NATO ASI Series

Advanced Science Institutes Series

A series presenting the results of activities sponsored by the NATO Science Committee, which aims at the dissemination of advanced scientific and technological knowledge, with a view to strengthening links between scientific communities.

The Series is published by an international board of publishers in conjunction with the NATO Scientific Affairs Division

A Life Sciences B Physics	Plenum Publishing Corporation London and New York
C Mathematical and Physical Sciences D Behavioural and Social Sciences E Applied Sciences	Kluwer Academic Publishers Dordrecht, Boston and London
F Computer and Systems Sciences G Ecological Sciences H Cell Biology I Global Environmental Change	Springer-Verlag Berlin Heidelberg New York London Paris Tokyo Hong Kong Barcelona Budapest

PARTNERSHIP SUB-SERIES

1. Disarmament Technologies	Kluwer Academic Publishers
2. Environment	Springer-Verlag/Kluwer Acad. Publishers
3. High Technology	Kluwer Academic Publishers
4. Science and Technology Policy	Kluwer Academic Publishers
5. Computer Networking	Kluwer Academic Publishers

The Partnership Sub-Series incorporates activities undertaken in collaboration with NATO's Cooperation Partners, the countries of the CIS and Central and Eastern Europe, in Priority Areas of concern to those countries.

NATO-PCO DATABASE

The electronic index to the NATO ASI Series provides full bibliographical references (with keywords and/or abstracts) to about 50 000 contributions from international scientists published in all sections of the NATO ASI Series. Access to the NATO-PCO DATABASE compiled by the NATO Publication Coordination Office is possible in two ways:

- via online FILE 128 (NATO-PCO DATABASE) hosted by ESRIN, Via Galileo Galilei, I-00044 Frascati, Italy.

- via CD-ROM "NATO Science & Technology Disk" with user-friendly retrieval software in English, French and German (© WTV GmbH and DATAWARE Technologies Inc. 1992).

The CD-ROM can be ordered through any member of the Board of Publishers or through NATO-PCO, Overijse, Belgium.

Series I: Global Environmental Change, Vol. 51

Springer
Berlin
Heidelberg
New York
Barcelona
Budapest
Hong Kong
London
Milan
Paris
Santa Clara
Singapore
Tokyo

Sediment Records of Biomass Burning and Global Change

Edited by

James S. Clark
Department of Botany
Duke University
Durham, NC 27708, USA

Hélène Cachier
Centre des Faibles Radioactivités
CNRS/CEA
Avenue de la Terrasse
91198 Gif-sur-Yvette Cedex
France

Johann G. Goldammer
Arbeitsgruppe Feuerökologie
Max-Planck-Institut für Chemie
c/o Universität Freiburg
D-79085 Freiburg, Germany

Brian Stocks
Great Lakes Forestry Center
Canadian Forest Service
Sault Ste Marie
Ontario, Canada P6A 5M7

With 141 Figures and 7 Colour Plates

Springer
Published in cooperation with NATO Scientific Affairs Division

Proceedings of the NATO Advanced Study Institute "Biomass Burning Emissions and Global Change", held in Praia de Alvor, Algarve, Portugal, October 1994.

Cataloging-in-Publication Data applied for

Library of Congress Cataloging-in-Publication Data

Sediment records of biomass burning and global change / edited by James S. Clark ... [et al.].
p. cm. -- (NATO ASI series. Series I, Global environmental change ; vol. 51)
"Proceedings of the NATO advanced study institute 'Biomass burning emissions and global change,' held in Praia de Alvor, Algarve, Portugal, October 1994"--T.p. verso.
"Published in cooperation with NATO Scientific Affairs Division."
Includes bibliographical references.
ISBN 3-540-62434-1 (hardcover)
1. Lake sediments--Congresses. 2. Burning of land--Environmental aspects--Congresses. 3. Fuelwood--Burning--Environmental aspects--Congresses. I. Clark, James Samuel. II. Series: NATO ASI series. Series I, Global environmental change ; no. 51.
GB1601.2.S43 1997
551.51'1--dc21 96-29756
CIP

ISBN 3-540-62434-1 Springer-Verlag Berlin Heidelberg New York

Printed in Germany

Typesetting: Camera ready by the authors/editors
Printed on acid-free paper
SPIN: 10466169 31/3137 - 5 4 3 2 1 0

ACKNOWLEDGMENTS

First, and foremost, I wish to thank NATO for the bulk of the funding. The NATO office responded rapidly to all requests for assistance, and they showed great patience during book preparation. NSF provided an additional contribution to support student participation, and the Quaternary Center of Duke University contributed travel support.

Bill Schlesinger suggested I approach NATO and provided advice and support throughout the planning and book preparation. Deborah Lawrence and Jean Porter coordinated travel and organization before and during the meeting. Brenda Cleveland provided invaluable editorial assistance. Chapter reviews were provided by workshop participants supplemented by anonymous outside reviewers.

Finally, I'd like to thank two mentors for many of the workshop participants, neither of which could attend the workshop itself. Ed Goldberg inspired much of the interest in this subject, and his presence was felt in many of the talks, some delivered by his former students. He expressed his regret at being unable to attend. I am particularly indebted to H.E. Wright Jr. on personal and professional levels. I thank him for his support during my student days at Minnesota, where my research on fire got off the ground. Much of the workshop agenda followed research directions instigated in collaboration with Herb, and his name came up often in casual discussion about fire.

TABLE OF CONTENTS

IV - STRATIGRAPHIC RECORDS

V - FUTURE DIRECTIONS

LIST OF PARTICIPANTS

Brigitta Ammann, Systematisch-Geobotanisches Institut, Universitat Bern, Altenbergain 21, CH-3013 Bern Switzerland

Scott Anderson, Center for Environmental Sciences & Education, PO Box 5694, Northern Arizona University, Flagstaff AZ 86011-5694, USA

John Birks, Botanical Institute, University of Bergen, Allegaten 41, N-5007 Bergen Norway

Platt Bradbury, US Geological Survey Box 25046, Federal Center, Denver, CO 80225, USA

Richard Bradshaw, Southern Swedish Forest Research Univ, Swedish University of Agricultural Sciences, Box 49, S-230 53 Alnarp Sweden

Helene Cachier, Centre des Faibles Radioactivites, CNRS/CEA, Avenue de la Terrasse, 91198 Gif-sur-Yvette Cedex France

Cristina Carapeto, Ministerio da Educaco, Universidade Aberta, Lisbon, Portugal

William Chaloner, Department of Biology, Royal Holloway/University of London, Egham, Surrey TW20 9LP United Kingdom

Jim Clark, Department of Botany, Duke University, Durham, NC 27706, USA

Randy Cofer, Atmospheric Sciences Division, NASA, Langley Research Center Hampton, Virginia 23665, USA

George Dubay, Department of Chemistry, Duke University, Durham, NC 27706

Augustin Ezcurra, Universidad del Pais Vasco, Exsuela Universitaria de Engeneria, Nieves Cano 12, 01006 Vitoria Spain

Mike Fosberg, USDA Forest Service–Forest Fire and Atmospheric Sciences Research, PO Box 96090, Washington, D.C. 20090, USA

Mike Garstang, Dept Environmental Sciences-Clark Hall, Univ Virginia, Charlottesville, VA 22903, USA

Annie Gaudichet, Universite Paris XII URA CNRS 1404, ave du General De Gaulle, 94010-CRETEIL, Cedex France

Johann Goldammer, Fire Ecology Research Group – Max Planck Inst for Chemistry, c/o Freiburg University, PO Box D-79085, Freiburg Germany

Tim Jones, Inst und Museum fur Geologie und Palaeontologie, Universitat Tubigen, Sigwartstrasse 10, D-72076 Tubigen Germany

Boone Kauffman, Dept Rangeland Resources, Oregon State University, Corvallis, OR 97331, USA

Peter Kershaw, Department of Geography, Monash University, Clayton Melbourne, Victoria 3168 Australia

Alexey Kokorin, Institute of Global Climate and Ecology, Glebovskaya 20B, 107258 Moscow, Russian Federation

K. Koutzenogii, Inst Chemical Kinetics & Combustion, Academy of Sciences, Institutskaia St 3, RU - 630090 Novosibirsk Russia

Thomas Kuhlbusch, Atmospheric Chemistry Department, Max Planck Institut für Chemie, Postfach 3060, D-55020Mainz Germany

Jean-Pierre Lacaux, Centre de Recherches Atmospheriques, 65300 Lannemezan, France

Deborah Lawrence, Department of Botany, Duke University, Durham, NC 27706, USA

Sam Levis, Atmospheric & Ocean Sciences, 1225 W. Dayton St, University of Wisconsin, Madison WI 53706, USA

Bo Lim, Centre des Faibles Radioactivites, CNRS/CEA, Avenue de la Terrasse, 91198 Gif-sur-Yvette Cedex France

Glen MacDonald, Department of Geography, McMaster University, Hamilton, Ontario L8S 4K1 Canada

Stephen Macko, Environmental Sciences, Univ Virginia, Charlottesville, VA 22903, USA

Pierre Masclet, Universite de Savoie Esigec, LESAM Campus Scientifique, Bat Mont Blanc, 73379 Le Bourget du Lac France

Sarah Millspaugh, Department of Geography, 107 Condon Hall, University of Oregon, Eugene OR 97403, USA

Tica Novakov, Lawrence Berkeley Laboratory, 1 Cyclotron Rd, Berkeley, CA 94720, USA

Bill Patterson, Department of Forestry, University of Massachusetts, Amherst, MA 01003, USA

Jean Porter, Department of Botany, Duke University, Durham, NC 27708, USA

Stephen Pyne, American Studies Department, Arizona State University West–PO Box 37100, Phoenix, AZ 85069, USA

Larry Radke, NCAR Research Aviation Facility, PO Box 3000, Boulder, CO 80307, USA

Ingemar Renberg, Department of Environmental Health, Umeå University, S-901 87 Umeå, Sweden

Jenny Robinson, NCGIA-SUNY, Department of Geography, 301 W. Wilkeson Quad. Box 610023, Buffalo New York 14261-0023, USA

Robert Sanford, Department of Biological Sciences, University of Denver, Denver, Colorado 80208, USA

Andrea Stevens, Department of Forestry, University of Massachusetts, Amherst, MA 01003, USA

Brian Stocks, Great Lakes Forestry Center, Canadian Forest Service, Sault Ste Marie, Ontario Canada P6A 5M7

Shinya Sugita, Department of Ecology, University of Minnesota, St Paul, Minnesota 55455, USA

Ilgonis Strauss, Ministry of Environmental Protection and Regional Development, 25 Peldu St, LV 1494, Riga Latvia

Daniel Suman, Division of Marine Affairs–RSMAS, University of Miami, 4600 Rickenbacker Causeway, Miami FL 33149, USA

Robert Swap, Department of Environmental Sciences, Clark Hall, University of Virginia, Charlottesville VA 22903, USA

Kimmo Tolonen, Department of Biology, University of Joensuu, PO Box 111, SF-80101 Finland

Mirjami Tolonen, Department of Botany, University of Helsinki, PO Box 7 Unioninkatu 44, SF-00170 Helsinki Finland

Vaughan Turekian, Dept Environmental Sciences, Univ Virginia, Charlottesville, VA 22903, USA

Karl Weiss, Fire Ecology Research Group – Max Planck Inst for Chemistry, c/o Freiburg University, PO Box, D-79085 Freiburg Germany

Cathy Whitlock, Department of Geography, University of Oregon, Eugene, OR 97403, USA

Marjorie Winkler, IES–Center for Climatic Research, 1225 W Dayton St, University of Wisconsin, Madison WI 53706, USA

Richard Zepp, EPA, Environmental Research Laboratory, Athens, GA 30605-2720, USA

An Introduction to Sediment Records of Biomass Burning

James S. Clark
Department of Botany
Duke University
Durham, NC 27708

Biomass burning has become recognized as a key element of global change. Best estimates of current and past emissions suggest fire now profoundly affects atmospheric chemistry, the carbon cycle, and climate and may have done so for millions of years. Literature values place carbon sequestration during biomass burning in excess of 20% of the total anthropogenic C releases to the atmosphere, which, failing oxidative losses, would tie up the globe's surface C in less than 10^6 yr (Goldberg 1985). And the relative importance of biomass burning may have been greater in the past, before large increases in fossil fuel combustion (Clark et al. 1996). The potential importance of these burning rates for biodiversity, C and O cycles, air quality, and climate have impressed a diverse community of ecologists and atmospheric chemists and served as a theme for a recent Dahlem workshop some of us organized on biomass burning (Crutzen and Goldammer 1993). That workshop together with subsequent discussions identified key challenges that require multidisciplinary efforts i) to better characterize combustion products that occur in sediments, including char, soot/fly ash, and polycyclic aromatic hydrocarbons (PAH's), ii) to calibrate these constituents against atmospheric measurements from wildland and prescribed fire emissions, and iii) to estimate the spatial and temporal patterns in combustion emissions at scales of individual burns to the globe. We are in need of baseline estimates of preindustrial burning rates in order to understand impacts of current emissions. Many sediment studies exist, but interpretation requires calibration against ground, airborne, and satellite estimates being completed by forest ecologists and atmospheric chemists.

The impetus for this workshop was a belief that calibrations needed to interpret the sediment record are possible given the right interdisciplinary mix of fire scientists. Sediment studies of wildfire and industrial emissions are abundant. Fire behavior studies provide the source-strength variables, and atmospheric measurements are beginning to yield the important details on emission factors and transport. Each discipline is now in a position to offer rough parameter values that could be used to estimate carbon sequestration during biomass burning and particulate emissions back into pre-industrial/pre-land clearance time.

NATO ASI Series, Vol. I 51
Sediment Records of Biomass Burning
and Global Change
Edited by James S. Clark, Hélène Cachier,
Johann G. Goldammer, and Brian Stocks

Several independent groups were identified that, together, might fully or partially resolve the critical questions regarding the role of fire during global changes of the past and present:

Paleoecologists: A network of sedimentary records show geographic patterns of past charcoal accumulation for boreal, temperate, and tropical biomes. These data sets are the evidence for past emissions, but they require calibration against fuel loads, burn efficiencies, and atmospheric transport during and after burns in order to be used to address the C cycle. Additional data sets from Pacific Ocean cores and PAH's also require calibration. The task for paleoecologists at this workshop was to begin assembling the mass flux estimates of charcoal in space through the past.

Fire ecologists: Ecologists have the best estimates of fuel loads and burn efficiencies, critical components for calculations of emissions and of C sequestration as elemental C. They have management goals that require knowledge of the role of fire with global change. Fire ecologists had the task of estimating fuel loads and burn efficiencies for principle vegetation types on modern landscapes.

Atmospheric chemists: A growing knowledge from airborne measurements of gaseous and particulate emissions and of C isotopic ratios in smoke during fire aid understanding of modern emissions. Atmospheric processes provide a link between fuel dynamics and sediment records and thus a means for calibration. They require a temporal perspective on direct and remotely sensed measurements in order to determine the importance of 20th century land-clearance activities. Atmospheric chemists were asked to provide the best estimates of emission factors for different types of burns in different vegetation and climate settings.

Organic chemists: Sediment records of PAHs have been used to argue that wildfire contributes small quantities of PAH to the atmosphere. This issue requires further consideration, because none of the analyses come from regions where wildfire was common, and they generally do not include flux estimates. Comparisons of charcoal and PAH sediment records together with measurements in smoke are needed.

The objective of our Advanced Research Workshop was to assemble the paleo records of past combustion and the atmospheric and terrestrial data needed to interpret emissions with global changes of the recent past. We attempted to better establish what constituents each group is measuring, what they represent, and to use these discussions as the basis for beginning the process of calibration against wildfire emissions. We anticipated new, collaborative research directions to be a principle outcome of this workshop.

This volume is organized under several themes. The issues that are addressed by sediment records is the topic of Part I. Jones et al. begin this section with a review of the many ambiguous terms that have been applied to charred particles, especially those that occur in sediments. Many definitions of "charcoal" are operational, and such terms as "black carbon", "elemental carbon", and "graphite" often depend on the method used to quantify it. This theme reappears in the chapter of Novakov et al., which demonstrates that different methods yield very different estimates of black carbon. Jones et al. suggest a standardization of terms and propose how those terms might be applied. Clark and Patterson's review of the spatial scales contained in stratigraphic records suggests promise that we can further refine interpretation of local and regional fire importances in the past. Fosberg and Levis suggest that model applications to past biomass burning hold potential for better understanding fire/climate relationships. In the final chapter of Part I, Pyne and Goldammer place long term changes in fire within the broader context of human cultural change. Through use of many examples they demonstrate that alteration of fire regimes is a ubiquitous feature of social change. Fire is a tool for intentional modification of human environments, but side effects are inevitable and, often, highly undesirable. Sediments provide a record these changes from Stone Age swidden agriculture to industrialization to fire suppression.

Characterization is treated in two chapters of Part II. Novakov et al. review methods for quantifying particulate emissions. Because most paleoecologists focus exclusively on particulates this review is especially relevant to the stratigraphic analyses in Part IV. Zepp and Macko examine the potential for PAH records from marine and lacustrine sediments. An unusually thorough treatment of transport effects on the PAH spectrum should prove valuable for reinterpretation of previous data. While analyses to date demonstrate the rise of fossil fuel combustion in industrial areas, this review suggests promise for analysis of wildfire, particularly through application of compound-specific stable isotopes.

Interpretation of biomass burning records requires knowledge of how combustion products are produced and transported to sediments. The three chapters of Part III examine emissions from different fuel types and how they are transported in the atmosphere. Stocks and Kauffman provide estimates of fuels and particle emissions from boreal, temperate, and tropical burns. Cofer et al. and Garstang et al. synthesize emissions and transport from a number of experimental burns. Insights for paleoecologists here include actual numbers for quantities and size distributions of particles at the source and how those abundances and size distributions can be modified under various atmospheric conditions. Especially noteworthy are the complex transports possible under different circulation patterns reviewed by Garstang et al.

The stratigraphic records of fire in Part IV provide an overview of the scales represented in sediment combustion profiles and some of the considerations needed to extract accurate interpretation. Robinson et al. provide convincing arguments that paleofire records extend at least as far back as the Devonian. The records themselves span changes in atmospheric composition, fuels, climate, and ignition rates. The Phanerozoic record of fusain may place bounds on atmospheric oxygen concentrations and, thus, help constrain models of the geologic carbon cycle. Broad interpretation of pre-Quaternary climates is not yet possible from such records. Robinson's hypothesis that past fire shaped evolutionary change (and vice versa) with the rise of lignin and fungal decomposers is mentioned here, as are the potential effects of fire on vegetation structure under altered climates and oxygen levels.

Suman et al. review the evidence for elemental C deposition in marine sediments, much of it from former students and associates of Ed Goldberg. An advantage of common methodology employed by this group from Scripps is potential for comparisons among cores. Suman et al. update the geographic and temporal patterns, and they provide new estimates of C sequestration rates.

The remaining chapters in Part IV provide an overview of Quaternary evidence from lake and mire deposits. Birks suggests a broad application of statistical methods to aid interpretation of profiles. Three chapters by Anderson and Smith, Winkler, and Bradshaw et al. provide new data and synthesis of fire evidence in montane, temperate deciduous, and boreal biomes, respectively. Whitlock et al. consider some of the processes that affect charcoal sedimentation in lakes. Monitoring the time course of accumulation in Yellowstone and Minnesota lakes, they show that charred particle deposition may continue well after a burn due to circulation of the water column. Swain (1978) was among the first to popularize the idea that fires might be interpreted from fluctuations in pollen curves, arguing for rather high rates of burning in northern Wisconsin (Clark and Royall 1996). Sugita et al. revisit the question with a more complex model of pollen transport. Predictions suggest sizes and distances of disturbances that might be recorded in pollen profiles.

The final two chapters of Part IV share the theme of human impacts recorded in lake sediments. Kershaw et al. provide the first comprehensive overview of long-term biomass burning in the Tropics. This is sure to be a valuable point of departure for future efforts to understand trends in human impacts. Lim and Renberg emphasize the records available in spherical carbonaceous particles and discuss how they relate to fossil fuel vs biomass burning sources. Together, the chapters of Part IV demonstrate that charcoal analysis has progressed substantially since its inception, with routine consideration of factors affecting particle production and transport and attention to implications for biome function and the atmosphere.

Part V contains an overview of working group discussions. This was the first such discussion involving paleoecologists, fire ecologists, and atmospheric scientists that focused solely on the theme of better understanding long term biomass burning. The working group report provides participant views of important new directions, including possible interdisciplinary research efforts. In the final chapter Goldammer reports on the international infrastructure that might support such efforts.

Literature references:

Clark JS, Royall PD (1996) Local and regional sediment charcoal evidence for fire regimes in presettlement northeastern North America. *J of Ecol* 84:365-382

Clark JS, Stocks BJ, Richard PJH (1996c) Climate implications of biomass burning since the 19th century in eastern North America. *Global Change Biology* (in press)

Crutzen PJ, Goldammer JG (1993) *Fire in the Environment: The Ecological, Atmospheric, and Climatic Importance of Vegetation Fires.* Wiley, New York

Goldberg ED (1985) *Black Carbon in the Environment.* Wiley, New York

Swain AM (1978) Environmental changes during the past 2000 years in north-central Wisconsin: analysis of pollen, charcoal and seeds from varved lake sediments. *Quat Res* 10:55-68

PART I:

THE ISSUES ADDRESSED BY SEDIMENT COMBUSTION RECORDS

Proposed Bio-geological and Chemical Based Terminology for Fire-altered Plant Matter

T.P. Jones [1], W.G. Chaloner [2], and T.A.J. Kuhlbusch [3]
Institut und Museum für Geologie und Palæontologie
Universität Tübingen
Sigwartstrasse
Tübingen 72076 Germany

Summary

A standard terminology is proposed applicable to all disciplines that produce, use, or examine fire-altered plant matter. The different materials are defined by macroscopic, microscopic, or chemical characteristics that can be determined using generally available equipment. No new terms are introduced. It is recommended that certain terms that have precise meanings in other fields should be either used with caution or discarded. Several existing terms should be disregarded, because they pre-suppose a mode of formation or structure for which the evidence is poor or incorrect.

Introduction

Within the diverse scientific disciplines interested in fire products (e.g., palaeobotany, fire ecology, atmospheric chemistry) the same term is often used for different materials, and different terms are sometimes used for the same materials. General usage terms to describe fire-altered plant material, such as 'charcoal', 'ash', 'smoke', and 'soot', do not appear in Biology (1970) or Geology (1985) dictionaries (Penguin reference books). Nor do definitions exist within the atmospheric science literature. In English language dictionaries these materials are usually described in terms of the processes that produce them (often inaccurately or imprecisely), rather than by their physical and chemical properties. In coal petrological descriptions (fire-altered plant matter is a common component of coal) coal components, or 'macerals' (substances found in coal analogous to 'minerals' in rocks), are defined primarily by physical and chemical characteristics and, to a lesser extent, by supposed modes of formation.

---

[1] Present address: Department of Earth Science, Unviersity of Wales Cardiff, CF1 3YE, Wales

[2] Department of Geology, Royal Holloway, University of London, Egham Hill, Egham, Surrey, TW20 0EX, U.K.

[3] Max-Plack-Institut fur Chemie, Abteilungen Luftchemie, Saarstrasse 23, 55122 Mainz, Germany

NATO ASI Series, Vol. I 51
Sediment Records of Biomass Burning and Global Change
Edited by James S. Clark, Hélène Cachier, Johann G. Goldammer, and Brian Stocks

This paper reviews the diverse current usage of terms for fire–altered plant materials, and it proposes a standardized nomenclature. Our proposed nomenclature consists of altered or extended revisions to pre-existing definitions intended to clarify or elucidate. We believe this standardization will foster communication and understanding within and among disciplines.

Current nomenclature

This first section contains existing nomenclature for fire-altered materials in general/geo-biological usage, and it summarizes some of the problems inherent in many commonly-used terms.

Terms in common use

Charcoal. The carbonaceous material obtained by heating wood or other organic substances in the absence of air (Webster's Encyclopaedic Unabridged Dictionary of the English Language 1989). Synonyms are char, chars.

For common usage Webster's definition is adequate, but for scientific descriptive purposes it is clearly inadequate. The degree of heating 'wood or other organic substances' is not specified. Furthermore, a forest does not burn 'in the absence of air', and forest fires certainly produce 'charcoal'; thus, strictly speaking, the definition is incorrect as well as inadequate. The Encyclopedia Britannica (1974) offers a different definition of 'charcoal'. "Charcoal is an impure form of graphitic carbon obtained as a residue when carbonaceous material is partially burned or heated, with limited access of air". It is noted that not all 'charcoal', especially charcoal produced at lower temperatures, has a measurable graphitic structure (Jones 1991). Thus, these two definitions differ significantly, and neither is truly valid for describing 'charcoal' obtained from wildfires.

The term 'char' is commonly used in the science of "thermochemical biomass conversion" for solid residues obtained by pyrolysis in absence or limited access of oxygen. The term 'chars' (or charcoals) was introduced by Medalia and Rivin (1982) for small fragments of carbonized wood or coals, ranging in size from microns to fractions of a metre. The use of the term 'fossil charcoal' to describe ancient fire-altered plant material found within rock is widespread, and is reviewed in Cope (1984) and Jones (1991). Early workers also used the expression 'mineral charcoal' for charcoal found in clastic rocks or coal, but this has fallen out of general usage.

Ash. The powdery residue of matter that remains after burning (Webster's Encyclopaedic Unabridged Dictionary of the English Language 1989).

In geological nomenclature ash has also a totally different and separate meaning, and is defined as "the unconsolidated fine-grained material formed as a result of volcanic explosions" (Penguin Dictionary of Geology 1985). For coal analysis, ash is defined as the powdery, mineral matter that remains after removing all organic and elemental carbon by combustion, e.g., 'ash content of coals'. It is noted that 'ash' (fine powdery residue) remaining after a wildfire can contain a significant organically-derived carbon component (up to 20% of dry mass), primarily controlled by the efficiency of combustion (Kuhlbusch et al., in press).

Fly ash. Fine, not combustible component of the fuel being carried with the emissions (translated from Brockhaus Enzyklopædie 1968).

Smoke. The visible vapor and gases given off by a burning or smoldering substance, especially the (white), grey, brown, or blackish mixture of gases and suspended carbon particles resulting from the combustion of wood, peat, coal, or other organic matter (Webster's Encyclopaedic Unabridged Dictionary of the English Language 1989).

Soot. The black carbonaceous substance produced during the imperfect combustion of coal, wood, oil, etc., rising in fine particles and adhering to the sides of the chimney or pipe conveying the smoke; also conveyed in the atmosphere to other locations (Webster's Encyclopaedic Unabridged Dictionary of the English Language 1989).

The word 'soot' originates from the old English word 'to sit', and possibly refers to domestic fire products 'sitting' on the chimney walls (Webster's Third New International Dictionary 1965). The use of 'soot' in the literature is twofold. It normally comprises just those carbon particles that form via gas–phase processes. The stages forming soot being nucleation and heterogeneous reactions occurring on the surface of the nuclei, followed by agglomeration or coagulation (summarized from Glassman 1977). Sometimes it is used for all particulates emitted by a fire, a usage that should be discontinued.

Wolbach et al. (1988) describe Cretaceous-Tertiary boundary clays that "are 10^2 - 10^4 –fold enriched in elemental C (mainly soot)". They define 'soot' as having characteristic 'grape bunch' morphology, and differentiate it from 'charcoal' which is 'Irregularly shaped, platy or pitted coarse carbon, presumably charcoal -- Unlike soot, which forms only in flames, charcoal forms at lower temperatures, by charring and devolatilization of fuel'. Soot contains not only elemental carbon, but also has an organic carbon component. Wolbach and Anders (1989) removed the organic component by wet chemistry oxidation, and quantified the remaining, more refractory carbon as elemental carbon.

Black carbon. 1) An impure form of the element produced by the incomplete combustion of fossil fuels or biomass. It contains over 60% carbon with the major accessory elements hydrogen, oxygen, nitrogen, and sulfur (proposed by Goldberg 1985). 2) A fire-derived highly aromatic to graphitic carbon fraction that is resistant at 340°C in pure oxygen and has a molar H/C-ratio of < 0.2 (proposed by Kuhlbusch and Crutzen 1995).

Relationships among terms

The current general usage of the terms is reasonably clear (Fig. 1). In a fire where a solid fuel is incompletely combusted (i.e. converted to a gas; CO_2, CO, etc.). The solid residue at the site of the fire includes 'charcoal', which consists of carbonaceous material, and 'ash', which is a white powdery mineral residue (mostly H_2O soluble). All visible material, either gaseous or solid, given off into the atmosphere is 'smoke', which includes small 'charcoal' fragments and 'soot'. The two latter components are determined by their microscopic morphologies; charcoal is irregularly shaped, and soot has a 'grape bunch' shape and both are black. 'Black carbon' is a certain carbon fraction being present in charcoal, ash, and smoke particulates.

Nomenclature of fire-altered matter in atmosphere science

Black carbon. The term black carbon was introduced to atmospheric chemists by Novakov (1984) as "combustion-produced black particulate carbon having graphitic microstructure." Synonyms and analogous terms include 'elemental carbon', 'polymeric carbon', 'soot carbon', 'carbon soot', 'free carbon', 'charcoal', 'fusain', 'microcrystalline carbon', 'graphitic carbon', and 'graphite'. Goldberg's (1985) descriptions of black carbon include the following: fire produced polymeric, aromatic to graphitic carbon; carbon fraction in ash, charcoal, partially-charred, and smoke particulates; absorbs and scatters radiation; absorption coefficient ranging form 2-25 $m^2 g^{-1}$; relatively inert to chemical, microbial, and oxidative degradation; carrier of toxic substances, e.g., Benzo[a]pyrene; porous; black.

"Environmental scientists often describe this material [black carbon] on the basis of their techniques of isolation or of measurement" (Goldberg 1985). Thus, a great variety of terms are used by atmospheric scientists for black carbon. Goldberg (1985) concludes that scientists from various disciplines only agree on one point, "the material has a highly variable nature". Both proposed definitions for black carbon are based on the chemical composition. Whereas Goldberg (1985) defines black carbon in general terms, Kuhlbusch and Crutzen (1995) try to be more precise, using the molar H/C-ratio as an indicator of degradability of black carbon in the environment The use of some synonyms for black carbon leads to confusion, because they may imply a chemistry of 100% carbon, a crystalline well-known structure, or an incompletely burned biomass associated with various organic phases.

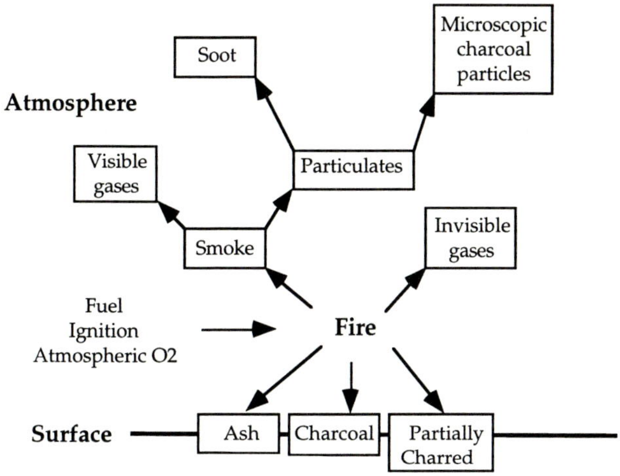

Figure 1. Relationships among fire products.

The term 'black carbon' has been used for a great variety of carbon compounds with the common property of absorbing radiation, one of the properties used to identify it. This carbon fraction is the main aerosol constituent absorbing light. There are two main analytical techniques for quantifying black carbon in aerosol sciences. One uses optical properties of black carbon (blackness) and the other uses chemical and physical stability of carbon compounds. The determination of 'black carbon' in sediments is mainly based on its stability against oxidation. Because of these different techniques and lack of a standard, data sets have to be carefully evaluated before comparison. 'Carbon black', 'lampblack', 'furnace black', 'channel blacks', 'active carbon', and 'activated charcoal' are commercially produced forms of carbon. Black carbon is one component, most probably the main, of these industrial products.

Nomenclature of fire-altered matter in coal petrology

The following is based on the International Commission for Coal Petrography (ICCP 1963, 1971, 1975), and on Stach's textbook of coal petrology, 3rd edition (1982). A requirement for these definitions is that the material must be sufficiently large for identification under standard coal petrology microscopes, ranging from 10 μm (e. g. fragment of a single cell wall) upwards. Definitions are usually based on optical and chemical properties and are not related to supposed modes of formation (Fig. 2).

Fusain. The term fusain was introduced by Grand'Eury in 1882 to designate the black silky, lustrous bands recognized macroscopically in coal. Fusain derives from the Latin word *Fusus*, for spindle or fibre, and is used in French for charcoal and drawings done with charcoal. Synonyms and analogous terms include 'faserkohle', 'charbon fibreux', 'mineral charcoal', 'mother of coal', 'russkohle'. In the Thiessen-Bureau of Mines (TBM) system, fusain is a 'component' of coal. In the Stopes-Heerlen (SH) system, fusain is a 'macroscopic constituent' of coal.

The ICCP (1963) description includes: macroscopically closely resembles wood charcoal; soft; friable; black; disintegrating readily into a black powder when roughly handled; hard varieties exist impregnated with mineral matter; silky lustre; blackens objects with which it comes into contact; microscopically usually opaque (black), in thin section may be translucent (dark red); shows well-preserved plant cell structure in which lumen may be empty; with carbonization, gives very low yields of by-products; does not lend itself to hydrogenation; does not readily oxidize, no tendency to spontaneous combustion. To constitute fusain under the TBM system, a unit of this material must have a minimum thickness of 37 mm, when used for coal petrology, seam analysis .

The ICCP (1963) system agrees with the Stopes-Heerlen system in the following two ways:

1) Fusain is bands and lenticles of fusinite and semifusinite > 37 mm thick.

2) Bands of fusain > 50 mm thick correspond to microlithotype (monomaceral microlithotype) fusite.

Fusite. "The term fusite was used for the first time by Potonie in 1924. In 1955 the Nomenclature Sub-Committee resolved to use this term for the microlithotype consisting principally of the macerals fusinite, semifusinite and sclerotinite. The Nomenclature Sub-committee resolved in 1964 to use the term fusite to denote a monomaceral microlithotype consisting of the maceral fusinite" (ICCP 1971). An analogous term is xylovitrain (Duparque 1933). Fusite must contain at least 95% of fusinite (ICCP 1971).

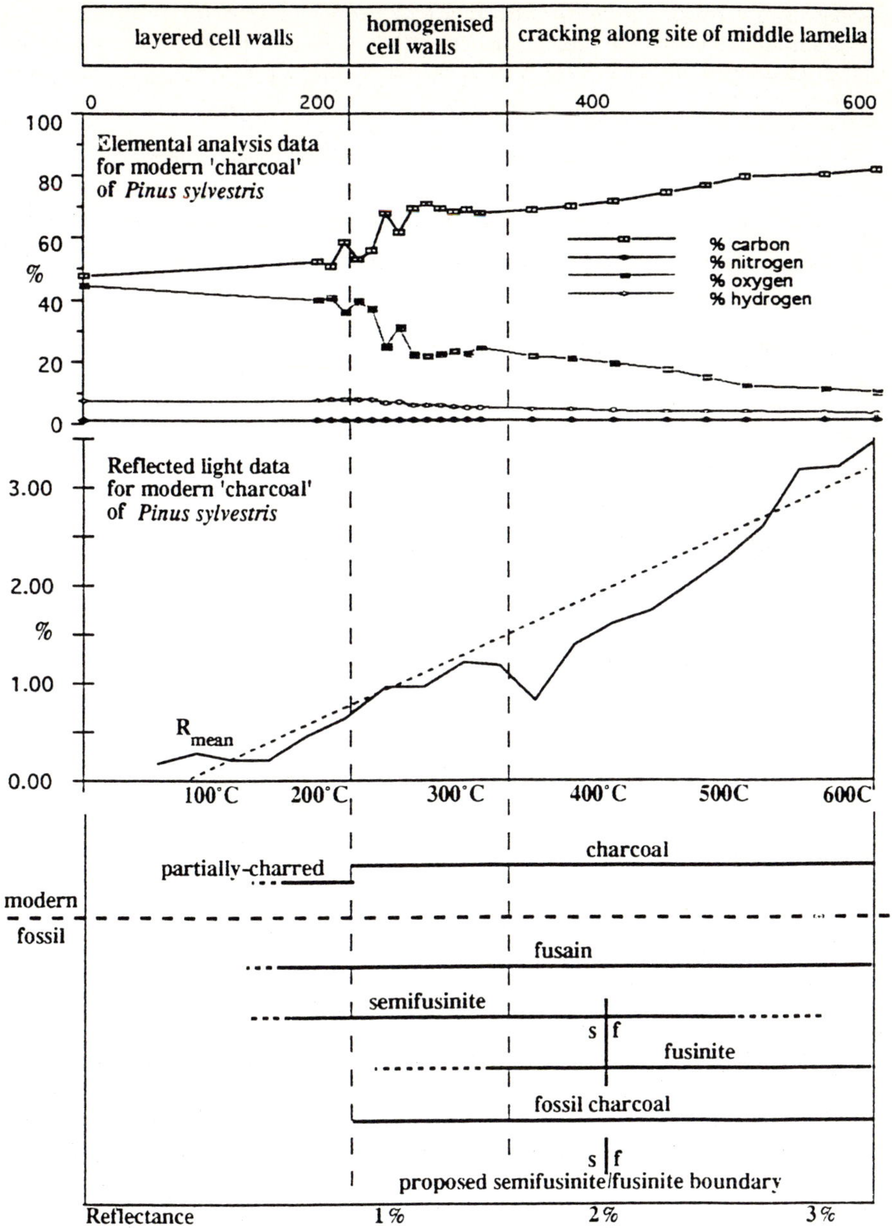

Figure 2. Changes in several properties along a heat-treatment gradient from 0°C (left side) to 600°C (right side) of figure. Lower panel gives terms recommended here.

Fusinite. The term fusinite was introduced by Stopes in 1935 for an opaque coal constituent that displays cell structure. It is a maceral (ICCP 1975). Synonyms and analogous terms include fusain (in part); xylinito-fusinite, vitrinofusinite (Mincev 1965); fusoid-group (Spackman 1958); fusinite-telinite (Timofeev and Bogoliubova 1964); alpha, beta, gamma and delta fusinite (Waltz 1956); xylenofusain, xylenovitrenofusain (Zemcuznikov and Ginsburg 1960).

The ICCP (1975) system summarizes fusinite as follows: consists of highly reflecting cell walls, usually thinner than the cell walls of corresponding semifusinite; bogen-structure is common (a discrete mass of small fragments, where an intact fragment has been shattered *in situ* as a result of compressive forces); cell lumina empty, but occasionally filled with gelinite or mineral matter; isolated fragments of cell walls with fusinite-like reflectivity are classified as inertodetrinite; color with transmitted light, black; color with reflected light, white, often with yellowish tinge; fusinite does not fluoresce; reflectance always high, irrespective of coalification (values Rro quoted 1.4% - 4.7%); density high (values quoted 1.47 - 1.64 SG, but note that cell lumina are often empty and this will affect the bulk values obtained); microhardness varies over wide range; characterized by relatively high carbon content, low hydrogen, oxygen and volatile components.

Semifusinite. The term semifusinite was first used by Jongmans et al. (1938) for a material transitional between vitrinite and fusinite (ICCP 1975). Synonyms and analogous terms include seminite (Ammosov 1956); Xylain (Duparque 1933); vitrofusite (Jongmans, 1938); semixylinitofusinite (Mincev, 1965); intermediates (Seyler 1948); semifusinoid group (Spackman 1958); vitrofusite (Stach 1935); Fusain (in part); semifusinite-telinite, gelifusinito-telinite (Timofeev and Bogoliubova 1964); Ubergangsstufen (ICCP 1963); alpha-, beta-, gamma-, delta-semifusinite (Waltz 1956); semifusain, semixylofusain, semi-vitrenofusain (Zemcuznikov and Ginsburg 1960).

ICCP (1975) summarizes semifusinite as consisting of cell walls whose reflectance lies between huminite and fusinite of the same coal. It is therefore a transition stage of textinite to fusinite formed of tissues with more or less well-preserved cellular structure. Cell walls of semifusinite are usually thicker than those of fusinite. As opposed to fusinite, semifusinite displays ill-preserved cell structures. Transitions of semifusinite to fusinite occasionally appear in the same cellular tissue. The color in transmitted light is red-brown to dark-brown, occasionally black. The color under reflected light is gray to white. It does not fluoresce; reflectance is stronger than huminite, weaker than fusinite (values Rro quoted 0.3% to 2.9%). The density lies between gelinite and fusinite (values quoted 1.35 to 1.50 SG).

Microhardness is less than that of fusinite. Chemical characteristics are between huminite and fusinite.

Fusinite and semifusinite both originate mainly from cellulose and lignin in plant cell walls. There are insufficient published empirical data to support the view that the preserved cell walls of semifusinite are generally thicker than those of fusinite. Charring experiments on modern secondary wood indicate that the controlling factor that determines reflectance and therefore whether the material should be classified as fusinite or semifusinite is temperature, and this does not affect cell wall thickness (Jones et al. 1991).

The suggested mechanisms of fusinization include, fire (incomplete carbonification), actions of bacteria and fungi (decomposition) and other 'unknown' causes (ICCP 1975). The expression "incomplete carbonification" is misleading, as it implies that material is replaced by externally-derived carbon; this is not the case. The heating results in the removal of volatiles, generally resulting in a carbon-enriched residue.

Pyrofusinite. "A part of fusinite and semifusinite can certainly be related to wood or to peat fires. It is well known that with large fires there is an oxygen deficiency and so frequently there is incomplete combustion and incineration; additionally, in the swamps, fires must be halted by moisture. Charring consequently takes place and so pyrofusinites represent fossil charcoal" (Stach 1982).

Petrology textbooks do not define any specific characteristics to distinguish 'pyrofusinite' from either 'fusinite' or 'semifusinite'. The decision to describe material as 'pyrofusinite' is therefore an arbitrary one chosen by the researcher, and often dependent on whether they support or oppose a fire-related origin for 'fusinite' and 'semifusinite'. There is still debate on the modes of formation of some fusinites and semifusinites, whether they could have formed by a non-fire related process, but the majority of palaeobotanists now support a wildfire-related origin. Because there appears to be no advantage in the use of the term 'pyrofusinite', and it pre-supposes a mode of formation (where a degree of uncertainty exists), it is suggested that this term be discarded.

Fusain-vitrinite transition fossils. Single fossil fragments that grade from fire-altered plant material (e.g., fusain) to non-fire-altered plant material (e.g., vitrinite), are termed transition fossils (Cope 1984, Jones et al. 1993). They represent the fossilized remains of a single plant fragment, which was only partially fire-altered; typically the outer surface was converted to charcoal, while the inner portion remained relatively unaltered.

Nomenclature of non-fire-altered '- fusinites' in coal petrology

Degradofusinite. "Probably because of desiccation and a relatively high oxygen supply at the peat surface, these degradofusinites (Zersetzungsfusinite) apparently form through dehydration and oxidation. Therefore they are also termed 'oxyfusinites'." (Stach 1982). Oxyfusinite is a synonym.

As yet no dehydration or oxidation processes have been recorded in modern peat-forming swamps that could produce a material resembling degradofusinite. Winston (1993) observes that "the presently accepted criteria for recognizing degradofusinite are inadequate even if degradofusinite does exist. Some evidence does exist but the key illustration, as published (Teichmüller 1950, 1989), does not demonstrate the existence of degradofusinite". Because there appears to be no advantage in the use of the term 'degradofusinite', and it pre-supposes a mode of formation (where a large degree of uncertainty exists), it is suggested that this term be discarded.

Rank fusinite. "There is considerable support for the opinion that some part of the fusinite of bituminous coals only forms during geochemical coalification of huminitic-liptinitic cell tissues. These fusinites are referred to as 'rank fusinites' "(Stach 1982).

There is now considerably less support for theories of geochemical coalification processes resulting in material resembling fusinite. There are no specific characteristics to distinguish 'rank fusinite' from 'fusinite'; the use this term is again arbitrary. Because, again, there appears to be no advantage in the use of the term 'rank fusinite', and it pre-supposes a mode of formation (where considerable uncertainty exists), it is suggested that this term be discarded.

Primary fusinite. "Just as in the case of fungo-sclerotinites, which forms from the black cell walls of fungi, the high reflectivity of some fusinites is clearly established in the original plants. Thus, the living fertile shoots of the horsetail, *Equisetum arvense*, are already remarkably brown coloured and display strongly reflecting cell groups" (Stach 1982).

The evidence for primary fusinite was originally based on color analysis in thin-sections of coal balls (calcium carbonate nodules formed in peat and preserved in coal seams), however this method is now known to be unreliable. The "strongly reflecting cell groups" in horsetails do not achieve fusinite reflectance levels (Winston 1993). This material is not fire-altered and does not achieve high enough reflectance to qualify as 'fusinite'. As it is usually associated with fungal activity it should be classified as 'sclerotinite'.

Summary of current coal petrology nomenclature

A number of problems exist with the current nomenclature system with respect to the study of inertinite macerals. The most important problem is the lack of precision when defining macerals, best seen when attempting to distinguish between fusinite and semifusinite. A maceral displaying reasonable cell structure, white coloured reflectance, Rro of 2.0%, opaque under transmitted light, no fluorescence, could be either fusinite or semifusinite (Table 1). Other problems include i) application of a nomenclature system that is designed for coal petrology to describe dispersed phytoclasts of fire-altered material in rock, and ii) regular misuse of nomenclature in published works alternating the terms, fusain, fusinite, fossil charcoal, and (even worse) fossil wood.

Table 1. Comparison of the main characteristics of fusinite and semifusinite, from ICCP (1963, 1971, 1975).

	Fusinite	Semifusinite
	Elemental analysis	
%C (af)[1]	79.34	74.13
%H (af)	3.31	3.94
%O (af)	11.23	N.D.
%N (af)	0.90	7.30
%S (af)	0.88	6.60
	Proximate analysis	
% moisture (af)	7.50	9.00
% ash	4.74	13.06
% humic acid (af)	4.00	10.30
Reflectance	High	Moderate
Rro[2]	1.4% - 4.7%	0.3% - 2.9%
Reflected light color	White, occasional yellow tinge	Gray to white
Cellular structure	Well-preserved	Ill-preserved
Cell walls	usually thin	usually thick
Bogen-structure	common	less common
Transmitted light	opaque, black	red, brown, occasionally black
Fluorescence	none	none
Density	1.47 - 1.64 SG	1.35- 1.50 SG
Microhardness	varies	usually less than fusinite

[1]Ash Free. [2]Random reflectance under oil. N.D. Not determined.

Proposed nomenclature

Fire-altered. Plant material that has had its chemical composition altered as a result of heating in a fire. Organic carbon is progressively decreased in the residue with increasing combustion efficiency, leaving non-combustible mineral matter. The heating of woody material typically results in an enrichment of carbon relative to other elements. The enrichment with carbon due to heat supplied by fire is different from carbon-enrichment as a result of diagenetic or maturation processes and can be recognized by physical and chemical characteristics of the respective end products.

Partially-charred. (a term to be applied to macroscopically described material) A single fragment or plant part that has been altered in part by fire, consisting of both burned and unburned material. An example is a standing trunk after a wildfire, where the external surface is blackened and consists of 'charcoal', grading through 'charred' material to the inner part which still consists of unaltered wood.

Charred. (a term to be applied to microscopically or chemically described material). Any plant material that is partially blackened and has had its chemical composition altered as a result of heating in a fire, but has not had its ultrastructure significantly altered. For example, secondary wood which is chemically altered, but still has stratified cell walls.

Charcoal. Any black-colored plant-derived material that has had its chemical composition and ultrastructure significantly altered as a result of heating in a fire, and retains recognizable anatomic structure of the parent plant, even if only in a fragmentary form. For example, secondary wood which is chemically altered, has homogenized (previously-stratified) cell wall ultrastructure, but retains its three-dimensional anatomical structure.

Char. A solid matter artificially produced by pyrolysis in inert, non-oxidizing atmospheres.

Ash. The mineral-rich powdery residue remaining on-site after a fire.

Fly ash. A fine, incombustible component of fuel carried with emissions.

Smoke. All visible matter emitted by a fire, comprising visible gases, volatiles, and smoke particulates.

Smoke particulates. Particles emitted with smoke. Particle size ranging from submicrometer ($< 10^{-6}$ m) up to a few millimetres. The latter are sometimes referred to as 'supergiant particles'. Consists of soot, small charcoal particles, fly ash, and sometimes soil particles in a wildfire.

Soot. Particles emitted with smoke and formed via gas phase processes; gas-to-particle conversion. Particle size ranging from submicrometer (mainly) up to less than 1 μm.

Small charcoal particles. Particles produced in fuel by fire and carried with smoke emissions. Particle size ranging from approximately 1 μm to several mm.

Black carbon. A fire–derived polymeric, aromatic to graphitic carbon fraction in fire-altered material or smoke particulates. It absorbs radiation , and is relatively chemically resistant to biodegradation.

Fusain. A general term for all fossilized fire-altered plant material.

Fossil charcoal. The fossilized remains of any black-colored plant-derived material that has had its chemical composition and ultrastructure significantly altered as a result of heating in a fire and retains the recognizable anatomic structure of the parent plant even if only in a fragmentary form.

Fusinite. As described in the ICCP (1963, 1971, 1975), but with minimum Rro of 2%.

Semifusinite. As described in the ICCP (1963, 1971, 1975), but with maximum Rro of 2%.

Literature references

Ammosov II (1956) Microscopie luminesente des charbons de mine. *Proc Internat Comm Coal Petrol* 2:14-17

Brockhaus Enzyklopædie (1968) 6:367

Cope MJ (1984) *Some studies of the origin, nature and occurrence of charcoalified plant fossils*. PhD Thesis University of London

Duparque A (1933) Structure microscopique des charbons du Bassin Houiller du Nord et du Pas-de-Calais. *Mem Soc Geol Nord* 11:83-96

Encyclopaedia Britannica (1974) Micropaedia 2:749

Goldberg ED (1985) *Black carbon in the environment*. Wiley New York

Glassman I (1977) *Combustion*. Academic Press New York

Grand'Eury FC (1882) Memoir sur la formation de la Houille. *Ann Mines* 1 Ser 8:106

International Committee for Coal Petrology (1963) *International Handbook of Coal Petrology* Second Edition CNRS Paris

International Committee for Coal Petrology (1971) *International Handbook of Coal Petrology* Supplement to the Second Edition CNRS Paris

International Committee for Coal Petrology (1975) *International Handbook of Coal Petrology*. 2nd Supplement to the Second Edition. CNRS Paris

Jones TP (1991) *The nature, origin and recognition of fusain*. PhD Thesis University of London

Jones TP, Scott AC, Cope MJ (1991) Reflectance measurements and the temperature of formation of modern charcoals and implications for studies of fusain. *Bull Soc Geol France* 162(2):193-200

Jones TP, Scott AC, Mattey DP (1993) Investigations of 'fusain transition fossils' from the Lower Carboniferous: comparisons with modern partially charred wood. *Intern J Coal Geol* 22:37-59

Jongmans WJ, Koopmans RG, Roos G (1938) Nomenclature of Coal Petrology. *Compt Rendu 2nd Congr Strat Geol Carbonif Heerlen* 3:1730-1737

Kuhlbusch TAJ, Andreae MO, Cachier H, Goldammer JG, Lacaux J-P, Shea R, Crutzen PJ Black carbon formation by savanna fires: measurements and implication on the global carbon cycle. *J Geophys Res* (submitted)

Kuhlbusch TAJ, Crutzen PJ A global estimate of black carbon in residues of vegetation fires, representing a sink of atmospheric CO_2 and a source of O_2. *Global Biogeochem Cycles* (in prep)

Medalia AI, Rivin D (1982) Particulate carbon and other components of soot and carbon black. *Carbon* 20:481-492

Mincev D (1965) Nomenclature groups of coal petrology sitz. *Ber Karpatho-Balkanische Geol Ass 7 Congr* 4:117-123

Potonie R (1924) *Einführung in die Allgemeine Kohlenpetrographie*. Berlin Verlag Gebrüder Borntraeger 34
Seyler CA (1948) The past and future of coal - the contribution of petrology. *Proc South Wales Inst Engin* 63:213-236
Spackman W (1958) The maceral concept and the study of modern environments as a means of understanding the nature of coal. *Trans New York Acad Sci* 20 Ser II Nr5:411-423
Stach E (1935) *Lehrbuch der Kohlenpetrographie*. Berlin Gebrüder Borntraeger
Stach E, Mackowski M-Th, Teichmüller M, Teichmüller R, Taylor GH, Chandra D (1982) *Stach's Textbook of Coal Petrology*. Gebrüder Borntraeger Berlin-Stuttgart 3rd edition
Stopes, MC. (1935) On the petrology of banded bituminous coal. *Fuel (London)* 14:4-13
Teichmüller M (1950) Zum petrographischen Aufbau und Werdegang der Weichbraunkohle (mit Berücksichtigung genetischer Fragen der Steinkohlenpetrographie). *Geol Jb* 64:429-488
Teichmüller M (1989) The genesis of coal from the view-point of coal petrology. *Int J Coal Geol* 12:1-87
Timofeev PP, Bogoliubova LI (1964) Main factors in the formation of generic types of humic coals and the specific features of their distribution in the different coal-bearing formations of the USSR. *CR 5e Congr Strat Geol Carbonifere* 3:1031-1038
Waltz IE (1956) On a new terminology of elements of the raw materials of bituminous coals. *Trudi Labor Geol Ugla Akad Nauk SSSR* 4:31-41
Webster's Third New International Dictionary (1965)
Winston RB (1993) Reassessment of the evidence for primary fusinite and degradofusinite. *Org Geochem* 20, 2:209-221
Wolbach WS, Gilmour I, Anders E, Orth CJ, Brooks RR (1988) Global fire at the Cretaceous - Tertiary boundary. *Nature* 334:665-669
Wolbach WS, Anders E (1989) Elemental carbon in sediments: determination and isotopic analysis in the presence of kerogen.*Geochim Cosmochim* Acta 53:1637-47
Zemcuznikov JA, Ginsburg IA (1960) *Basis of Coal Petrology*. Moscow-Leningrad Izdat Akad Nauk SSSR

Background and Local Charcoal in Sediments: Scales of Fire Evidence in the Paleorecord

J.S. Clark and W.A. Patterson III[1]
Department of Botany
Duke University
Durham, NC 27708 USA

Summary

Stratigraphic charcoal data are used to interpret combustion at a range of spatial and temporal scales. Interpretation is based on assumptions regarding how charred particles are transported to and deposited in sedimentary basins and how the variance structure of stratigraphic profiles is affected by transport. Here we examine evidence for "background" and "local" signals in stratigraphic data based on total abundances of particles and on distributions of particle sizes. Distributions of particle sizes in sediments and in the atmosphere show remarkable consistency, with a ~ 2% decrease in frequency for a 1% increase in particle diameter. Relatively large differences in source distance are required to produce differences in particle size distributions. Thus, stratigraphic samples with disproportionate representation of large particles are likely to represent a nearby source. Simple transport models suggest particles 10^0 to 10^1 μm have substantially longer atmospheric residence times than do particles > 10^2 μm. Several factors can produce a "dichotomy" (background vs local) in total abundances of charred particles. Transport during a single experimental burn shows abrupt decline in accumulation at a burn edge and relatively constant values out to 10^2 m. Saltation and redistribution of particles in surface runoff following fire are expected to focus accumulation within a catchment (local scale). The amount of transport that occurs by these modes is unstudied. The stratigraphic record appears to support a distinction between local, catchment sources vs. fires burning at greater distance. Charcoal profiles often show distinct peaks when fires burn within a lake catchment but are unaffected by fires that occur at greater distance. Broad subcontinental scale patterns in particle accumulation indicate shifts in regional importance of fire. The low frequency variance can be extracted from charcoal profiles and used to differentiate local from background changes in burning. Although the empirical data are still far too few to permit good characterization of particle transport, the evidence suggests utility in the concepts of background and local signals in profiles.

Introduction

Because charcoal stratigraphic data record emissions that can originate over a range of spatial scales, they can be difficult to interpret in terms of biomass burning at any one scale. Some of the earliest studies of fossil charcoal in palynological investigations of mires and lake sediments focused on local disturbance, both anthropogenic burning (Iversen 1941) and wildfire (Waddington 1978, Swain 1973,

[1] Department of Forestry and Wildlife Management, University of Massachusetts, Amherst, MA 01003-4210

NATO ASI Series, Vol. I 51
Sediment Records of Biomass Burning
and Global Change
Edited by James S. Clark, Hélène Cachier,
Johann G. Goldammer, and Brian Stocks

1978). Such analyses attempted to identify when fires occurred near the sample area as a basis for understanding how fire might have been used as a tool to manipulate vegetation or its role in forest succession. It was reasonable to expect that recent portions of these profiles might contain clean peaks matching those dated from catchment fire scars or historic documents. With this "validation" of the record in hand, one might then continue back into the past for a long-term view. Initial optimism for sediment charcoal records may have subsided somewhat as the first detailed records, including some from varved sediments, failed to reveal the neat patterns that might permit long-term interpretation of fire regimes. The high hopes for varved records in areas with independent records of fire were largely unrealized (Swain 1973, 1978, Cwynar 1978), perhaps because particle production, transport, and sedimentation were too poorly understood. Meanwhile, it became apparent that combustion products in sediments contained information on fire importance at much broader scales (Smith et al. 1973, Herring 1985). If charred particles accumulating in sediments originate from large regions, then sediment records might be used to understand how biomass burning affected atmosphere/biosphere interactions in the past. Analyses of elemental carbon in ocean cores revealed patterns that might result from changing regional fuel distributions and climate. As with lake and mire deposits, however, these records were difficult to interpret, because many poorly understood processes affect particle movement between production and sedimentation in the deep sea (Suman et al., this volume).

Speculation as to why charcoal profiles appear the way they do has focused attention but little research on methodology (how particles are quantified), basin characteristics, and transport of particles (Swain 1973, 1978, Cwynar 1978, Patterson et al. 1987, Clark and Robinson 1993). Concern over particle transport, particularly transport in the atmosphere, has led to a concept of "background" and "local" components of sediment charcoal abundances (Tolonen 1978, Clark 1988b). The local component is attributed to fires near a lake and background to particles transported long distances in the atmosphere. There is growing support for the utility of such a notion, but problematic examples remain. Until recently, there has been little attempt to find evidence for two general classes of source regions as implied by the "local/background" dichotomy.

In this chapter we examine evidence suggesting that sediment records of biomass burning contain information at a range of spatial and temporal scales. We review recent evidence from several sources that suggests there can be utility in the notion of background and local sources. In addition to factors already suggested as important for spatial resolution (e.g. basin size), we show that disparity among studies in how they emphasize different scales can result from methodologies that emphasize different particle sizes. We further point out that empirical evidence for water transport is lacking and potentially important for interpretations of scale. As part of our discussion on how methodology affects scale we offer suggestions on ways to increase comparability of studies and to emphasize the scales of most relevance to a given investigation.

What particle properties and transports would produce "local" and "background" signals?

A dichotomization of sediment charcoal signals into two source components implies a "threshold" response in the movement of particles from a burn to a lake. This notion implies high deposition close to the source, with abrupt declines at greater distance. A threshold distance could result because there exists i) a discontinuity in particle diameter distributions (with attendant consequences for transport), ii) a spatially discontinuous dispersal process, or both. A bimodal particle size distribution (as occurs with aerosols) is an example of a distribution that could produce a threshold in transport, with the large-diameter mode remaining nearby and the small mode transported further. In this section we examine the evidence for these two possibilities. We first discuss distributions of particle sizes in the size range quantified in sediment studies. We then consider whether transport processes might be expected to produce something like "local" and "background" signals.

The distribution of particle sizes in sediments

Do distributions of particle diameters in sediments show evidence of discontinuities between large and small that might produce local and regional signals? Diameter distributions for large numbers of samples in North America and Siberia are surprisingly conservative, with a 1.5 to 3 % decline in frequency for each 1% increase in particle size (Fig. 1). This slope of –1.5 to –3.0 on a log/log plot is observed in sediments from Minnesota, North Dakota, Siberia, Florida, and several sites in the northeastern United States, it is obtained in samples prepared by different methods, and it spans several orders of magnitude in particle diameter (Clark and Hussey 1996, Clark et al., in review). Figure 1 compares diameter distributions determined on slides prepared for pollen analysis and in samples sieved with a 180 μm mesh from Bor Lake, Siberia with those obtained on thin sections (Clark 1988a) from northwest Minnesota. Different diameter ranges represent different magnifications (15× for sieved samples, 64× for thin sections, 400× for pollen slides) and different methods of analysis. The size ranges of particles are given on Figure 1 and include thin sections: all particles > 20 μm, sieve samples: particles > 180 μm, and pollen slides: particles > 5 μm). When normalized for different diameter ranges included by each method (Clark and Hussey 1996), slopes are comparable. Thus, distributions are rather conservative and lack clear discontinuities. There is no sign of multiple modes as can occur in aerosol size classes.

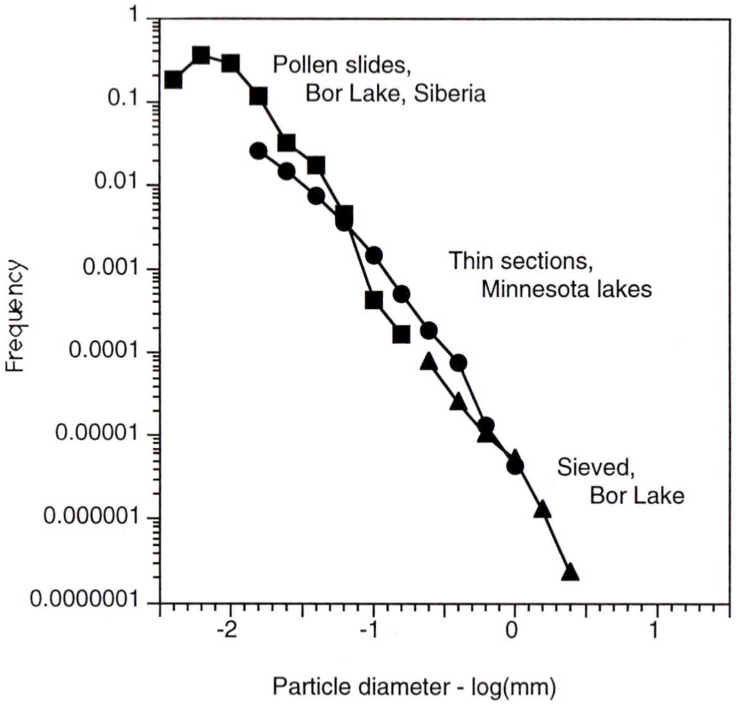

Figure 1. Comparison of normalized charred particle diameter distributions by two methods, pollen slides and sieved, from Bor Lake, Siberia and a third method, thin sections, from northwest Minnesota. Distributions are consistent across sites with different diameter ranges determined by magnification and method. Data from Clark and Hussey (1996) and Clark et al. (in review).

The smooth distribution of particle sizes without multiple modes means that particle properties are not expected to result in two distinct signals, e.g. a small diameter mode that remains aloft for extended intervals and a larger mode of rapidly-settling particles. Clark and Hussey (1996) found that particle shapes and settling rates also varied continuously across this diameter range (Fig. 2b). On the basis of particle properties extracted from sediments there is no reason to expect a natural break between locally dispersed large particles and regionally dispersed small particles within the size range of 5 μm to 3 mm.

Lack of clear discontinuities in diameter distributions does not preclude refinement of source region based on particle-size information. First, if transport depends on particle size, then categories of particle sizes might be lumped in ways that emphasize differences in transport distance. For example, although there is not an abrupt change in settling velocities at, say, 50 μm, there is a large (order of magnitude) difference in the expected fall velocities of particles in samples restricted to 10 to 50 μm diameters vs. those > 50 μm (Fig. 2b). The 10-to-50 μm size class is expected have a larger source region than the greater-than-50 μm size class. Second, if transport processes are highly non-linear in diameter, then there is also potential for separation of sources. Suppose atmospheric lifetimes of particles decrease abruptly at a particular diameter. Then analyses of diameter classes above and below this threshold would be expected to maximize separation of source area. It is further possible that different transport mechanisms dominate near and distant from a particle source. In the next section we consider modes of transport in the atmosphere and assess how they compare to the sediments distributions in Figure 1. Saltation and water transport are mechanisms for highly localized transport we consider. We then follow with evidence from sediments for local and background signals.

Transport in the atmosphere

Information on particle movement in the atmosphere from empirical analyses of air pollution and of fires is largely restricted to aerosol (< 2 μm) size classes, substantially smaller than the particles typically quantified by optical microscopy of sediments. These aerosols are widely dispersed. Observations from space show that smoke plumes can traverse whole continents (Andreae 1991). Empirical data come from such diverse sources as emissions estimates from individual burns (Sandberg et al. 1979, Radke et al. 1978) and regional data bases of particulate content of the atmosphere (e.g. Ogren 1982, Crutzen and Andreae 1990, EPA 1994). Individual burns provide estimates of how much of consumed fuels is represented by airborne particles. These "emission factors" vary among fuel types and with fire behavior. The data are spotty at best, because each estimate requires a well-instrumented experimental burn with pre-burn fuel measurements. The most up-to-date synthesis is by Stocks and Kauffman (this volume). Regional studies show how changes in combustion practice affect air quality. There are few such data on larger particles extracted from sediments.

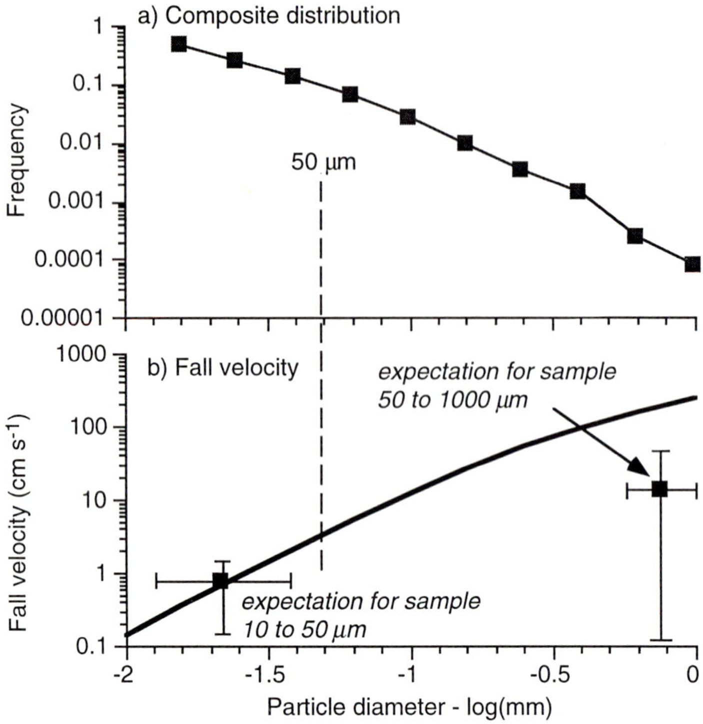

Figure 2. A composite distribution of particle size distributions from eastern North American lakes (a), fall velocities (b), and expected diameters and fall velocities for samples of 10 to 50 μm and of 50 to 1000 μm particles (Clark and Hussey 1996). Error bars are root mean square deviations calculated from the second moment of the particle size distribution. Although there is no sharp change in fall velocity at any one diameter value, the differences in expected values for whole samples taken over different diameter ranges are large.

A key factor influencing transport is fire intensity, which affects plume buoyancy (Clark 1988b). Turbulence lofts suspended particles, depending on particle size. As thermal buoyancy of the plume is lost, particles increasingly begin to settle out. Larger particles have higher settling velocities and so are preferentially lost sooner from the plume. Fire intensity and the wind field (which largely determines fire intensity) are extremely variable, resulting in a broad range of heights, from aerosols in intense convection columns that reach > 10^3 m to particles too large to be entrained in low-intensity surface fires.

Insights into how large particles move in the atmosphere are provided by transport models that assume forces acting on individual particles to predict how concentrations change over time. The source region for particles recovered in lake sediments depends on how long particles lofted during burns reside in the atmosphere. Once entrained in the atmosphere particles are removed by diffusion, settling, and precipitation. Each of these processes has an associated time scale, one of which (settling) is highly sensitive to particle size. Müller (1984) relates residence time τ for particles evenly distributed over a troposphere mixed layer of height H m as

$$\frac{1}{\tau} = \frac{v_0}{2H}\left[\left(\frac{d}{d_0}\right)^2 + \left(\frac{d_0}{d}\right)^2\right] + \frac{v_I}{H} + \frac{1}{\tau_{wet}}.$$

The vertical concentration profile is ignored here, an assumption relaxed in the next example. Three terms on the right-hand side are settling, impaction, and wet removal in precipitation. Settling rate depends on particle diameter d and is estimated relative to particle diameters having the longest residence time $d_0 \approx 0.3$ μm with fall velocity $v_0 \approx 0.002$ cm s^{-1}. v_I is a mean deposition velocity taken by Müller (1984) to be 0.018 cm s^{-1}. τ_{wet} depends on precipitation frequency, being roughly a week in many temperate regions. Müller's analysis is simplistic, but the differences in time scales it predicts overwhelms the approximate nature of parameter values and simplifying assumptions. For particles larger than 10 μm settling is the term that dominates atmospheric lifetime. Wet removal becomes the dominant term for aerosol size classes, meaning that particles remain suspended until washed out by precipitation. Thus, for particles in the size ranges quantified by optical microscopy (10 to 1000 μm), settling is likely to yield the most useful information on transport.

The time scales for removal processes imply transport distances that have been explored using simple Gaussian plume models (Clark 1988b). These models make the simplifying assumptions that particles are released at some height h (m) from a buoyant convection plume (as opposed to being evenly distributed throughout the troposphere), are wafted along by winds of velocity u (m s^{-1}), and eventually settle to the ground at a distance that depends on particle diameter, which determines settling velocity, and diffusion (Sutton 1953). Because settling velocity increases with particle diameter (Fig. 2), the model predicts an "evolution" of the diameter distribution, depending on three length scales in the model, height of particle release, distance downwind from the source, and wind speed (Fig. 3). We refer to wind speed as a "length" dimension in this context, because it serves to translate the plume downwind. Principle conclusions from this simple analysis are i) particles exceeding 10^0 mm in diameter released close to the ground are lost within 10^2 m from the source, ii) particles < 10^{-2} mm remain aloft well beyond 10^2 m, even when released from heights as low as 10 m, and iii) particles as large as 10^1 mm lofted to moderate (10^2 to 10^3 m) altitudes can be transported long distances from the plume, even at low to moderate wind speeds (0.5 to 3 m s^{-1}).

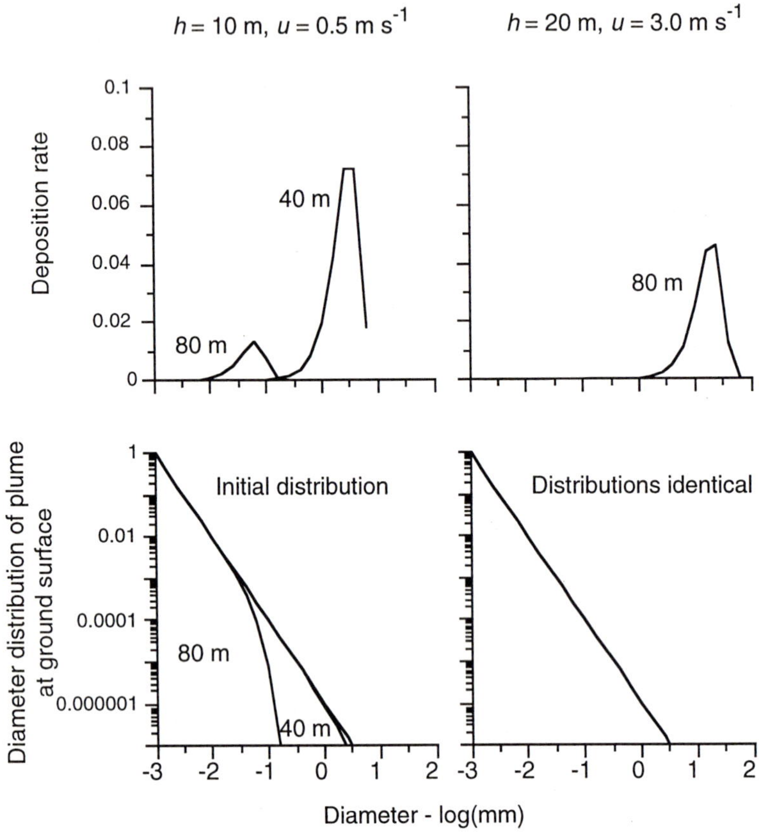

Figure 3. Distributions of particle diameters deposited at the ground (above) and in the plume at the ground surface (below) at two distances from the point of release (40 m and 80 m) based on a Gaussian plume model (Clark 1988). For particles released within 10 m of the ground in low winds (left) almost all particles > 10^0 mm are deposited within 40 m and particles > 10^{-1} mm are deposited within 80 m. At somewhat higher release heights and moderate winds (right) particles > 10^0 mm are deposited within 80 m, but smaller particles remain aloft.

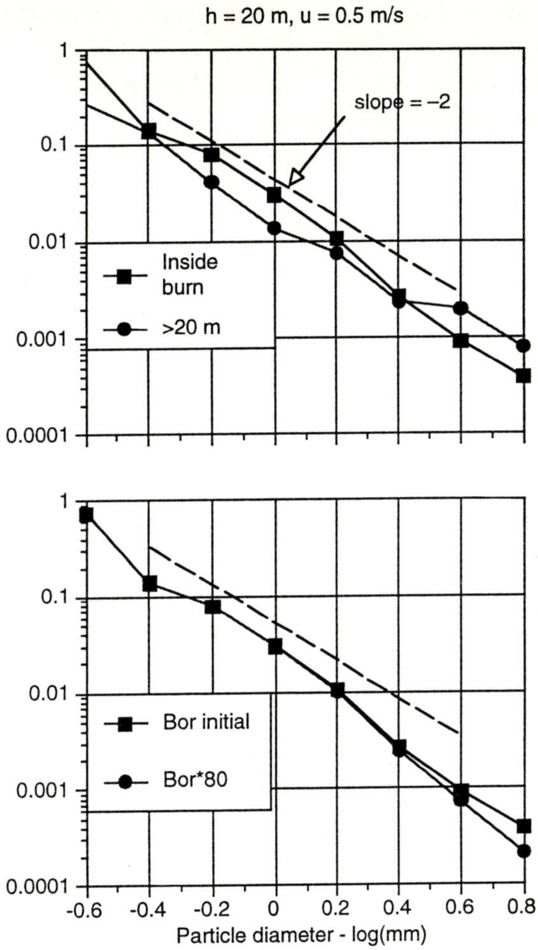

Figure 4. Above: Comparison of particle size distributions from samples within the Bor Island burn with those obtained in particle traps at a distance of 80 m. Slopes of the distributions are close to –2 (Clark et al., in review). Below: Using the distribution inside the burn as an initial distribution, the Gaussian plume model predicts only a slight decline restricted to the largest size classes at 80 m.

Direct tests of these predictions are not available, because the assumption of a point source is unrealistic for fires. Each burn has a vertical distribution of release heights and locations as a burn expands from the point of ignition. To gain some rough perspective for how a real burn compares to the predictions of a Gaussian plume we plotted initial particle size distributions from an experimental burn in Siberia (Clark et al., in review) against model results based on the observed initial particle size distribution. The experimental burn consumed surface *Cladonia* and some of the Scots pine crown fuels on a 50 ha island. The buoyant convection column from this intense fire reached > 5000 m. Particles were collected in particle traps on the surrounding fen surface to a distance of 80 m.

The intensity of the burn and column height precluded an evolution of particle size distributions with distance from the burn. The distributions from particles trapped within the burn did not differ significantly from those recovered at 80 m (Fig. 4, above). Model results assuming a release height of 20 m predicted no real change from the initial distribution, consistent with empirical data (Fig. 4, below). Because the bulk of the plume reached well above 20 m used in Figure 3 (> 5000 m), model predictions of a conservative diameter distribution are consistent with observed particle distributions.

The pattern of particle accumulation with distance demonstrates a clear nonlinearity of the sort implied by a notion of background and local sources (Fig. 5). There is a sharp decline within 5 m of the burn edge and no further trend in accumulation rate with distance outside the burn to 80 m; values are variable but the average is well below that at the burn edge. The non-linearity occurs at the burn edge; thus, if there exist two signals in charcoal profiles between "local" and "background", these data suggest the separation between the two lies between fires that burn to a lake edge (local) vs. those that do not. Fires that do not extend to within relatively short distances of a lake edge might still contribute substantial emissions to the atmosphere. But these emissions might be distributed rather evenly in space compared to the steep decline that occurs at the burn edge. Below we examine examples from stratigraphic studies.

Particle transport by wind may occur after the burn. Saltation of particles across burned landscapes is expected to be greatest for charred particles having diameters of roughly 100 μm. These particles have the lowest critical velocities for suspension (~ 10 m s^{-1}) and so are lifted by moderate winds (Clark 1988b). They are too large to remain suspended in light winds and so move by series of small leaps across the landscape. Because particles remain close to the ground, saltation is expected to move particles at catchment scales, with trapping in surrounding unburned vegetation. Saltation might therefore most emphasize a "local" component of charcoal profiles. Charred particles have been observed to diminish on the ground surface and to collect in topographic lows in days following fires (W.A. Patterson, personal observation). Saltation is the likely mode of dispersal for these macroscopic fragments. Surface flow that occurs soon after a burn might also redistribute particles locally.

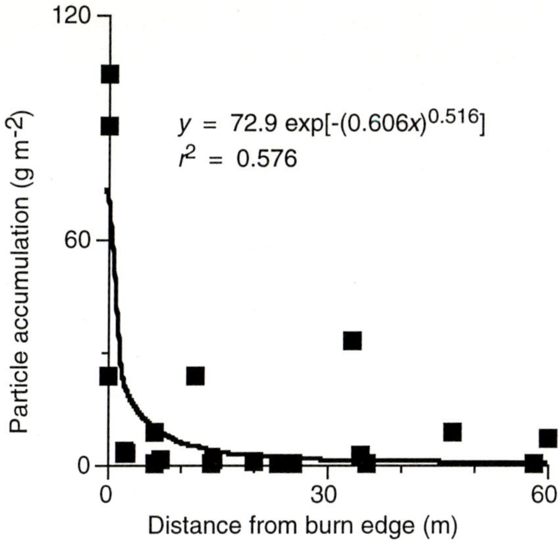

Figure 5. Pattern of particle accumulation with distance from the Bor Island burn. From Clark et al. (in review). Three samples along the vertical axis at within 3 m of the burn edge.

Particle transport models and a single experimental burn where large particles have been measured suggest that total accumulation rate and particle size distributions may be sensitive to sources at different spatial scales. Total particle accumulation may decline precipitously at a burn edge, marking a possible distinction between "local" and "background" near this point. Saltation might emphasize the local scale, because saltating grains are readily trapped in unburned vegetation around lakes. Thus, total accumulation of charred particles can be highly sensitive to "local" vs "background" sources, i.e. fires that burn into a lake catchment and, especially, to a lake edge vs. those that do not. Size distributions of suspended particles evolve gradually due to continuous change in settling velocity with diameter and a range of particle release locations and heights. Particle size distributions appear to be conservative and, thus, differences among distributions might only be apparent when there are large differences in source location.

Transport in flowing waters

Although water transport of charred particles is frequently invoked by paleoecologists, there is little research on the subject. Charred particles are buoyant with a density of about 0.5 g cm^{-3} (Clark 1988b). They are thus expected to be readily transported by streams or by any surface flow that may enter a collecting basin. Once on a lake surface they are expected to move with surface currents until they become waterlogged. We thus expect an important role for water transport in lakes having large inflowing streams and in catchments where surface flow is important (low infiltration rates, shallow soils, and steep slopes). We further expect winds to result in particle concentrations in littoral areas.

Clark (1988b) examined the argument that water transport may be the dominant vector for particle transport to lakes. These arguments are typically based on the assumption that large amounts of particles are moved in surface flow. The assumption does not apply to "typical" catchments. Surface flow does not occur following most fires, and it is generally uncommon in uncompacted forest soils (e.g. Waring and Schlesinger 1985). Analysis of lake waters in catchments with modest relief show little change following fire and rapid regrowth of vegetation (Wright 1976, McColl and Grigal 1977). Few fires identified by sediment charcoal in varved sediments in Minnesota (Clark 1990) and New York (Clark et al. 1996) are attended by increased varve thickness that might indicate erosion. The examples cited for increased erosion following fire come almost exclusively from burns that are especially intense (including logging slash fires and following blow downs)(e.g. Schindler et al. 1980, Swanson 1981), on steep slopes, or on soils with especially low infiltration rates (DeBano and Conrad 1978). Arguments for transport of pollen in surface flow come from a few studies from compacted agricultural soils or from lakes with inflowing streams (Peck 1973, Bonny 1976). There appears to be little appreciation for the substantially higher surface flows on compacted agricultural soils than occur in uncompacted forest soils. Inflowing streams constitute continuous surface flow capable of transporting large amounts of buoyant particles. Inflowing streams should be a potential important source of particle inputs where they occur. Obviously, they cannot used to argue for the importance of surface flow generally.

Where surface flow is important, it can transport particles at a watershed scale. If large inputs do move in water, the difference between "local" and "background" can be no larger than the size of the watershed. Experiments and mass balance studies of particle transport during and after burns are needed to assess the relative importance of water transport for charred particles.

Sedimentation processes

We do not review sedimentation processes here, because an excellent analysis of secondary deposition following fire is provided by Whitlock et al. (this volume). It is important to mention, however, that much charcoal analysis has been done in peats and other organic deposits, and interpretation of these records should consider differences between deposition here vs that in lakes, including i) deposition on subaerial surfaces is mostly atmospheric, and ii) burns can occur at the site itself, when dry conditions lower water tables or when dense emergent macrophytes carry fire above standing water (E. Grimm, pers. comm.). The first difference means that secondary deposition processes that might concentrate (e.g. focusing) or minimize (e.g. flotation to littoral zones) particles in lakes do not occur on subaerial peats. A peat surface may better reflect atmospheric fallout than lakes, provided the record is not obscured by water level fluctuations. There may be some vertical migration of particles in lose *Sphagnum* canopies; movement of charred particles on peat surfaces has not been studied. Deposition within lakes is expected to be less subject to water table fluctuations. Fires that occur directly on a site often leave distinct lenses, thus identifying where fire occurred. Obviously, some of the record is lost when the peat surface burns. Because particle accumulation rates in lakes of eastern North America (Clark and Royall 1994, 1995, Clark et al. 1996c) and Siberia (Clark et al., in review) are higher than expected based on current knowledge of fuel loads, fuel consumption, and emission factors, we suspect that sediments in the centers of lakes (where cores are taken) overestimate particle production. If so, organic peats and soil horizons would be expected to yield lower accumulation rates than lake sediments.

Evidence for local and background signals

The sediment evidence itself contains support for the utility of the concept of local and background signals. First is evidence for an inhomogeneous distribution of accumulation rate values in sediments, with extreme values concentrated at times when independent evidence suggests fire near a lake. We began this chapter by pointing out the charcoal profiles that failed to show evidence of documented nearby fire. These studies generally did not sample continuously, so hitting a "peak" is a chance event, and they focused on small (including 10 μm and lower diameter) particles, so samples might respond more to regional burning rather than local fires (Clark 1988b). It is not impossible to find evidence for fire in profiles that are discontinuous and that emphasize small particles, but it may be less likely than in continuously sampled profiles of large particles. Several studies are now available where samples are closely spaced within a core, and quantification is restricted to mostly large particles. Although these studies do not show one-to-one correspondence between all charcoal peaks and all nearby fires, they do indicate a much closer match than did the early sediment charcoal analyses.

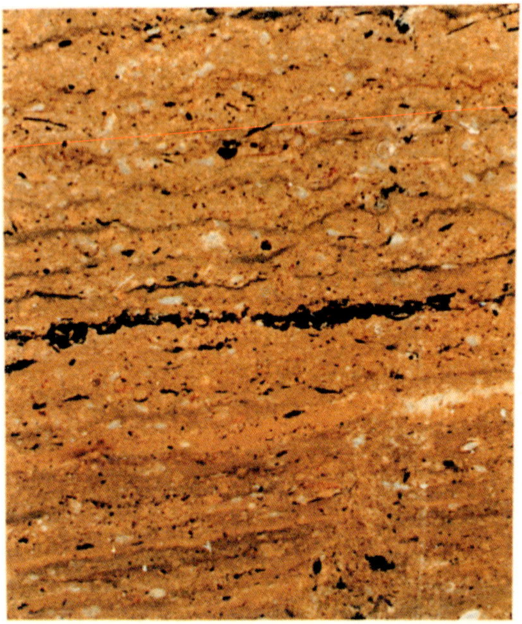

Figure 6. Photograph of a thin section of Deming Lake sediment showing annual laminations and a distinct lens of charred particles when the catchment of the lake burned.

Analysis of large particles on thin sections from varved sediments of three lakes in northwest Minnesota (Clark 1990) revealed distinct lenses of charcoal during years when fires occurred in the lake catchment and lenses generally did not occur otherwise, even during years when neighboring catchments burned (Fig. 6). A steep decline in charcoal followed with fire suppression that became effective after 1920. Patterson (1978) also found that charcoal declined in the 20th century at nearby Squaw Lake. The 1947 fire on Mt Desert Island, Maine produced peaks in charcoal series from catchments that burned, whereas series from other parts of the Island do show increases at this time (Fig 7). Charcoal lenses were absent over the last 2000 yr in the catchment of Crawford Lake Ontario, except when Iroquois occupied the catchment, a time when local fire is expected (Clark and Royall 1995b). Even when European settlement caused large increases in the average accumulation of charcoal since the 20th century, there was no occurrence of the large peaks that characterized local fires at Minnesota sites and the peak at Crawford during Iroquois occupation. Rather than a homogeneous distribution of accumulation rate values in the sediments of Devil's Bathtub, NY, Clark et al. (1996b) showed clear separation between large peak values that occurred at an average interval of 80 to 90 yr in early Holocene spruce/jack pine forest. This inhomogeneity in the distribution of values could not be explained by episodic erosional events; annual sediment accumulation was uncorrelated with charcoal accumulation. Large peaks in profiles of sieved (125 μm) charcoal from contiguous sediment samples of Wyoming lakes showed large peaks, many of which corresponded to independent evidence of fire (Millspaugh and Whitlock 1995). Peaks were substantially higher than intervening "background" values.

A detailed analysis of sediment charcoal from a lake in Wood Buffalo National Park, Alberta included contiguous five-yr samples (MacDonald et al. 1991). This study was one of the first to compare a range of particle size class information to fire importance at different spatial scales. The stratigraphic data are difficult to interpret, because the independent record of fire is limited to fire scars on jack pines, which are not abundant around the lake, and they are often killed by fire. Fifteen fire scars were located within several km^2 surrounding the sample lake. The charcoal series from this study show many changes in abundance, but the numbers and sizes of fires burning around the lake is largely unknown. A comparison of different particle classes with modern fire statistics is limited to the scale we term "background", because even the finest spatial scale extends to 20 km. Some of the fires recorded in fire scars on jack pine match peaks in the charcoal record, while others do not. Charcoal values were higher in the 19th century times when fires appear to have entered the lake catchment and spruce was established.

A second line of evidence supporting a concept of local and background signals is a tendency for particle sizes to increase when independent evidence suggests nearby fires. The proportion of large particles declined in 20th century samples from Squaw Lake following fire suppression (Patterson 1978). Clark and Royall (1995a) asked whether charcoal analysis by methods that emphasize different particle sizes would show evidence of altered geographic patterns in combustion since the 19th century in eastern North America. Pollen slide and thin section methods of analysis (Fig. 8) were available from a suite of lakes spanning the upper Midwest to northeastern North America. Because thin section analyses emphasize larger particles, Clark and Royall hypothesized i) that Midwest sites would show a sharp decline in charcoal accumulation with fire suppression and tillage of prairie following cultivation and ii) that Northeastern sites, where local fires were less common, would show little response to European settlement. Pollen slides were predicted to record these transitions differently. Because of small size and potentially broader dispersal, northeastern sites were predicted to show large increases with settlement due to regional increases in burning in the Northeast. Although local fires in forests surrounding individual sample lakes might not occur, use of fire for heating, fuel, and industry increased throughout the Northeast. Because of their small size, pollen-slide particles were not predicted to show the large decreases with fire suppression predicted for thin sections. Again, broad dispersal might record continued combustion on a regional scale following fire suppression. The patterns were consistent with these predictions based on particle size (Fig. 6), supporting that different particle sizes emphasize signals of different spatial scale. There are many factors that affect calculated fluxes of charred particles in sediments, so we do not expect every site to display the geographic pattern suggested by this analysis. Changing patterns of deposition within lakes affect calculated fluxes, so much variability is expected. Smoothed patterns of charcoal accumulation across eastern North America (see below), however, suggest a broad scale pattern consistent with changes in burning from prairie in the Midwest to eastern temperate forests.

If large accumulation of particles occurs when fires are nearby, and if large particles have a more restricted source area than small particles, then we predict positive correlation between mean particle diameter (or the fraction of large particles) and total accumulation rate. Clark et al (in review) tested this hypothesis in sediments from Bor Lake, Siberia, and obtained results consistent with this hypothesis for one method (sieved particles) and not for another (pollen–slide particles). Although some correlation is expected between particle size and total accumulation rate, failure to obtain correlation indicates that the practice of interpreting local fires when large particles are found is questionable. At Bor Lake, average diameter of sieve particles was positively correlated with accumulation rate, consistent with the notion that higher accumulation rates also mean closer fires (Fig. 9, lower panel). There was no such relationship between accumulation rate and particle size for particles counted on pollen slides. Thus, the size information from the small (~ 10 μm), pollen–slide particles may be less useful than thin-section particles for interpreting source regions. Expected fall velocities were used to calculate expected atmospheric lifetimes for particles. For particles lofted 10^2 m above the ground these time scales are on the order of minutes (Fig. 9, upper panel). For particles lofted in large convection columns that reach 10^3 m residence times are hours.

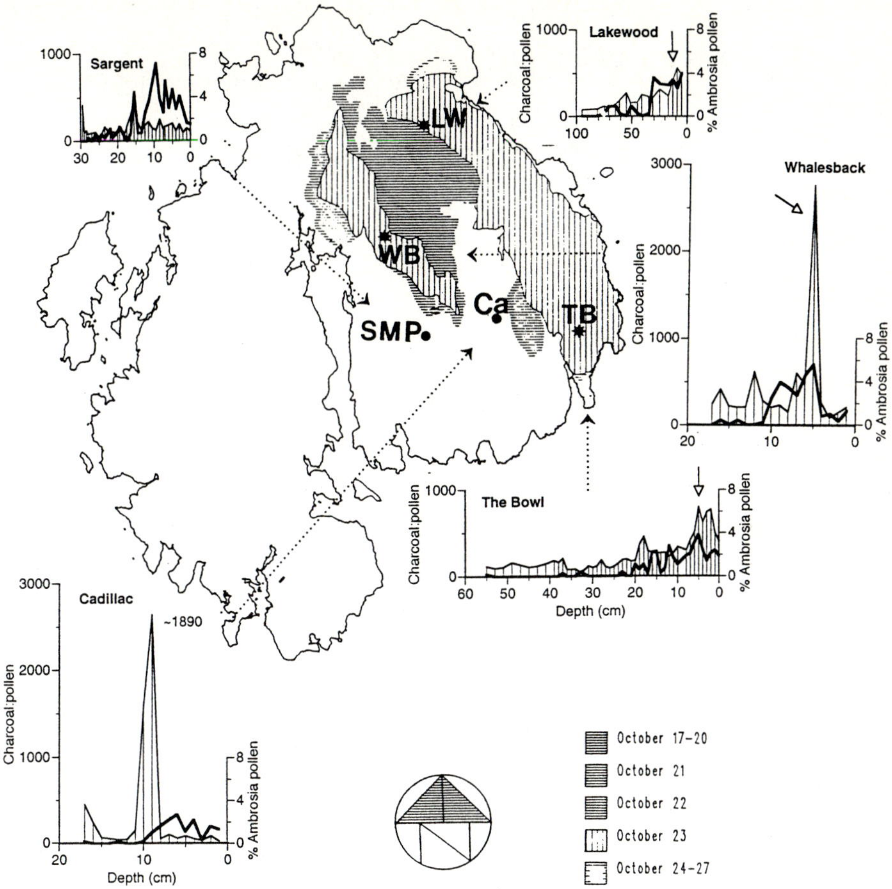

Figure 7. Charcoal profiles from sites within (Lakewood, Whalesback, The Bowl) and outside (Sargent, Cadillac) the 1947 burn on Mt. Desert Island. Profiles are dated by pollen cultural horizons, including *Ambrosia* (thick line), which increases at ~ A.D. 1790 and ^{210}Pb (not shown). All sites within the burn show increased charcoal (thin lines) after 1947, whereas sites outside the burn show no such increase. The large peak at Cadillac occurs earlier and corresponds to burn that occured there at ~1890.

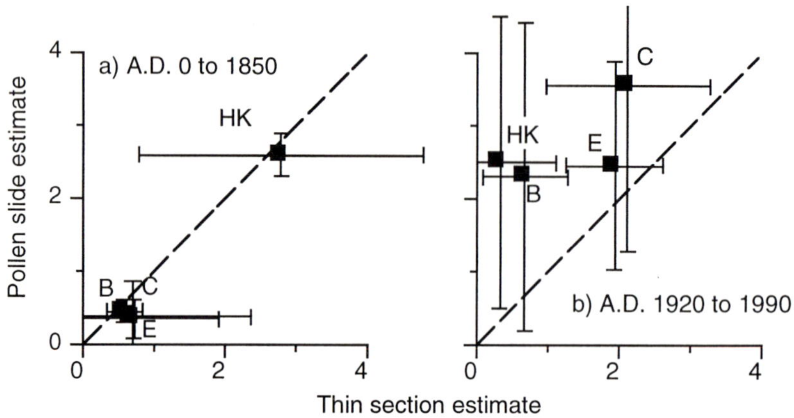

Figure 8. Comparison of thin section ("large" particles > 50 μm) and pollen slide ("small" particles > 5 μm) charcoal for pre and post settlement times. The dashed line of agreement is indicated in both plots. The two methods show similar values for presettlement time, but small pollen slide particles increase disproportionately in the 20th century. Included are three northeastern US sites (B - Basin Pond, Maine, C - Conroy Lake, Maine, E - Ely Lake, Pennsylvania) and one midwestern site (HK - Hell's Kitchen, Wisconsin). Values are $mm^2cm^{-2}yr^{-1}$. Calculated from data in Clark and Royall (1995a).

Composite diameter distributions for samples with different accumulation rates clearly show a tendency for more large particles when accumulation rates are high (Fig. 10). Four portions of the Bor Lake core were selected to represent "High", "Intermediate", and "Low" accumulation rates. "High" accumulation rates also mean greatest frequency of large particles and vice versa. The correlation between total accumulation and the fraction of large particles was also obtained for pollen-slide charcoal (Patterson 1978) at Squaw Lake, Minnesota.

A third pattern consistent with local and background signals comes from geographic patterns in charcoal accumulation. Patterson and Backman (1988) and Clark and Royall (1994) noted general decline in charcoal accumulation in presettlement samples from the Midwest eastward to the Northeast. Clark et al. (1996a) showed that the frequency distribution of charcoal accumulation rates across the same region followed underlying vegetation patterns, with the exception that charcoal showed especially sharp decline within Great Lakes pine forests between Minnesota and Wisconsin sites. More striking, however, are geographic patterns illustrated by maps from a recent compilation of charcoal accumulation rate data across eastern North America (Clark et al 1996c). Patterns before European settlement clearly respond to regional climate (Fig. 11). Despite large peaks in Midwest samples that record individual burns, the smoothed values that are controlled by a majority of background values clearly show the prairie/forest border, further decline that roughly follows the intrusion of the Pacific Air stream into the Lake States in summer, and the lowest values of the eastern deciduous and northeastern boreal forests, where fires were rare. This is the pattern expected if broad patterns in climate control the background component of charcoal profiles.

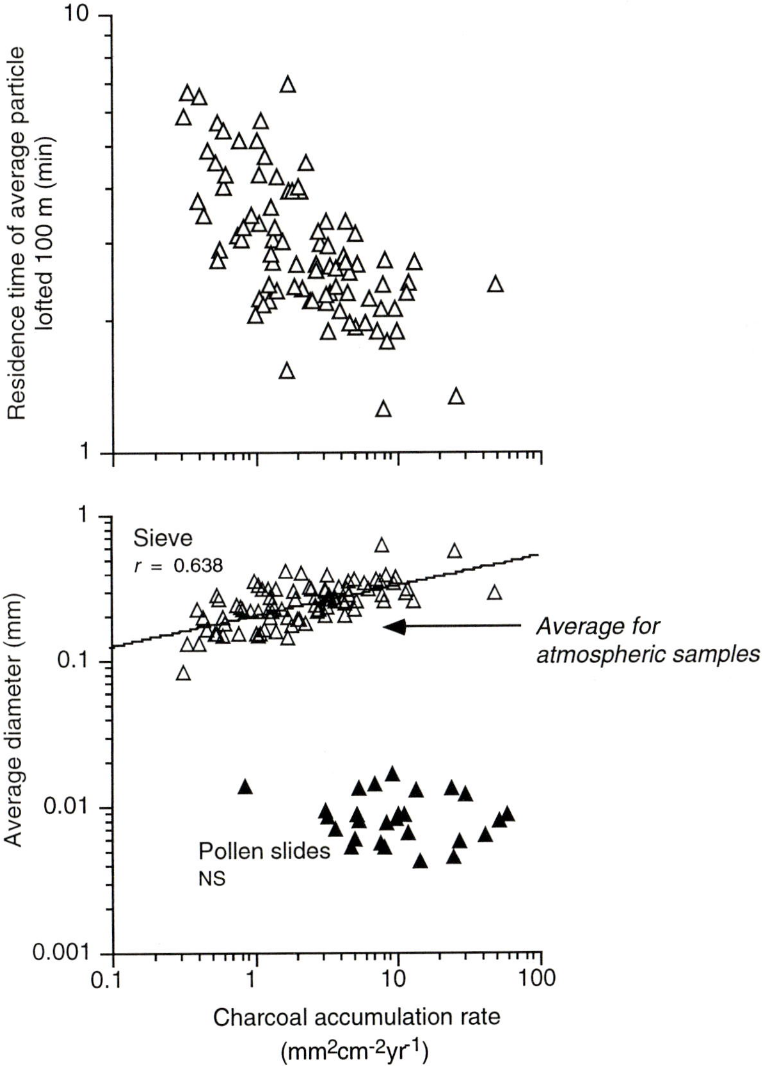

Figure 9. The relationship between average particle size and accumulation rate for large "sieve" and small "pollen–slide" particles (below) and atmospheric lifetimes for sieve particles (above) from Bor Lake, Siberia. (From Clark et al., in review)

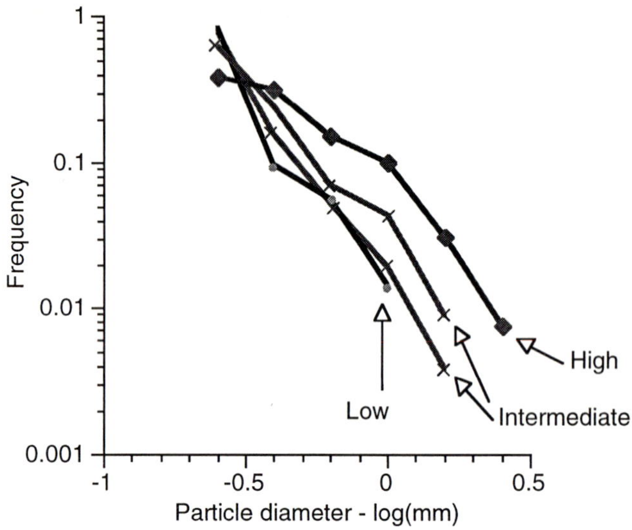

Figure 10. Particle size distributions for four sections of the Bor Lake core having "high", "intermediate", and "low" accumulation rates (From Clark et al., in review).

Separating the signals

Although "background" and "local" signals may exist in some charcoal profiles there is sure to be overlap. How might we emphasize these two signals to better separate regional patterns from events occurring nearby? One candidate for background is a simple average of accumulation rates across samples in a core. This "temporal" average tends to be dominated by the many years when no fires occur nearby. Unless fires are especially frequent we expect this average to be minimally influenced by the few peaks that may occur during nearby burns. We might subtract this average from a series and view the resultant series of peaks as something closer to a local signal. A limitation of a simple average, however, is that the controls on background inputs, e.g. regional climate, may vary on many time scales. By simply subtracting the mean we may ignore much "regional" variability that occurs at higher frequencies than the length of the series used for averaging.

Low frequency variability expected to characterize "background" can be emphasized by smoothing a series. A smoothed series can be created and then subtracted from the original profile. The smoothed profile emphasizes low frequency variability. The residual variance emphasizes high frequency variability, i.e. the variability associated with short time scales and, thus, primarily, local sources. A limitation of such an approach is the fact the high-frequency variance is not really removed, it is just smoothed out.

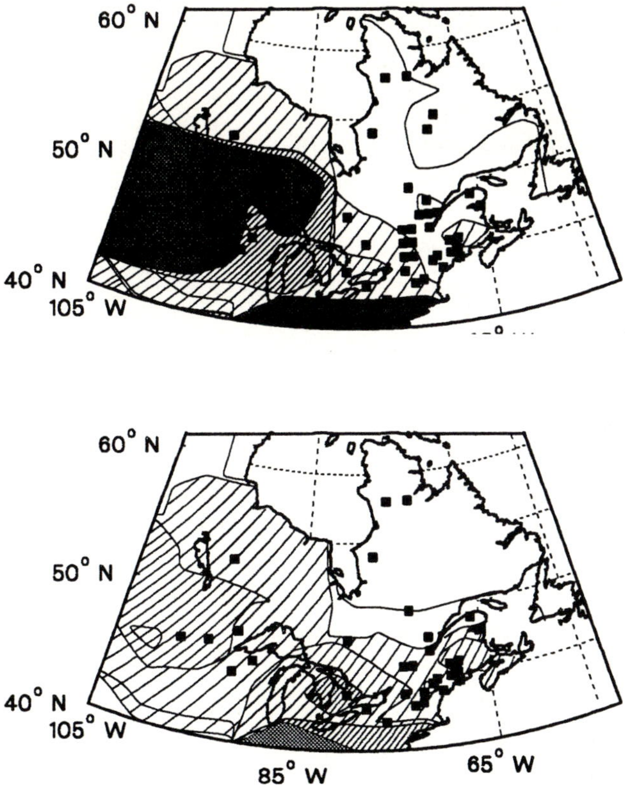

Figure 11. Contour maps of charcoal accumulation rate for presettlement (A.D. 0 to 1850, upper) and 20th century (lower). Symbols indicate locations of sample lakes. Accumulation rates are calculated as the averages for samples falling within the two time periods (Clark et al. 1996c).

Clark and Royall (1996a) emphasized regional and local sources by removing the high frequency variability from charcoal series and subsequently examining it independently. The method involved producing a Fourier transform of the series. The low frequency was then modeled from the transform, and the modeled transform was inverted. The resultant series emphasizes variability at scales of decades to centuries. The series of residuals emphasizes frequencies less than a decade.

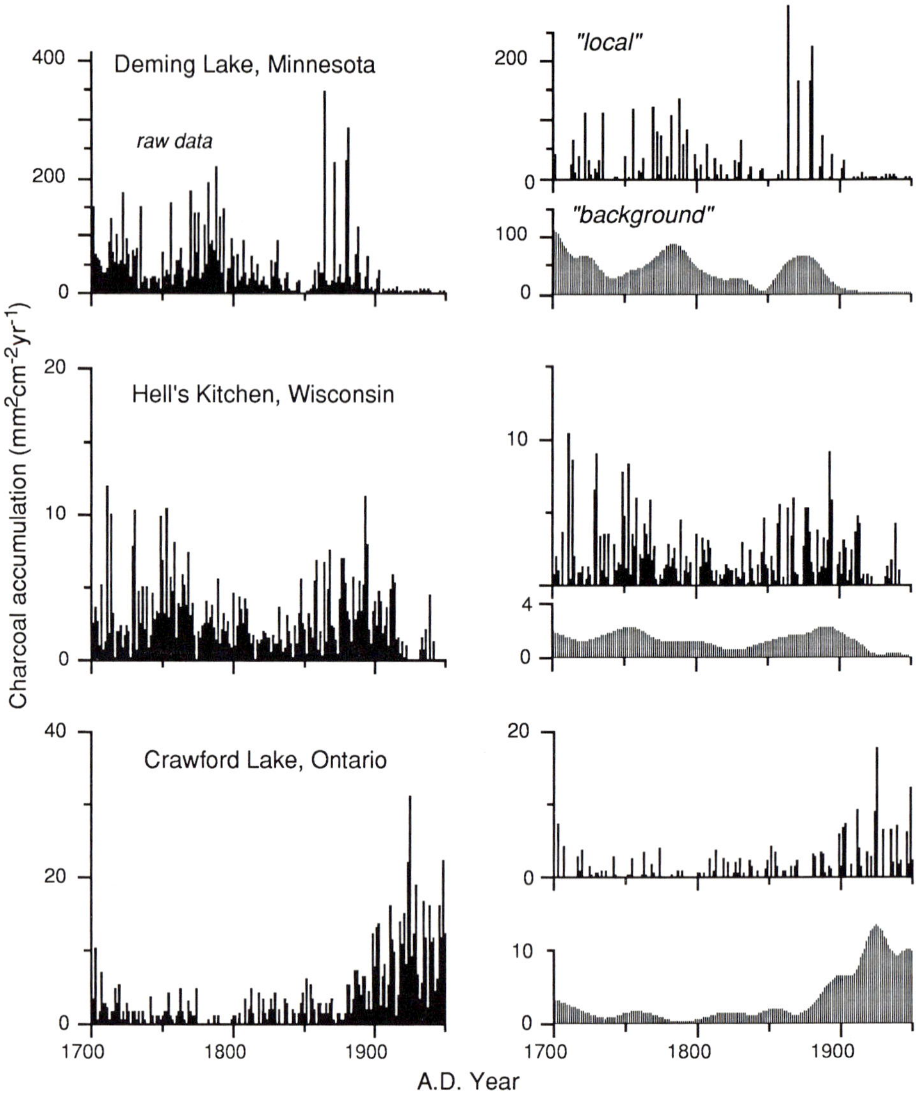

Figure 12. Series filtered to emphasize ("background") and remove ("local") 20-yr components of charcoal series using the method of Clark and Royall (1996a). Fourier transforms of raw series (left) were modeled and then inverted. The "background" is this inverse, and the "local" is the residuals of this inverse.

In most cases local and background signals were similar. An example is Deming Lake, where local fires disappeared with region-wide fire suppression after A.D. 1920 (Fig. 12). Presettlement fires within the catchment may have been controlled by climate variability that is present in "background" series; both series show similar patterns. Several exceptions to this general pattern were observed. Hell's Kitchen lies within a mosaic of forest types. Northern hardwoods around the lake are not expected to burn frequently, but nearby are stands that include substantial jack pine. This series was dominated by low frequency (> 10 yr) variability (Fig. 12); large peaks that might signal nearby fire are not obvious. Variability at the low frequency includes the decline at fire suppression. The decline is clear in the background diagram, but less obvious in the local diagram. An interpretation consistent with this analysis is that fires may have occurred in the region, but were infrequent locally. The background series records this regional decline, and the local series records little change in the catchment.

Crawford Lake lacks the dominance of low frequency variability, suggesting little effect of climate fluctuation on fire occurrence in the region. There is a significant increase in charcoal accumulation during Iroquois occupation (A.D. 1360 to 1640) and a larger increase after A.D. 1850 (Fig. 12), probably associated with regional combustion activities. The catchment of Crawford Lake is wooded and probably not the source of the recent charcoal increase. The background series is dominated by this increase, which is large and prolonged. The local series is less dominated by this increase. The comparison of components is consistent with the interpretation that late 19th century increases were regional, and little change occurred near the lake itself.

Conclusions

Although we find no evidence for a strict dichotomy in source regions for particles extracted from lake sediments, we identify several transport mechanisms expected to emphasize different source areas. Settling is the dominant mode of particle removal from the atmosphere. Neither particle diameter distributions nor settling velocities suggest discontinuities of the sort expected to produce clear separation of signals. Particle diameter distributions are continuous and conservative among lakes and similar to those in smoke columns. They have a single mode, likely below the smallest diameters quantified by optical microscopy, suggesting a continuous gradation of responses to removal processes from air and flowing waters. Atmospheric transport processes do not predict threshold responses with particle diameter or with distance. Factors that can tend to segregate by spatial scale are saltation of particles across burned areas following fire with accumulation in hollows (including lakes) and water transport that might move particles within, but not beyond, catchments.

Empirical data suggest a strongly non-linear decline in particle accumulation rate at a burn edge, which could explain high accumulation rates in lake sediment when a fire burns to its shore and low values otherwise. Particle transport sampled from a Siberian burn suggests abrupt transition from local to more regional levels (Fig. 5). This abrupt transition is consistent with the observation of clear peaks in thin sections of sediments dated to times when burns reach a lake's shores, and low values when fires remain outside a lake catchment. There is thus reason to expect that limited wind dispersal beyond the burn edge, saltation, and, perhaps, water transport of large (> 50 μm diameter) particles are all mechanisms to emphasize local burns in sediment profiles. Broad geographic patterns among lakes, consistent with climate controls on fire and with changing human use of fire, lend support to these interpretations.

Literature references

Andreae MO (1991) Biomass burning: its history, use, and distribution and its impact on environmental quality and global climate. In Levin J (ed) *Global Biomass Burning: Atmospheric, Climatic, and Biospheric Implications* pp 3-21 MIT Press Cambridge Massachusetts

Bonny AP (1976) Recruitment of pollen to the seston and sediment of some Lake District Lakes. *J of Ecol* 64:859-888

Clark JS (1988a) Charcoal-stratigraphic analysis on petrographic thin sections: recent fire history in northwest Minnesota. *Quat Res* 30:67-80

Clark JS (1988b) Particle motion and the theory of charcoal analysis: source area, transport, deposition, and sampling. *Quat Res* 30:81-91

Clark JS (1990) Fire and climate change during the last 750 years in northwestern Minnesota. *Ecol Mono* 60:135-159

Clark JS (1996) The baseline biomass burning emissions of eastern North America. In Levine JS (ed) *Biomass Burning and Global Change* (in press)

Clark JS, Hussey TC (1996) Estimating the mass flux of charcoal from sediment records: the effect of particle size, morphology, and orientation. *The Holocene* 6:129-144

Clark JS, Hussey TC, Royall PD (1996a) Presettlement analogs for Quaternary fire regimes in eastern North America. *J of Paleolimnol* 16:79-96

Clark JS, Robinson J (1993) Paleoecology of fire. In Crutzen PJ, Goldammer JG (eds) *Fire in the Environment: The Ecological, Atmospheric, and Climatic Importance of Vegetation Fires* pp 193-214 Wiley New York

Clark JS, Royall PD (1994) Pre-industrial particulate emissions and carbon sequestration from biomass burning in North America. *Biogeochemistry* 23:1-17

Clark JS, Royall PD (1995a) Particle-size evidence for source areas of charcoal accumulation in Late Holocene sediments of eastern North American lakes. *Quat Res* 43:80-89

Clark JS, Royall PD (1995b) Transformation of a northern hardwood forest by aboriginal (Iroquois) fire: charcoal evidence from Crawford Lake, Ontario, Canada. *The Holocene* 5:1-9

Clark JS, Royall PD (1996a) Local and regional sediment charcoal evidence for fire regimes in presettlement northeastern North America. *J of Ecol* 84:365-382

Clark JS, Royall PD, Chumbley C (1996b) The role of fire during climate change in an eastern North American forest at Devil's Bathtub, New York. *Ecology* (in press)

Clark JS, Stocks BJ, Richard PJH (1996c) Climate implications of biomass burning since the 19th century in eastern North America. *Global Change Biology* (in press)

Crutzen PJ, Andreae MO (1990) Biomass burning in the tropics: impact on atmospheric chemistry and biogeochemical cycles. *Science* 250:1669-1678

Cwynar LC (1978) Recent history of fire and vegetation from laminated sediment of Greenleaf Lake, Algonquin Park, Ontario. *Can J of Bot* 56:10-21

Debano LF, Conrad CE (1978) The effect of fire on nutrients in a chaparral ecosystem. *Ecology* 59:489-497

EPA (1994) *National air pollution trends, 1900 - 1993.* EPA-454/R-94-027 Research Triangle Park NC

Herring JR (1985) Charcoal fluxes into sediments of the North Pacific Ocean: The Cenozoic record of burning. In Sundquist ET, Broecker WS (eds) *The Carbon Cycle and Atmospheric CO2: Natural Variations, Archaen to Present* Geophysical Monograph 32 American Geophysical Union pp 419-442 Washington DC

Iversen J (1941) Land occupation in Denmark's Stone Age. *Danmarks Geologiske Forenhandlungen II* 66

MacDonald GM, Larsen CPS, Szeicz JM, Moser KA (1991) The reconstruction of boreal forest fire history from lake sediments: a comparison of charcoal, pollen, sedimentological, and geochemical indices. *Quat Sci Rev* 10:53-71

McColl JG, Grigal DF (1977) Nutrient changes following a forest wildfire in Minnesota: effects in watersheds with differing soils. *Oikos* 28:105-112

Millspaugh SH, Whitlock C (1995) A 750-yr fire history based on lake sediment records in central Yellowstone National Park. *The Holocene* 5:283-292

Müller J (1984) Atmospheric residence time of carbonaceous particles and particulate PAH-compounds. *The Science of the Total Environment* 36:339-346

Ogren JA (1982) Deposition of particulate elemental carbon from the atmosphere. In Wolff GT, Klimisch RL (eds) *Particulate Carbon: Atmospheric Life Cycle* pp 379-392 Plenum New York

Patterson WA (1978) *The effects of past and current land disturbances on Squaw Lake, Minnesota and its watershed.* Ph.D. thesis University of Minnesota Minneapolis MN

Patterson WA, Backman AE (1988) Fire and disease history of forests. In Huntley B Webb III T (eds) *Vegetation History* pp 603-632 Kluwer Academic Dordrecht

Patterson WA, Edwards KJ, Maguire DJ (1987) Microscopic charcoal as a fossil indicator of fire. *Quat Sci Rev* 6:3-23

Peck RM (1973) Pollen budget studies in a small Yorkshire catchment. In Birks HJB, West RG (eds) *Quaternary Plant Ecology* pp 43-60 Blackwell Scientific

Radke LF, Stith JL, Hegg DA, Hobbs PV (1978) Airborne studies of particles and gases from forest fires. *J of Air Poll Control Assoc* 28:30-34

Sandberg DV, Pierovich JM, Fox DG, Ross EW (1979) *Effects of fire on air.* USDA Forest Service Gen Tech Report WO-9 Washington DC

Schindler DW, Newbury RW, Beatty KG, Prokopowich J, Ruszczynski T, Dalton JA (1980) Effects of a windstorm and forest fire on chemical losses from forested watersheds and on the quality of receiving streams. *Can J Aquatic Sci* 37:328-334

Smith DM, Griffen JJ, Goldberg ED (1973) Elemental carbon in marine sediments: a baseline for burning. *Nature* 241:268-270

Sutton OG (1953) *Micrometeorology.* McGraw-Hill New York

Swain AM (1973) A history of fire and vegetation in northeastern Minnesota as recorded in lake sediment. *Quat Res* 3:383-396

Swain AM (1978) Environmental changes during the past 2000 years in north-central Wisconsin: analysis of pollen, charcoal and seeds from varved lake sediments. *Quat Res* 10:55-68

Swanson FJ (1981) Fire and geomorphic processes. *In Fire and Ecosystem Processes.* USDA Forest Service General Technical Report WO-26 pp 401-420 Washington DC

Tolonen M (1978) Palaeoecology of annually laminated sediments in Lake Ahvenainen, S Finland I. Pollen and charcoal analyses and their relation to human impact. *Annales Botanicae Fennici* 15:177-208

Waddington JCB (1978) *Vegetational changes associated with settlement and land clearance in Minnesota over the last 125 years: a comparison of historical and sedimentary records.* Ph.D. Thesis University of Minnesota Minneapolis MN

Waring RH, WH Schlesinger (1985) *Forest Ecosystems: Concepts and Management.* Academic Press Orlando FL

Wright RF (1976) The impact of forest fire on the nutrient influxes to small lakes in northeastern Minnesota. *Ecology* 57:649-663

Reconstruction Of Paleo-Fire Through Climate And Eco-System Models

Michael A. Fosberg and Samuel Levis[1]
USDA Forest Service
Washington DC 20090

Summary

This chapter addresses two aspects of paleo-fire and carbon deposits. The first is that the bio-physical system in climate determines the ecotype - biomass structure and function, and as a result, the carbon available for release in the atmosphere. The second part defines the relationship between ecosystem structure, weather, and fire behavior - the emission of carbon particulates. Modeling paleo - emissions of carbon from fire requires an understanding of emissions from current fire behavior in current ecosystems. Modeling fire from past ecosystems requires, first, that we can reconstruct the ecosystem, and second, requires that we assume that current models of fire behavior will apply. These key assumptions can be restated as hypotheses: can current modeling capability reproduce paleo-fire? and, can climate models, coupled with ecosystem models and fire models, replicate the spatial patterns of fire emissions?

Introduction

Biomass burning has clearly been established as an important factor in the global carbon budget (Crutzen and Andreae 1990) and as a contributor to greenhouse gas forcing of climate (IPCC 1990). Despite the current uncertainties in biomass burning emissions of greenhouse gases (carbon dioxide, carbon monoxide, methyl bromide and others), it is clear that biomass burning accounts for approximately 2 to 5 X 10^{15} grams of carbon release annually. Society is currently concerned with the uncertain future of our planet. How will changes in climate and vegetation affect future fire patterns and emissions of greenhouse gases from biomass burning? Rind et al. (1989) have projected increased variability of precipitation and reduced variability of temperature as a result of increased greenhouse forcing of climate. These future scenarios are based on large number of assumptions of population growth, per-capita consumption of energy, and policy measures that might be taken to mitigate change. Sorting out the natural and anthropogenic components of this forcing is difficult. Climate has changed naturally, vegetation has responded to those climate changes by redistributing

[1]Atmospheric and Ocean Sciences, University of Wisconsin, Madison WI 53706

NATO ASI Series, Vol. I 51
Sediment Records of Biomass Burning
and Global Change
Edited by James S. Clark, Hélène Cachier,
Johann G. Goldammer, and Brian Stocks

biomass and structure of the landscape. Land cover and land use have changed as a result of increasing populations and incidence of fire has changed from both land use/land cover changes and anthropogenic use of fire. Current assessment of biomass burning provides only a benchmark from which to quantitatively estimate past and future contributions of biomass burning to greenhouse forcing. There needs to be an understanding of the processes in which changes in biomass burning take place, and in the causes and mechanisms of those changes. In order to gain confidence in scenarios of the future, we need to look to the past, where we can compare those models used to construct future scenarios with their capability to describe past events.

There are several ways in which a framework for analysis of biomass burning contributions can be structured. The framework we have chosen is based on natural sciences, with social sciences as an exogenous factor. We recognize that alternate approaches are equally valid. This natural science approach is taken because we are more able to quantify the natural system than the social system. Alternate approaches, based on a social science focus, with natural science as exogenous would be more comprehensive, but would lead to greater uncertainty, and an approach based on a fully integrated natural and social science system would be so complex that it would be unanalyzable. Our approach is one of expedience and ability to quantify components of the system rather than one which fully mimics the real world.

We have chosen to structure our framework with climate as the driver of change. This climate driver results in vegetation and soil systems which are in quasi-equilibrium with climate. Soil formation from parent rock is influenced by climate. Similarly, vegetation is a result of site suitability (water, light, temperature, nutrients) for a particular species and for competition between species. This framework places human induced land cover and land use change as an exogenous force. The spatial distribution of climate and vegetation, along with cause of fire, determine the source of biomass burning carbon, and the frequency and intensity of fire. Again human caused fire is exogenous to this framework, and only natural, or lightning caused fire is considered explicitly. Given fire, airborne particulate carbon is transported from the fire or source region to deposition sites and incorporated into sediments. This framework is illustrated in Fig. 1.

Our focus in this chapter is to model fire probability and fire behavior given a climate driving force. The core of this framework and the structure of the modeling system is based on linking general circulation models of climate to ecosystem models and fire behavior models in order to predict fire severity and risk. We use the climate forcing of the first box of Fig. 1 to drive the ecosystem models in the second box and fire climate in the third box. We also use the ecosystem composition and structure defined in the second box to drive the potential fire behavior. Transport of carbon particles (Garstang and others, this volume) and deposition (several chapters in this volume) are included in this framework in order to illustrate where this chapter fits into the overall system.

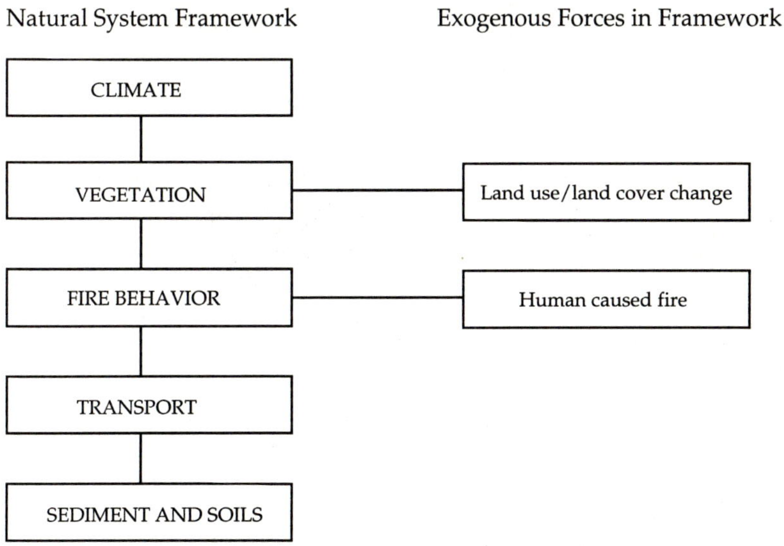

Figure 1. Framework For Relationship Between Climate, Vegetation, Biomass Burning, Transport Of Carbon Particulate And Deposition In Sediments. Arrows between boxes indicate flow of influence. Feedbacks are not included in this simple diagram.

Our approach to evaluation of climate change on fire is to examine the past 18,000 years, a period during which the northern hemisphere emerged from an ice age to the current interglacial warm period. This period provides a basis in which to evaluate our ability to construct future scenarios in that pollen record of past vegetation distribution are known, carbon deposits from paleo-fire have been documented, and there are useful data to reconstruct climate from a variety of sources - pollen records from plants with known climate ranges, tree ring climate reconstructions of precipitation, and in some cases, changes in human settlement patterns. When we overlay our understanding of how vegetation patterns change with climate, and our understanding of how fire intensity and risk change with climate, we can estimate the frequency and intensity of paleo-fire.

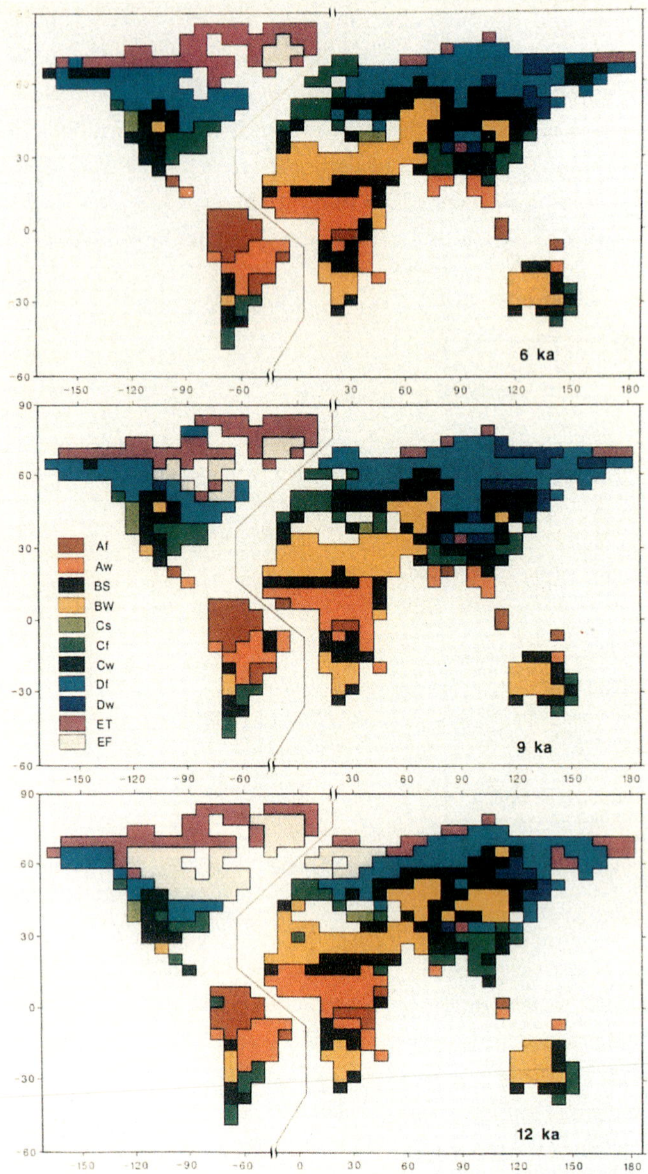

Figure 2. Climate of the past 12,000 years, expressed as vegetation distribution through the Koppen climate system (from Guetter and Kutzbach 1990).

In this paper, we base our discussion on the output of a single atmospheric general circulation model which has been used to simulate the climate of the past 12,000 years (Fig. 2, from Guetter and Kutzbach 1990). We have focused on three intervals, 12,000 to 6000 BP, when glacial retreat resulted in major change in vegetation patterns, the period around 6000 BP, an exceptionally dry period and also when the vegetation patterns began to stabilize to something resembling the present, and on the present itself, our reference case for quantitative evaluation.

Climate Change - The Driver

The fact that climate has varied between glacier-dominated cold periods and warm interglacial periods with warm - dry, or warm - wet interglacial periods is not controversial. The primary cause, the Milankovitch cycles (Berger 1988), has dominated the climate record of geologic history. Equally, the greenhouse effect of trapping infrared radiation by atmospheric gases such as water vapor and carbon dioxide, and by reflected incoming solar radiation in the visible range by cloud cover has controlled climate is not controversial. What is controversial is our ability to quantify that change and predict the future change. Oceans, which cover three-quarters of the earth's surface and have significantly higher heat capacity than the atmosphere, dominate the regional climate distribution. Major above-ocean mountain systems modify that global system and result in regional climate distribution. During glacier periods, the great bulldozer, the continental scale glaciers, ground the mountain system to level plains. Their small cousins, the mountain glaciers, carved deep valleys in those mountains not laid flat by the big bulldozer. The results are an ocean-dominated climate modified first by the major mountain ranges of the world, the Himalayas, the Alps of Europe and the western cordillera of North and South America, and, to a smaller extent, the lower elevational changes of older mountain systems.

Climate Change - The Forcing Factor

Our current capability to forecast weather and to make projections on climate variability a year in advance is based on high temporal and spatial observations of the present weather and climate. There are more than 10,000 permanent weather observation sites around the world, supplemented by a large number of ship and aircraft observations, and, most importantly, continuous high-resolution satellite observations on a global scale. When we attempt to model the past in order to evaluate our understanding of processes, or when we attempt to forecast the future on decadal to century scales, we do not have the benefit of high time and space resolution data. Instead, we must begin with the spatially and temporally averaged data. This is true whether we analyze the past or project scenarios of the future. A second problem arises in modeling climate. Our climate models are based on well-established laws of physics and chemistry. As example, conservation of energy (the second law of thermodynamics) conservation of mass, and Newton's second law (force equals mass times acceleration) to name but a few. When each of these and all of the other fundamental laws of science are expressed mathematically, we have a higher non-linear system of equations which cannot be solved exactly. We know in advance that there is no correct answer that we can look up in the back of the book. Instead, our complex model of the atmosphere can only pass tests that do not lead to nonsensical outcomes. The models are only testable in terms of reasonable solutions, but not testable in terms of correct solutions. Because of the mathematical complexity of the models, namely through maintaining stability by minimizing error growth rates, the solutions of these mathematical models on large computers require months of real time to simulate century to millennia time intervals. In order to increase spatial resolution in these models, time required to make the necessary calculations would increase dramatically. As an example, if the spatial resolution were doubled, the computational effort would increase tenfold. In addition to the spatial resolution problem of atmospheric general circulation models, we need to add the problems of incomplete physics and chemistry. The incomplete physics problem is somewhat more straightforward, meaning that small scale (less than the scale of the node spacing) processes are expressed in statistical form rather than in physical or chemical process form. The chemical and/or phase state of chemicals in the atmosphere are far more difficult to model. These problems are both one of understanding the processes and one of statistically representing these processes in the atmospheric general circulation models. The remaining major problem is that, while the ocean is the driver of the heat engine of the atmosphere, most of the climate models do not include a physics-based interactive ocean and atmosphere, nor an interactive biosphere and atmosphere. The atmosphere, the definer of terrestrial climate, has limited feedback to the ocean, the flywheel of the heat engine, and to the change in the terrestrial biosphere in most climate models. This feedback problem is somewhat overcome by use of asynchronous solutions in earth system models by modeling the ocean, the atmosphere, and the biosphere independently, but periodically bringing the three systems into equilibrium by using predicted state of one solution to initialize the other models in order to approximate an interactive system. These three problems in projecting the earth system, that of incomplete representation of the physical, biological and chemical processes, a system which is only partially interactive, and the fact that enormous computer time is required to make a single calculation, have resulted in a limited number of projections of either the past, which we must use for verification of the models, or in projections of the future from which we will base our future actions. We add to this problem the fact that subtle difference in assumptions and choice from a range of physical, biological, or chemical parameters have resulted in markedly different outcomes in model climate and biosphere results (Cess et al. 1989, Joyce et al. 1990). As a result, projections of the future or evaluation of the past climates and biospheres should be considered as plausible scenarios rather than precise statements of events. Ideally, a large number of model projections would be available, with probabilities attached to the range of assumptions and numerical values of parameters in order to establish a probability distribution of past or future scenarios.

Given the current state of modeling, we need to identify what can be expected from models, what resolution of spatial scales are realistic, and what

Figure 3. Current ecotype distribution at 2° by 2° resolution (from Dorman and Sellers 1989).

confidence we have in climate information contained in the model projections. We know that the model climate qualitatively differs from real climate, but that spatial patterns are similar between the model and observed climate (Fosberg et al. 1993). We finesse our way around this problem by placing higher confidence in the spatial patterns, and determining the difference between the modeled current climate and the projected difference of past or future climates. The changes between the model climates is then applied to the current climate in order to estimate the real change in our past or future scenario (Fosberg et al. 1995). We have greater confidence in temperature change than in the hydrologic change. We also know that the models do not depict interannual or interdecadal variability with an acceptable degree of utility. Specifically, the interannual and interdecadal variability demonstrated in tree rings (Stahle et al. 1988) are not duplicated in model projections. This weakness is particularly important in fire. While major biome shifts are driven by climate change, the more subtle difference in stand structure and composition are driven by interannual and interdecadal variability, and fire occurrence and behavior are driven by daily and weekly weather. Finally, we address the spatial resolution of these projections. Typically, the models are capable of stimulating climate of somewhere between 2 and 5 degrees of latitude and longitude - a resolution that reduced the Rocky Mountains and Himalayas to big mounds (removing all micro climate aspects of ecosystems) and totally eliminated the Southern Alps of New Zealand.

Even with these caveats and limits on projection capability, much can be done to begin constructing the implications of past and future climate change. While a single model run will not forecast or hindcast the fire potential, severity, or, more importantly, the contributions to the global carbon budget, each calculation and projection begins to define the order of magnitude of impact of biomass burning, each additional projection establishes the range of variability, and a large number of projections will establish the probability of emissions of carbon from biomass burning.

A common way in which regional and seasonal depictions of climate are illustrated is through climate space analogs. Most common of these are the correlation of climate to vegetation patterns. The two most commonly used are the Koppen climate classification system which is based on classifying seasonal climate and climate extremes to vegetation, and the Holdridge system which evaluates temperature (as energy), precipitation, and potential evapotranspiration as definitions of vegetation composition and structure, and then maps that composition and structure to climate. These climate space models of vegetation provide the zero order approximation of how ecosystems respond to climate change. This correlation method assumes equilibrium between climate and ecotype. This zero order approximation has great utility because of its simplicity. For this analysis, we have chosen a depiction of potential vegetation with somewhat higher level of information (Fig. 3, from Dorman and Sellers 1987). We can add to this correlation structure, a correlation model of potential fire severity which related to our current depiction of these ecosystems analogues to climate and climate change (Fosberg 1989, Fosberg et al. 1990, Fosberg et al. 1993). This gives us a zero order approximation of potential carbon emissions from fire in both the gas and particulate phases as driven by climate and climate change.

Ecosystem Response To Climate

Zero order approximations of ecosystem response and interaction with atmospheric general circulation models are slowly being replaced with models which contain ecosystem structure and process. Ecosystem structure is particularly important in fire behavior. Leaf area index provides a basis for estimating surface area of fuels to volume of the fuels in foliage. This is the area most exposed to heat transfer in forest fires, and if there is small volume, the fuels will reach pyrolysis temperatures more rapidly (Rothermel 1972). Also if the vegetation is dry, it will reach pyrolysis temperature more rapidly (Fosberg 1973). The physical structure of forest fuel distribution and size are crucial to fire intensity and energy output (Rothermel 1972, Fosberg et al. 1993). The models of ecosystem response that are replacing the zero order models contain leaf area in several levels of the canopy, net primary productivity, and in a few instances, forest composition. Model results have shown that there is a high degree of sensitivity to projections of climate to land cover (Dickinson and Henderson-Sellers 1988, Henderson-Sellers et al. 1993, Shukla et al. 1990). These results have clearly demonstrated a need for a higher level of process interactions between the terrestrial biosphere and the atmosphere. Development of first-order approximation ecosystem models capable of interacting with the atmosphere are underway. These models are being developed from both the atmospheric interaction perspective (Foley 1994) with both multiple canopy layers and soil moisture regimes, and from the perspective of plan response to nutrients, energy, and water availability on plant growth and net primary productivity (McGuire et al. 1992). What characterizes these first generation models is that they are climate feedback models or are responsive to climate. They also contain a higher level of ecosystem structure than the correlation models.

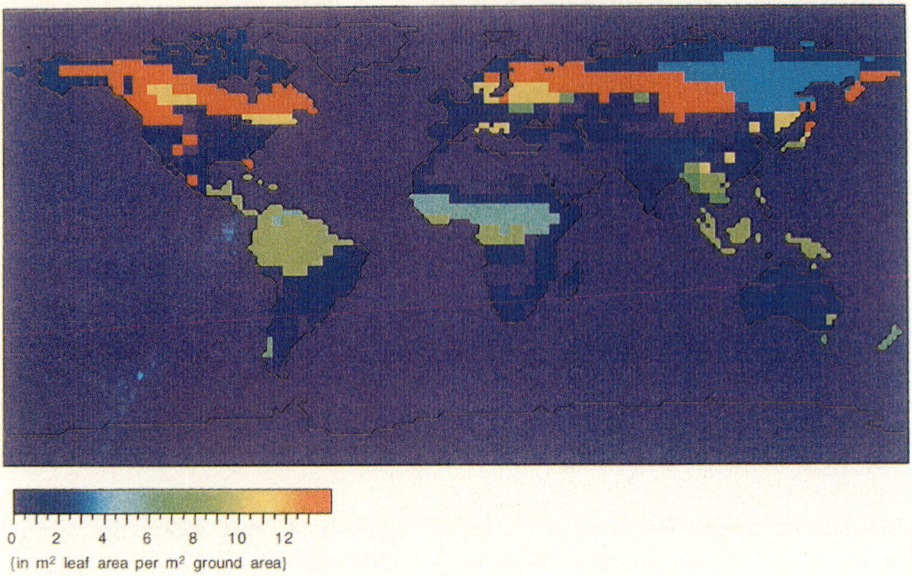

Figure 4a. Current climate two-sided leaf area index in August for upper canopy (from Dorman and Sellers 1989)

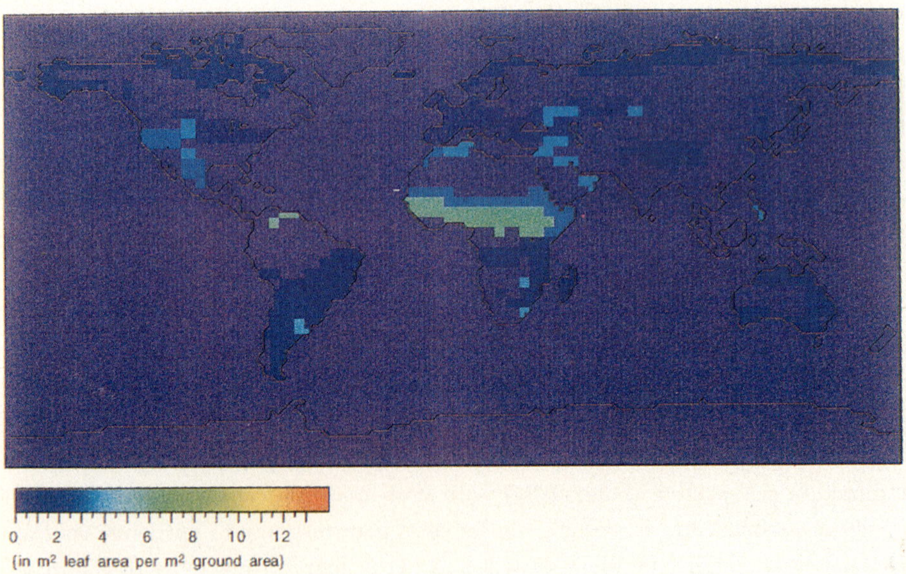

Figure 4b. As in 4a but for lower canopy (from Dorman and Sellers 1989).

In addition to providing interactions between the atmosphere and biosphere through water balance, energy exchange and frictional drag, some of these models simulate net primary productivity of ecosystem biomass as constrained by heat and water. A few of these models also include growth limiting factors such as available mineral nutrients (McGuire et al. 1992). Most land surface models rely on fixed distribution of vegetation. Numerous simulations have demonstrated that the atmospheric general circulation models show high sensitivity to changes in ecosystem - atmosphere interactions (Dickinson and Henderson-Sellers 1988, Henderson-Sellers et al. 1993, Shukla et al. 1990). As example, Foley et al. (1994) simulated high latitude warming through northward expansion of the boreal forest during the mid Holocene exceeding that explained by Milankovitch cycles. There are significant shortcomings in these first order approximations. First, and foremost, is that vegetation distribution is fixed by climate. This approximation does not recognize the potential for ecoregions to spring from refugia instead of migration. Second, this first order approximation does not include disturbance such as fire, insect and disease in ecosystems (Fosberg 1988).

Second-order approximation models do not exist. It is anticipated that they will come from the ecological community through the gap models of Botkin et al. (1972) and further developed by numerous others. While these gap models are not connected to the atmosphere through momentum, energy and matter at the moment, they provide a more complete depiction of the ecosystem, and are driven by energy and material exchange with the atmosphere. These stand and landscape gap models, when coupled interactively, either directly or in asynchronous mode, will provide the basic science with which to assess the impacts of climate change.

As in the climate scenario, which were based on a single model, we have chosen a particular land surface interactive model from the first level of approximation. Atmospheric general circulation models (AGCMs) require the flux of energy, momentum and moisture from the oceans, soils and vegetation as boundary conditions. Spatial variability of these fluxes, as well as the interaction of the atmosphere to provide surface boundary conditions of the AGCMs have been developed (Dickinson et al. 1986, Sellers et al. 1986, Pollard and Thompson 1995). We have based our analysis on the Pollard and Thompson Land Surface Model (LSX). LSX differs from earlier models in enhanced resolution of the biosphere rather than conceptual differences in physics. LSX contains 6 soil layers and 2 layers of vegetative canopy. This model includes radiative flux of energy between the canopy layers as well as net flux between the soil, plant, atmosphere system. The model determines turbulent flux of energy, mass and momentum by coupling mixing length regions above and between canopy layers with simple momentum - diffusive regions within each layer. The model also includes interception of precipitation by the canopy, with drip, stem flow and evaporation. Water reaching the ground is distributed to surface run-off, infiltration, plant evapotranspiration, and sub-surface flow. Vegetation is defined in terms of both structural and physiological characteristics, leaf area index (Fig. 4), canopy height, fractional canopy cover, phenology and stomatal resistance (Dorman and Sellers 1989).

The Climate - Ecosystem - Fire Relationship

The micro climates created by small, but significant differences in elevation result in major differences in water and temperature distribution. The resultant effect is pockets of vegetation in the current interglacial climate which reflect the glacial period distribution of that species. As an example, the remnant pockets Douglas fir (*Pseudo tsuga*) in the Rocky Mountains are genetically different from the Douglas fir of the west coast of North America. These isolated pockets of Douglas fir exist in cool, moist ravines in forests dominated by ponderosa pine (*Pinus ponderosa*) in lower elevation sites and lodgepole pine (*Pinus contorta*) at higher, cool and moist sites. This micro mosaic of vegetation distribution is also seen at the landscape to global scales. We see this same climate - vegetation relationship in wet-tropical forests and species mix, the wet-cold forest of the northern Pacific Coast of North America and in Scandinavia, the dry savannah of oak, pine and scrub of Mediterranean climates, and all of the other climate regions. In each of these regions, we see the influence of microclimatic variation on mosaics of vegetation. The importance of these vegetation mosaics are clearly illustrated in the fire associated with them. Grass fires spread fast with low intensity, while mature deciduous forests, with fire occurring during fall or spring, will burn with high intensity only when driven by strong winds, and not burn during winter leafless or summer full green foliage periods. Mature, old growth coniferous forests, which depend on fire for continuance, will burn with high intensity and spread rapidly. These latter forests depend on fire and are genetically adapted to fire. As example, lodgepole pine has cones which will only release seed under heat and moisture stress. Lodgepole pine has been chasing the glacier retreat at a pace set by the fire recurrence interval, a rate much slower than climate change. Other species, spruce (*Picea sp*), true firs (*Abies sp*), birch (*Betula sp*) and others have migrated to new regions more rapidly. The successional advance of individual species into new regions, the abundance of individual species in that region, and the age (or biomass) determine the structure of the ecosystem and potential fire behavior.

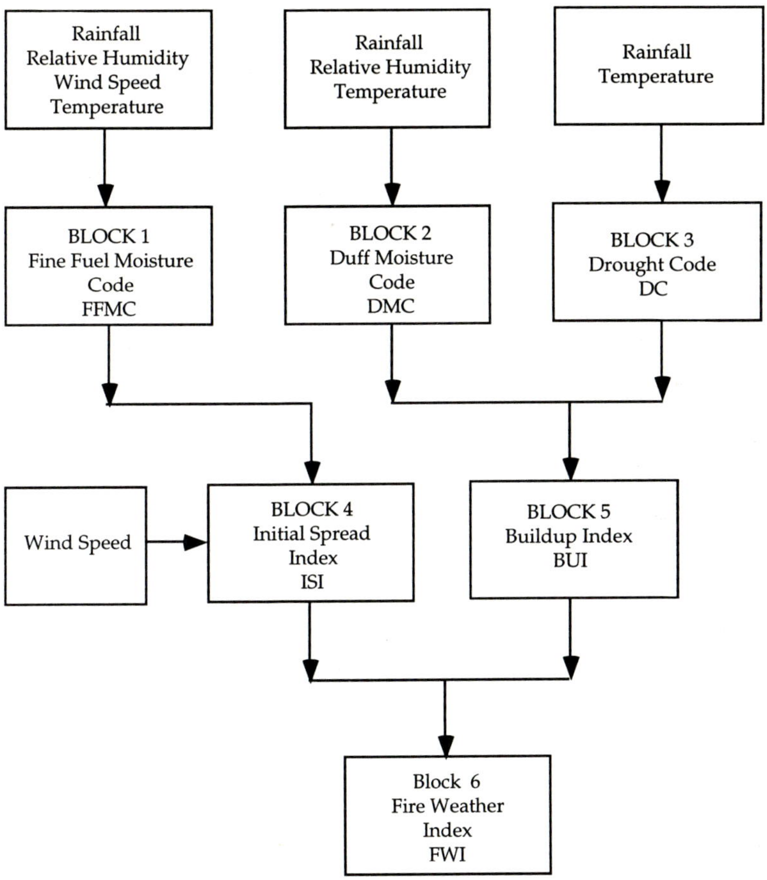

Figure 5. Schematic structure of fire behavior models.

Ecosystems respond not only to the climate change driver, but also to interannual and interdecadal variations of climate. Competition between species for light, water and nutrients determine species abundance and growth. Species composition and biomass change over decadal periods because of this competition. Ecosystems are not static, nor are they in equilibrium with climate.

Add to this disturbance from fire, weather events such as drought and wind storms, or insect or disease outbreaks. These disturbances frequently result in shifting from one serel path to a new path. The best documented case is that of Mediterranean regions in which forests were removed by human action. The savannah forests of Mediterranean regions are now dominated by brush fields maintained through fire disturbance.

Fire Severity And Risk

Fire severity can be expressed in terms of ability to control fire spread and damage. Fires of low intensity can be controlled by personnel using hand tools, fire of moderate intensity can be controlled by direct attack of personnel using hand tools and mechanized equipment such as heavy tractors, while fires of extreme intensity can only be controlled by indirect means such as use of aircraft delivering fire retardants to the fire front and by construction of fuel breaks in advance of the fire front. The relationship between fire severity, fire behavior and biomass consumed by fire has been one of the main thrusts of forest fire research for the past 50 years. The need to estimate fire severity for fire control and planning has led to development of models which describe fire behavior, namely fire spread, fire intensity, flame length, consumption of organic soils, tree mortality and transition from ground and understory fire to crown fire. Fire behavior models have been developed from both empirical and theoretical frameworks. The empirical models such as those used in Canada (Van Wagner 1987) are theory based, but rely on data to define the functional relationship between fire behavior, forest biomass, and weather. The fire behavior models used in the United States, in contrast, are more strongly theory based and mimic process with a higher degree of fidelity (Rothermel, 1972). A generic structure of fire behavior models is shown in Fig. 5. This schematic has ecosystem structure, expressed as biomass by size and amount, porosity of the ecosystem, and weather as boundary conditions. The model framework requires that the ecosystem be classified by fuel size, surface area to volume radio, and biomass in each of these classes. Distribution of fuel throughout the ecosystem is important. A system which is connected from top to bottom will burn as a single fire. One in which there is a separation of an upper canopy and lower foliage will burn as a ground fire, a canopy fire or a mix of the two, depending on wind speed, heat transfer and interconnectedness of the levels. Knowing the structure of the canopy, as expressed by leaf area index (Fig. 4) is crucial to modeling low intensity understory fire and high intensity crown fire. The model also requires that the classes be identified by function, that is, dead or living biomass, foliage, or woody material. Moisture content of the fuels are an important determinant of fire behavior. Because moisture content of each of these elements are determined by both physical and biological processes, fire behavior will change from day-to-day. Weather and climate enters the fire equations through several paths. Day to day fire behavior enters through moisture content of the forest fuels, as example, the number of days with measurable precipitation (Fig. 6). Live fuel moisture is determined by water availability, transpiration and seasonal/phrenological state. Dead fuel moisture, on the other hand, reacts more directly through heat and mass transfer processes (Fosberg et al. 1993). The second path in which weather and climate enter fire behavior is through the process of determining ecosystem structure and composition. While long term climate will define ecosystem type, shorter term climate variations and weather will control the abundance and condition of smaller fuels such as foliage, forbs and grasses. While these control the biomass, actual fire behavior is highly dependent on daily changes in weather. As example, dead grasses will be highly combustible during mid-afternoon when they are dry, but are less flammable at night with high humidity. Large dead fuels respond much more slowly to weather, taking days or weeks, and sometimes months to dry out or acquire moisture.

Table 1. Correlation between ecotypes and fire behavior (from Fosberg, Mearns and Price, 1993)

Ecoregions	Spread rate	Flame length
Tropical rainforest	-	-
Subtropical moist forest	-	-
Dry forest	54	41
Warm temperate forest	23	45
Semiarid grasslands	100	13
Hot desert	-	-
Shrublands	52	100
Steppe	8	7
Cool desert	-	-
Cool temperate forest	11	19
Boreal forest	5	21
Cold parklands	-	-

This continuum can be modeled with three approaches as a first approximation. The first approach is structured to correspond to the climate space models of Koppen and Holdridge. This is a correlation model which related potential fire behavior, rate of spread and energy output, to ecosystem type (Table 1). This zero order approximation ranks ecosystem type (using the Koppen analog) under extreme moisture stress and high wind on an arbitrary scale in which the fire behavior model's results were scaled to the highest values of fire spread and energy output independently. Grass fires, in this case, had an arbitrary value of 100 for fire spread, but a low value of energy output. Mediterranean brush fields had the highest value of energy output and were set arbitrarily to a value of 100. Actual values of spread and energy were then scaled to these values in order to provide a ranking system. The second approach is to use net primary productivity models to capture interannual and interdecadal variability of fire. Land surface exchange models and second order models of ecosystem response to climate will play an important role here. Currently, this second approach has not been developed.

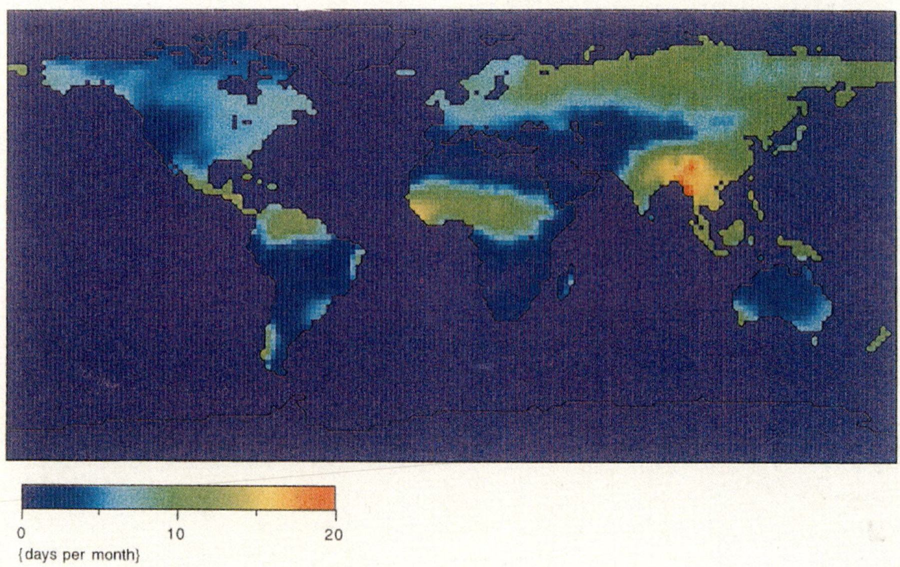

Figure 6. Number of days in August with precipitation greater than 0.1 inches for current climate (based on data from the World WeatherDisc 1994).

The third approach focuses on day-to-day variations of weather. Fire behavior is calculated from observed or modeled weather on a daily basis. We make the assumption that a fire exists, and then determine what the behavior of that fire would be each day. The statistical distribution of the resultant fire behaviors lead to an evaluation of risk of fire damage, or in the case of this paper, carbon emissions from fire. These three approaches capture long-term variations in climate, interannual and inderdecadal variability, and day-to-day variability in fire behavior and approximate the major forces that produce ecosystem - fire variability in charcoal records.

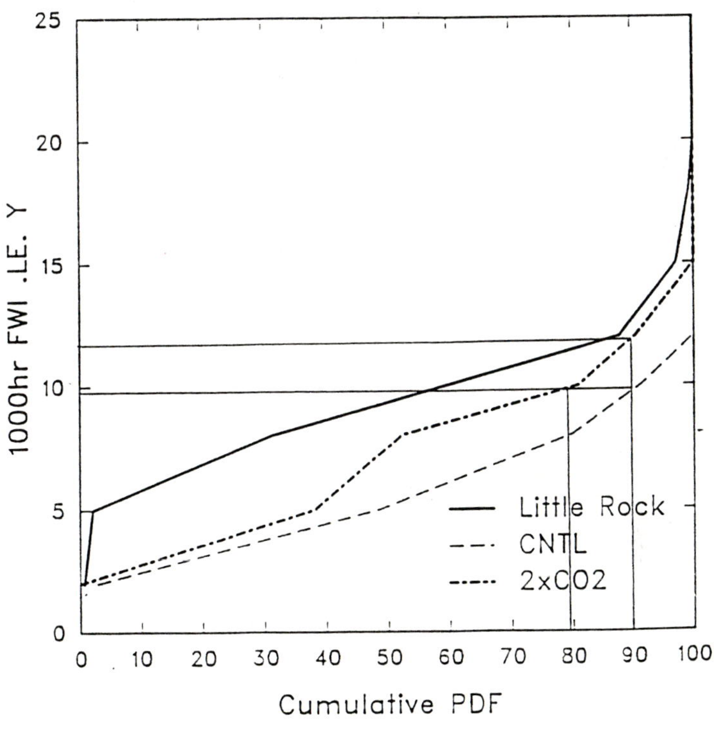

Figure 7. Probability Distribution Of Fire Severity, Based On Current Climate At A Single Location

In recent times, the fire regime has been dominated by a large number of small fires, with only a few fires per year, or a few years in which large, stand replacement fires occur. It is these infrequent, large fires that are responsible for large mass emissions of charcoal and are distributed over widespread areas. Small fires will leave a charcoal signature locally, but because of low emissions and atmospheric dispersion, will have small airborne concentrations and deposition rates outside the immediate fire area. The Yellowstone fires of 1988 resulted in widespread smoke and deposition, yet many small fires in other years, of much greater total area, spread out over a longer period of time and of lower intensity, will probably never be detected in charcoal deposits. What fires should we look to for charcoal sources? There is a general rule of thumb in fire management that states that only 10 percent of fires result in 90 percent of the damage, or in this case, 90 percent of the carbon emissions. We use this rule of thumb to guide our analysis of daily variation in fire severity. A cumulative probability distribution of daily fire behavior potential (Fig. 7) provides the basis for defining the spatial distribution of extreme fire behavior. Here, we first determine the cumulative distributions of fire flame length for an individual weather station and fuel structure. The 90th percentile in the cumulative distribution sets the lower limit for extreme fire conditions, and those most likely to result in high carbon emission (Fosberg, Mearns and Price 1993). In Fig. 7, the 90th percentile (bottom line) gives a fire index of approximately 10 under the current, or control run from the model. Under a doubled carbon dioxide scenario, the fire index would be 12, or roughly a 20 percent increase in resources needed to control fire. Alternative interpretation is that at the fire index of 10 and the current 90 percent fire preparedness level, a doubled carbon dioxide atmosphere would mean that we were prepared for fire control slightly less than 80 percent of the time.

When we map these risk values at the current and projected levels of fire severity (Fig. 8), we have a map of intense fire behavior potential and a map of potential carbon emissions from fire. The particular fire behavior system used in this analysis is the Canadian Fire Weather Index (FWI) System (Van Wagner 1987). The FWI System was applied to the region 50 N to 70 N to determine fire potential in the boreal forest (Fosberg et al. 1995). The FWI System contains all of the elements and precision described in a schematic of Fig. 6, but has been scaled to a 1 to 10 index for managerial and communication purposes. This index can be interpreted by a value of 2 having the potential for large carbon release and a value above 4 having a likelihood of large carbon emission.

Emissions of particulates from fire are related to fire intensity. They are also related to pyrolysis stage (Ward et al. 1991). The active flaming stage of the fire will produce fine particulate matter, while the incomplete combustion associated with smoldering fire will produce larger organic particulate. These heavy carbon particles, produced during the waning states of fire, and with little convective

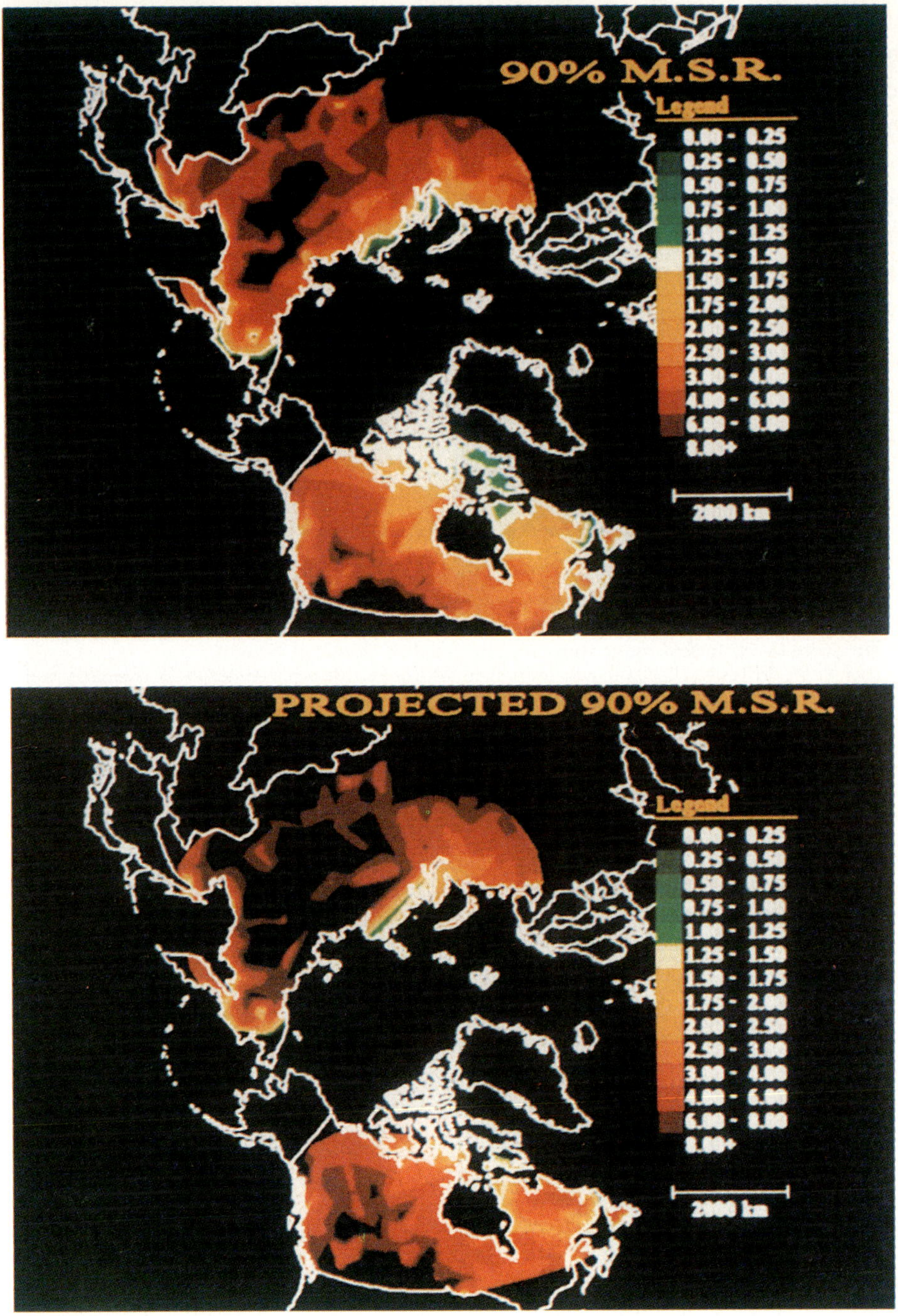

Figure 8. Spatial distribution of extreme fire severity.

lift, are likely to be deposited near the fire. This process is complicated by the fact that during the flaming stage, burning embers are carried aloft in the convection column. Deposition of these large particles of carbon can be far removed from the source. Emission of carbon particulate from fire and transport to deposition are addressed in other chapters of this volume. Emission, transport and deposition cannot be described in quantitative terms without a full understanding of the ecosystem structure, how that ecosystem evolved, and what role fire plays in that evolution.

Literature References

Berger A (1988) The spectral characteristics of pre-Quaternary climatic records, an example of the relationship between astronomical theory and Geo-Sciences. In Berger A, Schneider S, and Duplessy J-Cl (eds) *Climate and Geo-Sciences: A Challenge for Science and Society in the 21st Century.*NATO ASI Series C: Mathematical and Physical Sciences 285:47-76 Kluwer Academic Publishers Dordrecht

Botkin DB, Janak JF, Wallis JR (1972) Some ecological consequences of a computer model of forest growth. *J of Ecol* 60:849-873

Cess RD, Potter GL, Blanchet JP, Boer GJ, Ghan SJ, Kiehl JT, LeTreut H, Li Z-X, Liang X-Z, Mitchel JFB, Morcrette J-J, Randeall DA, Riches MR, Roeckner E, Schlese U, Slingo A, Taylor KE, Washington WM, Wetherald RT, Yagai I (1989) Interpretation of cloud-climate feedback as produced by 14 atmospheric general circulation models. *Science* 245:513-516

Crutzen PJ, Andreae MO (1990) Biomass burning in the tropics: Impact on atmospheric chemistry and biogeochemical cycles. *Science* 250:1669-1678

Dickenson RE, Henderson-Sellers A (1988) Modeling tropical deforestation: a study of GCM land-surface parameterizations. *Quat J Royal Meteor Soc* 114:439-462

Dorman JL, Sellers PJ (1989) A global climatology of albedo, roughness length and stomatal resistance for atmospheric general circulation models as represented by the simple biosphere model (SiB). *J Applied Meteor* 28:833-855

Fosberg MA (1988) Forest productivity and health in a changing atmospheric environment. In Berger A, Schneider S, Duplessy J-Cl (eds) NATO ASI Series C: Mathematical and Physical Sciences 285:681-688 Kluwer Academic Publishers Dordrecht

Fosberg MA (1989) Climate change and forest fires. In Topping Jr JC (ed) *Proceedings, Second North American conference on preparing for climate change* pp 292-302 Climate Institute Washington DC

Fosberg MA, Goldammer JG, Rind D, Price C (1990) Global Change: Effects on Forest Ecosystems and Wildfire Severity. In Goldhammer JG (ed) *Fire in the Tropical Biota: Ecosystem Processes and Global Challenges Ecological Studies* 84:463-486 Springer-Verlag Berlin

Fosberg MA, Mearns LO, Price C (1993) Climate Change - Fire Interactions at the Global Scale: Predictions and Limitations of Methods. In Crutzen PJ, Goldammer JG (eds) *Fire in the Environment: The Ecological, Atmospheric, and Climatic Importance of Vegetation Fires* pp 123-137 John Wiley and Sons New York

Fosberg MA, Stocks BJ, Lynham TJ (1995) Risk analysis in strategic planning: Fire and climate change in the boreal forest. In Goldammer JG (ed) *Fire in Northern European Ecosystems* John Wiley (in press)

Foley JA, Kutzbach JE, Coe MT, Levis S (1994) Feedbacks between climate and boreal forests during the Holocene epoch. *Nature* 371:52-54

Garstang M (paper in this volume)

Guetter PJ, Kutzbach JE (1990) A modified Koppen classification applied to model simulations of glacial and interglacial climate. *Climatic Change* 16:193-215

Henderson-Sellers A (1993) Continental vegetation as a dynamic component of a global climate model: a preliminary assessment. Climatic Change 23:337-377

IPCC (1990) *Climate change: the IPCC scientific assessment.* Houghton J, Jenkins GJ, Ephraums JJ (eds) Cambridge Univ Press Cambridge

McGuire AD, Melillo JM, Joyce LA, Kicklighter DW, Grace AL, Moore B III, Vorosmarty CJ (1992) Interactions between carbon and nitrogen dynamics in estimating net primary productivity for potential vegetation in North America. *Global Biogeochem Cycles* 6:101-124

Pollard D, Thompson SL (1995) Use of a land-surface-transfer scheme (LSX) in a global climate model (GENESIS): the response to doubling stomatal resistance. *Global and Planetary Change* (in press)

Rothermel RC (1972) A mathematical model for predicting fire spread in wildland fuels. *USDA For Serv* Res Paper INT-115

Sellers PJ, Mintz Y, Sud YC, Dalcher A (1986) A simple biosphere model (SiB) for use with general circulation models. *J Atmos Sci* 43:505-531

Shukla JL, Nobre C, Sellers PJ (1990) Amazon deforestation and climate change. *Science* 247:1322-1325

Stahle DW, Cleveland MK, Hehr JG (1988) North Carolina climate changes reconstructed from tree rings: AD 372 to 1975. *Science* 240:1517-1519

Van Wagner CE (1987) Development and structure of the Canadian Forest Fire Weather Index System. *Can For Serv* For Tech Report 35

Ward DE, Setzer AW, Kaufman YJ, Rasmusson RA (1991) Characterization of smoke emissions from biomass fires of the Amazon Region - BASE - A Experiment. In Levine, JS (ed) *Global biomass burning, atmospheric, climatic and biospheric implications* pp 394-402 MIT Press Cambridge

World WeatherDisc (1994) Worldwide airfield summaries. In *World WeatherDisc Seattle* WeatherDisc Associates Inc

The Culture of Fire: An Introduction to Anthropogenic Fire History

Stephen J. Pyne and Johann G. Goldammer[1]
American Studies Department
Arizona State Univ. West
Phoenix, AZ, USA

Summary

The capture of fire by the genus *Homo* changed forever the natural history of the Earth. The manipulation of fire, a species monopoly, defined humanity's ecological niche. Even today fire appears at the core of many popular scenarios for environmental apocalypse. Yet the larger history of fire—the varied ways human societies have sought to use and control fire, the trajectory of change over centuries if no millennia—is not well known.

A useful scheme to conceptualize that history fire is to frame anthropogenic fire with its two competitors—natural fire on one side and the industrial combustion of fossil fuels on the other. The introduction of humans is almost everywhere marked by a new fire regime, typically by an initial increase in charcoal. Increasingly, however, industrial combustion is substituting for traditional pyrotechnologies. The impact of fossil hydrocarbons, not only as fuel but as surrogate fallows, is profoundly rewiring the ecology of Earthly fire in ways not well understood.

Introduction

The capture of fire by the genus *Homo* marks a divide in the natural history of the Earth. Fire of course long predated hominids, and many organisms display adaptations, often highly specific, to fire. But from the time of *H. erectus* one species acquired the capacity to start and stop fires, a niche filled by no other organism. A uniquely fire creature became bonded to a uniquely fire planet.

The human capture of fire was thus a profound event for the Earth. It forced biotas to adjust to new fire regimes and new fuel complexes as shaped, deliberately or accidentally, by human societies. Equally, it forced humans to compete against natural sources of fire. If fire is a primary technology for making the world habitable, then it is essential that humans, not lightning, control the fire regime. It is no less imperative that humanity not allow any other creature to possess fire. None does, and it is unthinkable that humans would allow any other species to break its monopoly.

The fire-mediated relationship between humans and the Earth is fantastically complex. Not only can humans, within limits, control ignition and manipulate the

---

[1]Fire Ecology Group, c/o Freiburg University, Freiburg, Germany

NATO ASI Series, Vol. I 51
Sediment Records of Biomass Burning
and Global Change
Edited by James S. Clark, Hélène Cachier,
Johann G. Goldammer, and Brian Stocks

available fuels, but they exercise even greater range through fire's power as a catalyst, an enabling device for hunting, foraging, farming, pastoralism, heat engines, and fire-dependent technologies from fire-tempered spears and stones to ceramics and metallurgy. Accordingly the wholesale anthropogenic modification of the biosphere did not begin with the industrial revolution or with the Neolithic revolution but with the hominid revolution announced with promethean splendor by the capture of fire.

Nearly all fire origin myths identify the acquisition of fire as the means of passage from life among the beasts into special status as a human being. In ecological terms that mythology contains more than a kernel of truth. Humans have sought out fire-prone landscapes, and have tried to convert less fire-prone biotas into forms more susceptible to fire. But wherever they have gone, humans have carried the torch, even to the ice-steppes of Antarctica, even to the Moon. Anthropogenic fire is as much a cultural artifact as chopping stones and skyscrapers, and landscapes forged in those fires are as much a creation of human societies as marble sculptures and parking lots. If fire takes on the character of the landscapes within which it burns, so it also assumes the traits of the people who oversee it. Thus fire is at once cause, consequence, and catalyst.

These cultural differences extend to the understanding of fire. Europeans tend to see fire within a wholly human context; lightning fire is a freak of nature. Americans prefer to study "natural" fire, and see human manipulations as a kind of contamination. Scientists from temperate climates that lack clearly defined fire seasons--those from the northeastern U.S. and much of western Europe, for example, have difficulty appreciating the power of fire in landscapes for which fire comes as regularly as summer rains. Intellectuals housed in cities have difficulty imagining fire other than as urban conflagration or siege. The immense reconstruction of the planet through fossil-fuel combustion is changing not only fire practices but fire perceptions. What makes fire history particularly difficult is also what makes it especially fascinating, precisely this complicated fusion of nature and culture.

Sources

The difficulties of fire history as related to sedimentary records can be conceptualized through an examination of sources and sinks and the linkages that join them. This takes three forms. One is the question of linking fires with sedimentary deposits. It is not clear what record of fires is captured in lakes, soils, and bogs. Some fire regimes create good records, some poor ones; some environments preserve their fire products better than others. But for all of them the geologically archived record is inherently skewed and partial. Most critically, there is often no objective means by which to determine whether the fire recorded by the residue originated from human or natural causes.

The second question involves the linkages between source regions and sink regions. Environments that burn frequently do so because they are at least seasonally or episodically dry; environments that preserve charcoal do so because they are normally wet. The regions that produce the most fire do not always record that fact, while the regions that successfully retain fire records are generally not those that produce the most fires. Thus there is a regional bias in sampling.

The third question is even broader. The sources for fire history are not easily linked with the sinks—the understanding—demanded by science and society. Only a fraction of those records belong with the natural sciences and those, as explained above, have their biases. The richest sources reside elsewhere, but because they are primarily textual, they are not easily converted into a scientific epistemology. They are not readily quantified. In truth, anecdotal evidence gleaned from the historic record has its equivalent in anecdotal numbers generated in the absence of any valid historic baseline. The world did not begin with CO_2 measurements in East Anglia, fire-scarred pine in the 19th-century Sierra Nevadas, or the Icelandic sagas.

Linking data sources and conceptual sinks is not simple. Divining fire history solely from travelers' accounts or lacustrine charcoal or bog-preserved pollen is an exercise in scholarly pyromancy. Dismissing whole ranges of evidence is not a solution. Disciplines must hold to their standards and intellectual rigor, that is what makes them a discipline. But fire history demands interdisciplinary study. Only through corroborating evidence, analogy with contemporary practices, comparative histories, ecological reasoning, and omnivorous foraging through whatever data is available—sediments, fire-scarred trees, forest age-structures, charcoal dating techniques, official statistics, unofficial reports and explorers' observations, ethnographic studies, photographs, paintings, folklore, whatever speaks to the topic—will something like an accurate history be possible. Analysis by thermoluminescence can determine the time that clay was baked, whether that clay survived as a ceramic shard or as the hearth of an abandoned campfire. Tree-ring analysis can provide detailed reconstructions of past fires and even determine the season of burning (Figs. 1 and 2). Collectively such methods have an authority that any one of them independently lacks.

What follows is an introduction to anthropogenic fire. The essay tries to explain what anthropogenic fire is, how it derives its power, how it can be understood. It proposes a conceptual model for interpreting anthropogenic fire within the context of the Earth as a fire planet. And it sketches in rough outline the contours of human fire history, both qualitative and quantitative.

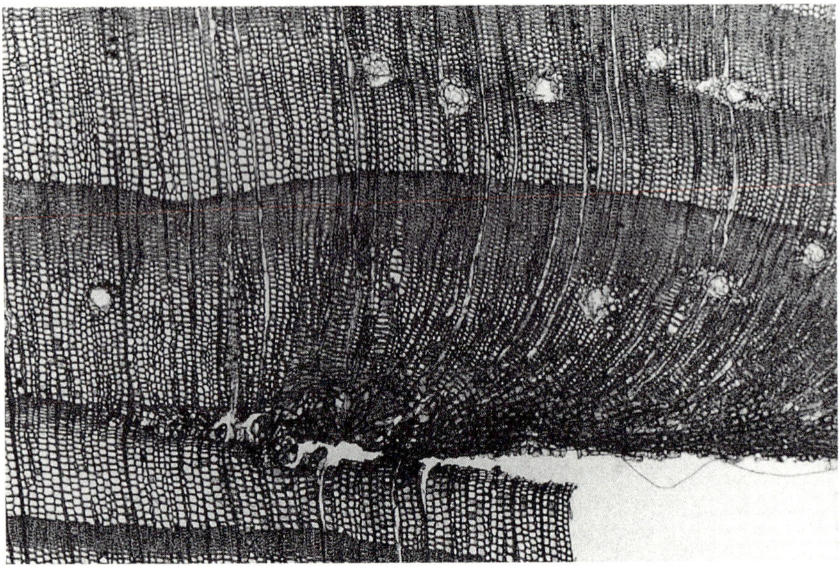

Figure 1. Microscopic photograph of a fire-scarred *Pinus ponderosa* from the Santa Rita Mountains, Arizona, USA. The position of the scar within the annual ring structure indicates the fire burned between August and September 1834. The pattern of rings over the age of the tree, and of fire-scarred trees over the region, are useful indications of past fire regimes. From Ortloff (1992).

The Ecology of Fire: A Primer

Fire is a creation of life. Terrestrial life provides the fuel and life everywhere furnishes the oxygen required for combustion. While ignition comes from various sources, all except lightning were trivial, at least until the Pleistocene. Evidence of fire is apparent in all the coal-bearing strata of the geologic record. For eons, then, fire has shaped ecosystems, informed evolutionary trends, and shared in the complexity of living systems.

The patterning of fire—the fire regime—resembles a two-cycle engine, oscillating between wet and dry conditions, with lightning supplying the spark. There has to be enough moisture to produce fuel and enough dormancy or drought to ready that fuel for burning. This wet-dry cycle can crack open a biota the way a frost-thaw cycle can split open rock. In many environments this informing cycle occurs annually as a well-recognized fire season. In other places it appears over the course of decades or even centuries, as climatic tides of wet and dry weather wash over the biota. In desert regions outbursts of exceptional rains create the fuels that otherwise do not exist in sufficient mass for fire. In rainforests periods of exceptional drought prepare ever-abundant biomass for burning. The sharper the contrasts between wet and dry—the more intense the gradient and the faster the transition—the more vigorous the fire regime. Hence Mediterranean climates, monsoonal forests, and landscapes subject to El Niño-Southern Oscillation-style bouts of drought and deluge are particularly prone to fire. But fire can be found nearly everywhere, and it appears more profusely during times of rapid and extreme climatic change, or when, through their technologies, humans can force conversions of equal magnitude.

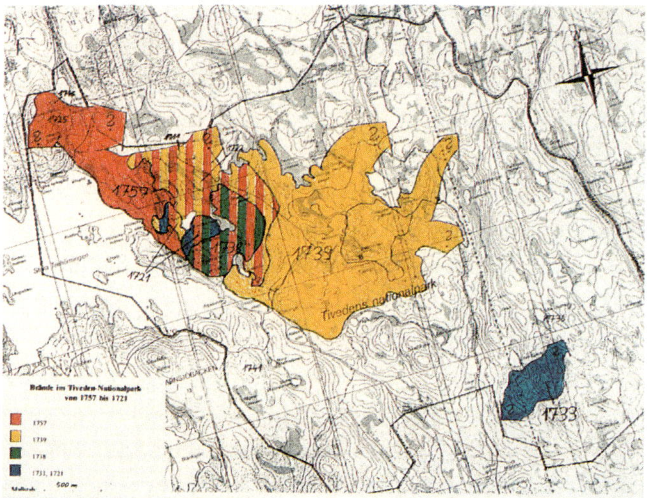

Figure 2. Reconstructions of past fires from fire scars and forest age structure at Tividens National Park, Sweden. Many fires do not replace stands wholesale; here fires from 1757, 1739, and 1738 all overlap in their geographic spread. From Page (1995).

Adaptations to fire are many, complex, and subtle. Rarely are adaptations fire-specific or unique to a particular type of fire. More typically organisms show a suite of traits that adapt to a suite of conditions; and they adapt not to fire in the abstract but to a particular regimen of fire. Thus drought, grazing, and fire, for example, form a complex of pressures that result in similar kinds of traits such as storing a larger proportion of biomass in root stocks or the capacity to resprout from top-killed stems. The ability to adapt to disturbances of many kinds also helps an organism adapt to fire.

What matters is not whether fire exists or not—the prospect of fire exclusion is an illusion of industrial societies—but the character of the fire regime. A change in regimes will cause a biotic shift within an ecosystem, favoring some organisms over others. Although fire is present in both instances, the changes in intensity, frequency, seasonal timing, size, and so on will affect an ecosystem differently. The choice is not to adapt to fire in the abstract but to adapt to various fire regimes. Most of the planet's terrestrial biotas, excepting perhaps tundra, stony deserts, and spring-fed grottos, have in fact adapted to anthropogenic fire, or had their biotic ore hammered in its forge.

The Culture of Fire: First Principles

Anthropogenic fire builds on natural conditions, and anthropogenic fire practices mimic natural fire processes. Most traditional fire uses have analogues in fires started by natural sources; most fire practices exploit conditions that already exist. Humans seek out fire-prone landscapes or sites that, with other technological leveraging, they can make more fire-prone. Thus in Mediterranean or monsoonal climates, aboriginal burning can effect great changes, while in temperate climates that lack a clearly defined fire season, major shifts in fire regimes require other preparations such as clearing and grazing. For the one, agriculture was unnecessary; for the other, a precondition to wholesale environmental manipulation.

The exercise of anthropogenic firepower takes many forms. The most obvious expression is through direct control over ignition itself. Humans can start fires at will, although the environment will not always accept them; and they can stop fires, although this is an even more troublesome task, and historically the most common way to contain wildfire is to replace it with domesticated fire, to base fire control upon fire use, which is to say, that fire is best controlled through controlled burning.

Only with industrialization have societies attempted outright fire suppression, in effect substituting for open burning the controlled combustion of fossil fuels embedded in firefighting technologies. In most regions this strategy, by itself, cannot be sustained. The long-term effect is to substitute infrequent, large fires for frequent, low-intensity fires. This shift in fire regimes has serious consequences for fuel complexes (and wildfire) as well as for biodiversity. Figure 3 shows the impact of suppression on the distribution of fires, as manifest in northern Ontario. Figure 4 documents the shift in fire regimes in the southern Sierra Nevadas of California and the ponderosa pine forests of northern Arizona as a result of attempted fire exclusion.

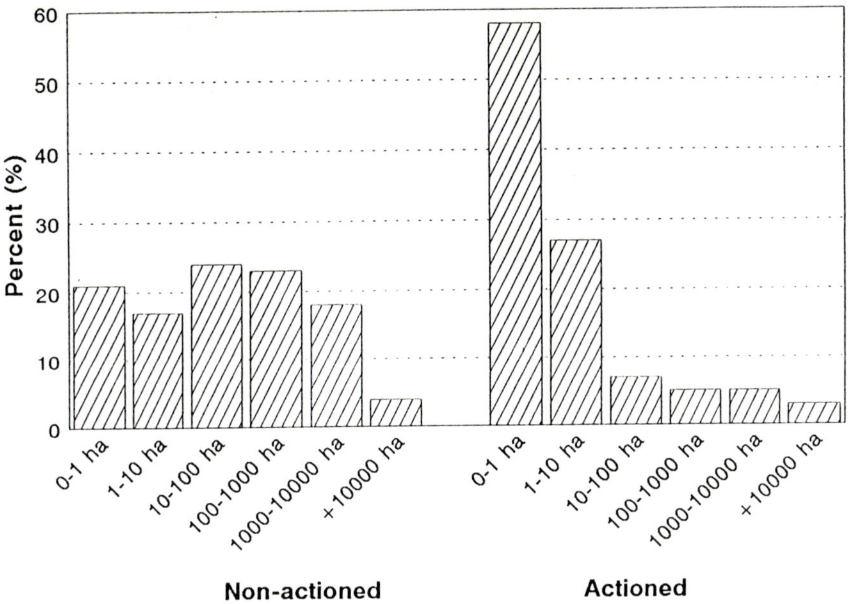

Figure 3. The effects of fire suppression as revealed by the size of fires for which fire control is attempted ("actioned") or not ("non-actioned"). The data derive from the removal of firefighting forces from Northern Ontario in 1976 to 1988. With suppression the average size of fires decreases not only because early attack is effective but because smoldering fires no longer reside on the land and experience repeated runs. Data from Stocks (1991).

No less importantly human societies can also influence a fire environment in indirect ways by restructuring its fuels. For this purpose fire is also a vital catalyst, making possible many other mechanical and biological technologies, from smelting to grazing. Farms, cities, metallurgy, and domestic herds may indirectly compete with lightning fire for a landscape's available biomass and thus shape the fire regime. Until very recently most of these technologies required fire, so even in these secondary forms anthropogenic fire practices still competed with natural fire. Anthropogenic fire is the quintessential interactive technology. Almost never does it occur by itself; almost always it occurs in association with other practices, each leveraging the power of the other.

Figure 4. The two photos from the Black Hills of South Dakota, identically positioned but taken a century apart (1874 and 1974), document the changes in forest structure and density that resulted in large measure from fire exclusion. Lightning is common in the Black Hills, but indigenes also burned widely. Grazing by livestock, patterns of fixed land ownership, and later active suppression by federal agencies all contributed to fire's extinction. The fire hazard has increased in proportion to the increased fuels. Source: Progulske and Shideler.

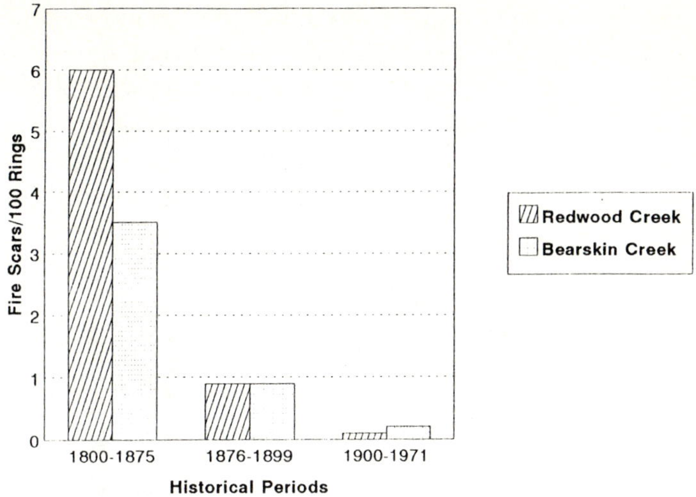

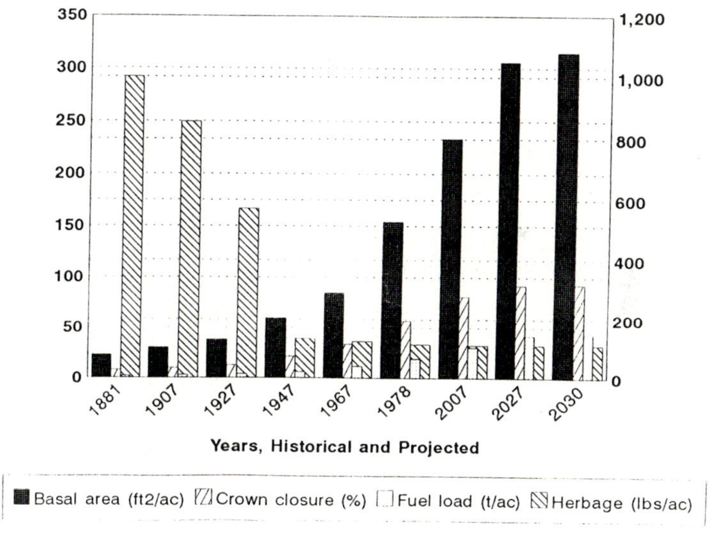

Figure 4 (cont'd). The two graphs show the same shift in fire regimes. The top graph traces the reduction in fire-scarring with the reservation of sites as public land. Redrawn from Kilgore and Taylor (1979). The bottom graph, from ponderosa pine forests in northern Arizona, calculates the actual and projected changes in grasses, trees, canopy, and fuel loads. Data from Covington and Moore (1992).

But while anthropogenic fire builds on natural conditions, it quickly transforms them. It redefines fire seasons, inscribes new fire regimes, and informs ecosystems according to very different fire dynamics. Anthropogenic fire is a kind of ecological virus that takes over the nuclear machinery and redirects it to new purposes. In some cases, fire continues; in some, it augments; in others, it diminishes. In first-contact landscapes it typically increases. In industrialized landscapes, it contracts. Where pastoralists enter humid environments, they escalate the fire load; where they meet arid environments, they depress it. In all the fire regime becomes intelligible only through reference to anthropogenic fire and the social context within which it resides.

The Competition for Combustion : A Conceptual Model

Figure 5 sketches the boundary conditions for anthropogenic fire as a "competitor" within the Earth's combustion environment. Initially humans competed with natural ignition sources for the available biomass, and the amount of burning was limited by the source—the fuels—accessible in common to both. Increasingly, however, anthropogenic fire is being forced into competition with another human combustion technology, namely, the burning of fossil fuels. In this case anthropogenic fire competes not over a common source, biomass, but over a common sink, the atmosphere, the amount of emissions that regional or global airsheds can absorb without major distortion. In neither case is the combustion frontier well understood.

Nature contains several ignition sources. Active volcanoes start fires from lava flows. Rockfalls have kindled fires in South Africa and India. Spontaneous combustion in drying bogs and caves have, reportedly, led to self-sustaining fires. Coal-seams can burn for millennia, and have periodically rekindled fires along their surface contours in North America, Australia, Siberia, and Borneo. But all of these sources are local, and their ecological consequences, parochial.

The dominant source of ignition is lightning. Estimates reckon that the Earth experiences 1,800 thunderstorms a day with collective cloud-to-ground discharges that exceed 8 million per day. Although only a minuscule fraction of this quantity leads to fires, the numbers are large. Fire-igniting lightning, moreover, does not occur randomly: it selects some places more than others, and some seasons more than others. But sooner or later it comes to nearly everywhere. Lightning accounts for the Earth's fundamental fire geography. It is against lightning that humans compete for control over fire.

Traditionally they competed by burning under their terms what was available for burning from all sources. Over the past century, however, fossil hydrocarbons have increasingly substituted for biomass, and industrial combustion for free-burning fire. The competition with lightning takes the form of direct suppression, made possible by the pyrotechnologies of internal-combustion engines. The more serious outcome is the abolition of deliberate burning. While planetary combustion remains high, planetary fire has decreased. As biomass burning and fossil-fuel combustion compete in the atmospheric sink, traditional burning is likely to diminish further. All this has serious ecological implications. Fire is as powerful by its removal as by its application. Not only air pollution but ecological deterioration is the ambiguous outcome.

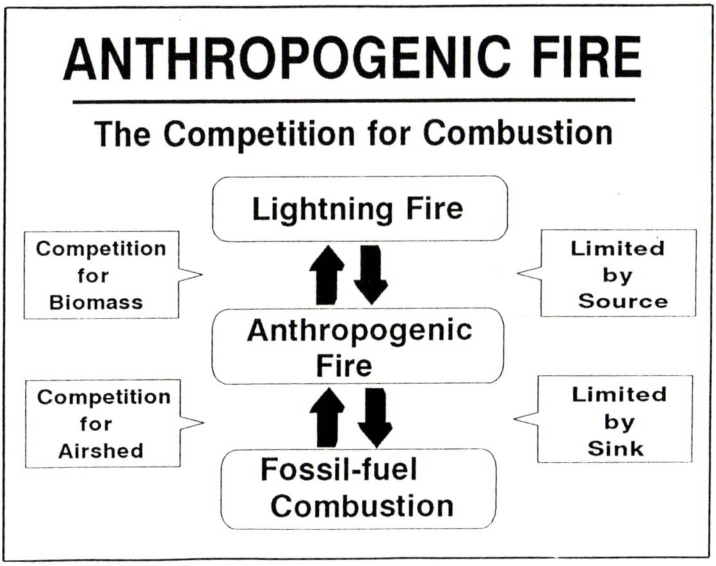

Figure 5. The Competition for Combustion. A simplified schema for identifying the major boundaries of anthropogenic fire.

The Competition with Lightning Fire

In most of the Earth the boundary between anthropogenic and lightning fires became obscured long ago. For thousands, and in some cases, for hundreds of thousands of years anthropogenic fire practices have defined the matrix within which combustion must proceed. It is likely that the initial encounter came as a shock wave, a flaming front of rekindled hominid migration as *Homo sapiens* added to the hominid tool kit and propagated around the globe.

This front involved burning on a major scale as humans sought through fire to restructure the new lands into forms more suitable to them. In many environments, Australia for example, colonization seems to be recorded in the form of a charcoal horizon (Fig. 6). In recent years Australian Aborigines have rekindled that tradition on islands in the Gulf of Carpentaria. After the shock of first contact the record shows a different presence for fire, typically one of routine, low-intensity burning characteristic of aboriginal societies. Similar records are found in more recent colonizations of previously uninhabited islands—the Maoris in New Zealand, the Norse in Iceland and Greenland, the Portuguese in Madeira, and so on.

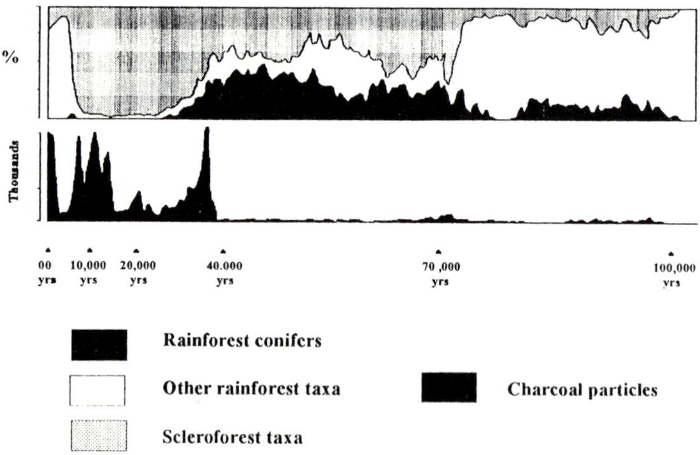

Figure 6. Fire front: the hominid colonization of Australia. Summary pollen diagram from Lynch's Crater, northeast Queensland, Australia. The sudden eruption of large charcoal corresponds closely with the earliest dated remains of humans. Redrawn from Clark (1994).

So thoroughly has anthropogenic fire replaced lightning fire that early explorers often dismissed the possibility of lightning as an ignition source in many landscapes for which it now exists. In the early 1950s Americans denied lightning fires in central Alaska; Europeans today continue to dismiss lightning fire as other than a freak of nature. Partly this reflects a failure to observe, partly it represents an unwarranted extrapolation that the magnitude of native burning witnessed by the observer must apply everywhere, and partly it testifies to the success with which human societies had replaced wild fire with a more domesticated fire. By farming, grazing, and burning human societies simply controlled the overall combustion environment (see Fig. 7).

Where humans have withdrawn, however, lightning fire has often returned. This has happened historically where wars, diseases, and migrations have

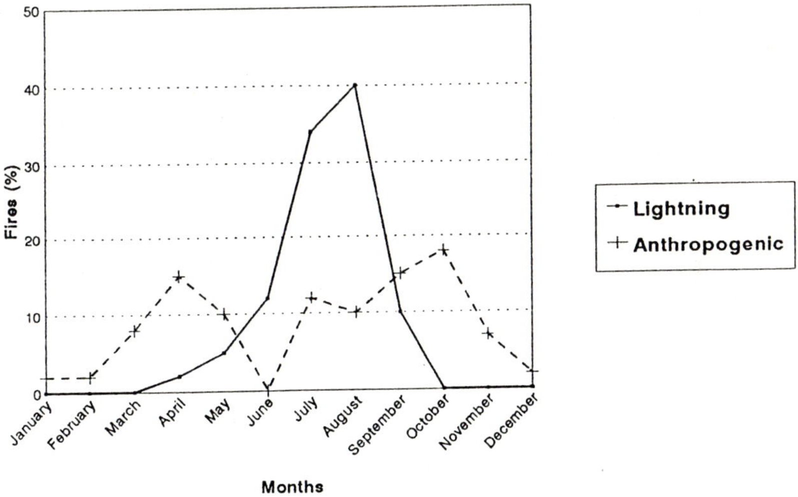

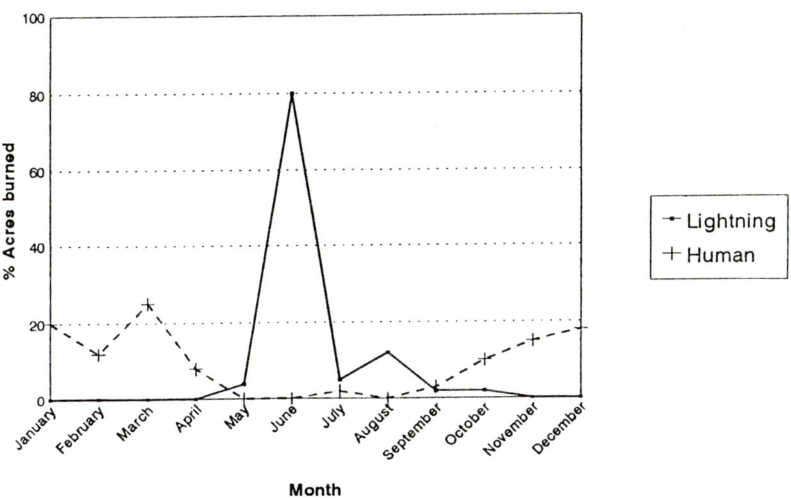

Figure 7. Examples of how anthropogenic fire practices redefine the dimensions of fire seasons, and how by off-set seasonal burning humans contain lightning fire. (a) Everglades. Data from Everglades National Park. (b) Northern Great Plains. Data from Higgins (1984).

temporarily depopulated a region, but there are a number of interesting contemporary examples as well in which people have voluntarily withdrawn from sites they once occupied. The most common are nature reserves where "natural" processes are left untrammeled. Thus lightning fire has reasserted itself in Australia, North America, and even Europe, not in its historical forms but in a new regime for which management philosophy (and environmental ideology) have dictated the acceptable regime (Fig. 8).

A dramatic illustration of lost contact is the revival of fire in the region around the Chernobyl nuclear plant. After the catastrophic meltdown of 1986, radioactive contamination forced the abandonment of the surrounding region. With no one to tend the vegetation, fuels rapidly build up, and in the summer of 1992 fires broke out for which control was virtually impossible. The fires in turn entrained and redistributed some of the radioactive soils and plants in a horrifying cycle of positive feedback. It is not clear whether lightning kindled these fires, but in the absence of a local population it is likely, and regardless, the revealed pattern of feral fire reclaiming once-domesticated land is a scenario that is becoming increasingly common.

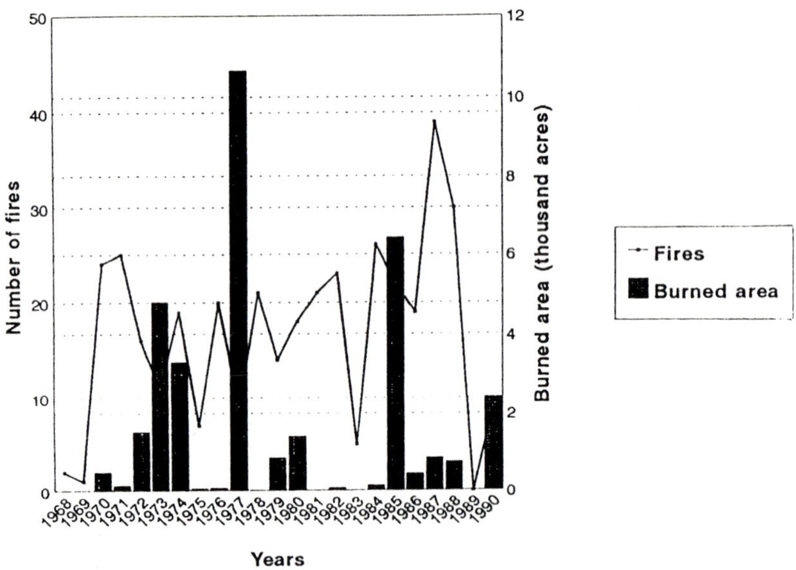

Figure 8. Lost contacts, restored contacts: the deliberate "restoration" of lightning fire. It would be fatuous to assert that humans can stop and start lightning fires at will. But it is equally absurd to claim that calculated attempts to reintroduce lightning fires (and only lightning fires) is not an artifact of human society. Compare to the overall reduction in fire load shown in Fig. 2. Data from Sequoia-Kings Canyon National Park.

Figure 9 shows the recent geography of fire in Alaska. In this instance land legislation has determined where humans can settle and what kinds of technologies they may use. The division of the land into various categories has effectively defined separate domains for lightning and anthropogenic fire. Lightning fire is confined to the interior valley; anthropogenic fire to fields of settlement and corridors of traffic. Only in a few areas do they overlap.

The lightning fire regime is far from "natural," however. Gone are most traditional anthropogenic fires. Gone, too, are the older patterns of fire. Through firefighting institutions like the Alaska Fire Service, the National Park Service, and the Fish and Wildlife Service, humans "manage" many lightning fires through decisions about what fires to control and in what forms. The larger fire regime remains very much an expression of human will. Humans have deliberately sought to segregate their presence (and fires) from that of nature (and its fires); the fire geography of Alaska expresses exactly this cultural decision.

Another illustration of how landscape ecology reflects anthropogenic fire (and vice versa) is the story of 20th-century Greece. Figure 10 traces the historical record of fire. Note, particularly, the acceleration in burning over the past 50 years. In this case the disintegration of traditional agriculture, even outright land abandonment, is encouraging a steady buildup of fuels that sustains an increase in fire. The spikes in burning correspond to periods of social unrest when ignition increases and the capacity for fire suppression decreases. Similar scenarios, the stockpiling of fuels due to the breakdown of intensive farming, characterize other northern Mediterranean landscapes.

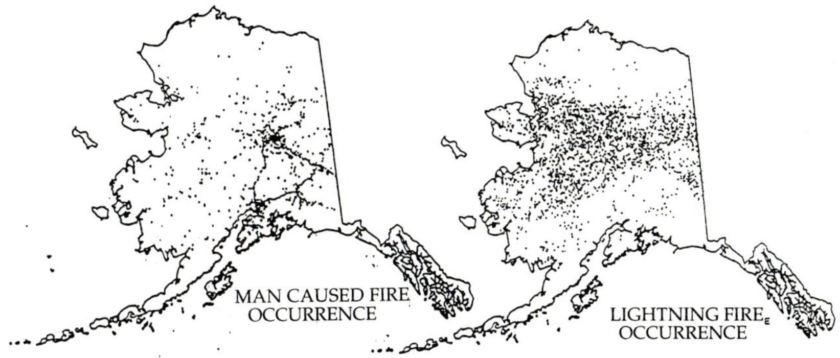

Figure 9. The modern fire geography of Alaska. The different domains of anthropogenic and lightning fire reflect decisions about land use. From Gabriel and Tande (1983).

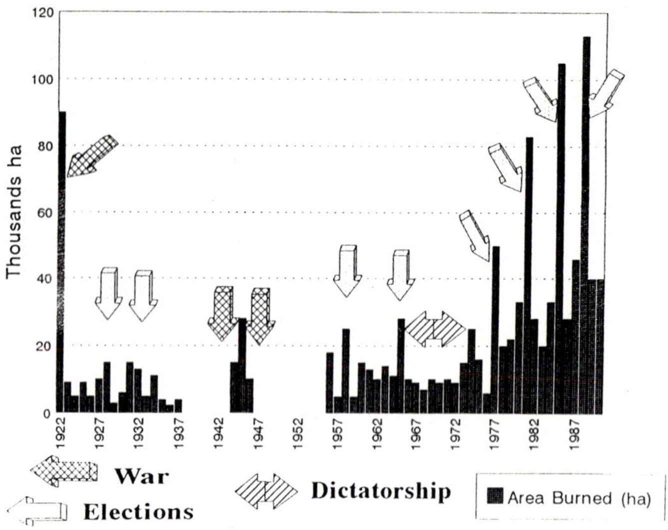

Figure 10. Greek fire: area of forests and scrublands burned. The overall increase in area burned relates to changing land use, particularly the decay of traditional agriculture; the spiked ignitions correspond to periods of social unrest. Data from Kailidis (1992).

While lightning has not apparently contributed to this revival—lightning fire is believed rare in the Mediterranean Basin, although a high fraction of fire causes are "unknown"—the potential exists, and it may be that lightning will find an environment more suitable than any it has known since the Pleistocene. The absence of lightning fire may have been another artifact of very long-term human presence like terraced olive orchards and hard-browsed maquis. Whether it is a prelude or a freak, it is worth noting that lightning kindled one of the largest wildfires in recent years in Spain, an event that would have been unthinkable before large quantities of fuel became generally available.

The Competition Among Anthropogenic Fires

The exchange of a lightning fire regime for one based on anthropogenic fire occurred long ago in most of the world. But this is not to say that regimes ceased to change beyond that. Torch met torch. Societies evolved, clashed, merged, and otherwise engaged in ways that have affected how fire has been applied or withheld from the environment. Various anthropogenic fires began to compete among themselves.

A full narrative of this competition is nothing less than the history of humanity and its interaction with the environment. This involves of course many peoples and many lands in many narratives played out through many climates. Moreover, the parts themselves change through time. Human societies evolve as well as migrate; ecosystems adjust to species gained or lost, and to processes added or subtracted; climate fluctuates both locally and globally. Empires assemble and then disintegrate; epidemics strip societies bare, like the Black Death in 14th century Europe and the pandemics that swept the Americas after the European Encounter; cattle and horses replace bison on the Great Plains of North America, and metastasize exponentially throughout the Argentinean pampas; the spread of metallurgy deforests local landscapes for charcoal; a fashion for beaver hats upsets the ecology of northern hemisphere forests; the list is endless. Each of these changes affects the local fire regime and the global fire budget. Equally, fire serves as an indicator, and integrator, of these other environmental reformations.

The major divisions of anthropogenic fire history thus follow the prominent divisions of human history—the initial colonization of continents; the Neolithic revolution; the expansion of Europe; fossil-fuel combustion associated with industrialization. For each of these processes there are unique historical examples. Aboriginal fire shaped Australia. The Americas mixed aboriginal and agricultural fire practices, but without domesticated livestock. Portions of central Eurasia and Africa relied more or less exclusively on pastoral burning. To varied degrees Africa, Europe, and Asia experienced mixed economies and displayed fire regimes appropriate to farming, herding, and, in more remote sites, hunting and foraging. Antarctica exhibits a continent without fire—and for that reason is unlivable without imported fire.

The magnitude of human impact varied widely, of course. Where environmental conditions were favorable, as in Australia and California, aboriginal peoples could reshape landscapes wholesale. The firestick became a lever that, suitably placed, allowed them to move continents. In the Great Plains of North America climate, landforms, and fire promoted vast grasslands; but through anthropogenic burning that biome could be perpetuated even after climatic changes nominally favored trees. The power of aboriginal fire is nicely illustrated by its removal. Everywhere the abolition of aboriginal burning has inspired dramatic ecological changes. For the most part, these involve shifts from grass to woods, from prairies or forest-steppes to scrub and forest.

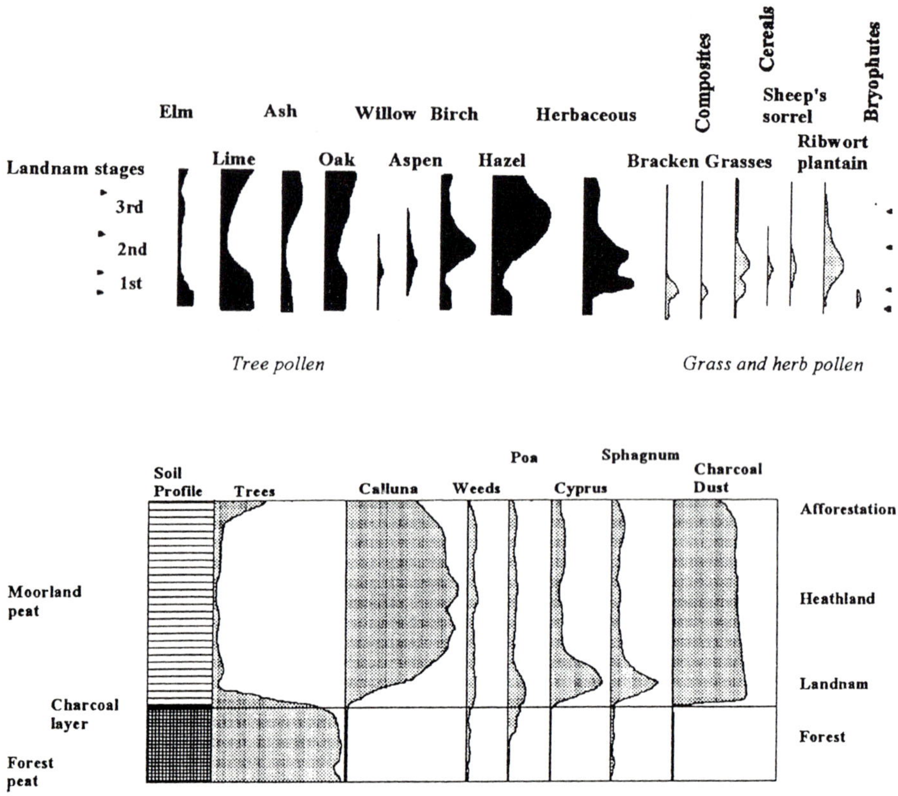

Figure 11. Landnam: Neolithic fire. With other tools, and especially with domesticated animals, Neolithic tribes in Europe could exploit fire to greater effect than could Mesolithic peoples. A first-contact swidden was the outcome, often resulting in significant floral shifts. (Top) Landnam in Denmark. Redrawn from Iversen (1973). (Bottom) Landnam in Norway. Redrawn from Behre (1988).

Other regions humans made more favorable through allied technologies, notably agriculture. Even simple swiddeners outfitted with stone axes, and particularly grazers with small flocks of goats, sheep, cattle, and swine, could induce dramatic reformations in the landscape. For the first, it was not necessary to fell all (especially large) trees, only the small ones. For both, it was not necessary to fell, only to kill the shade-producing species. This they could easily do by ring-barking and stripping. Once the forest surface was exposed to sun and wind, it rapidly dried and became available for burning. So did standing dead trees. livestock and human foragers could then select among the species that sought to reclaim the site. In this way, it was possible to restructure landscapes extensively. Extended over centuries (or millennia) whole landscapes became, in effect, anthropogenic artifacts.

In this way shifting agriculturists could break up putatively stable, old-growth forests. A consistent outcome was an increase in biodiversity. There were more niches, a tighter mosaic of biotic patches, more dynamism to the cavalcade of species successions. The classic description of course comes from Denmark through the work of the palynographer, Johannes Iversen. The entombing lime (*Tilia*) forest of Mesolithic Europe splintered before the diffuse *landnam* (literally, "land taking") of Neolithic tribes (Fig. 11). Species driven into sun-refugia along river banks and lakes now percolated throughout the landscape. Under the impress of clearing, grazing, and fire, heath and moor also proliferated. Interestingly, where those pressures have been removed, as in Sweden's Dalby Söderskog National Park, biodiversity has collapsed. Between 1925 and 1970 the park lost 41% of its species, largely through inattention, largely because the park excluded rather than promoted the chronic disturbances that had nurtured its nascent variety.

Of course, during any period of change, some species thrive and some falter. Some species cannot tolerate fire. But increasingly ecologists have discovered that fire exclusion (or the exclusion of fire-catalyzed practices) is a powerful destroyer of biodiversity, and more and more ethnographers are describing fire-forged landscapes for which, once the fire was extinguished, the biota turned cold, immobile, and monolithic. In the Brazilian *cerrado*, the South African *fynbos*, the Mediterranean *maquis*, North American tallgrass prairie, Swedish *heden*, southern pineywoods, Everglades sawgrass, Australian spinifex—in an endless register, the power of aboriginal and agricultural fire has been demonstrated and reconfirmed, not only for individual species but for habitats and landscapes. Earlier critics had assumed that fire, so visible and dramatic, was inevitably a force of destruction. The real issue, however, was not the presence of fire but its regime, its scale, and the fire-catalyzed technologies with which it was allied.

Fire and Fallow: Agricultural Burning

Sedentary agriculture too relied on some regimen of burning. Until very recently agriculture, after all, was rotational, either because it moved the farm through the landscape or because, in effect, it moved the landscape through the farm; both forms demanded fire. Agriculturists looked to fire to perform its twin duties, to purify and to fertilize. (The major exception was irrigation agriculture which substituted water for fire, although even here it was common practice to burn the stubble.) But in order to burn it was necessary to have fuel, or in agronomic language, fallow. "Swidden" is a poor description for the robust spectrum of such fire-fallow regimes. (In its origin the term derives from old Norse, meaning "to burn," it came to England with Norse invaders, and it referred specifically to the decadal burning of ling, or *Calluna* heather.)

One reason why the term fails is that most landscapes were compounds and mixtures of habitats, not just fields, and they experienced a medley of fire practices. There was slash-and-burn cultivation in outfields, stubble burning in infields, pastoral burning in abandoned swidden plots, wetlands, and rough fields, underburning to promote native tubers and berries for foraging, broadcast burning to sustain hunting grounds, smoking to harvest bees, patch burning to produce roofing thatch and branches for basketry. Landscapes encountered not one fire but many, and found those burns remarkably specific, subtle, and, in their sum, pervasive.

Especially among European agronomists the fallows were considered a badge of shame. This was land out of production, land that with proper management could grow food and fodder to sustain Europe's spiraling population. Ideally, agricultural savants sought to substitute a calculated succession of useful crops to replace those species which flourished opportunistically after abandonment. Ultimately they hoped it would be possible to dispense with fallows altogether, which is to say, with fire, for a primary purpose of the fallows was to produce the fuel without which fire was impossible. Agricultural systems raised fuel for field burning just as they cultivated oats for draft horses.

The fallows served other purposes, and were often rich in semi-cultivated (or at least selected) species that could be put to various uses. That they were the scenes of chronic disturbances by grubbing, grazing, cutting, harvesting, planting, and burning made them notoriously profuse in species diversity. But for agronomic critics biomass, not biodiversity, was the measure of success. Political theory, too, preferred sedentary to shifting cultivators. As Europeans expanded their dominion across the globe those notions of proper farming, good societies, and healthy landscapes dominated their understanding of the agricultural systems they encountered.

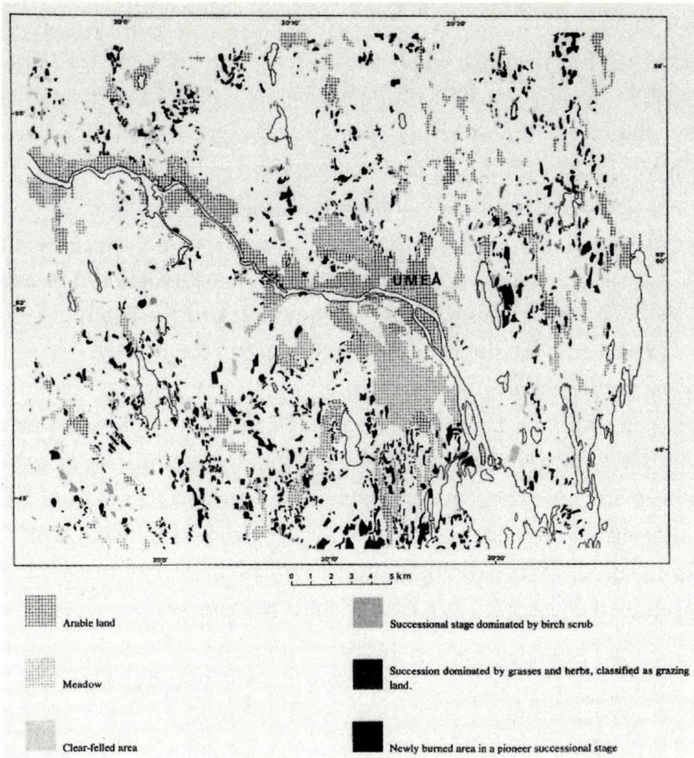

Figure 12. Swidden in Sweden: map of Umeå showing the mosaic of managed lands. The pattern is similar to traditional slash-and-burn cultivation elsewhere in the world. From Zackrisson (1977).

Vestal Fire: The European Exemplar

The dominant story over the past half millennia has been the expansion of Europe. Mariners reconnected in decades what geologic processes had separated over tens of millions of years. Plants, animals, microbes, diseases, and peoples mixed—invading, transplanting, eradicating, mingling—in what Alfred Crosby has proposed as the greatest biological event since Pleistocene glaciation. What resulted was not merely a new global economy but a reconstituted global ecology and, after the consolidation of modern science, a global scholarship by which to understand these events. European science would replace indigenous lore as the interpreter of fire, and Europe would declare itself the standard by which fire practices should be judged.

The expansion of Europe and the subsequent spread of industrialization based on fossil-fuel combustion thus offers a range of encounters. In Australia it met peoples who practiced neither farming nor herding. In the Americas it confronted hunters, foragers, and farmers. In Africa and Asia it interacted with mixed economies much like its own. The fire histories of these regions thus differ, not only because of the intensity of colonial rule but because of the shock wave of encounter itself. The greater the difference, the greater the fragmentation. In general, however, the amount of open burning has declined, woody biomes have replaced grasslands, and where it persists, fire has moved from the field to the furnace.

Because of Europe's influence, the fire history of Europe holds particular interest. Virtually every fire practice found in the world finds expression in European history. Europeans practiced fire hunting and fire fishing. They burned to improve berries and mushrooms. They practiced slash-and-burn cultivation in nearly every environment; the agricultural revolution that preceded the industrial revolution was based, in part, on the extensive conversion of heath, moor, and other organic soils through a regimen of slashing and burning (see Figs. 12 and 13). Swidden was still practiced in Scandinavia and the Midi region of France in the 1920s, and continued in Finland through the refugee resettlements after W.W.II. The agricultural geography of Tsarist Russia resembled a series of nested dolls, the scale of each sized by the length of its fallows (Fig. 14). There was no absence of fire.

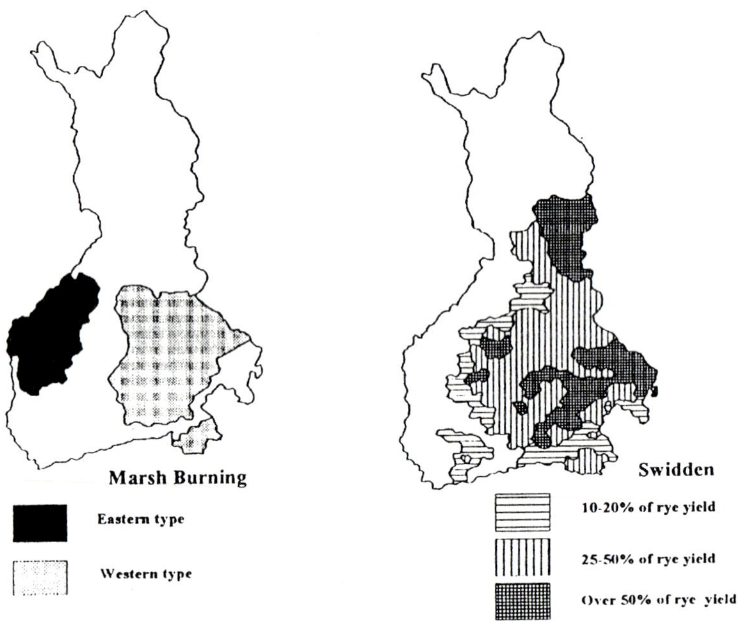

Figure 13. Finnish fire: two patterns of slash-and-burn cultivation still apparent in 1830s Finland. In the better drained sites forests were swiddened (left map); in the mires, the organic soils, a practice that characterized much of northern Europe during the agricultural revolution of the 18th and early 19th centuries (right). Redrawn from Soininen (1974).

Until the mid-18th century Europe was little different from other parts of the world subjected to similar economies. Its fire regimes were only marginally distinctive, a reflection of local environments and histories; its encounter with fire regimes outside Europe reflected more the indirect effects of its expansion in terms of plants and animals introduced an extirpated, and of diseases and the demographics of contact than of any deliberate policy aimed at restructuring those lands through fire. But officials and intellectuals, almost all of them urban, detested fire and denounced it with edict and epistle.

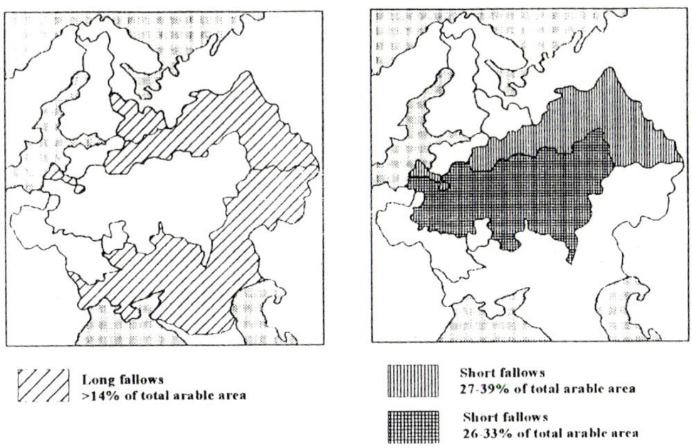

Figure 14. Fire and fallows: the Russian example. The pattern corresponds both to climatic gradients (the domain of the mixed European forest) and to the east-trending spread of Russian settlement. Even longer fallows characterize Siberian agriculture. Redrawn from Pallot and Shaw (1990).

Increasingly European intellectuals of the Greater Enlightenment sought to constrain fire. They saw abused landscapes—organic soils pared and burned down to rock, fallowing cycles shortened dangerously, expanding heaths, felled woods, marauding shepherds, eroded hillsides, mountain torrents. Reason demanded some way to protect humus and eliminate fallows. They saw agriculture, under the crush of increasing population, trying to square the circle instead of close it. They identified fire, ever present, as a cause of destruction. Even the legendary Linnaeus was forced to delete from his 1752 Skåne journal favorable comments regarding swidden in Småland and substitute a passage on manure.

They unambiguously separated primitive from rational agriculture according to whether or not it used fire. Intensification demanded an end to fallows; increased production required that biomass go to the slow combustion of metabolizing livestock and humans rather than the fast combustion of open flame. Agronomists had their reasons. National wealth depended on agriculture, agriculture on the state of the humus, the humus on whether or it not it was burned. Besides, broadcast burning works best where populations cycle through a landscape, and as enclosure and fixed ownership spread, fires were less desirable. Free-burning fires belonged with the flaming front of colonizers and pioneers. But most of Europe was beyond this. So too were Europe's intellectuals, almost all of whom were urban. Only those of rural backgrounds saw much value to fire—Virgil and Linnaeus among the most prominent. Others knew fire only through the terrors of urban conflagration, and that, as often as not, the outcome of hostile siege and sack.

The swelling population itself brought such pressures on the land that there was simply less and less free biomass available for burning; peasants raked up pine needles for stable bedding, seized every reed and blade, lopped off tree limbs for fuel and fodder; in response, agronomists became obsessed with restoring the organic component of the soil, primarily by manuring. The degradation of Mediterranean landscapes by promiscuous browsing and burning complemented Edward Gibbon's *The Decline and Fall of the Roman Empire* as an allegory of political economy and moral malfeasance. Encounters with tropical islands such as St. Helena, Barbados, St. Vincent, and Mauritius became semi-controlled experiments on the disastrous consequences of abusing lands, microcosmic warnings of macrocosmic doom. Cutting, grazing, and burning became the unholy trinity of environmental critics.

Counter Fire: The Engineering of Free-Burning Fire

The European hearth became a norm, a skewed vestal fire for the planet. Europe's experiences within its own borders were sharpened by the encounter with the "strange fire" of distant colonies and eventually became the basis for a more systematic understanding of biomass and burning. The exchange with overseas colonies, however, went only one way. Europe exported its science and fire control apparatus but never imported indigenous knowledge about fire ecology and practices. These experiences were institutionalized as the 18th century Enlightenment remade classical arboriculture into modern forestry.

Academic forestry grew out of European agronomic thinking, and thus accepted humus as a universal index of environmental health. Since fire burned humus, it was intrinsically evil. Besides, swiddeners burned forests before they could mature into sawtimber, and pastoralists routinely fired woodlands to encourage browse and grass and in so doing directly attacked trees. Foresters freely identified the weapon, fire, with the assailant, herders and shifting cultivators, and sought to control the users by controlling the use. Out of all the disciplines that might have addressed fire, the task fell to foresters with consequences that are still very much with us.

Foresters struggled to create institutions to suppress traditional burning and to explain the damages wrought by fire. In general indigenous lore, like indigenous fire practices, was extinguished. This was as true in Europe as elsewhere. Fire suppression became the norm, and the relatively fire-free forests of central Europe, an aberration, became a standard. Germany's forests stood to landscapes as Max Weber's Prussian bureaucracy did to social institutions, a Platonic type. Foresters in turn joined those other great corps of engineers, civil, hydraulic, and mining, that were so influential in projecting European influence throughout the world.

The presence or absence of forestry became in fact a defining criteria in the evolving geography of fire. Those nations that accepted it, or had it imposed, removed too much fire but created an infrastructure for the management of forests and wildlands that later served for the reintroduction of fire. Those that did not develop forestry continued traditional burning (but within an altered and much more densely populated world) and today generally lack an institutional means to contain those fires.

This exercise in technology transfer was easiest in North America and messiest in the southern hemisphere where biotas were dramatically different and where, as in Asia and Africa, indigenous peoples and livestock outnumbered European imports. There traditional burning persisted in defiance of European desires. In fact colonists often adopted native fire practices or hybridized with them in mutual hostility to the edicts of imperial proconsuls and the theories of European savants. Fire policy became an expression of colonial rule, a means of reordering the ruled landscapes. In response natives burned illicitly, at once a political protest and an attempt to restore traditional lands to traditional purposes. Regardless of whether local peoples retained legal access to state-reserved lands, unless they could use fire they lacked biological access to the potential resources those lands held. The character of the land changed, often irrevocably.

All this did not pass unnoticed. Typically a debate ensued, often formal and even published, that pitted European standards against local practices, fire control against fire use. The most dramatic confrontations flared in British and French colonies (or former colonies), where foresters nurtured in Franco-German traditions were most aggressive at imposing policies that aspired to fire exclusion. In North America fire control triumphed; in India, Australia, and South Africa awkward compromises resulted that first denounced controlled burning, only to recant later and ultimately to absorb them into official doctrine. In time controlled fire returned to North America as well although in a language ("prescribed burning") that denied the legitimacy of its earlier incarnation ("light burning").

Fire Ecology Reconsidered: Recycling a Concept

All this has reformed the meaning of fire ecology. The range of anthropogenic fire matches that of human society, no longer confined to particular sites or limited to a penumbral shadow of off-site effects. Of course ecosystems had long accommodated anthropogenic fire practices. An ecosystem's energy flows, its nutrient cycles, its structure, its pattern of disturbances, its fire regimes — all had been influenced if not dominated by humans. As human societies expanded and evolved, so did the nature of fire ecology. Increasingly Europe's widened contacts had rewired ecosystems, redefined their scales, and modified their tempo.

A global economy cycled nutrients on a global scale. It could for example mix guano from Pacific islands with forests in France, replace native eucalypts in Australia with American pines or substitute eucalypts for Indian deodar, trade Ukrainian wheat for British steel. More or less bounded ecosystems broke up and were reconstituted, at least in part, on different scales through the trophic chain of a money economy. Human institutions rapidly connected lands previously bound only through the longest and most tenuous of biogeochemical cycles.

The organization of distant colonies followed from the knowledge (or ignorance, or misinformation) of imperial powers. Science and engineering, no less than economic imperatives, created institutional valences that bonded India to Cape Colony, and Corsica to the Ivory Coast. What foresters, for example, learned in one part of the British Empire would be felt in other parts. Imperial foresters could transfer fire practices developed for Madhya Pradesh to Burma and Kenya and the Yukon.

In the United States fire effects similarly cycled through national institutions. The 1910 fires so traumatized the U.S. Forest Service that it would not consider controlled burning until its founding generation had passed from the scene thirty years later. By killing 15 firefighters, the otherwise obscure 1937 Blackwater fire in Wyoming stimulated both the smokejumper and the 40-man crew programs, each of which profoundly altered the fire regimes of America's public lands. By forcing the National Park Service and U.S. Forest Service to reevaluate their programs, the Yellowstone fires of 1988 had ecological consequences for cognate lands in Oregon and Minnesota and New Mexico. Thus physics, biology, and culture may evaluate fires different. A low-intensity fire may be biologically significant, and an ecologically trivial fire, as measured by area burned, may be culturally important. A true fire ecology must integrate them all.

While fire scientists (especially Americans) have sought to reconstruct the "natural" ecology of fire, landscapes in fact reflect their human histories and the dynamics of social institutions. The cycling of information through bureaucracies is as critical as the flow of carbon through lodgepole pine forests. Budget cycles become as important as the rhythms of the seasons; policy determinations as potent as evolutionary selection; the flow of money as vital as that of sunshine. The study of fire ecology without reference to the human presence is meaningless, like physicists studying perpetual motion machines because they choose to disregard entropy.

The Competition with Industrial Combustion

Not least of all industrial Europe began to sublimate the power of fire into machines. Controlled combustion began to replace controlled burning. By the end of the 18th century industrialization compounded imperialism into a new order. The encounter between Europe and the remainder of the planet broadened, and its tempo quickened. For fire history the critical events involved the exhumation of fossil biomass, the invention of new fire engines that could combust it, and the substitution of artificial herbicides, pesticides, and fertilizers for tasks once done by fire. Those engines redefined the character of pyrotechnologies, while coal and petroleum became, in effect, fossil fallows that allowed agriculture to transcend its closing ecological circles.

Fire Engines: From Flame to Combustion

In the beginning industrial combustion, and the industrial economy, depended on biomass burning or other sources of power such as watermills. But fossil hydrocarbons soon changed that, and imposed a change in the character of the Earth as a fire planet (Fig. 15). Fossil-fuel combustion burned without regard to the living environment. It literally stood outside the ancient ecology of fire, and outside the traditional social mores that had guided anthropogenic fire use. Humanity metamorphosed from the keeper of the flame to the custodian of the combustion chamber.

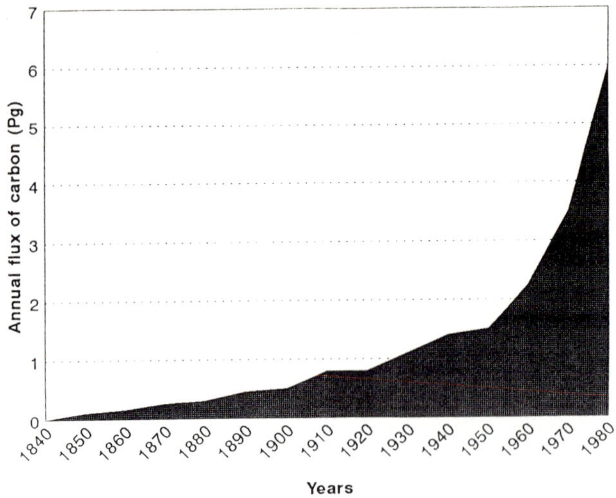

Figure 15. The escalation of fossil-fuel combustion, as measured by annual flux of carbon (Pg). Data from Houghton and Skole (1990).

Industrialization's reach exceeded its grasp. Fossil fuels burn beyond the parameters of traditional fire ecology. Their combustion proceeds with savage indifference to season, time of day, the biotic rhythms of nutrient capture, growth, and decay. They restructure fire regimes, both directly by redefining and controlling ignition and indirectly by reorganizing the biosphere and its fuels. They have redefined what are natural resources, how they might be exploited, and what lands will be used in what ways. Converted to petrochemicals and combustibles, fed into tractors, diesel trucks, and the turbines of commercial ships, fossil fuels have allowed agriculture to overcome, at least momentarily, the closed ecology of fallow farming.

What had long exploited—in fact, required—fallows now had an alternative. Fossil fallow could substitute for living fallow. Output was not restricted to available inputs. European agriculture, in particular, had long been obsessed with recycling waste, whether dung or fallow fields; only by increasing the biomass fed into a system could farmers and pastoralists hope to extract more out of it. That was the agronomic logic behind the European critique of folk burning. But when coal and oil could substitute for fuelwood, when diesel-powered tractors, trucks, and trains would replace draft animals and their fodder requirements, when petroleum-derived pesticides, herbicides, and fertilizers could do what only fallow-fueled fire had done in the past, then it was possible to break out of the limits of fire-fallow agriculture, or more properly to transcend its biotic geometry. Agriculture could stop chasing its tail around the field and leap over the fences that once confined it.

No longer, or at least for several centuries, does available biomass limit productivity. Fossil fallow is matter and energy added to, not extracted from, landscapes of cultivation. The new combustion regime is no longer restricted by what the land could grow. Instead fossil hydrocarbons are a kind of biotic bullion, plundered from the past. Unfortunately, fossil biomass is acting on nature's economy as the 16th-century flood of Inca gold and Mexican silver did on Spain's. It is inflationary, often ecologically corrupting, and shaped less by input than by output, by the capacity of the system to absorb the excess effluent. Increasingly the Earth's ecology traces the nutrient cycles and energy flows of a global economy. Increasingly the limits to productivity are set by available sinks rather than sources.

The full range of interactions is not well understood. There are not even formal terms to discriminate between traditional burning and industrial burning. Call the new regime "combustion," since it has disaggregated burning from flame. Reserve "fire" for traditional open burning. It is clear, however, that if lightning fire defines one border to anthropogenic fire, then fossil-fuel combustion defines another (see Fig. 5). This pyric revolution has demanded not only new combustion technologies for furnace and forge but new fire practices for field and forest. Perhaps the most critical difference ecologically is that fossil-fuel combustion competes with anthropogenic fire not over a common source, biomass, but over their common sinks, principally the atmosphere, the amount of emissions that regional or global airsheds can absorb without major distortion.

Again the nature of this boundary is not well understood. While industrial combustion can substitute for many human pyrotechnologies, it cannot supplant the full range of ecological effects that come with free-burning fire. It can replace older fire technologies but not fire ecologies; fire as a tool has surrogates, fire as a process may not. It is one thing to substitute an electrical wall heater for a fireplace, another to propose that chain saws and bulldozers can mimic the disturbance and fire regimes of a forest. But if the ecological processes are blurred, the historical border is remarkably precise. Fire follows fuel. Fire's modern age tracks exactly the exploitation of fossil hydrocarbons. Had industry merely burned biomass, the ancient fire-fallow cycle would have held. But industrial technology discovered, instead, a new fuel, outside the nagging limits of its grown biomes. In doing so industrialization inaugurated a new era of fire history.

The range of encounters between new and old is large, confusing, and often incommensurable. More and more, the Earth divides between two geographies, one of industrial combustion, another of open fire. With the exception of India (here, as in other matters, seeming to absorb new elements into its fire castes), lands belong to one or the other. The steady abolition of fire from industrialized landscapes reflects not only the removal of agriculture and its human tenders but the application of fire suppression technologies, all dependent on such combustion-powered machinery as portable pumps, wildland fire engines, and slurry-dropping aircraft; even fire-retarding chemicals derive from fossil hydrocarbons. Previously, the best — often the only—protection against wildfire was to burn under controlled conditions, which had the effect of keeping fire, some fire, domesticated fire, on the land. Paradoxically, the dramatic diminution of controlled fire has allowed biomass to bloom, and in many exurban landscapes, from Los Angeles to Sydney, these proliferating fuels have powered devastating wildfires. Even more curiously, industrial societies have cultivated an alternative to the biodiversity-rich fallow of agricultural landscapes. Nature reserves have become the fallow-surrogates of the industrial Earth.

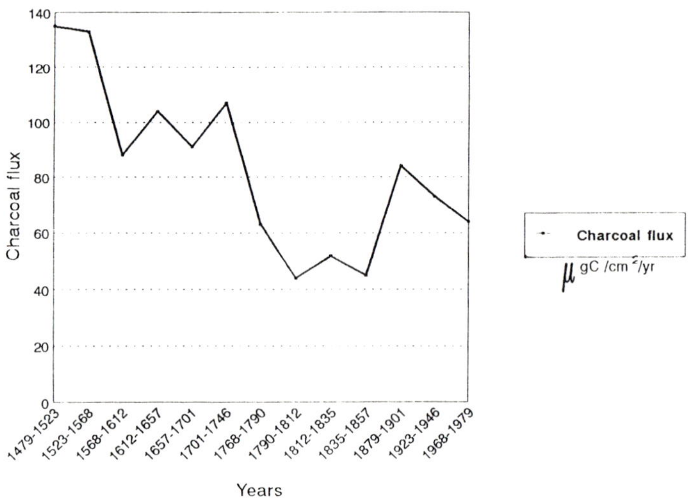

Figure 16. Lost contacts: the removal of anthropogenic fire. Charcoal fluxes embedded in Pacific sediments from off-shore Mesoamerica. Data from Suman (1991).

Making and Unmaking Agriculture: The Amazon and the Mediterranean

Since the late 1980s Amazonia has become an international symbol of combustion run amok. Slashing and burning on a colossal scale have, it is charged, degraded landscapes, impoverished biotas, and unhinged the climatic order of the planet. In fact fire is everywhere in Amazonia's natural history. Charcoal lenses underlie the soil; small-scale swidden accounts for perhaps 35-40% of the biodiversity of many sites; the grassy *cerrado* are fire-dependent, soon overwhelmed by species-poor woods if not regenerated routinely by burning. The rainforest appears to be an artifact of the Holocene, a biota that has coexisted with humans and thereby with anthropogenic fire. How much of this forest is a secondary product of depopulation following a violent, disease-brokered Encounter with Europe is unclear. In Mesoamerica the evidence suggests that burning has not yet returned to preconquest levels (Fig. 16). Undoubtedly this is also true for most of South America. Almost certainly it is true for North America.

Not enough is known to say whether this also characterizes Amazonia. Clearly, however, the reckless deforestation of the past two decades is a product of industrial combustion, either directly through internal transportation grids or indirectly through linkages to a global economy that determines the value of Amazonian land and makes profitable its production of commodities. Industrialization after all is the announced goal of the Brazilian state and stands behind its subsidies of development. Similar scenarios characterize Indonesia, Malaysia, Thailand, and parts of India.

The Mediterranean Basin presents the obverse scenario. Here a traditional agriculture that has in one form or another existed for several thousand years is disintegrating. The region has a fire-prone climate, a pyrophytic biota, and a very long history of anthropogenic burning for fuel, pastoralism, and farming. The intensity of traditional usage is such that wild fires are rare—there is simply not enough wild fuel to sustain them.

Industrialization and integration into the European Community are forcing changes, however. Without massive subsidies subsistence agriculture cannot endure. Accordingly cities are growing, and farm land is being abandoned or converted to second homes or other uses. The liberation of fuels has liberated fire. Figure 17 shows the relationship between the intensity of usage and the propensity to uncontrolled fire. This change, of course, underlies the record of renewed fire for Greece expressed in Fig. 9. Similar situations characterize North America and Australia where exurbanites are reclaiming land from rural economies, but because they do not engage the land—harvesting, grazing, gathering, burning—they are allowing fuels to build up. This "intermix" fire scene is generally recognized as the dominant problem of the American fire community. The Southern California conflagrations of 1993 demonstrate its volatility and destructive potential.

Hostile Hearth: The Energy Exchange

A related process characterizes the change in domestic fuels. The substitution of fossil fuels for biomass is allowing, in many places, for a buildup of woody vegetation. Figure 18 shows this exchange for Korea and the U.S., a transformation typical of industrialization.

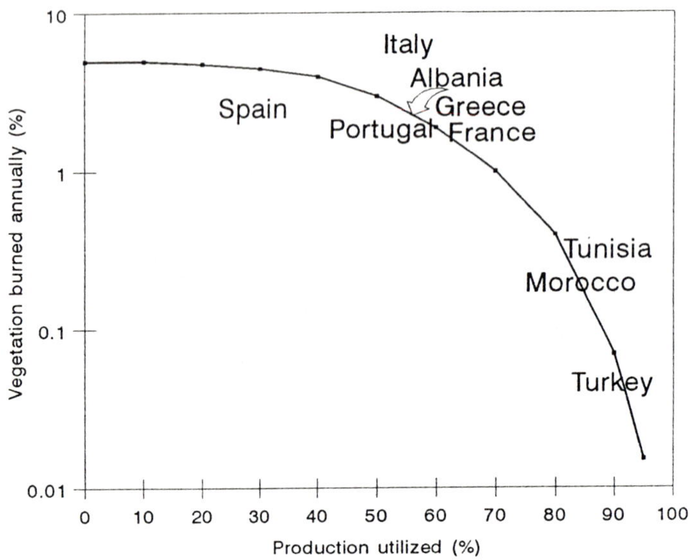

Figure 17. The competition for biomass, a Mediterranean example (1980-85). The greater the utilization of biomass the less the area burned. Redrawn from Castro Rego (1992).

Many observers consider this change desirable. Biomass burning releases greenhouse gases, fouls the air near human inhabitants, wastes energy, and subverts surrounding ecosystems. The explosive growth of human populations has made the hearth, long an emblem of human solidarity, into a symbol of human degradation of the environment. Larger and urban populations demand a shift in domestic energy from biomass to some more efficient combustion based on fossil fuels. In principle the amount of carbon released can be absorbed by the vegetation that is spared.

Of course the issue is more complicated than this, and among its unexpected side-effects is the prospect for wild fire. If nothing else consumes that renewed growth, fire typically will. This in turn upsets the fire regime of the ecosystem in ways that are often as damaging as direct harvesting of its flora. In this sense the substitution of fossil fuels for biomass does not abolish fire but argues the need for a new regimen of anthropogenic burning in which controlled burning can substitute for wildfire.

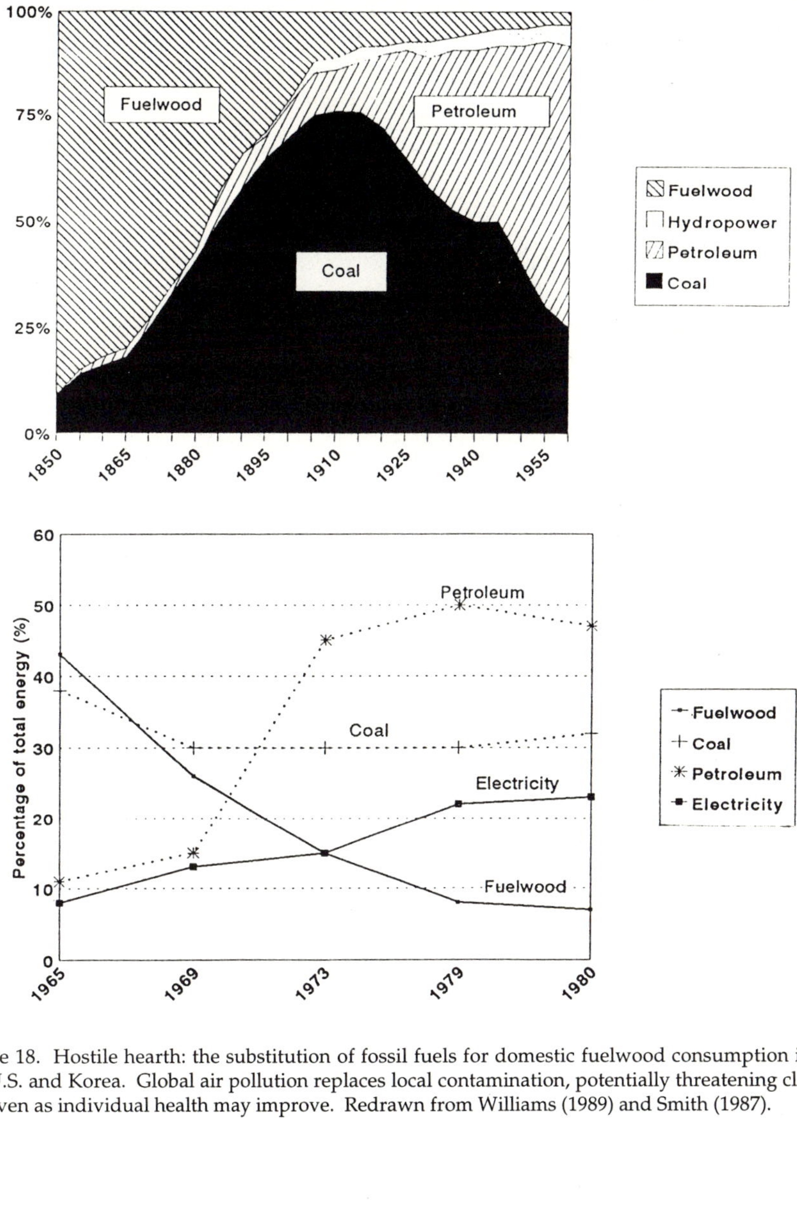

Figure 18. Hostile hearth: the substitution of fossil fuels for domestic fuelwood consumption in the U.S. and Korea. Global air pollution replaces local contamination, potentially threatening climate even as individual health may improve. Redrawn from Williams (1989) and Smith (1987).

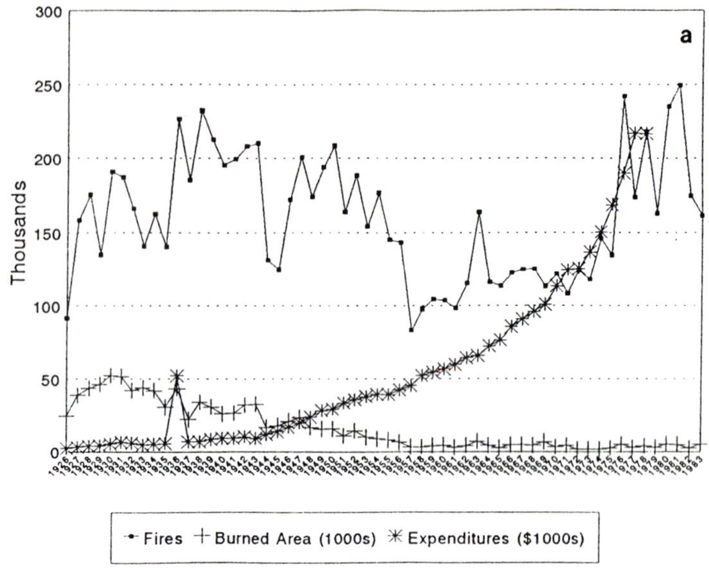

NB: Costs exclude emergency funds

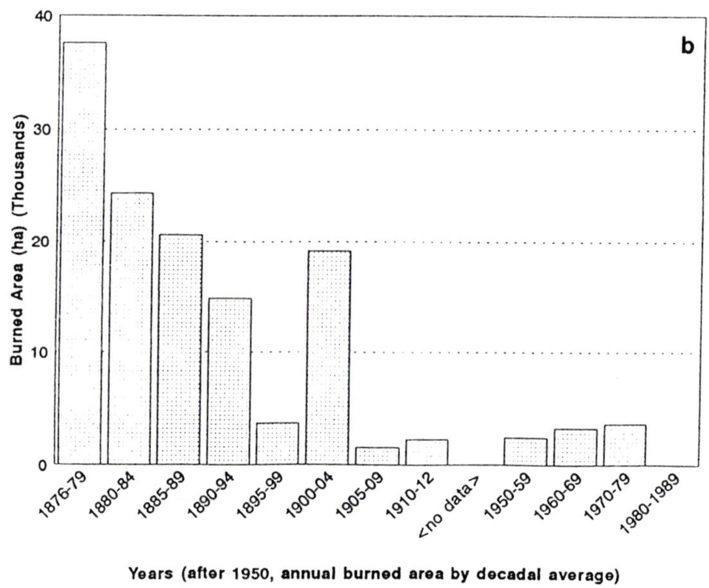

Figure 19. The expulsion of fire. (a) U.S. There are two logistic curves, one tracing the removal of free-burning fire, the other the costs of these acts; neither is sustainable. Reforms have sought, with limited success, to control costs and reintroduce fire. Data from *Historical Statistics of the United States*. (b) Sweden. In converting farm and woods to commercial forest Sweden treated fire as it would other agricultural pests, and intensive use (note the decline in dead wood) has virtually abolished fire. No coherent statistics exist for many decades, although large fires did break out in 1934. From Högbom (1934) and Stocks (1991).

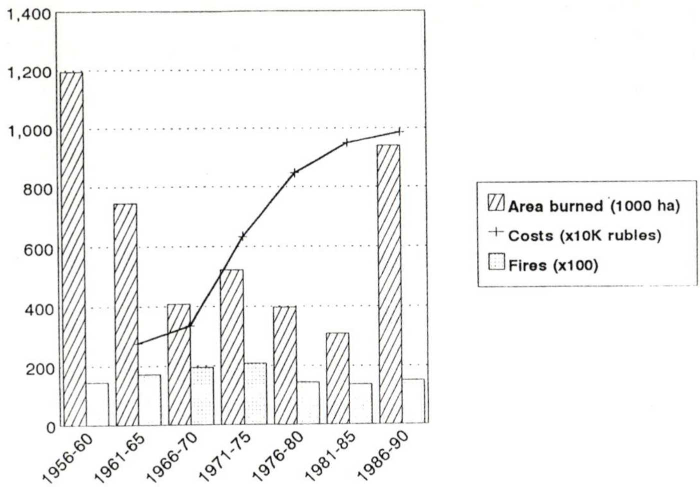

Figure 19. (cont.) (c) USSR. The figures are almost certainly false - they should be increased by a factor of 2-10. But the trend is probably correct, and shows remarkable similarities to that from the U.S. Data from Avialesookhrana.

Friendly Fire: Wildland Management

Among the curious consequences of industrialization is the attempt to suppress fire of all kinds in commercial forests and nature reserves, both landscapes of special concern to the kind of urban societies that industrialization has made possible. Breaking the fire cycle was deemed necessary to protect such sites for large-scale logging and, after the enthusiasm for wild nature became fashionable, for parks, wildlife sanctuaries, and nature preserves. In both cases traditional fire practices were prohibited; fire management was based on fire control, not on fire use. Thus forestry revealed its origins in European agronomic thought and park administrators their belief that a prelapsarian state of nature represented the natural order. For the first, administrators had to intervene to reshape a quasi-natural system to more suitable purposes; for the second, they had only to prevent human practices of any kind other than low-impact tourism and scientific research.

The experiment has been particularly vigorous in the northern hemisphere where it has progressed for almost a century. Figure 19 shows some examples of how dramatically the fire load has dropped. One explanation for this putative success is that state forests and nature reserves eliminated traditional burning. Comparative histories suggest that very large returns result from initially small investments in fire control. Even a first-order infrastructure can lead to dramatic reductions in burned area. The real dilemma comes from trying to sustain this condition through ever-escalating inputs of money, machinery, and manpower. The ecological crisis was not that wild fires were suppressed but that controlled fires were no longer set. A change in fire regimes, not fire exclusion, should have been the goal.

There were other reasons for the delayed appreciation of what industrial fire control meant, reasons peculiar to the geography of industrial firepowers. Almost all these nations resided in the north, where, indeed, the bulk of the world's landmass and forests also lie. It is characteristic of the boreal forest that a small number of large fires during drought years accounts for most of the acreage burned; rapid attack can control those critical ignitions at great savings. Still another explanation is that the ponderous rate of change in the northern landscapes disguised the full consequences of fire exclusion for decades whereas in the southern hemisphere the negative impact was felt within a handful of years.

But felt they were, even in the north. Landscapes changed, fuels built up, species responded. It became apparent that to withhold fire was as powerful an ecological act as to introduce it. Systems degenerated, biodiversity decayed, and lands became unstable with respect to fire. Fire control had succeeded in maintaining an incremental improvement by investing in elaborate technologies, by ultimately fighting fire with fire but through internal combustion engines rather than through torches. But this soon reached another point of diminishing returns. The triumphant profile of suppressed fire was equally a profile of an accumulating fire deficit.

For the past two decades the fire organizations of the United States, for example, have struggled to reinstate fire into state forests and wildlands. A grand experiment in letting nature determine the fire regime more or less ended with the debacle surrounding the Yellowstone fires of 1988. It is clear that the historic environment had adapted to anthropogenic fire, not to fire in the abstract, and that an anthropogenic fire regime must be reinstated if that landscape is to be preserved in something like its traditional state. Thus, even in lands from which people have been excluded, there is a compelling case for active fire management, which is to say controlled burning. There is a fire tithe owed most ecosystems, and it has been a human duty to pay it on behalf of the biota. If humans fail to do so, then lightning, arson, or accident will extract it in ways that will not always agree with human goals.

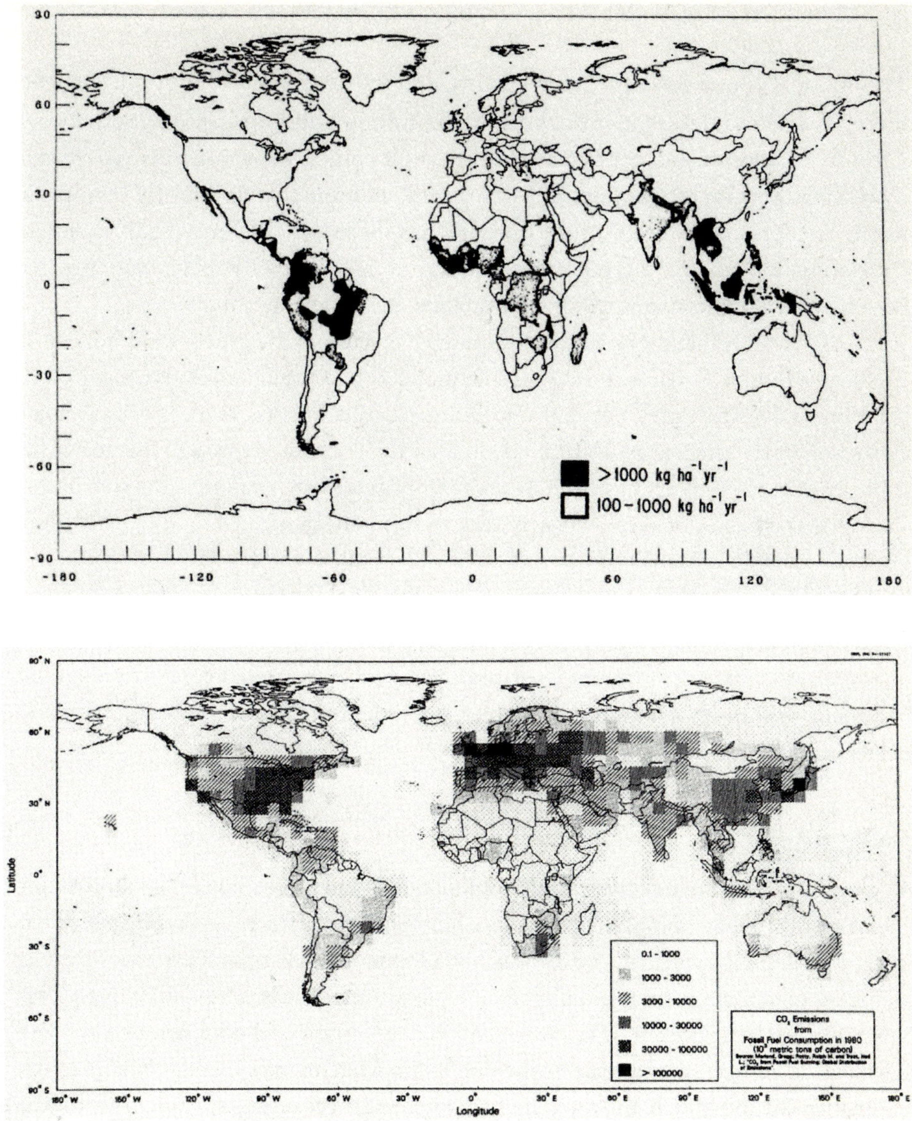

Figure 20. The two geographies of contemporary combustion: fossil fuel and biomass. Top map describes the geographic distribution of the net flux of carbon from terrestrial ecosystems to the atmosphere in 1980. The bottom map shows the geographic distribution of emissions of carbon from fossil fuel combustion. The one area of significant overlap is monsoonal Asia. From Marland et al. (1985) and Houghton et al. (1987).

Scorched Sky: The Iconography of Uncontrolled Combustion

The boundary between anthropogenic fire and industrial combustion differs from those that define the frontier between lightning fire and human fire, and between various regimes of anthropogenic burning. Of course their domains do express themselves geographically (Fig. 20). But they do not compete directly for biomass so much as for airshed; air, not humus, has become the universal index of environmental health, supplemented by measures of biodiversity; and fire has become part of an iconography of international environmentalism.

Nuclear winter, greenhouse summer, slashed-and-burned biodiversity in Brazil and Indonesia, Kuwait oil fires, the meltdown of the Chernobyl nuclear plant, wildfires in Manchuria, Borneo, and suburban California, even the parbroiled extinction of the dinosaurs—all have animated their concerns through images of fire as destroyer. One effect has been to make fire itself seem a cause of disaster rather than a catalyst for it, or even a consequence of mismanagement.

Sinks

That, in brief, is the overall cultural history of anthropogenic fire. The data for a quantitative restatement do not exist, and at the level that most researchers would like them, will never exist. It is, however, possible to trace general trajectories and block out the major trends.

Fire Report

A systematic inventory of the Earth's combustion history does not yet exist. It is not even decided what parameters should define such a history. Existing historical studies, ecological evidence, and atmospheric conditions all argue for a major shift in planetary combustion from biomass fires to fossil-fuel combustion. Intuitively one could argue that at present there is too much combustion and not enough fire. The fires that do persist are maldistributed -too much wildfire in the wrong places at the wrong times; too much burning in furnaces and not enough fire in the fields. Biomass burned is not equivalent to biota burned. Free-burning and confined fires have different biological consequences, and thus different atmosphere effects. Fossil fallows function differently than contemporary ones; eventually their added chemicals must go somewhere, and this typically involves some form of industrial combustion.

Over the last 500 years the Earth has probably experienced a net reduction in fire—how much is impossible to say with any confidence. This has been felt at different places at different times to different degrees. At what point industrial combustion has made up that net "loss" is difficult to say, if in fact it has done so. Figure 21 gives the best estimates of contemporary biomass burning. It is believed that the ratio of fossil-fuel to biomass combustion is now 60:40. These are measurements of mass consumed; they speak primarily to atmosphere chemists worried about emissions; they do not speak adequately to area burned, which is probably a more useful index of fire's biological impact.

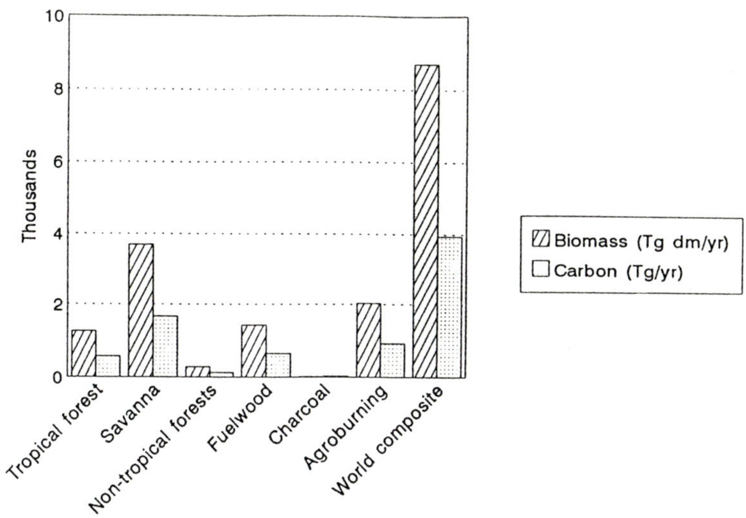

Figure 21. Estimates of global biomass burning. Data from Andreae (1991).

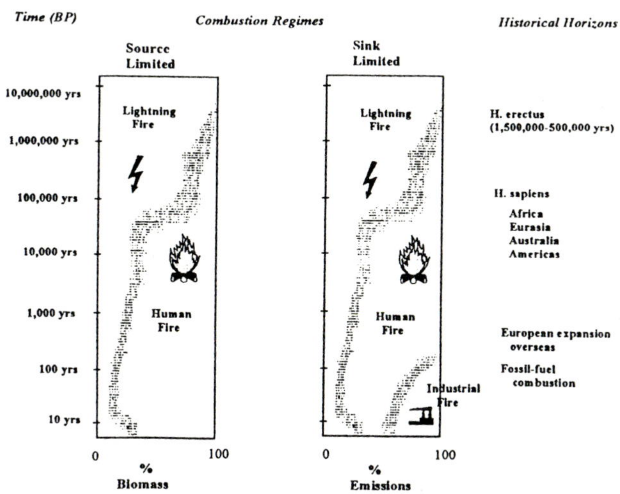

Figure 22. The competition for combustion: a schematic history. The graphs estimate only the percentage contribution from each source.

Figuring Fire

The following figures attempt to translate these gross events and more subtle intuitions into the framework for a quantitative fire history. Figure 22 assesses the relative contributions of lightning-caused, anthropogenic, and industrial fires within the context of a common global fire environment. They describe proportional effects on the consumption of available biomass and emissions without regard to the absolute quantities involved. In effect it traces the changing competitions described in Fig. 5.

Figure 23 sketches—with great trepidation and scant conviction—the actual quantities combusted. The left graph tracks the amount of biomass burned (by area), normalized against contemporary estimates. The right graph adds fossil-fuel combustion to create a fuller summary of total combustion. Again, past quantities are normalized against contemporary estimates. The maximum is about five times the present level of area burned.

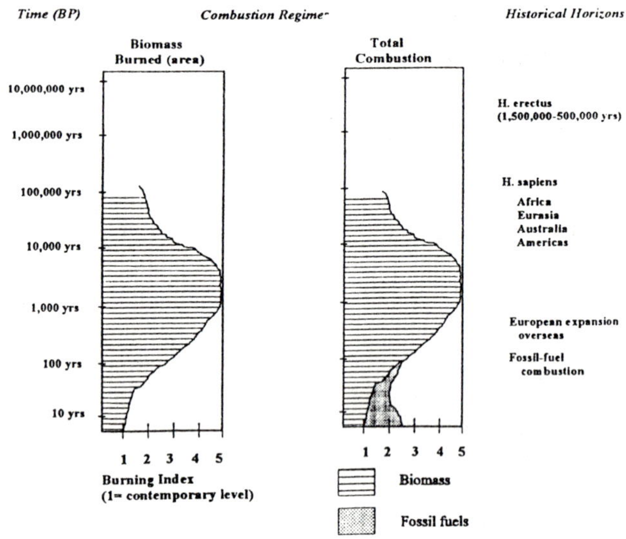

Figure 23. Combustion through the Holocene. An estimate of absolute values as normalized to contemporary burning.

The collapse of open burning is remarkably recent. In many regions, it has crashed only in the past fifty years. In others it appears to have accelerated, but that is relative to recent times and it has become possible only because a global economy, powered by fossil-fuel combustion, has penetrated tropical forests in Brazil,

Indonesia, and southeast Asia. These dramatic episodes, however, convey a distorted impression of fire history. They are equivalent to the outbreak of large, economically damaging fires that typically accompanies industrial logging and landclearing, for example, in 19th-century America. While such high-intensity events did increase, they occurred against a loss of background burning. Again, there is an important distinction between area burned and biomass burned. Forests were cut and burned, but fires in grasslands snuffed out. Overall the Earth is experiencing a sudden diminution of biomass fires and a concomitant acceleration of fossil-fuel combustion. It is likely that its overall fire load has decreased.

In describing the Earth's fire history it is far too easy to dismiss aboriginal or agricultural burning, especially in pre-Encounter landscapes. The Australian example is neither trivial nor exceptional. Perhaps a majority of the world's grasslands were created through a convergence of climate and fire, but they were subsequently maintained through routine anthropogenic burning. When those pressures were removed, the landscape spontaneously regenerated to forest or scrub. For much of the world, the European Encounter has meant an increase in woody vegetation at the expense of grasslands.

Excepting tundra, stony deserts, and spring-fed grottos, the terrestrial biotas of the planet had more or less adapted to anthropogenic fire, or at least they had until the expansion of Europe scrambled ecosystems and industrialization through the burning of fossil fuels began refiguring the combustion calculus of the Earth. In some locales the substitution of fossil hydrocarbons has liberated biomass for burning. In others, the power of industrial combustion has substituted for, or suppressed outright, traditional burning. The outcome is a grand maldistribution of fire—too much of the wrong kind of fire in the wrong times and places, not enough of the right kind of fire in the right times and places. Increasingly, the limitations on combustion of all kinds is less their source fuels than the sinks that must receive their byproducts.

The reduction and redistribution of biomass burning has thus created for many ecosystems a fire deficit. Industrial combustion can replace ancient pyrotechnologies; and fossil-fuel combustion may substitute for some lost processes —establish new energy flows, fashion novel nutrient cycles, substitute money and information for phosphorus and nitrogen, and so on. But for systems retained in quasi-natural conditions there is no real alternative except the restoration of anthropogenic fire according to something like its historical regime. Burning fuelwood in ovens does not have the same ecological effect as burning it in the woods. The full ecological consequences of this immense shift in anthropogenic fire practices are far from understood.

Literature References

Andreae, MO (1991) Biomass Burning: Its history, use, and distribution and its impact on environmental quality and global climate. In Levine JS (ed) *Global Biomass Burning* pp 3-21 MIT Press Cambridge

Artsybachev EV (1984) *Forest Fires and Their Control*. Russian Translation Service 15 AA Balkema

Balee W (1992) Indigenous History and Amazonian Biodiversity. In Steen HK, Tucker R (eds) *Changing Tropical Forests* pp 185-196 Forest History Society Durham NC

_____ (1993) Indigenous Transformation of Amazonian Forests. An Example from Maranhao, Brazil. *L'Homme 126-128 XXXIII* 231-254

Bartlett HH (1955-61) *Fire in Relation to Primitive Agriculture and Grazing in the Tropics: Annotated Bibliography*, 3 vols Ann Arbor: University of Michigan Botanical Gardens

Batchelder R, Hirt H (1966) Fire in Tropical Forests and Grasslands. Technical Report 67-41-ES US Army Natick Laboratories

Behre KE (1988) The role of man in European vegetation history. In Huntley B, Webb III T (eds) *Vegetation History* pp 633-672 Kluwer Academic Publisher Boston MA

Booysen PdvV, Tainton NM (eds) (1984) *Ecological Effects of Fire in South African Ecosystems.* Springer-Verlag Heidelberg New York

California Dept. of Forestry and Fire Protection and Fire Research Insitute (1990) Bibliography and Directory of Remote Sensing of Wildland Fires

Castro Rego F (1992) Fuel Management. In U.N. FAO et al. *Seminar on Forest Fire Prevention, Land Use, and People* Athens: Ministry of Agriculture pp 209-221

Clark L (1994) The Prehistory of Bushfires. In Stanbury P (ed) *Bushfires. Their Effect on Australian Life and Landscape* 2nd ed pp 61-74 International Association Wildland Fire

Conrad CE, Oechel WC (tech coords) (1982) Proceedings of the Symposium on Dynamics and Management of Mediterranean-Type Ecosystems General Technical Report PSW-58 US Forest Service

Covington WW, Moore MM (1992) Postsettlement Changes in Natural Fire Regimes: Implications for Restoration of Old-Growth Ponderosa Pine Forests. In Old-Growth Forests in the Southwest and Rocky Mountain Regions Proceedings of a Workshop US Forest Service General Technical Report RM-213 pp 81-99

Crosby A (1986) *Ecological Imperialism. The Biological Expansion of Europe 900-1900*. Cambridge University Press

Crutzen PJ, Goldammer JG (eds) (1992) *Fire in the Environment: The Ecological, Atmospheric, and Climatic Importance of Vegetation Fires*. Dahlem Workshop Reports Environmental Sciences Research Report 13 John Wiley & Sons Inc Chichester England

diCastri F, Mooney H (eds) (1973) *Mediterranean-type Ecosystems. Origin and Structure.* Springer-Verlag Heidelberg New York

diCastri F, Goodal DW, Specht R (eds) (1981) *Mediterranean-type Shrublands.* Springer-Verlag Heidelberg

Dusha-Gudym, SI (1992) "Forest Fires on the Areas Contaminated by Radionuclides from the Chernobyl Nuclear Power Plant Accident," *International Forest Fire News 7* (August 1992) pp 4-6

Gabriel HW, Tande GF (1983) "A Regional Approach to Fire History in Alaska," BLM-Alaska Technical Report 9.

Gill AM, Grove RH, Nobel IR (eds) (1981) *Fire and the Australian Biota.* Australian Academy of Science Canberra

Goldammer JG (1993) *Feuer in Waldökosystemen der Tropen und Subtropen*. Birkhäuser

Goldammer JG (ed) (1990) *Fire in the Tropical Biota.* Ecological Studies 84 Springer-Verlag

Goldammer JG, Jenkins JJ (eds) (1990) *Fire in Ecosystem Dynamics. Mediterranean and Northern Perspectives*. SPB Academic Publishing bv The Hague

Goudsblom J (1992) *Fire and Civilization*. Viking

Harris DR (ed) (1980) *Human Ecology in Savanna Environments.* Academic Press New York

Hallam S (1979) *Fire and Hearth: a Study of Aboriginal Usage and European Usurpation in South-western Australia.* Australian Institute of Aboriginal Studies Canberra

_____ (1985) The history of Aboriginal firing. In Ford J (ed) *Fire Ecology and Management in Western Australian Ecosystems* pp 7-20 Perth Western Australian Institute of Technology Canberra

Heikinheimo O (1915) *Kaskeamisen vaikutus Suomen metsiin.* Helsinki Helsinki

Higgins KF (1984) Lightning fires in grasslands in North Dakota and in pine-savanna lands in nearby South Dakota and Montana. *J. Range Management 37*:100-103

Hough W (1926) *Fire as an Agent in Human Culture.* US National Museum Bulletin 139 Smithsonian Institution

Houghton RA, Skole DL (1990) Carbon. In Turner II et al. *The Earth as Transformed by Human Action* pp 393-408

_____, Boone RD, Fruci JR, Hobbie JE, Milillo JM, Palm CA, Peterson BJ, Shaver GR, Woodwell GM, Moore B, Skole DL, Myers N (1987) The flux of carbon from terrestrial ecosystems to the atmosphere in 1980 due to changes in land use: geographic distribution of the global flux. *Tellus 39B*:122-139

Huntley BJ, Walker BH (eds) (1982) *Ecology of Tropical Savannas.* Ecological Studies 42 Springer-Verlag New York Berlin

Högbom AG (1934) *Om skogseldar förr och nu och deras roll i skogarnas utvecklingshistoria.* Norlanddkst Handbibliotek 13 Uppsala

International Association of Wildland Fire. *International Bibliography of Wildland Fire* (IAWF 1992 updated quarterly) Nearly 35,000 entries

Jones R (1969) Fire-stick farmers. *Australian Natural History* 16:224-228

_____ (1975) The Neolithic, Palaeolithic and the hunting gardeners: man and land in the Antipodes. In Suggate RP, Cresswell MM (eds) *Quaternary Studies* pp 21-34 Royal Society of New Zealand Wellington

Kailidis DS (1992) Forest Fires in Greece. In UN FAO et al. *Seminar on Forest Fire Prevention, Land Use and People* pp 27-40

Kardell L, UN Food and Agriculture Organization, UN Economic Commission for Europe, UN International Labor Organization (1980) Svedjebruk förr och nu. Rapport 20 Avdelning för landskapsvård SLU

Kilgore B, Taylor D (1979) Fire History of a Sequoia Mixed Conifer Forest, *Ecology 60*(1):29-142

Kozlowski TT, Ahlgren CE (eds) (1974) *Fire and Ecosystems.* Academic Press

Kunstadter P, Chapman EC, Sabhasri S (1978) *Farmers in the Forest.* University Press of Hawaii

Larsson LJ (1989) Svedjebruk i Värend och Sunnerbo. In Nordström O et al. *Skogen och Smålänningen* pp 61-99 Historiska föreningens i Kronobergs län skriftserie 6 SmpTRYCH AB

Levine J (ed) (1991) *Global Biomass Burning.* MIT Press Cambridge

Lewis HT (1985) Burning the 'Top End': kangaroos and cattle. In Ford J (ed) *Fire Ecology and Management in Western Australian Ecosystems* pp 21-32 Western Australia Institute of Technology Perth

Lewis HT, Ferguson T (1988) Yards, corridors, and mosaics:how to burn a boreal forest. *Human Ecology* 16(1): 57-77

Luke RH, McArthur AG (1978) *Bushfires in Australia.* Australian Government Printing Service

Marland G et al. (1985) CO_2 from fossil fuel burning: global distribution of emissions. *Tellus 37B*:243-258

Minnich RA (1983) Fire Mosaics in Southern California and Northern Baja California. *Science* 219:1287-1294

Montelius S (1953) The Burning of Forest Land for the Cultivation of Crops. *Meddelanden från Uppsala Universitets Geografiska Institution* Serie A No 87

Mooney HA, Bonnicksen TM, Christensen NL, Lotan JE, Reiners WA (eds) (1981) *Proceedings of the Conference, Fire Regimes and Ecosystem Properties.* US Forest Service General Technical Report WO-26

Mooney HA, Conrad CE (tech coords) (1977) *Proceedings of the Symposium on the Environmental Consequences of Fire and Fuel Management in Mediterranean Ecosystems.* General Technical Report WO-3 US Forest Service

Nodvin SC, Waldrop TA (eds) (1991) *Fire and the Environment: Ecological and Cultural Perspectives.* Proceedings of an International Symposium Southeastern Forest Experiment Station US Forest

Service General Technical Report SE-69
Ortloff W (1992) *Holzanatomische Untersuchungen zur Rekonstruktion der Feuergeschichte eines Bestandes von Pinus ponderosa Laws. in den Santa Rita Mountains, Arizona.* Diploma Thesis Freiburg University Dept Forestry with Max Planck Institute for Chemistry
Östlund L , Linder P (1993) Orsa Forest Common 1885-1991, The Transformation of an Old-Growth Boreal Forest in Sweden. In Östlund, *Exploitation and Structural Changes in the North Swedish Boreal Forest 1800-1992*. Dissertations in Forests Vegetation Ecology 4 Department of Forest Vegetation Ecology Sveriges Lantbruksuniversitet
Pallot J, Shaw DJB (1990) *Landscape and Settlement in Romanov Russia 1613-1917.* Oxford University Press
Perles C (1977) *Prehistoire du Feu.* Masson Paris
Peters WJ, Neuenschwander LF (1988) *Slash and Burn. Farming in the Third World Forest.* University of Idaho Press Moscow
Pyne S (1991) *Burning Bush. A Fire History of Australia.* Henry Holt and Co, Inc New York
_____ (1982 1988) *Fire in America: A Cultural History of Wildland and Rural Fire.* Princeton University Press Princeton
_____ (1990) Fire conservancy: the origins of wildland fire protection in British India, America, and Australia. In Goldammer JG (ed) *Fire in the Tropical Biota* pp 319-336 Springer-Verlag Berlin
_____ (1990) Firestick history. *J of American History* 76(4):1132--1141
_____ (1984) *Introduction of Wildland Fire*. Wiley-Interscience New York
_____ (1995) *World Fire. The Culture of Fire on Earth*. Henry Holt and Co, Inc
Raumolin J (ed) (1987) Special Issue on Swidden Cultivation. *Suomen Antropologi 4*
Robinson J (1987) The Role of Fire on Earth: A Review of the State of Knowledge and a Systems Framework for Satellite and Ground-based Observations. PhD dissertation University of California Santa Barbara
Sauer C (1963) Fire and early man. In Leighly J (ed) *Land and Life* pp 288-299 University of California Press Berkeley
_____ (1975) Man's Dominance by Use of Fire. *Geoscience and Man X*:1-13
Shervakov IP (ed) (1979) *Forest Fires in Yakutia and Their Influence on the Nature of Forests*. Novosibirsk Nauka
Singh G (1982) Environmental Upheaval. In Smith JMB (ed) *A History of Australasian Vegetation* pp 90-108 McGraw-Hill Sydney
Smith KR (1987) *Biofuels, Air Pollution, and Health. A Global Review.* Plenum New York
Society of American Foresters and American Meteorological Society (1969-present) *Proceedings, Conference on Fire and Forest Meteorology* (alternate years) Vol 1-10
Soininen AM (1959) Burn-beating as the Technical Basis of Colonisation in Finland in the 16th and 17th Centuries. *Scandinavian Economic History Review 7*(2):150-166
Soininen AM (1974) Vanha Maataloutemme. *Historiallisia Tutkimuksia 96* Helsinki
Stocks BJ (1991) The Extent and Impact of Forest Fires in Northern Circumpolar Countries. In Levine JS (ed) *Global Biomass Burning* pp 197-202 MIT Press Cambridge MA
Suman DO (1991) A Five-Century Sedimentary Geochronology of Biomass Burning in Nicaragua and Central America. In Levine JS (ed) *Global Biomass Burning* pp 512-518 MIT Press Cambridge MA
Tall Timbers Research Station (1962-1976 1991) *Proceedings, Tall Timbers Fire Ecology Conference* 17 vols. Tall Timbers Research Station Tallahassee
Thomas Jr WL (ed) (1956) *Man's Role in Changing the Face of the Earth,* 2 vols. University of Chicago Press
Turner II BL, Clark WC, Kates RW, Richards JF, Mathews JT, Meyer WB (eds) (1990) *The Earth as Transformed by Human Action. Global and Regional Changes in the Biosphere over the Past 300 Years.* Cambridge University Press
Wein R, MacLean DA (eds) (1983) *The Role of Fire in Northern Circumpolar Ecosystems.* SCOPE 18 John Wiley & Sons New York
Williams M (19890 *Americans and Their Forests. A Historical Geography*. Oxford University Press
Wright H, Bailey A (1982) *Fire Ecology.* Wiley-Interscience New York
Zackrisson O (1977) Vegetation Dynamics and Land Use in the Lower Reaches of the River Umeälven. *Early Norrland* 9:7-74

PART II:

CHARACTERIZING COMBUSTION PRODUCTS IN SEDIMENTS

Characterization of Particulate Products of Biomass Combustion

T. Novakov, H. Cachier[1], J.S. Clark[2], A. Gaudichet[3], S. Macko[4], P. Masclet[5]
Lawrence Berkeley National Laboratory
1 Cyclotron Rd. MS-73
Berkeley, CA 94720 USA

Summary

Biomass burning, one of the most important global sources of particulate matter, produces both airborne particles that may influence global and regional climate, and particles incorporated into sediments that provide records of past local, regional, and global impacts of biomass burning. Determining the mass concentrations, chemical compositions, and physical properties of these particles is a prerequisite for assessing the environmental effects of biomass burning. In this Chapter we describe methods for chemical and physical characterization of airborne and sedimentary particles used in biomass burning studies, with particular emphasis on the carbonaceous material. For each method we review the principles, problems and uncertainties and, when possible, discuss analytical features presently or potentially applicable to characterizing both airborne and sediment particles.

Introduction

Biomass burning produces copious amounts of gaseous and particulate species. CO_2 is the main gaseous constituent of biomass smoke, representing on average 90% of volatilized carbon (Crutzen and Andreae 1990). The particulate component of smoke is dominated by carbonaceous species. Gaseous and particulate smoke composition strongly depends on burn efficiency of fuels, physical parameters of combustion (i.e. flaming or smoldering combustion) and, to a lesser extent, the type of vegetation burned (Ward et al. 1996).

Burning of biomass fuels produces three kinds of airborne particles. (1) Carbonized vegetation residues found mainly in the coarse (>2 μm diameter) range. These particles, commonly referred to as charcoal, largely retain the original vegetation cell structure. Although charcoal particles constitute a significant fraction of the total mass produced in biomass burning, quantitative data are sparse. (2) Mineral particles (salts and metal oxides), found in both micron and submicron sizes ranges, are produced from trace element constituents of vegetation. Additionally,

---

[1] Centre des Faibles Radioactivites, CNRS-CEA, av de laTerrasse, Gif sur Yvette, (France)
[2] Departments of Botany and Geology, Duke University, Durham, NC 27708 (USA)
[3] LISA, UA-CNRS 1404, Faculte des Sciences, av du General de Gaulle, Creteil (France)
[4]Environmental Sciences, University of Virginia, Charlottesville, VA 22903 (USA)
[5] LESAM, Universite de Savoie, Chambery (France)

NATO ASI Series, Vol. I 51
Sediment Records of Biomass Burning
and Global Change
Edited by James S. Clark, Hélène Cachier,
Johann G. Goldammer, and Brian Stocks

some exogenous particles (e.g. terrigenous dust or anthropogenic and/or marine aerosols), previously deposited and trapped at the surface of the vegetation, may be re-entrained during the burn, thus altering the chemical "fingerprint" of some fires. (3) Carbonaceous particles are produced by pyrolysis followed by condensation of organic species and by gas-to-particle conversion of reactive organic gases. These particles, found in the submicron size range, are the main part of biomass burning aerosols. Emission measurements have shown that carbonaceous particles account for 1–15% of volatilized carbon, depending on burn type (Cachier et al. 1995a).

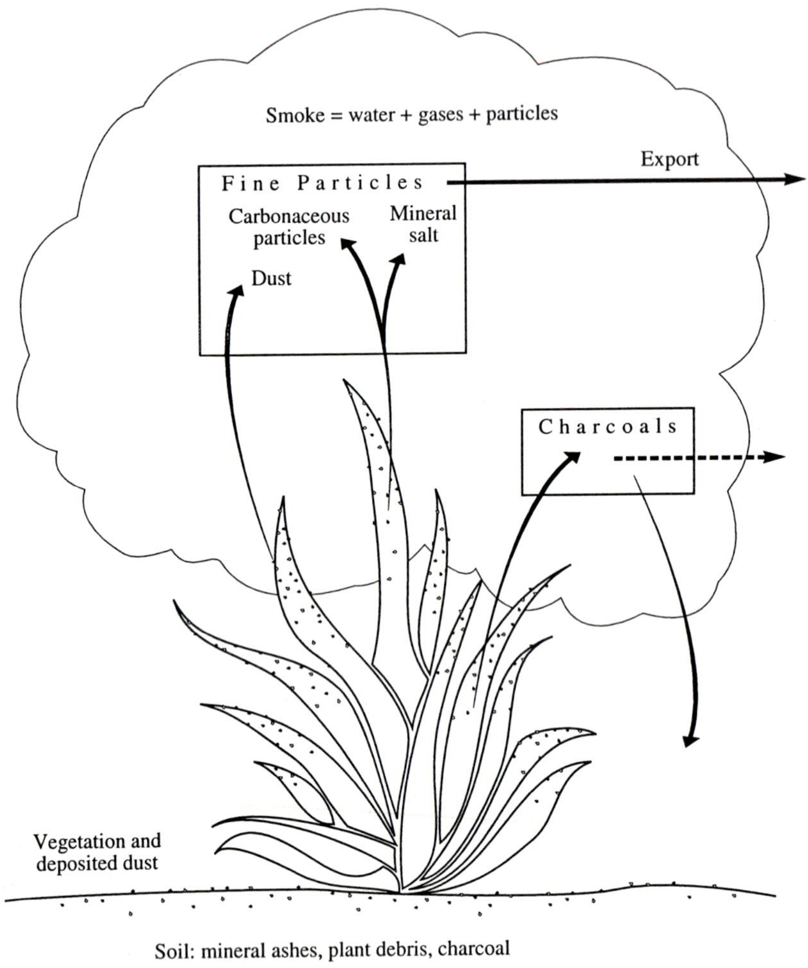

Figure 1. Schematic illustration of particle formation from biomass burning.

After a fire, the burn site is a source of particulate matter, composed principally of mineral ash, charcoal, and partially burned debris. Postburn effects involve secondary remobilization by wind and runoff of the material remaining on the ground. The translocation area of this material will be mostly local, although no quantitative estimates of its magnitude and extent are available. These particle generation processes are schematically illustrated in Fig. 1.

Airborne particles from biomass burning are lifted into the atmosphere by turbulent ascends and then entrained in regional circulation. During their lifetimes in the atmosphere they are subjected to processes that affect their size and chemical composition. These particles are removed from the atmosphere through incorporation in hydrometeors (cloud droplets, rain, or snow) and deposited on continents or oceans. Chemically inert components of these particles may therefore be incorporated into sediments and provide records of past local, regional, and global biomass burning (Goldberg 1985).

Determining mass concentrations, chemical compositions, and physical properties of particulate products of biomass burning is a prerequisite for assessing effects of biomass burning on the composition of the atmosphere and its influence on global and regional climate change. Because biomass burning is the most important global source of carbonaceous particles, the following discussion emphasizes characterization of airborne and sedimentary carbonaceous components. In this chapter we describe methods for chemical and physical characterization of particles commonly used in biomass burning studies. For each method we review the principles involved and, as much as possible, discuss the problems and uncertainties encountered in its application.

Airborne Particles

Total, organic, and black carbon determinations

The carbonaceous component of aerosols consists of broad classes of materials commonly referred to as organic carbon (OC), and black carbon (BC) (also known as elemental or graphitic carbon)(Jones et al., this volume), and total carbon (TC = OC + BC). Accurate determination of OC and BC concentration is important, because the relative amounts of BC or OC in biomass burning aerosols are highly variable, and BC is partly quantified as "charcoal" in sediment records, whereas OC is not. The BC fraction of total particulate carbon in biomass smoke samples ranges from 4 to 25% (Cachier et al. 1995a,b). The BC/TC mass concentration ratio is directly related to the fraction of total particulate carbon in biomass smoke samples ranges from 4 to 25% (Cachier et al. 1995a,b). The BC/TC mass concentration ratio is directly related to the quality of combustion. Efficient combustion (high CO_2 yield) results in lower BC/C production than inefficient combustion. During flaming biomass burning, backing fires, which expand slowly against the wind, produce aerosols with the highest BC content. Conversely, violent heading fires may create pockets depleted of oxygen that decrease the relative BC content. Smoldering processes produce smoke particles with the highest OC content and the lowest BC content (Patterson et al. 1986, Cachier et al. 1995a,b). Generally, the BC fraction in biomass aerosols is considerably lower than in industrial and urban particles. This makes quantitative determination of BC more difficult, as shown below.

Methods for carbon determination can be divided into two categories—thermal (for TC, OC, and BC) and optical (for BC). The definitions of organic carbon (OC) and black carbon (BC) used in conjunction with these methods are based on presumed volatilization, combustion, and optical properties of these classes of carbonaceous materials and are therefore operational and method-dependent. These analytical limitations are compounded by a lack of accepted standard materials and terminology, e. g. terms such as "black," "elemental," and "graphitic" carbon are sometimes used interchangeably. For simplicity, in this Section we will use the term "black carbon" as a synonym for both elemental and graphitic carbon. Jones et al. (this volume) summarize relationships among some of the more common terms.

Thermal Methods. Methods for total aerosol carbon determination rely on combusting a sample in oxygen, thereby converting all carbon to CO_2. The CO_2 is measured either directly as CO_2 or after methanation over a suitable catalyst to CH_4. Such methods were first applied to urban aerosols by Mueller et al. (1972), variants of which are still in use. The carbonate fraction is determined by comparing total carbon from untreated samples with HCl-treated samples to remove the carbonate carbon. Separation of organic and black carbon by thermal methods relies on the assumption that these components can be distinguished by their volatilization and combustion properties. Violations of these assumptions depend on method and on overall sample composition. Thus, definitions of OC and BC are operational.

The first thermal method used to determine TC, OC, and BC content of urban particulate matter was developed by Malissa et al. (1976). The procedure consisted of progressively heating a sample at a constant rate of 10°C min^{-1} in a pure O_2 or N_2 atmosphere and measuring the rate of evolution of gases as a function of sample temperature. When O_2 was used as the carrier gas, the evolved CO_2 was detected by a conductometric method. When N_2 was employed, the hydrocarbons released from the sample were detected by a flame ionization detector (FID). The plot of the rate of CO_2 (or hydrocarbon) evolution vs. temperature gives the "thermogram" of carbonaceous material evolved from the sample. The peak appearing at the lowest temperature (150–350°C) and observed with both O_2 and N_2 was defined as OC; the peak at ~425°C, seen only in runs with O_2 but absent in N_2 thermograms, was assigned to black carbon. The peak appearing at the highest temperature (~600°C) was attributed to inorganic carbonate carbon.

The above approach was subsequently modified by including an oxidation catalyst (CuO at 900°C) to assure complete conversion of evolved gases to CO_2, by detecting the CO_2 with a nondispersive infrared (NDIR) analyzer to increase sensitivity, and by incorporating simultaneous measurement of optical transmission through the sample collected on a quartz filter (Novakov 1981, 1982). Such combined optical and thermal measurements showed that the loss of optical absorption (or sample blackness) occurred simultaneously with the appearance of the high temperature (~420–550°C) thermogram peak (Rosen et al. 1982, Gundel et al. 1984). The aerosol carbon component giving rise to this peak was therefore termed "black carbon" (Novakov 1981, 1982). The carbon concentrations corresponding to the rest of the thermogram were defined as OC. Combined optical and thermal measurements also demonstrated that the combustion temperature of the black carbon peak was not constant but dependent on the kind of sample.

The method developed by Huntzicker et al. (1982) employs stepwise ramping of the sample temperature and two carrier gases—inert (pure He) and oxidizing (2% O_2 + 98% He mixture). With both gases the evolved gases were converted to CO_2 over a MnO_2 catalyst, reduced to CH_4, and detected with a flame ionization detector (FID). The sample darkness was continuously monitored by reflectance. This method defines OC as the sum of concentrations of species that volatilize at 350°C in a He/O_2 mixture and the fraction volatilizing at 600°C in He. BC is defined as carbon oxidized in three temperature steps at 400, 500, and 600°C in a He/O_2 mixture. The detection limit for total carbon determination by this method is about 0.2 μg per cm^2 of filter sample. The decrease in reflectance observed when the temperature is increased in an inert atmosphere, attributed to sample charring by pyrolysis, is used as the correction for black carbon.

A modified version of the Huntzicker et al. apparatus was constructed by Chow et al. (1993). In this method organic carbon is defined as the sum of the carbon component evolving in a He atmosphere at four discrete temperature steps (from ~25–120°C, 120–250°C, 250–450°C, and 450–550°C). The BC component is taken as the integrated carbon concentrations released in a 2% O_2 + 98% He atmosphere at 550, 700, and 800°C. The detection limit achieved with this method is 0.2 μg (C) cm^{-2}. Both these methods monitor the changes in sample darkness to correct for sample charring.

In addition to the above methods that use either continuous or stepwise increase of the sample temperature, several simplified two-step methods have been developed. Mueller et al. (1982) placed a sample in contact with MnO_2, which serves as the oxidizing agent. OC is defined as the carbon evolved from the sample at 550°C. BC is determined by the difference between total carbon (measured on a second sample aliquot) and organic carbon. In both steps the evolved CO_2 is detected by a FID after conversion to methane. The detection limit achievable with this method is about 0.8 μg C per sample.

Another two-step thermal analysis procedure developed by Cachier et al. (1989). In this method a sample is first exposed to 340°C for 2 hr in pure O_2 to remove the organic material. BC is defined as the total carbon, determined by coulometric titration of the CO_2 produced during the complete combustion at 1100°C in O_2 that remains on the sample after the precombustion treatment. OC is derived by the difference between the TC concentration, determined by combustion of another aliquot of the untreated sample, and the BC concentration. The accuracy of the carbon analyzer and the reproducibility of the carbon content of blank filters allow the determination of 10 μg C.

Wolff et al. (1982) used a two-step volatilization/oxidation procedure at a constant temperature (650°C). The concentration of carbon-containing gases evolved in a He atmosphere (oxidized to CO_2 and detected by NDIR) during the first step is defined as OC. In the second step, the carrier gas is changed from He to O_2, and the detected CO_2 is attributed to BC.

Comparison of thermal methods. From the foregoing, it is evident that OC and BC determinations by all commonly used methods are based on operational definitions. The consequences of these operational definitions on reported organic and black carbon concentrations are clearly illustrated by the results of the Carbonaceous Species Methods Comparison Study (Shah and Rau 1990). In this intercomparison study, similar sets (i.e. portions taken from a larger filter sample) of particulate samples were prepared and distributed to a number of laboratories for analysis by a variety of methods, mostly thermal. The samples were ambient urban samples, source samples, and—of particular relevance to the topic of this Chapter—two ambient samples heavily influenced by residential wood burning.

The intercomparison results show that most methods can accurately measure total carbon (within 5% of the common average). Organic and black carbon determinations of urban ambient samples had coefficients of variation of 6–9% and 28–31%, respectively. The discrepancies between OC and BC concentrations for the two wood smoke samples obtained by 11 thermal and optical methods were significantly greater. The results and the operational definitions used by different methods, compiled from Shah and Rau (1990), are summarized in Table 1. These data show that total carbon concentrations (TC = OC + BC) range from 63 to 101 μg cm^{-2} and from 228 to 469 μg cm^{-2} for the two sample sets. If the TC concentrations obtained by Method 2 are excluded as obviously too low, the remaining TC values agree within about 13–17% (concentration ranges of 86 to 101 μg TC cm^{-2} and 328 to 469 μg TC cm^{-2}).

Table 1. Methods and results of OC and BC concentration (μg cm^{-2}) determinations in wood smoke influenced ambient samples, compiled from Shah and Rau (1990).

Method[a]	Operational Definitions		Sample #1			Sample #2		
	OC	BC	OC	BC	OC/BC	OC	BC	OC/BC
1	600° C, He	750° C, 2% O_2	88.3	10.4	8.5	374.5	21.7	17.2
2	600° C, He	TC - OC	47	16	2.9	195	33	5.9
3	600° C, He	650° C, 2% O_2	86.1	5.3	16.2	257.6	77.6	3.3
4	950° C, He	950° C, 2% O_2	82.4	21.8	3.8	390.5	78.7	5.0
5	610° C, He	750° C, 10% O_2	71.8	26.6	2.7	299.2	94.7	3.2
6	550° C, He	650°-800° C, 2% O_2	62.5	35.8	1.7	253.5	74.8	3.4
7	700° C, He	900° C, 2% O_2	82.7	8.4	9.8	320.5	11.7	27.4
8	525°C,He(MnO_2)	850° C, He(MnO_2)	86.1	3.8	22.6	364.7	1.6	228
9	400° C, O_2	700° C, O_2	92.9	6.5	14.3	392.1	11.9	32.9

[a]Laboratories: 1) Environmental Monitoring Services, Inc., 2) Oregon Department of Environmental Quality, 3) US EPA, 4) General Motors Research Laboratory, 5) Oregon Graduate Institute, 6) Desert Research Institute, 7) Sunset Laboratory, 8) Environmental Research and Technology, 9) Global Geochemistry Corporation.

Discrepancies in OC and especially BC determinations are much larger than TC determinations. OC concentrations differ by about 50% (concentration ranges 62 to 93 μg OC cm^{-2} and 253 to 392 μg OC cm^{-2}). The reported BC concentrations range from 3.8 to 36 μg BC cm^{-2} and from 79 to 1.6 μg BC cm^{-2} for the two samples, resulting in 10- to 50-fold differences among individual determinations. Consequently, the OC/BC ratios differ by up to two orders of magnitude (OC/BC ranges 1.7 to 23 and 3.2 to 228). Furthermore, OC/BC ratios differ significantly, even when obtained by seemingly similar analytical procedures. For example, the OC/BC ratios obtained by two-step procedures (Methods 1 through 4) differ by a factor of about 5, and methods that account for sample charring (Methods 5 through 7) differ by factors of 5.7 to 8.5.

Sources of uncertainties . From the above discussion it follows that only total carbon can be reliably determined by thermal methods. The differences in TC concentrations determined by individual methods can most likely be attributed to the nonuniformity of the deposits on large filter samples from which the aliquots were taken. The substantial differences in OC concentrations measured in wood smoke samples demonstrate the method dependency of the results. The much larger differences in BC concentrations reflect the uncertainties in OC determinations. Even relatively small under- or overestimation of OC concentrations results in much greater uncertainty in the BC concentrations, especially in samples with the large organic fraction characteristic of biomass smoke particles.

The authors of the intercomparison study referred to above identified the following possible reasons for the observed discrepancies in OC and BC determinations: (1) different temperatures used for OC volatilization and BC oxidation; (2) trace amounts of O_2 in the inert carrier gas partially oxidizing BC during the OC volatilization step; (3) pyrolysis corrections for samples containing light-absorbing species other than BC, such as colored organic materials and metal oxides; and (4) interference of carbonate with BC determination.

Several additional factors influencing OC and BC determinations in redwood (*Sequoia sempevirens*)-needle and eucalyptus smoke samples were recently identified (Novakov and Corrigan 1995). They analyzed both untreated samples and the same samples after acetone and water extraction by evolved gas analysis. Acetone extraction was used to remove most of the organic material that may interfere with BC determination. Water extraction was used to examine the possible role of water-soluble cations, such as K^+ and Na^+, that are known to lower the combustion temperature of carbonaceous materials because of their catalytic activity (Lin and Friedlander 1988).

The results of these experiments show that the apparent BC concentrations inferred from analyzing untreated samples significantly exceed the BC concentrations determined on acetone-extracted samples. Such differences were observed for all samples except those with the lowest (about 5%) organic fraction. For samples with high organic content (>80%), the apparent BC concentrations can be up to 50 times greater than the more realistic BC concentrations, determined on acetone-extracted samples. These results indicate that the organic fraction of biomass smoke particles includes a relatively nonvolatile component with a combustion temperature close to that of black carbon. Furthermore, a substantial fraction of the organic material in biomass smoke particles is practically nonvolatile in an inert atmosphere when heated to about 550°C in a manner comparable to that used in many thermal methods. (A similar non-volatile organic component was also observed in urban aerosols [Ellis et al. 1984]). Both these materials may erroneously be assigned to black carbon and result in an overestimation of its actual concentration. Finally, the combustion temperature of black carbon in biomass smoke samples is significantly lower when K and Na are present in the sample. Removal of these species, an integral part of biomass smoke particles, by water extraction causes the combustion temperature of black carbon to increase by more than 100°C. This observation suggests that a portion of BC in samples containing K or Na may be oxidized at a temperature at which it would not combust in samples free of these catalytic species.

In summary, the results and discussion presented above clearly show that the determination of organic and black carbon concentrations is not only method dependent but also highly dependent on the content and kind of other organic and inorganic constituents of the sample. Therefore, only a combined approach using several methods (for example, thermal, optical, elemental analysis, and solvent extraction) can give actual black carbon and organic concentrations.

Optical Methods. Optical methods for estimating black carbon concentrations evolved from techniques originally developed for measuring aerosol absorption coefficients. These methods rely on measuring the intensity of visible light transmitted through a blank filter I_0 and through a loaded filter sample I. A convenient way of relating the attenuation of light through a filter deposit and surface concentration X (μg cm^{-2}) of the absorbing species is through the relationship ATN = $s_x X$, where

$$\mathrm{ATN} = 100 \ln\left(\frac{I_0}{I}\right) \tag{1}$$

and the proportionality factor s_x is the specific mass absorption cross section of light-absorbing material in units of m^{2}g^{-1}.

The first method of this kind, the integrating plate method, was developed by Lin et al. (1973). Blank and loaded Nuclepore filters are placed on a light-diffusing opal glass support. Field tests of this method showed agreement with other independent techniques to within about 25% (Weiss and Waggoner 1982). A more sensitive method, the integrating sandwich, was developed for very low concentrations of absorbing material (Clarke 1982). In this approach the filter is placed between two diffusely reflecting plates. Light penetrating one side of the sandwich assembly experiences multiple diffuse reflections while passing through the filter, thereby amplifying net absorption by the aerosol layer. This approach resulted in a sensitivity increase of about a factor of 40 over the integrating plate method for transmission changes of less than a few percent.

Somewhat similar techniques, using quartz filters as both collection substrates and light diffusers, are routinely used for estimating BC concentrations. These are either applied to precollected filters (Rosen et al. 1980) or in a continuous mode for near real-time monitoring (Hansen et al. 1984). The instrument for continuous monitoring, known as the aethalometer, measures the real-time attenuation of light transmitted through particles accumulating on a quartz filter.

Application of optical transmission measurements to BC estimation is based on empirical results that assume the attenuation of visible light by the filter deposit (eqn 1) is proportional to the surface concentration of black carbon M_{BC} ($\mu g\ cm^{-2}$),

$$ATN = s_{BC}\, M_{BC}. \qquad (2)$$

A number of experiments using different optical measurements and analytical methods for BC determination obtained s_{BC} values ranging from 2 to 24 m^2g^{-1} (see Liousse et al. 1993). Such a large spread demonstrates that the apparent s_{BC} values obtained in this manner should not be taken as representing the actual BC absorption cross section. At best, these should be viewed as empirical calibration factors relating ATN to BC, determined for specific instruments, analytical procedures, and sampling locations.

Because optically measured BC mass concentrations are determined from the relationship in eqn 2, the accuracy of such determinations critically depends on the choice of the s_{BC} value, the accuracy of the ATN measurement, and the assumption that contribution of species other than BC to light absorption is small. The proportionality factor s_{BC} is empirically determined from ATN measurements and BC analyses (for example by a thermal method) performed on the same filters. As mentioned above, there is a large variability in s_{BC} values derived by different investigators (see Liousse et al. 1993). A significant part of this variability is undoubtedly caused by errors in BC determination. There is also evidence that the ATN vs. BC proportionality may depend on the kind of aerosol, even when the same transmission measurements and procedures for BC determination are used. Niessner and Petzold (1994) showed that the light absorption measured by an aethalometer correlates well with BC concentrations defined as the nonextractable (in a 1:1 mixture of toluene and 2-propanol) fraction of total particulate carbon at all measurement sites (urban, suburban, and rural). However, the coefficient s_{BC} varied from about 5 m^2g^{-1} for rural to 18 m^2g^{-1} for urban aerosols. Liousse et al. (1993), also using an aethalometer and BC concentrations measured by a thermal method (Cachier et al. 1989), obtained s_{BC} values that were similarly variable, depending on sample location. The lowest value of 5 m^2g^{-1} was obtained in remote areas, and the highest value of 20 m^2g^{-1} was measured in African savannah.

Finally, species other than black carbon, such as mineral dust and some colored organic materials, may contribute to light absorption. Derivation of specific mass absorption coefficients or cross sections requires knowledge of mass concentrations of all absorbing species. Therefore, if a s_{BC} value derived from measurements on samples whose only light-absorbing species is BC were to be applied to estimate the BC concentration in aerosols with substantial concentrations of other light-absorbing species, the BC concentrations would be overestimated.

Schnell et al. (1994) estimated the contribution of mineral dust to light absorption measured at Mauna Loa during a large "Asian Dust" event. Their approach was to measure the ATN values on quartz filters before and after these were exposed to 600°C in air to remove all combustible material. The percent of light absorption contributed by dust ranged from 1 to 19%, with an average < 10%. These authors concluded that mineral dust has a mass absorption cross section for visible light 3 to 5 orders of magnitude smaller than BC.

The contribution of organic species (some of which are yellow or brown) in biomass smoke to light absorption can be roughly estimated from optical transmission and thermal analysis measurements of samples whose carbonaceous fraction is 98.5% organic material (Novakov and Corrigan 1995). The ATN values and OC concentrations for two such samples were 23.8 and 55.3, and 118 and 443 μg cm^{-2} respectively, resulting in apparent mass absorption cross sections of the total organic material in these samples of 0.20 and 0.12 m^2g^{-1}. These values are two orders of magnitude lower than the average s_{BC} for pure black carbon of ≈20 m^2g^{-1}. If we assume that the least volatile (presumably highest molecular weight) organic fraction is the light-absorbing organic material, then their mass absorption cross sections would range from 1.50 to 0.7 m^2g^{-1}.

Molecular characterization

Organic geochemical characterization of extractable materials from either a sedimentary layer or aerosol trapped on a glass fiber filter represents a powerful approach for identifying sources and history. Procedures usually call for organic solvent (typically dichloromethane) extraction of the dry sediment or filter. This extract is usually evaporated to concentrate the compounds. The mixture can then be saponified using heat, a strong base, and an alcohol. This reaction ruptures the triglycerides and waxes. Compounds of interest (hydrocarbons, PAHs) can now be separated using gas chromatography (GC). Others must be derivatized (alcohols, fatty acids) to increase their volatility and thus amenability for the GC separations. Through analysis of the compounds first separated by GC, using fragmentation of the compounds following collision with high energy electrons, the authentication of the identity of a component can be achieved (GC/mass spectrometry [MS]). By selectively identifying specific mass fragments (for example, m/z 217, steranes), trace level components—the biomarkers—can now be identified in mixtures of organic materials. Biomarkers are of particular interest because of their specificity to source materials.

Isotopic Characterization

Stable isotopes offer a unique perspective on the sources of organic materials. Through biochemical reaction, processes alter the carbon isotope ratio (ratio of ^{13}C to ^{12}C) from that of carbon dioxide to one that can be used to discriminate vegetation types as being either C_3 or C_4 terrestrial, or resolve terrestrial source materials from marine. The basis for this difference is the existence of enzymatic pathways for carbon fixation—ribulose bisphosphate in C_3 plants, and phosphoenol pyruvate in C_4 plants. At this level, the required instrumentation is simply a stable isotope ratio mass spectrometer, which makes determinations on the order of fractions of an approximately parts per thousand basis. The reproducibilities should be better than 0.005 per mil (‰), and the sensitivity is on the order of a few microliters of CO_2. To introduce the bulk material into the mass spectrometer, it must be converted to a gas of suitable purity for analysis, without altering the isotopic composition.

Of recent interest is the separating of the mixture of components using a gas chromatograph (GC) and then shunting the components through a combustion (C) furnace, with the products being CO_2 pulses (GC/C/IRM). Following water trapping, the pulse of CO_2 is introduced directly into the source of the isotope ratio mass spectrometer (IRMS). As a result as little as 0.5 nM of a compound can be isotopically established, with a precision typically better than 0.3 ‰ (Macko 1991). It appears that the isotopic compositions of individual molecular components are capable of surviving the pyrolytic process and reflect the source materials from which they were generated (Ballentine et al. 1995, Turekian et al. 1995).

Trace elements, anions and cations

The principal methods for determining trace element concentrations in biomass smoke samples are proton-induced X-ray emission (PIXE) and X-ray fluorescence (XRF) spectrometry. As PIXE requires that the sample be in high vacuum, analyses of light and volatile elements (such as Cl) are difficult. Water-soluble anions and cations are analyzed by ion chromatography (IC).

Concentrations of trace elements in biomass smoke aerosols range from 5% of the total particulate mass in flaming conditions to 1% in smoldering condition (Echalar et al. 1995). Relative concentrations of trace elements (i.e. the trace element "fingerprint") are principally influenced by burning conditions and, to a much lesser degree, by the trace element content of the biomass fuels (Table 2).

Table 2. Relative amount of some trace elements (%) in vegetation and aerosols (From Echalar et al. 1995).

Element	Vegetation	Aerosol
K	savanna (0.4)	Flaming (0.3)
	rain forest (0.2)	Smoldering (0.3)
S	savanna (0.08)	Flaming (0.6)
	rain forest (0.08)	smoldering (0.3)
Ca	savanna (0.4)	flaming (0.3)
	rain forest (0.3)	smoldering (0.1)

The principal trace elements identified in biomass burning aerosols are listed in Table 3, according to the fire phase (flaming vs. smoldering), together with their contribution to total particulate mass, and their mass fraction found in fine particle (<1 μm diameter) mode (Gaudichet et al. 1995, Echalar et al. 1995). Of these, K, Cl, P, Br, and Zn originate primarily from vegetation fuels. Al, Ti, and Fe are mainly soil dust-derived elements, probably originating from particles predeposited on vegetation. Si, S, and Ca are of mixed origin. Most trace elements are associated with the fine aerosol (<1 μm diameter) mode, with the exception of P, Si and Ca, found preferentially in the coarse charcoal particles (>5 μm). In contrast to combustion-derived industrial and urban particles, biomass burning aerosols are characterized by a significant potassium but relatively low sulfur content. A striking feature of biomass smoke is the presence of appreciable amounts of zinc or bromine, traditionally used as tracers of industrial emissions. IC data show that K is largely water soluble. Other major ions found in notable amounts are NH_4^+, Na^+, Ca^{2+}, Mg^{2+}, NO_3^-, SO_4^{2-}, and Cl^-. It is interesting to note that IC analyzed data show detectable amounts of oxalate ($C_2O_4^{2-}$), which may serve as a tracer for biomass burning.

Table 3. Relative amount of the main biomass burning trace elements and their fraction in the fine particle mode.(From Echalar et al. 1995)

Element	% of total particle mass		% in fine fraction
	Flaming	Smoldering	
K	3.5	0.3	84
Cl	2.4	0.05	71
S	0.6	0.3	71
Si	0.8	0.1	16
Ca	0.2	0.1	15
P	0.1	0.03	0
Br	0.07	—	100
Zn	0.03	0.02	7

An important transformation of the trace-element fingerprint may occur during particle aging. As an example, the Cl/K ratio in fresh biomass burning particles is higher (~0.8) than in aged particles (~0.2) (Gaudichet et al. 1995). Conversely, the K to BC ratio is found to be relatively constant and allows discrimination of biomass burning aerosols from other anthropogenic sources. K enrichment is not applicable to sediment record studies because of its solubility and presence in mineral sediment grains. It appears, however, that Zn enrichment in sediments may be investigated as an indicator of biomass burning.

Size and morphology

Electron microscopy is the primary tool for investigating the size and morphology of airborne particles. The choice of microscopic methods depends primarily on the size of particles, which covers a diameter range from millimeter to submicron. Coupling morphological and chemical analyses (mostly by energy dispersive x-ray spectrometry) provides more complete characterization of particles, including information about their sources and origins. These methods require a good dispersion of single particles onto suitable collection supports. A summary of microscopic methods and their application to biomass burning particles is given in Table 4.

Scanning electron microscopy (SEM) is preferentially used to investigate medium-size particles such as charcoal or fly ash (Fig. 2a). Fly ash from coal combustion is composed mainly of Si and appears as smooth spherical particles. Fuel oil combustion results primarily in porous or sponge-like carbonaceous spheres. SEM images can be used to infer specific surface area of particles, an indicator of their capability for adsorbing gaseous species. SEM or optical studies of charcoals show the altered cell structure and may help identify parent vegetation.

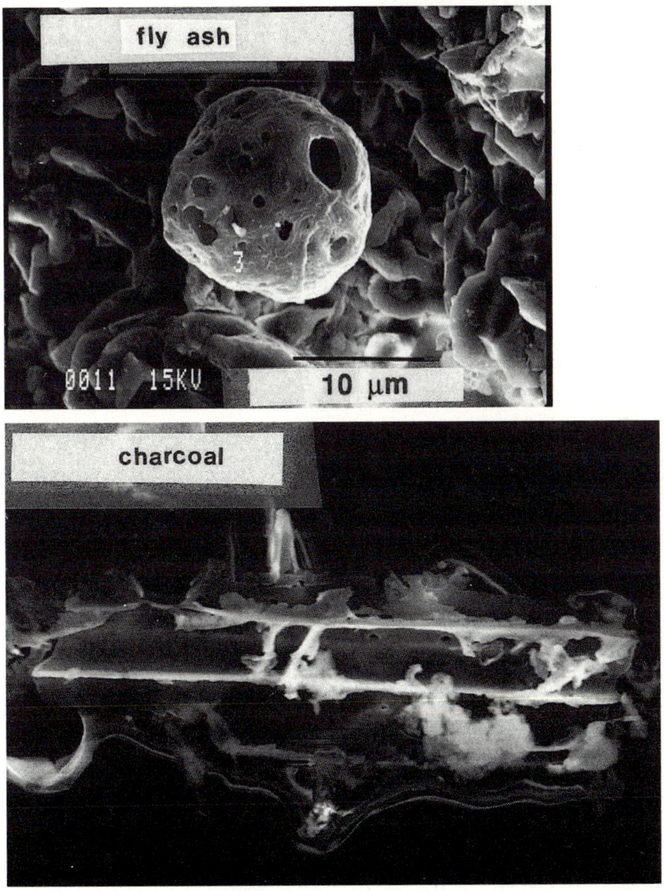

Figure 2a. Scanning electron microscope (SEM) pictures of Above: porous fly ash; Below: charcoal fragment (from Amazon forest fire).

Table 4. Microscopic methods and their applications to combustion-generated particles.

Method	Size range (μm)	Applications	Complementary analyses
Optical microscopy	1 - 1000	charcoal	refractive index
Scanning electron microscopy (SEM)	1 - 100	charcoal, fly-ash	elemental analysis, X-ray spectroscopy
Transmission electron microscopy (TEM)	0.01 - 10	fly-ash, micro-soots	elemental analysis, cristallography, electron diffraction

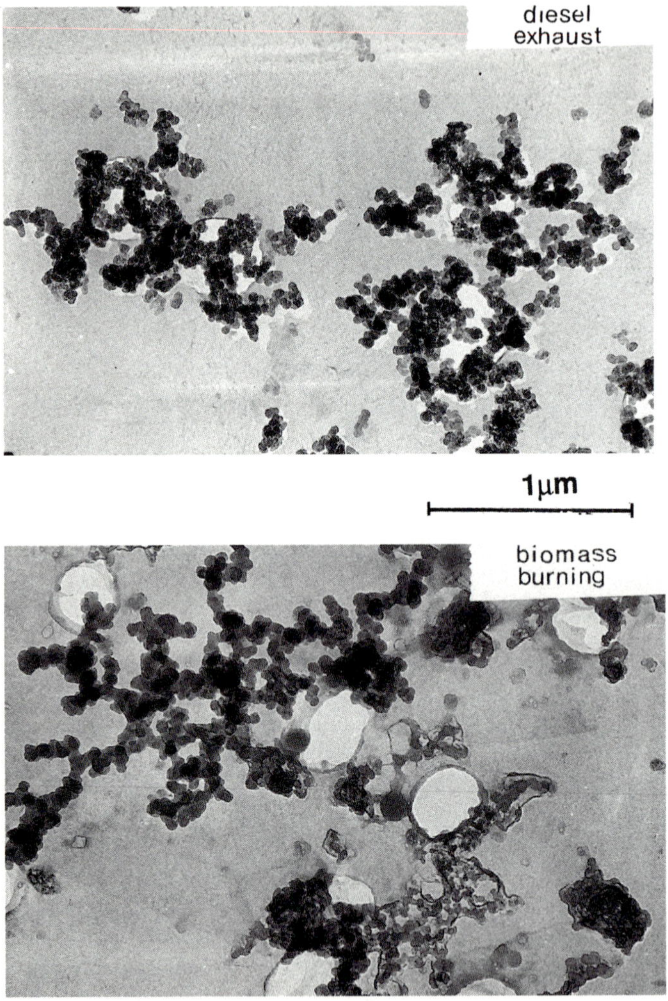

Figure 2b. Morphological features of microsoots from transmission electron microscopy (TEM) of Above: diesel exhaust particles; Below: biomass burning plume particles (from savanna fire).

Transmission electron microscopy (TEM) is mostly applied to studies of small spherical carbonaceous clusters or aggregates ("microsoots") (Fig. 2b). There is no unambiguous morphological criterion to distinguish diesel or biomass burning soot aerosols. However, carbonaceous aggregates from biomass burning may be qualitatively distinguished from those derived from diesel emissions. The former appear in the 20 to 30 nm size range, while diesel spheres are generally smaller. Biomass burning particles include some hollow carbonaceous spheres with a mean diameter of 200 to 400 nm, similar in appearance to fly ash particles although much smaller. These spheres are mostly produced during smoldering combustion. An interesting facet of the electron microscopy data is the suggestion that individual biomass smoke particles can be traced by their ubiquitous potassium content, detected by systematic X-ray analysis of single particles.

Sediment Particles

Carbonaceous particles are extracted from sediments of lakes, mires, and oceans using a variety of coring methods. Samples removed from cores are dated by radiometric techniques (e.g. ^{14}C or ^{210}Pb), by annual laminations (varves), or by stratigraphic indicators. Particle concentrations can be transformed into flux estimates using a time model established by dating methods.

Wet chemical methods

Chemical characterization generally involves wet oxidation of organic constituents followed by quantification of residual (presumably black) carbon, BC. Because it is inexpensive, gravimetry has been the preferred method of BC quantification by paleoecologists (Winkler 1985, Burney 1987). Limitations to the method include problems detecting low amounts of BC, broad overlap in OC and BC, and unknown matrix effects. Detection limits are probably undesirably high, but they have not been systematically examined. Because of the low BC content, this method is best suited to sediments from regions where emissions are relatively high. Detection is also difficult, in part, because OC and BC are not chemically distinct. The wide range of types and amounts of OC that occur in sediments makes it difficult to factor out the effects of organic matter on BC estimates (see above). The background sediment matrix is often dominated by algal material but can also contain terrestrial materials high in lignin, e.g. leaves. The potential effects of hydrated clays on gravimetric estimates of BC and the presence of catalysts for oxidation may also affect estimates. Despite these drawbacks, rapid analysis and the ability to analyze for total carbonaceous emissions, rather than simply those particles that fall in the largest diameter classes, suggest continued promise. Systematic characterization studies are needed to assess detection limits, oxidation losses of different organic materials, and confounding matrix effects.

Thermogravimetric analysis

Thermogravimetric analysis (TGA) is used to characterize C-containing compounds based on their potential to volatilize and/or oxidize over a range of temperatures. As described above, thermogravimetry involves subjecting a sample to a program of controlled temperatures and atmospheres to separate components that differ in volatility and/or reactivity. Mass is monitored over the course of the program to identify fractional losses as a function of temperature and/or time. BC oxidizes in the presence of O_2 but does not volatilize in an inert atmosphere; heating in an inert atmosphere will not cause loss of mass unless O_2 is introduced. In sediments, OC consists primarily of cellulose and lignin that volatilize and oxidize over a range of temperatures. Thus, OC and BC are believed to differ in that BC does not volatilize in the absence of O_2, but OC does. The advantages of TGA include high reproducibility of results and relatively good differentiation between cellulose and lignin (Weidemann et al. 1988).

In addition to the drawbacks of TGA for separation of OC and BC mentioned above (potential O_2 traces in the inert carrier gas, interference of carbonates, nonvolatile organic smoke constituents that contribute to "BC" estimates, presence of water soluble cations that lower combustion temperatures) are several concerns more specific to sediment samples. These include potentially large amounts of lignin-containing organics, which have temperature ranges for oxidation that overlap with BC, charring of OC that occurs in an inert atmosphere, and hydrated clays. The high OC content of sediments means that some of the mass loss that occurs in the range of BC could include nonvolatile organics. Heating in N_2 does not solve this problem because, in addition to volatilization, some char is produced from OC in an inert atmosphere. Thus, the method produces the constituent it aims to quantify.

Some common sediment constituents are illustrated with several examples (Figs. 3, 4). Thermograms for fuels such as leaves, fine twigs, and grass in O_2 show the typical overlapping peaks of cellulose and lignin with peaks at ~280 and 400°C, respectively (Fig. 3a). Burning causes preferential loss of cellulose and the charring of lignin (Fig. 3b, c). Organic sediments contain these same two peaks, together with an additional carbonate peak near 600°C, depending on the rate of temperature increase (Fig. 4a). Clays lose water over a broad temperature range that can span both the BC and carbonate peaks (Fig. 4b). Because of these overlapping temperature ranges for the main constituents in sediments, further characterization work is necessary before TGA can be used to confidently quantify BC in sediments.

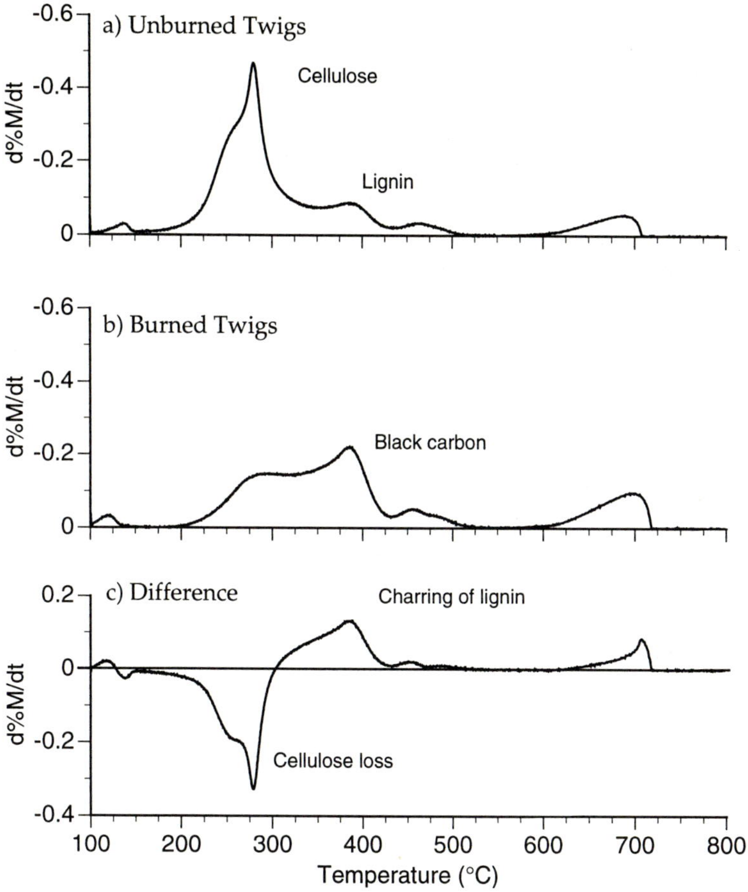

Figure 3. Thermograms for fuels and sediments in an O_2 atmosphere. The vertical axis is rate of mass loss in a sample heated from 30 to 800 °C. Unburned fuels (a) show typical cellulose and lignin peaks. Charring (4 hr at 280°C in air) causes preferential loss of cellulose and charring of lignin to produce "black carbon" (b). The difference between curves in a and b demonstrates volatilization of cellulose and lignin charring.

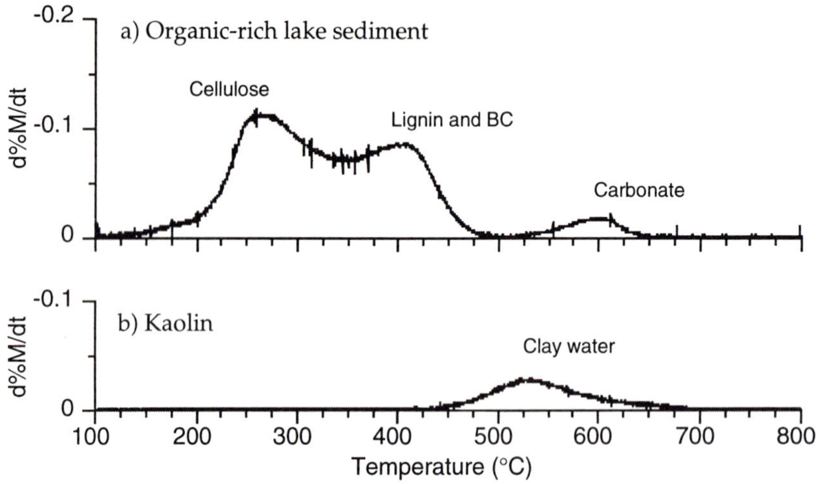

Figure 4. Sediments from Round Pond, NY show the cellulose peak and a lignin peak that may include BC. Carbonates and lignin/BC may overlap with clay water (b), which can be important in inorganic sediments.

Optical microscopy

Optical microscopy is a favorite method for sediment charcoal analysis, because optical properties are among the most diagnostic for charred particles, and the method involves equipment and expertise already available in many paleoecology labs. Although microscopy is always used for quantification by this method, sample preparation and analysis can vary substantially. We summarize the principal approaches here.

Particles are either concentrated by chemical methods, extracted from the sediment matrix by sieving, or quantified *in situ* on petrographic thin sections of embedded, undisturbed sediments. Chemical digestions are the method used when charcoal quantification is to proceed from slides prepared for pollen analysis. Processing involves steps to deflocculate sediments (hot KOH) and to remove carbonates (HCl), silicates (HF), and organics (acetolysis). Following these treatments only the most recalcitrant organics (including pollen), charcoal (which contains OC and BC), and some miscellaneous minerals (e.g. pyrite) remain. The residue is suspended in silicone oil and mounted on glass slides.

Sieving is accomplished by gently washing sediments through one or several sieve(s). Most studies have used sieves with 100 to 200 μm diameter openings

(Whitlock et al., this volume). The particles remaining on the sieve must then be transferred to a suitable background or suspended (in water, for example) for identification and counting. Because charred particles span a diameter range of several orders of magnitude (see below), a series of sieve sizes can decrease subsequent counting time by limiting the range of particle sizes on a given sample.

Thin sections are prepared by embedding sediments in a medium suitable for sectioning. Before embedding can take place, pore waters must be replaced by a low viscosity material that is miscible with the embedding medium. Media that have been used for sediment charcoal include Carbowax (Allison et al. 1986) and epoxy (Clark 1988a). Carbowax has the advantage of being easier to embed; however, it is soft and unsuitable for sediments containing clastic material. Epoxy yields a hard block suitable for embedding any sediment, but it requires a more laborious process. Dehydration is accomplished by freeze drying or with successive acetone treatments (Clark 1988a). If freeze dried, the epoxy resin must be introduced under vacuum to insure good penetration. Acetone can be replaced with a low viscosity formulation such as Spurr's (Clark 1988a). The embedded block is cured under moderate heat and then sectioned. Carbowax preparations are sectioned with a microtome; epoxy preparations are sectioned by usual petrographic methods.

Quantification involves counting and/or measuring individual particles or a sampling of the processed sediment. Charcoal particles on pollen slides are counted, measured, and/or sampled by a point-count method (Clark 1982). Simultaneous counts of a spike of known concentration are needed to calculate concentrations. The method usually quantifies particles at a magnification of 400× and includes particles in the 5 to 50 μm diameter range. The method is rapid, but there have been concerns that particles may be fragmented and/or lost during processing (Patterson et al. 1987).

Microscopic analysis of particles on thin sections has been completed mostly on varved sediments and involves quantifying the area of sediment subtended by charcoal particles (Clark 1988a, 1990) or the geometric mean area of charcoal per sediment area (Clark and Royall 1995, Clark and Hussey 1996). Using stereology relationships, the flux of particle area per sediment area can be determined from cross sectional measurements of particles relative to length of varve sampled (Clark 1988). Because particles are quantified *in situ*, there is no spike involved in calculating concentrations. The method is best applied at 20 to 100× and quantifies particles >50 μm diameter.

Sieved samples are generally limited to large (> 100 μm) particles counted on a stereoscope. Determination of particle concentration requires that all particles in the volumetric sample be counted or measured. Many of the counts include only particle numbers, so it is difficult to relate results to emissions that are based on mass or volume fluxes (Clark and Hussey 1996).

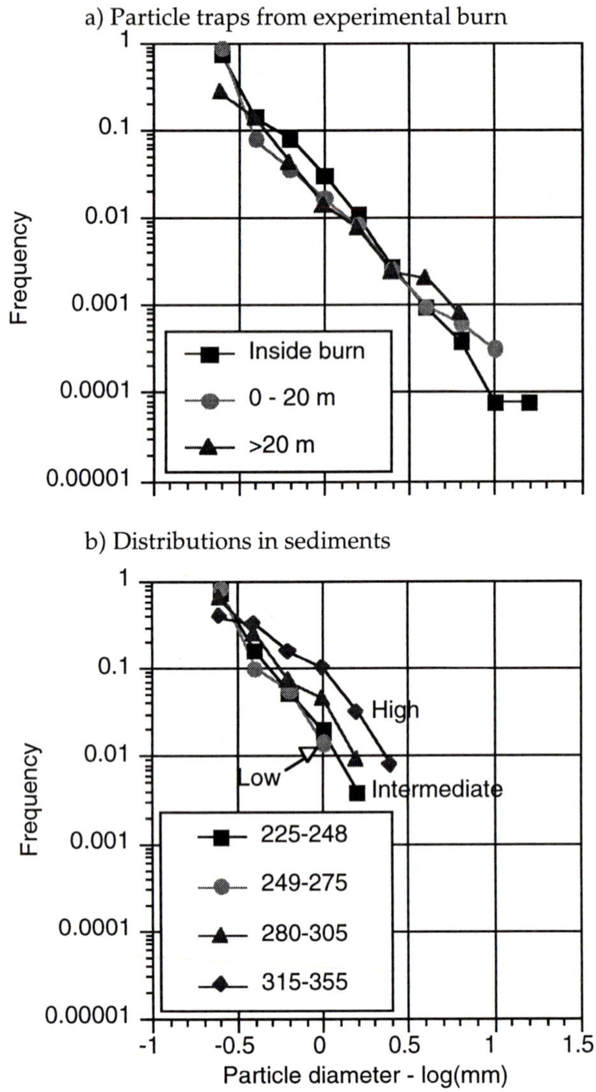

Figure 5. Size distributions for charred particles from traps at the ground surface at three distances during an experimental burn (a) and from sediments in a nearby lake (b). Distributions from sediments come from parts of core showing high, intermediate, and low particle accumulation rates (see Figure 6).

Advantages of optical microscopic analysis include the potential for particle size analysis, which may aid interpretation of source distance. Large particles have higher settling velocities and so might reflect local sources (Tolonen 1983, Patterson et al. 1987, Clark 1988b, Clark and Royall 1985). Despite much speculation on the potential importance of particle size for atmospheric transport, the size distribution of carbonaceous particles in sediments is poorly understood. Because optical microscopy is the only method that is routinely used to quantify individual particles in sediments, we know little about the bulk of combustion emissions that fall in smaller particle classes (Radke et al. 1991). And, until recently, no sediment studies have sought to obtain detailed distributions of particle size classes because size distributions may change during processing, and the studies are laborious. A number of analyses have classed particles in more than one diameter category, under the assumption that large particles remain closer to the source than do small particles. Regional comparisons of methods that quantify different size classes support this view (Clark and Royall 1995).

Clark et al. (in review) used image analysis to compare diameter distributions from sediments of Bor Lake, Siberia, with distributions obtained from a burn in a nearby *Pinus sylvestris/Cladonia* forest. For both sample types, samples were sieved with a 180-μm mesh, particles were identified individually under a compound microscope at 15×, and areas were determined using image analysis. They found that airborne and sediment size distributions shared some attributes and showed some differences. Particle traps arrayed along transects from the experimental burn to a distance of 60 m all showed similar particle size distributions (Fig. 5a). The largest particles were >10 μm diameter and occurred within the burn, but particles nearly as large occurred along the entire transect. The modes of distribution were consistently near the smallest sieve diameter, 180 μm (i.e., $10^{-0.75}$ μm on Fig. 5a). Particle frequencies declined rapidly above this diameter, with diameters >$10^{0.5}$ μm being rare.

The sediment samples showed greater size variability than the atmospheric samples, and that variability was largely explained by differences in charcoal accumulation. Charcoal accumulation in the core showed the highest values between 4000 and 5000 yr BP, declined to a minimum by 2000 yr BP, and increased thereafter (Fig. 6). Particle size distributions in Fig. 5b are the composite distributions from four different parts of this core, corresponding to "high" (315–335 cm), "intermediate" (280–305 and 225–248 cm), and "low" (249–275 cm) charcoal accumulation (Figs. 5b, 6). The greatest representation of large particles was found when total accumulation was high, and vice versa. Intermediate accumulation rates were likewise attended by intermediate diameter distributions.

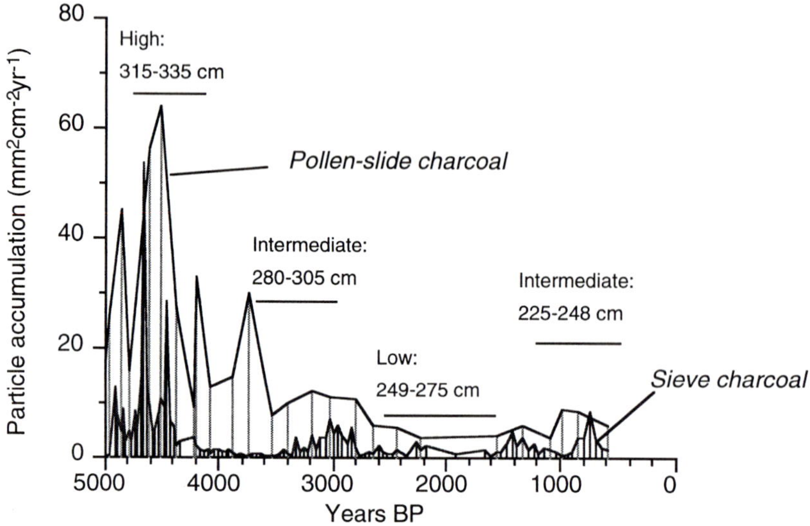

Figure 6. Accumulation of charcoal in Bor Lake, Siberia over the last 5000 yr. Accumulation of particles on pollen slides is greater than in sieve samples, probably because small particles are counted on pollen slides, and small particles are more abundant than large particles. The segments labeled "High", "Intermediate", and "Low" are those for which particle size distributions are presented in Figure 5.

The single data set comparing atmospheric particle size distributions with those in Bor Lake sediments suggests that calibration of sediment charcoal records may facilitate interpretation. Highly similar particle-size distributions from airborne samples during fire and sediments suggests that sediments may provide an accurate representation of particle sizes produced by fire. The fact that large particles are most abundant at times of high accumulation suggests that particle sources may indeed have been closer at those times. These relationships require further study to allow for more detailed calibration.

In summary, existing methods for sediment charcoal analysis represent tradeoffs between the time-consuming but time-tested microscopic methods (optical, SEM, and TEM) versus a variety of methods from atmospheric sciences and geochemistry that are potentially rapid and diagnostic but not well tested in sediments. There is much potential for application of chemical and thermal methods to sediments, but much characterization is needed to fully understand how each distinguishes black carbon from organic carbon in different types of sediment matrices.

Acknowledgments

The work at Lawrence Berkeley National Laboratory was supported by the U.S. Department of Energy, Office of Health and Environmental Research, Environmental Research, Environmental Sciences Division under contract DE-AC03-76SF00098.

Literature References

Ballentine DC, Macko SA, Turekian VC, Gilhooly WP, Martincigh B (1996) Chemical and isotopic characterization of aerosols collected during sugar cane burning in South Africa. In Levine J (ed) *Chapman Conference on Biomass Burning* (in press)

Burney DA (1987) Late Holocene vegetational change in central Madagascar. *Quat Res* 28:130-143

Cachier H, Bremond M-P, Buat-Ménard P (1989) Determination of atmospheric soot carbon with a simple thermal method. *Tellus* 41B:379-390

Cachier H, Liousse C, Buat-Ménard P, Gaudichet A (1996) Particulate content of savanna fire emissions. *J Atmos Chem* special DECAFE issue (in press)(a)

Cachier H, Liousse C, Gaudichet A, Echalar P, Kuhlbusch T, Lacaux JP (1996) Particulate emissions during savanna fires in Kruger National Park (South Africa). In Levine JS (ed) *Biomass Burning and Global Change* MIT Press Cambridge (in press) (b)

Chow JC, Watson JG, Pritchett LC, Pierson WR, Frazier CA, Purcell RG (1993) The DRI thermal/optical reflectance carbon analysis system: description, evaluation and application in US air quality studies. *Atmos Envir* 27A:1185-1201

Clark JS (1988a) Charcoal-stratigraphic analysis on petrographic thin sections, recent fire history in northwest Minnesota. *Quat Res* 30:67-80

Clark JS (1988b) Particle motion and the theory of charcoal analysis, source area, transport, deposition, and sampling. *Quat Res* 30:81-91

Clark JS (1990) Fire and climate change during the last 750 years in Northwestern Minnesota. *Ecol Mono* 60:135-159

Clark JS, Royall PD (1995) Particle-size evidence for source areas of charcoal accumulation in Late Holocene sediments of eastern North American lakes. *Quat Res* 43:80-89

Clarke AD (1982) Integrating sandwich: a new method of measurement of the light absorption coefficient for atmospheric particles. *Appl Optics* 21:3011-3020

Crutzen PJ, Andreae MO (1990) Biomass burning in the Tropics: Impact on atmospheric chemistry and biogeochemical cycles. *Science* 250:1669-1778

Echalar F, Gaudichet A, Cachier H, Artaxo P (1996) Aerosol emissions by biomass burning in Africa and in the Amazon basin: characteristic trace elements and fluxes. *Geophys Res Lett* (in press)

Ellis EC, Novakov T, Zeldin M (1984) Thermal characterization of organic aerosols. *Sci Total Envir* 36:261-270

Gaudichet A, Echalar F, Chatenet B, Quisefit JP, Malingre G, Cachier H, Buat-Ménard P, Artaxo P, Maenhaut W (1996) Trace elements in Tropical African savana biomass burning aerosols. *J Atmos Chem* special DECAFE issue (in press)

Goldberg ED (1995) *Black Carbon in the Environment* Wiley New York

Gundel LA, Dod RL, Rosen H, Novakov T (1984) The relationship between optical attenuation and black carbon concentration for ambient and source particles. *Sci Total Envir* 36:271-276

Hansen ADA, Rosen H, Novakov T (1984) The aethalometer—an instrument for the real-time measurement of optical absorption by aerosol particles. *Sci Total Envir* 36:191-196

Huntzicker JJ, Johnson RL, Shah JJ, Cary RA (1982) Analysis of organic and elemental carbon in ambient aerosols by a thermal-optical method. In Wolff GT, Climish RL (eds) *Particulate Carbon: Atmospheric Life Cycle* pp 79-88 Plenum New York

Lin C, Friedlander SK (1988) A note on the use of glass fiber filters in the thermal analysis of carbon containing aerosols. *Atmos Envir* 22:605-607

Lin CL, Baker MB, Charlson RJ (1973) Absorption coefficient for atmospheric aerosols: a method for measurement. *Appl Optics* 12:1356-1363

Liousse C, Cachier H, Jennings SG (1993) Optical and thermal measurements of black carbon aerosol content in different environments: variation of the specific attenuation cross-section, sigma (σ). *Atmos Envir* 27A:1203-1211

Macko SA (1991) Compound specific approaches using stable isotopes. In Lajtha K, Michener RH (eds) *Stable Isotopes in Ecology and Environmental Science* pp 241-247 Blackwell Scientific Publications London

Malissa H, Puxbaum H, Pell E (1976) Simultane Kohlenstoff— und Schwefelbestimmung in Stäuben. *Fresenius Z Anal Chem* 273:109-113

Mueller PK, Mosley RW, Pierce LB (1972) Chemical composition of Pasadena aerosol by particle size and time of day: carbonate and noncarbonate carbon content. *J Coll Interface Sci* 39:235-240

Mueller PK, Fung KK, Heisler SL, Grosjean D, Hidy GM (1982) Atmospheric particulate carbon observations in urban and rural areas of the United States. In Wolff GT, Climish RL (eds) *Particulate Carbon: Atmospheric Life Cycle* pp 343-370 Plenum New York

Niessner R, Petzold A (1994) Comparison study on thermal, optical, and photoelectrical methods for elemental carbon analysis. Paper presented at the *Fifth International Conference on Carbonaceous Particles in the Atmosphere* Berkeley CA 23-26 August 1994

Novakov T (1981) Microchemical characterization of aerosols. In Malissa H, Grasserbauer M, Belcher R (eds) *Nature, Aim and Methods of Microchemistry* pp 141-16 Springer-Verlag Vienna

Novakov T (1982) Soot in the atmosphere. In Wolff GT, Climish RL (eds) *Particulate Carbon: Atmospheric Life Cycle* pp 19-41 Plenum New York

Novakov T, Corrigan CE (1996) Thermal characterization of biomass smoke particles. *Mikrochemica* Acta (in press)

Patterson WA, Edwards KJ, Maguire DJ (1987) Microscopic charcoal as a fossil indicator of fire. *Quat Sci Rev* 6:3-23

Patterson EM, McMahon CK, Ward DE (1996) Absorption properties and graphitic carbon emission factors of forest fires aerosols. *Geophys Res Lett* 13:129-132

Radke LF, Hegg DA, Hobbs PV, Nance JD, Lyons JH, Laursen KK, Weiss RE, Riggan PJ, Ward DE (1991) Particulate and trace gas emissions from large biomass fires in North America. In Levine JS (ed) *Global Biomass Burning: Atmospheric, Climatic, and Biospheric Implications* pp 209-224 MIT Press Cambridge

Rosen H, Hansen ADA, Dod RL, Novakov T (1980) Soot in urban atmospheres: Determination by an optical absorption technique. *Science* 208:741-744

Rosen H, Hansen ADA, Dod RL, Gundel LA, Novakov T (1982) Graphitic carbon in urban environments and the Arctic. In Wolff GT, Climish RL (ed) *Particulate Carbon: Atmospheric Life Cycle* pp 273-294 Plenum New York

Schnell RC, Kuniyuki DT, Bodhaine BA, Hansen ADA (1994) The dust component of aerosol light absorption measured at Mauna Loa Observatory. Paper presented at the *Fifth International Conference on Carbonaceous Particles in the Atmosphere* Berkeley CA 23-26 August 1994

Shah JJ, Rau JA (1990) Carbonaceous species methods comparison study: Inter laboratory round robin interpretation of results. *California Air Resources Board* Research Division Contract No. A832-154 (final report)

Tolonen K (1983) The post-glacial fire record. In Wein RW, Maclean DA (eds) *The Role of Fire in Northern Circumpolar Ecosystems.* pp 21-44 Wiley Chichester

Turekian VC, Macko SA, Gilhooly WP, Ballentine DC, Swap RJ, Garstang M (1996) Bulk and Compound specific isotopic characterization of the products of biomass burning: laboratory studies. In Levine JS (ed) *Chapman Conference on Biomass Burning* (in press)

Ward DE, Hao WM, Susott RA, Babbitt RA, Shea RW, Kauffman JB, Justice CO (1996) Effect of fuel composition on combustion efficiency and emission factors for African savanna ecosystems. *J Geophys Research* special SAFARI issue (in press)

Weiss RE, Waggoner AP (1982) Optical measurements of airborne soot in urban, rural and remote locations. In Wolff GT, Climish RL (eds) *Particulate Carbon: Atmospheric Life Cycle*, pp 317-325 Plenum New York

Wiedemann HG, Riesen R, Boller A, Bayer G (1988) From wood to coal: a compositional thermogravimetric analysis. In Earnest CM (ed) *Compositional Analysis by Thermogravimetry* pp 227-244 American Society for Testing and Materials Philadelphia

Winkler MJ (1985) Charcoal analysis for paleoenvironmental interpretation: A chemical assay. *QuatRes* 23:313-326

Wolff GT, Groblicki PJ, Cadle SH, Countess RJ (1982) Particulate carbon at various locations in the United States. In Wolff GT, Climish RL (eds) *Particulate Carbon: Atmospheric Life Cycle* pp 297-315 Plenum New York

Polycyclic Aromatic Hydrocarbons in Sedimentary Records of Biomass Burning

Richard G. Zepp and Stephen A. Macko[1]
U.S. Environmental Protection Agency
960 College Station Road
Athens, Georgia

Summary

Lacustrine and coastal marine sediments often contain polycyclic aromatic hydrocarbons (PAH) of combustion origin. Many higher molecular weight PAH, like charcoal, are hydrophobic and resistant to biodegradation. Recent studies are presented here which indicate that the stratification of PAH in bottom sediments can provide a useful historical record of combustion activity, ranging back millions of years. The distribution of individual PAH in environmental samples often has been used to infer their source. However, this PAH "fingerprint" is affected not only by the combustion conditions and fuel, but also by transport and transformation processes that affect PAH as they runoff from land or move through the atmosphere and water to the sediments. Recent studies, though sparse, suggest that non-linear, higher-molecular-weight PAH sorbed on aerosol black carbon are most resistant to alteration during long-range environmental transport. During the past century there has been a large increase in sedimentary PAH derived from fossil fuel combustion. The latter qualitatively produces PAH distributions similar to biomass burning. The use of new techniques, such as compound-specific isotope analysis (CSIA), which allow for the isotopic determinations of individual compounds, has shown great promise for apportioning PAH sources, e.g. fossil fuel or biomass burning.

Introduction

Fire has long had a major impact on the Earth's biogeochemical cycles as well as functioning and dynamics of ecosystems (Crutzen and Andreae 1990, Andreae 1991, Goldammer 1991, Clark 1990, Payette and Gagnon 1985, Anderson et al. 1986, Payette 1992, Clark and Robinson 1993). This impact includes feedbacks that affect carbon and oxygen cycles over time scales of millions of years (Chaloner 1989, Robinson 1989). Fires affect species composition, climate and atmospheric composition at scales of years to decades. Changes in climatic parameters (e.g. lightning frequency, drought, temperature, humidity), as well as human actions such as fire suppression and ignition, strongly influence fire frequency and intensity. For example, long periods without fire can result in a buildup of biomass that can enhance the intensity of eventual fires, e.g. the recent fires in Yellowstone National Park (Whitlock et al. this volume).

Fires have important effects on the trace gas composition of the

[1]Department of Environmental Sciences, University of Virginia, Charlottesville, VA 22903.

NATO ASI Series, Vol. I 51
Sediment Records of Biomass Burning
and Global Change
Edited by James S. Clark, Hélène Cachier,
Johann G. Goldammer, and Brian Stocks

atmosphere. On a global basis, biomass burning results in direct emissions of greenhouse and chemically active gases and aerosols that are approximately equal to those produced by fossil fuel combustion (Crutzen and Andreae 1990, Andreae 1991). Man-made fires in tropical biomes are major contributors of these direct emissions. Natural fires in temperate and high-latitude biomes may have long-lasting, post-burn impacts on carbon storage and net fluxes of greenhouse gases (Cofer et al. 1990, Smith and Shugart 1993, Kasischke et al. 1995). Fire intensity affects the nature of greenhouse and chemically active gas emissions (Crutzen and Andreae 1990, Andreae 1991). Hot fires with good oxygen supply efficiently convert organic matter to carbon dioxide. The production of other partially-oxidized trace gases, such as carbon monoxide and methane, occurs much less efficiently. Smoldering fires, on the other hand, release carbon monoxide and other partly oxidized gases much more efficiently. Therefore, data concerning the frequency and intensity of past fires can provide useful insights to their impacts on past atmospheric composition.

In addition to greenhouse gases, fires produce other semivolatile gaseous and particulate species that are emitted to the atmosphere, as well as a substantial amount of charred material that remains on the ground (Kuhlbusch and Crutzen 1995, Lobert and Warnatz 1993, Ward and Radke 1993, Crutzen and Andreae 1990, Andreae 1991). These substances are either pyrogenic, i.e., formed by combustion of the biomass, or they are organic constituents of plants that are released to the surrounding environment during conflagration (Macko, this volume). Semivolatile pyrogenic substances include polycyclic aromatic hydrocarbons (PAH), discussed further below, and various other substances derived from the alteration of long-chain aliphatic hydrocarbons at higher temperatures, e.g. through oxidation or cracking to shorter or odd chain lengths. The products of smoldering, or low temperature fires, include fatty acids, waxes and alcohols, which are transported to the atmosphere through the process of distillation at temperatures around the boiling point of water. For low temperature fires, there appears to be less of a pyrogenic signal [i.e., polycyclic aromatic hydrocarbons (PAH)] and more of a biogenic signal (see below). [This in itself presents a chemical problem of distinguishing between biomass that has been released and transported through the environment in association with dusts (Simoneit et al. 1977, Gagosian et al. 1982) and those compounds that are emitted to the environment through "steam-distillation" by fires].

Charcoal and black carbon are pyrogenic substances that have been widely used as atmospheric and sedimentary tracers of past fire activity. In this paper, we refer to charred material blackened by fire as "charcoal." "Black carbon" is a component of charcoal that is defined as "...a fire-derived highly aromatic to graphitic carbon fraction ..." (Kuhlbusch and Crutzen 1995). Of particular interest here are PAH, pyrogenic compounds that are widely distributed in the environment, including lacustrine and marine sediments. PAH are aromatic hydrocarbons that comprise two or more fused aromatic rings. Like charcoal and black carbon, sedimentary PAH also have been used as tracers of past vegetation fires (Youngblood and Blumer 1975, Venkatesan and Dahl 1989, Killops and Massoud 1992). The presence of PAH in million-year-old sediments has been used to suggest the prehistoric occurrence of fire. PAH also have been used as indicators of pollution derived from fossil fuel burning during the past century (Laflamme and Hites 1978, Gschwend and Hites 1981, Charles and Hites 1987, McVeety and Hites 1988).

Although the chemical distribution of PAH in sediments or atmospheric particles is generally indicative of the source of these compounds, there exist several factors that may limit their use as indicators of past fire activity. Both natural and anthropogenic combustion produces similar PAH distributions, and there may need to be some historical interpretation of a sediment in order to clarify which of these sources may have produced the PAH (Hites et al. 1980). Since various combustion sources qualitatively produce similar distributions of PAH (Laflamme and Hites 1978), the use of new techniques, such as compound-specific isotope analysis (CSIA) which allow for the isotopic determinations of individual compounds, has shown great promise for identifying PAH sources (O'Malley et al. 1993, O'Malley et al. 1996a).

Our objectives in this paper are to discuss sources of PAH in the environment, with emphasis on biomass burning, to describe factors that affect the transport and transformation of PAH from source to sediments, and, finally, to compare techniques that are being used to characterize PAH in lake and marine sediments, including new isotopic methods that can be employed to evaluate PAH sources (O'Malley 1994, O'Malley et al. 1994).

Sources of Polycyclic Aromatic Hydrocarbons

Fossil fuel combustion and vegetation fires are major sources of polycyclic aromatic hydrocarbons in aquatic sediments (Neff 1979, McElroy et al. 1989). Human activities are likely the dominant source of sedimentary, pyrogenic PAH in populated tropical and temperate regions. For example, in the United States a significant fraction of benzo[a]pyrene (BaP) emissions, a PAH that is mainly derived from combustion, are derived from wood combustion (Wilber et al. 1992). In the less-developed high latitude regions, natural fires in the boreal forest possibly may be a significant source of sedimentary PAH. Fire frequency and extent is likely to be linked to the climate change that is occurring in these regions (Kasischke et al. 1995).

In addition to combustion, there are other known sources of PAH. These other sources, however, can be distinguished from pyrogenic PAH, because they have different PAH composition or "signatures." Of these other sources, by far the most important are oil spills and other releases of petroleum hydrocarbons (Neff 1979). PAH from petroleum hydrocarbon spills can be readily recognized, because they contain substantial amounts of alkyl-substituted 2- and 3-ring PAH (Neff 1979). In addition to petroleum hydrocarbons, one other less important natural source of PAH in sediments is early diagenesis of sedimentary organic matter (Youngblood and Blumer 1975, Laflamme and Hites 1978, Wakeham et al. 1980). Diagenetic PAH include readily recognizable compounds, such as perylene, that are not present in pyrogenic PAH. Retene, a PAH often found in sediments, is produced both by combustion and diagenesis (Ramdahl 1983). A few early studies indicated that a small amount of PAH is synthesized by certain microbiota and other organisms, but current research has shown that microbial processes generally provide net sinks rather than sources for PAH in aquatic environments (Cerniglia and Heitkamp 1989).

It should be emphasized that most data on PAH in aquatic environments have been obtained near urban centers in developed countries, where fossil fuel burning is clearly the major source. Fossil fuel combustion also is a major source of charcoal in aquatic environments (Griffin and Goldberg 1981, Lim and Renberg, this volume). PAH and charcoal profiles in recent temperate lacustrine and marine sediment cores have been used to evaluate historical trends in fossil fuel combustion (Laflamme and Hites 1978, Wakeham et al. 1980, Gschwend and Hites 1981, Charles and Hites 1987, McVeety and Hites 1988). Hites and coworkers have provided ample evidence of the importance of fossil combustion as a major source of PAH in recent temperate sediments of lakes in industrialized countries.

Comparatively little is known about PAH in sediments of regions where intense vegetation burning occurs. It should be kept in mind that, on a global basis, biomass burning is approximately as important as fossil fuel combustion (Crutzen and Andreae 1990, Andreae 1991). Biomass burning is more likely to be the primary PAH source in the tropics and in remote areas such as the boreal biome. Savanna or boreal forest fires are common events. Most vegetation burning takes place in the tropics, and these fires are usually started by man. Burning of biomass as a fuel also can be a significant PAH source in some regions. Other studies indicate that forest fires have provided the major input of PAH to sediments that are not close to fossil combustion sources (Youngblood and Blumer 1975, Venkatesan and Dahl 1989). Grass fires in African savannas also produce PAH (Masclet et al. 1995). Moreover, as noted earlier, recent studies have shown that significant amount of PAH that may have been derived from vegetation fires are present in ancient sediments (Venkatesan and Dahl 1989, Killops and Massoud 1992).

Generally, as biomass combustion efficiency increases, emission factors decrease for particles and most gases, other than CO_2. Emission factors for particles have been quantified by a number of recent studies (Cachier et al. 1996), but emission factors for PAH are poorly understood. The sparse data available on PAH production from vegetation fires indicate that the nature and amount produced are influenced by temperature, fuel-to-oxygen ratio, moisture content and other factors (Howard and Longwell, 1983; McMahon and Tsoukalas, 1978; Mazurek et al. 1991). Temperatures close to 800°C appear to be optimal for PAH formation. Lower-temperature combustion (< 300°C) favors formation of alkylated PAH (Neff 1979) and organic particles that do not strongly absorb visible radiation (Patterson et al. 1986, Ward and Radke 1993), whereas higher temperatures (> 800°C) promote formation of unsubstituted 4-and 5-ring compounds (Neff 1979) and black carbon (Patterson et al. 1986, Lobert and Warnatz 1993).

Experiments under controlled conditions in the laboratory indicate that the pyrolysis of almost all organic compounds can produce PAH, but that the nature of the organic compound and combustion conditions affect PAH yield (Neff 1979). Organic substances with cyclic structures produce PAH much more efficiently than branched or straight-chain species (Halaby and Fagerson 1971). Because a wide array of PAH are produced by burning, the possibility exists that PAH composition in environmental samples can be used to "fingerprint" combustion sources. Field studies have provided conflicting data on this possibility. Studies of PAH composition in the sediments of Buzzards Bay, Massachusetts suggested that the composition of unsubstituted PAH produced by energy-related combustion of wood or fossil fuels are similar (Laflamme and Hites 1978). However, PAH composition in urban atmospheres impacted by fossil fuel combustion (Paris, France) reportedly differed significantly from that in atmospheres impacted by biomass burning in central Africa (Masclet et al. 1995). These workers have argued that PAH composition may facilitate characterization of tropical biomass burning effluents. Isotopic composition of individual PAH may provide another valuable tool for characterization of their sources (see below).

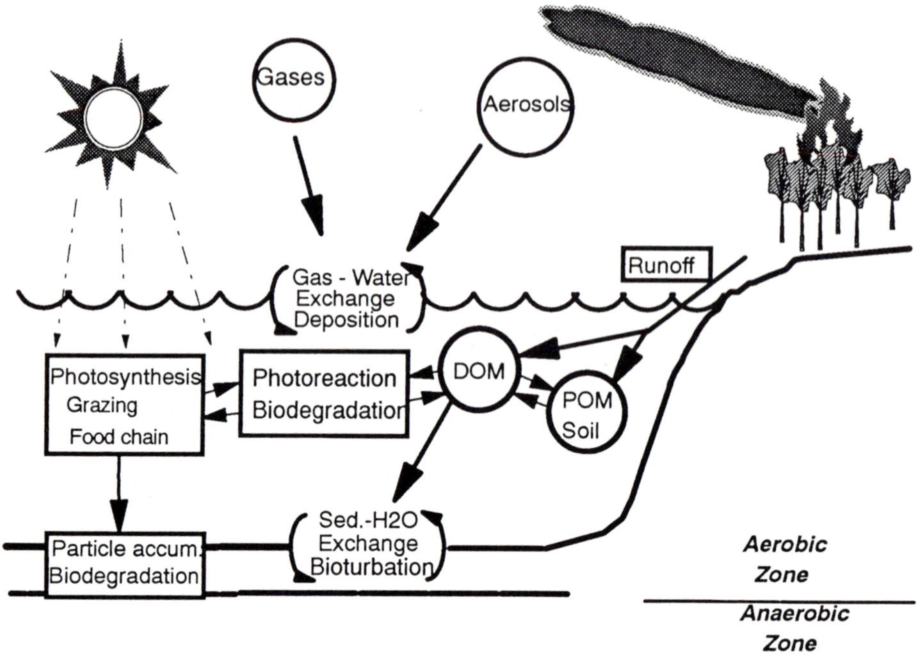

Figure 1. Environmental processes affecting PAH transport and transformation.

Results discussed in this section indicate that PAH provide a convenient tracer of combustion activity. PAH composition is quite sensitive to changes in combustion conditions. Combustion conditions in the environment are related to climate and the nature of the vegetation. For example, the fuel-to-air ratio and temperature for burning grass in African savannas is typically lower than that observed in intense crown fires in the boreal forest. Thus, the composition and amount of PAH archived in sediments may provide a useful tool for the elucidation of past climatic conditions and vegetation types. Isotopic analyses of individual sedimentary PAH can further strengthen this approach. As discussed in the next section, however, the array of PAH produced from a given source may be affected by transformation processes as the PAH are transported through the atmosphere and water column into sediments.

Transport and Transformation of PAH in the Atmosphere

Various transport and transformation processes that can affect PAH as they move from point of emission to sediment are depicted in Fig. 1. Aerial deposition and surface runoff are the two predominant pathways by which combustion-derived PAH enter water bodies (Neff 1979). PAH inputs from runoff are primarily derived from fires in the immediate vicinity of a water body, e.g. the basin that drains into a lake or estuary. On a regional or global basis, atmospheric deposition is the predominant pathway for entry of combustion-derived PAH into the aquatic environment (Neff 1979). In the following section we discuss the atmospheric component of the journey from point of emission to sediment, with emphasis on gas-particle partitioning, transformation, and deposition. Discussion of atmospheric transport processes relevant to regional to global distribution of the biomass burning emissions is presented elsewhere in this volume (Garstang et al. this volume). Following the atmospheric transport section, we further consider factors that affect PAH composition in water bodies.

PAH are generally classified as semivolatile organic compounds (SOCs). Such compounds are present in the atmosphere in both gaseous and particulate phases (Junge 1977, Yamasaki et al. 1982, Bidleman and Foreman 1988, Pankow 1987). Because particle-associated and gas-phase SOCs have different mechanisms for transformation and deposition, recent studies have focused on developing an improved understanding of the partitioning process. Mathematical relationships have been derived that describe gas-particle partitioning of SOCs in the atmosphere in terms of partition coefficients that correlate with subcooled liquid vapor pressures of the SOCs (Pankow 1987, Bidleman and Foreman 1988). Such partition coefficients increase with decreasing temperature and with increasing PAH molecular weight. Under equilibrium conditions at 25°C, PAH with 5 rings, such as benzo[a]pyrene, are predominantly partitioned in the particle phase, whereas 3-ring PAH such as phenanthrene are mainly in the gas phase (Kamens et al. 1995, Bidleman and Foreman 1988, McVeety and Hites 1988, Masclet et al. 1995). Field studies of gas-particle partitioning of a series of PAH showed that the fraction sorbed on particulates increased over an order of magnitude during winter months in the atmosphere over northern Lake Superior (McVeety and Hites 1988). Recent observations suggest that PAH desorption from the particle phase can become sufficiently retarded under cool outdoor conditions that equilibrium considerations may not apply (Kamens et al. 1995). Other studies also have suggested that PAH can be occluded in combustion-derived particles in such a form that the PAH are not equilibrated or readily transformed on entry into the aquatic environment (Readman et al. 1984).

In addition to atmospheric particulates, gaseous PAH partition to plants and soil at the Earth's surface. Equilibrium relationships have been employed to describe partitioning of PAH to vegetation, and air-to-plant partitioning correlates with lipid content of the plant matter (Simonich and Hites 1994a, Simonich and Hites 1994b, Simonich and Hites 1995). As expected, PAH partitioning between atmospheric gas phase and vegetation increases with decreasing temperature and thus exhibits seasonal trends in temperate latitudes with the greatest sorption occurring during winter.

PAH concentrations in the atmosphere also are affected by transformations that take place in both the gas phase and particle phase. In the gas phase, PAH oxidation occurs mainly via attack by hydroxyl radicals (Atkinson et al. 1990). Hydroxyl radicals are highly reactive and thus not very selective in their attack on aromatic compounds. Thus, gas-phase oxidation of PAH should not have a major effect on the composition of a PAH mixture. It is interesting to note that vegetation fires indirectly affect atmospheric PAH transformations by producing chemically-active species, such as nitrogen oxides and carbon monoxide, that affect tropospheric hydroxyl radical concentrations (Crutzen and Carmichael 1993). The situation is more complicated in the case of PAH photoreactions on atmospheric particles (see Wehry and Mamantov 1993, Odum et al. 1994 and references therein). Thus, PAH photoreactions have been shown to be rapid on particles such as alumina, silica gel, and certain wood soot and other atmospheric particles collected by high-volume filtration. Organic aerosol constituents such as methoxyphenols may enhance the photodegradation of PAH on atmospheric organic aerosols (Odum et al. 1994). On the other hand, PAH sorbed on black carbon and many coal stack ash particles have been found to be resistant to photo-oxidation (Daisey et al. 1982, Readman et al. 1984, Behymer and Hites 1985, Behymer and Hites 1988, Wehry and Mamantov 1993).

Atmospheric deposition of PAH is affected by gas-particle partitioning (Eisenreich et al. 1992). Both gas-phase and particle-phase PAH are introduced into aquatic environments through wet and dry deposition processes. Wet deposition proceeds through scavenging of gas- and particle-phase PAH by falling rain. Particle-associated PAH are scavenged more efficiently by this process than gases. In the case of dry deposition of particle-associated PAH, coarse particles (> 2 μm) are deposited more rapidly than fine particles (< 2 μm). Generally, wet deposition of PAH in the particle phase is the most important deposition pathway in water bodies that are distant from the source, such as the ocean. Dry deposition of PAH can be a dominant contributor to lakes close to the combustion source. For higher-molecular weight PAH, deposition of particle-associated PAH is the dominant mode of entry, whereas air-water exchange can be an important pathway for return of the PAH to the atmosphere (see below).

The above discussion indicates that atmospheric transport and deposition of higher molecular weight PAH occurs mainly in the particle (or aerosol) phase. Gas-phase oxidation by hydroxyl radicals is likely to be most significant for low-molecular-weight PAH (2-and 3-ring), because they exist predominantly in the gas phase and are removed most slowly from the atmosphere by deposition. Because higher-molecular-weight PAH are mainly sorbed on particles, their atmospheric residence times are shortened by deposition processes, compared to the 2- and 3-ring compounds. This factor may reduce the degree that such PAH are transformed in the atmosphere. Transformation of higher-molecular weight PAH is retarded on certain carbonaceous particles, although it is clear that photoreaction can occur on other forms of atmospheric particles. We conclude that the higher-molecular-weight fraction of PAH produced by biomass burning are most likely to move through the atmosphere into aquatic environments with little change in composition.

Transport and Transformation of PAH in the Aquatic Environment

Once deposited in a water body, the distribution of PAH in the system is affected by a variety of competing processes (Fig. 1):

(1) The PAH are transported within the water body via advection and dispersion processes as discussed below.

(2) PAH move to the bottom sediments through sorption or sedimentation on soil or detrital particles, such as zooplankton fecal pellets. Resuspension of bottom sediments can return PAH to the water column.

(3) PAH are exchanged back to the atmosphere by volatilization of dissolved PAH.

(4) PAH are transformed by photoreactions and microbial degradation.

A detailed discussion of transport within lake systems is beyond the scope of this paper. Advection involves the movement of dissolved PAH or PAH sorbed to fine particles at the current velocity. Dispersion and turbulent (or eddy) diffusion mix the PAH within the water column. These transport processes are affected by climatic factors such as wind shear at the surface of the water or density gradients caused by temperature differentials in the water column. Molecular diffusion is too slow to contribute significantly to PAH transport in most parts of the water column, but it may be important in bottom sediment pore water.

During warm periods, these density gradients result in regions of transport resistance that are referred to as "thermoclines." Thermoclines greatly retard the exchange of water between the upper part of lakes or the ocean and water at greater depths. Stratification also can be caused by density gradients attributable to differences in salt concentrations in estuaries.

By analogy to the above discussion of atmospheric processes, water-particle partitioning is an important determinant of the transport and transformation of PAH in the aquatic environment (Karickhoff 1984, Westall 1987, Zepp and Wolfe 1987). As illustrated by the data in Table 1, the aqueous solubility of PAH generally decreases as molecular weight increases. Note also that the sorption partition coefficient, K_{OC}, which is normalized to the fractional organic content of aquatic particles, increases with increasing molecular weight. Thus, PAH with higher molecular weights have the greatest tendency to sorb to settling soil particles or to biota that eventually settle to the bottom sediments. In larger lakes algae and detritus are the main sources of particles responsible for sedimentation of SOCs, including PAH. Thus, peak sedimentation in temperate and high latitude lakes occurs during summer and spring algal blooms (Imboden and Schwarzenbach 1985). Zooplankton that graze on algae accumulate SOCs then release them packaged in rapidly sinking fecal pellets (Eisenreich et al. 1992), thus providing a particularly effective pathway for sedimentation.

Recent studies have further indicated that PAH may sorb to the dissolved organic matter (DOM) in natural waters (Schlautman and Morgan 1993, Backus and Gschwend 1990, Landrum et al. 1984). Such sorption, by reducing the concentration of PAH dissolved in water, can reduce the bioavailability, volatilization rates, and possibly photodegradation rates of high-molecular-weight PAH such as BaP.

Table 1. Physicochemical properties of selected PAH (McVeety and Hites 1988)

PAH	Molecular Weight	Water Solubility (μM)	Henry's Law Constant, 25°C[a]	K_{oc}[b]
Naph[c]	128	263	3.9×10^{-2}	1.3×10^{3}
Phen[c]	178	7.2	1.1×10^{-4}	2.6×10^{4}
Bz[a]A[c]	228	0.048	6.3×10^{-5}	2.6×10^{5}
B[a]P[c]	252	0.016	7.5×10^{-6}	6.9×10^{5}

[a]atm m^3 mol^{-1}; [b]Sediment-water partition coefficient, normalized to organic content of sediment; [c]Naph=naphthalene; Phen=phenanthrene; Bz[a]A=benz[a]anthracene; B[a]P=benzo[a]pyrene

Volatilization has been proposed as a major process for loss of PAH from lakes (Mackay and Yuen 1983, Eisenreich et al. 1992). Gas-air exchange is described by equations that have been developed by Mackay and Yuen (1983), Liss and Merlivat (1986) or Wanninkoff (1992). The lower Henry's Law constants (H) for 4-and 5- ring PAH (Table 1) indicate that these compounds have lower volatilization rates than the 2- and 3-ring PAH.

A variety of transformation pathways can affect the concentrations of SOCs in lakes and marine environments. In the case of PAH, two major classes of processes, microbial degradation (Cerniglia and Heitkamp 1989) and photodegradation (Zepp 1980, Zepp and Schlotzhauer 1979), are the most dominant transformation pathways.

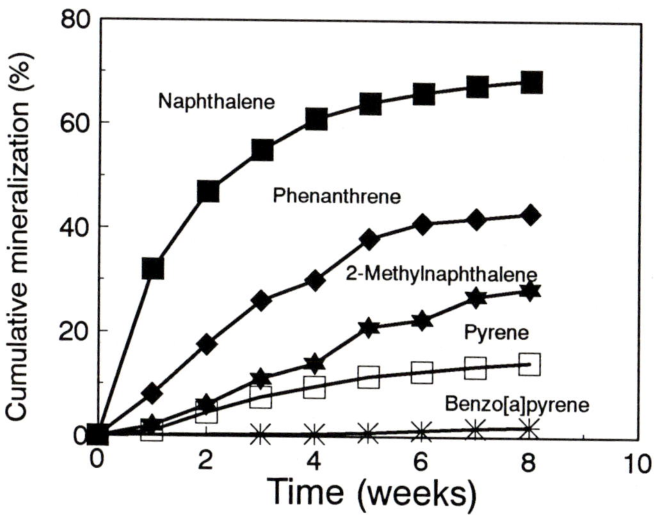

Figure 2. Microbial degradation rates for selected PAH in sediments from Redfish Bay, an oil-exposed estuarine marine ecosystem (*Source:* Cerniglia and Heitkamp 1989).

In general, PAH biodegradation rates decrease with increasing number of fused rings (i.e., molecular weight) and with increased ring substitution. This trend is illustrated by data for microbial degradation of several PAH in sediments from Redfish Bay, an estuarine ecosystem (Fig. 2) (Cerniglia and Heitkamp 1989). Oxygen plays an important role in the microbial degradation of PAH. Under controlled conditions in the laboratory, PAH biodegradation becomes generally undetectable over a period of months under anaerobic conditions in sediments obtained from freshwater or intertidal marine systems (Cerniglia and Heitkamp 1989, Bauer and Capone 1985). Field studies of PAH vertical profiles in sediments further indicate that biodegradation must be extremely slow, if it occurs at all (see discussion below).

Photodegradation of certain PAH dissolved in the aqueous phase is quite rapid with full exposure to solar radiation (Zepp and Schlotzhauer 1979, Mill et al. 1981, Lee and Ryan 1983, Landrum et al. 1984, McElroy et al. 1989). Photodegradation rates in sunlight tend to increase with increasing number of linear fused rings with particularly large increases in going from 2-ring to 3-ring compounds. As with all photochemical processes, PAH photodegradation exhibits seasonal trends in temperate and high latitude aquatic systems with the lowest rates during winter and highest in summer (McElroy et al. 1989). Unfortunately, the photoreactivity of PAH sorbed on aquatic particles, atmospheric particles that enter water bodies (such as carbonaceous particles from fires), or on DOM is poorly defined. Indeed, it is known that sorption on sediments reduces the photoreactivity of SOCs such as the chlorinated hydrocarbon, DDE (Zepp and Wolfe 1987). Moreover, based on similarities observed in the composition of PAH on suspended sediments and bottom sediments in an English estuary, it has been suggested that PAH may be occluded in aquatic particulates and thus not susceptible to photodegradation (Readman et al. 1984). Finally, it is conceivable that larger combustion-derived particles that are deposited from nearby fires or PAH sorbed on fecal pellets may rapidly settle to a lake bottom with no photodegradation of associated PAH.

Mathematical simulations (Imboden and Schwarzenbach 1985, Zepp 1980 using EXAMS model described by Burns et al. 1982), that take into account light attenuation effects and competing processes such as sedimentation, microbial degradation and volatilization, indicate that photodegradation is likely the major loss process for BaP and other 4- and 5- ring PAH in freshwater lakes and ponds. These simulations assume that PAH are not photoreactive when sorbed on aquatic particles, and that photoreaction takes place completely in the aqueous phase. The models also assume that particle-water partitioning of PAH is at equilibrium. Other mass balance models have indicated that the inputs of BaP and other PAH into the Great Lakes' water columns are balanced primarily by outputs that include sedimentation, volatilization, and transformation (Eisenreich et al. 1992, McVeety and Hites 1988). Based on the simulations noted above, photodegradation of the PAH is likely to be a much more significant loss process than volatilization.

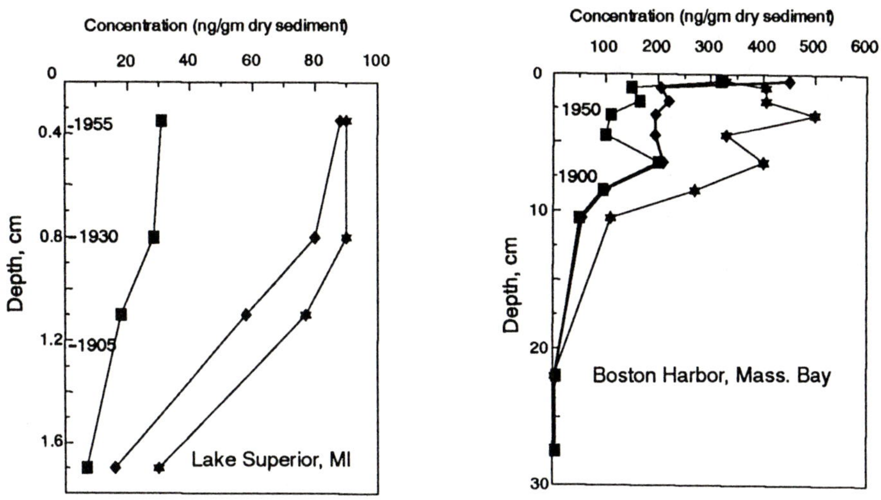

Figure 3. Vertical profiles of PAH in sediments of selected U.S. lakes (*Source:* Gschwend and Hites 1981).

Compositional differences in PAH sedimentary profiles from urban and remote sites are consistent with the expected effects of photodegradation between combustion source and sediment (Gschwend and Hites 1981). Compositional changes observed by Killops and Massoud (1992) in comparisons of PAH profiles from ancient sediments and recent urban sediments exhibit interesting similarities to those noted by Gschwend and Hites (1981), such as the preferential loss of linear PAH. These similarities suggest that the changes may have involved pre-sedimentation transformations of the PAH, rather than sedimentary transformations.

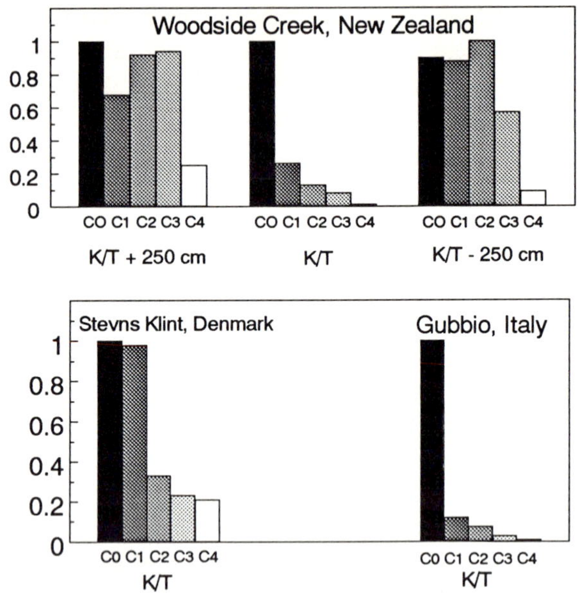

Figure 4. Comparison of alkyl homologue distributions of PAH in samples extracted from Woodside Creek, New Zealand (*Source:* Venkatesan and Dahl 1989).

Characterization of PAH in the Sedimentary Record

One possibility of deciphering the sources of sedimentary biogenic materials, which co-occur in sediments with PAH would be the isotopic analysis of both of these compound types. Because there only appears to be a small fractionation associated with combustion and production, the biogenically derived materials should be isotopically "identical" to isolated PAH. In this section, we briefly discuss procedures that can be used to analyze PAH in sediments, ancillary data that are used to help interpret vertical profiles of sedimentary PAH, and, finally, isotopic techniques to characterize PAH sources.

Methods used to isolate and separate sedimentary PAHs are well-established. For example, the U.S. Environmental Protection Agency uses standard methods for PAH extraction from sediments and subsequent analyses by gas chromatography. Deuterated standards are generally spiked into the sediments prior to extraction to evaluate recovery efficiencies. Ultrasonic extraction of the sediment, best by a polar organic solvent, is now a widely used procedure preferable to soxhlet extraction (Grimwalt et al. 1984). Recent advances in instrumentation have provided reliable, sensitive desktop combined gas chromatography-mass spectrometry (GC-MS) systems at reasonable costs for analysis of organic mixtures. With GC-MS the mass spectrum for each individual component of the PAH mixture can be obtained. As well, homologous series of a chemical family can be measured using the selected ion mode (SIM) on the GC/MS.

In addition to gas chromatographic procedures, PAH mixtures also can be analyzed by high performance liquid chromatography (HPLC). Microbore columns afford high resolution separations, and fluorescence detection provides high sensitivity. HPLC can also be combined with mass spectrometry (LC/MS) to enhance analytical capabilities.

Vertical profiles of PAH concentrations in lake sediments, in conjunction with ancillary data that quantify deposition rate, can be used to estimate PAH inputs as a function of time. Useful ancillary data include distributions of lead-210 and cesium-137 in the sediments. These distributions can be used to help date the sediment layers over the past few decades to centuries and to determine the extent to which they may have been disturbed by bioturbation (Charles and Hites 1987).

Like charcoal, many higher molecular weight PAH are hydrophobic and biologically refractory, especially under the anaerobic conditions that often prevail in freshwater sediments. Thus, the stratification of PAHs in the bottom sediments can provide a useful, long-term record of combustion activity during previous times in the surrounding area. For example, the stratification of PAH in lake sediments (Fig. 3) indicates that PAH inputs from fossil fuel combustion have greatly increased during the past century and are now the dominant source (Gschwend and Hites 1981, Charles and Hites 1987, McVeety and Hites 1988, Eisenreich et al. 1992). Mass balance studies indicate that atmospheric deposition is the main source of PAH in the Great Lakes, U.S.A. However, comparatively few studies have been conducted in regions where vegetation fires are the major source of sedimentary PAH. The higher abundance of non-alkylated PAH homologues (CO) (compared to alkylated homologues, C_1, C_2, C_3, C_4) at the Cretaceous/Tertiary (K/T) boundary (Fig. 4) indicates a predominant combustion source and provides evidence for widespread fires during this period (Venkatesan and Dahl 1989).

In earlier sections of this chapter, we discussed various approaches to elucidating the sources of PAH based on the distribution of individual PAH in the PAH mixture. This approach is somewhat uncertain, in particular because the pre-sedimentation transformations of PAH, such as photodegradation, are not well understood. Determination of the isotopic composition of individual PAH, compound-specific isotope analysis (CSIA), provides the investigator with a powerful new tool for evaluating PAH sources (O'Malley et al. 1993, O'Malley 1994, O'Malley et al. 1996a). This technique involves combining a stable isotope ratio mass spectrometer (IRMS) with a gas chromatograph (GC) that separates PAH mixtures into their individual components (Fig. 5). The mass spectrometer allows for the determination of the stable-isotope ratios of CO_2 gas generated from the oxidation of the individual PAH, following separation on the gas chromatograph (O'Malley 1994, O'Malley et al. 1996b). The isotope ratios are relative measurements and are expressed as such in the "delta" notation relative to the international standard, Pee Dee Belemnite (PDB):

$$\delta^{13}C_{PDB} = \left(\frac{\left({}^{13}C \div {}^{12}C\right)_{sample}}{\left({}^{13}C \div {}^{12}C\right)_{standard}} - 1 \right) * 10^3.$$

As shown in Fig. 6, the isotopic composition for individual PAH in carsoot (fossil combustion) and fireplace soot (biomass combustion) differed significantly. Comparison of these data with corresponding data for PAH extracted from sediments in Newfoundland water bodies indicated that the sedimentary PAH were primarily derived from wood burning (Fig. 6). These results indicate that the PAH fingerprint in combination with the isotopic composition of individual PAH isolated from environmental samples may greatly facilitate the identification of PAH sources.

Unknowns and Future Research Areas

Sedimentary PAH have not been fully utilized as a means of elucidating the history of fire on Earth. The analytical tools for measuring PAH at low concentrations in sediments are widely available. However, although PAH profiles in the sediments of lakes and rivers of urban, industrialized areas have been well-characterized, there is a paucity of data in developing parts of the world where biomass burning has been intense. Likewise, there are few PAH measurements in high latitude regions experiencing increasing fire activity that may be linked to climate change.

In addition to the need for additional field research concerning PAH in poorly investigated regions exposed to intense burning, the following are other potentially fertile areas for future investigations:

(1) New techniques should be developed to better assign the sources of sedimentary PAH. Techniques such as CSIA show great promise for making such assignments.

(2) Additional information should be obtained on the factors that control PAH distributions from biomass burning and how PAH emission factors relate to those of other substances produced by fire. How are PAH and charcoal formation related? How do they depend on burning conditions (fuel-to-air ratios, combustion temperatures) and vegetation type? What are the relationships between the PAH production and the production of radiatively-active and chemically-active trace gases?

(3) How are PAH altered as they move through the atmosphere? In particular, which PAH are most stable to abiotic degradation? Available sedimentary data suggest that non-linear, high-molecular-weight PAH may be particularly resistant to decomposition, especially on highly carbonaceous particles, but only a very limited sample size is available. Is photodegradation most efficient in the gas phase or particle phase? How does gas-particle and gas-vegetation partitioning affect transport and transformation of PAH through the atmosphere?

(4) What are the relationships between particle size and the transport of PAH through the atmosphere into water bodies? Are PAH on coarse particles generated by biomass burning rapidly deposited in nearby water bodies? What is the relative importance of runoff and atmospheric deposition? Can PAH that are transported great distances on fine particles be used as regional indicators of fire activity?

(5) How are PAH transported and transformed in the water columns and bottom sediments of lakes and marine environments? PAH photodegradation rates in water appear to increase with increasing molecular weight, but so does their tendency to sorb to aquatic particles. How susceptible are PAH to photodegradation when sorbed on soil derived and detrital particles in lakes and the sea? On the other hand, biodegradation rates of PAH generally decrease with increasing molecular weight. Anaerobic conditions retard PAH biodegradation in sediments, but can other electron acceptors besides oxygen oxidize PAH over geologic time scales?

Although many questions remain, current research has established that PAH profiles in sedimentary records provide useful information about past fires, and possibly, in conjunction with other data such as charcoal and pollen distributions, about historical climatic conditions and vegetation. Moreover, because trace gas emissions from fires also are sensitive to combustion conditions, PAH profiles could possibly be used to help estimate the impact of past fires on atmospheric composition.

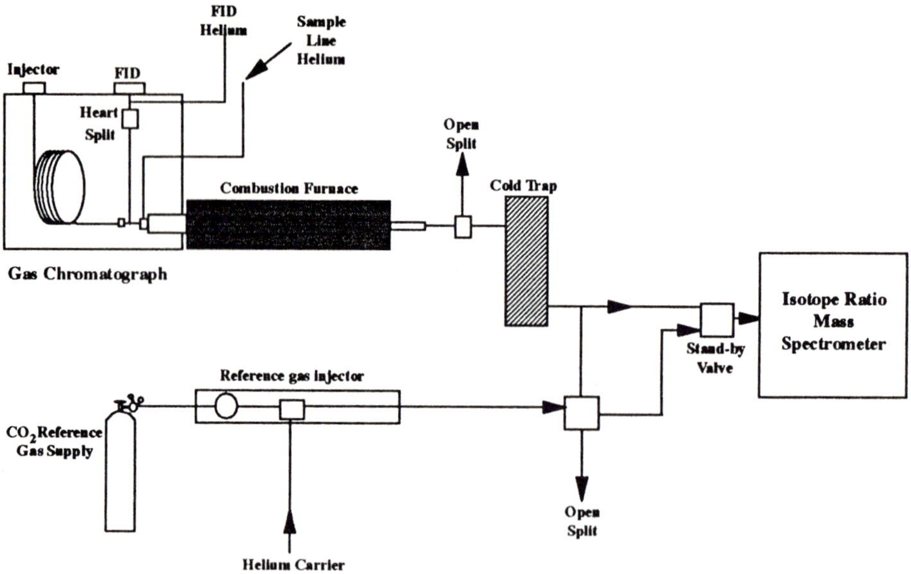

Figure 5. Schematic of the combined gas chromatograph/stable isotope ratio mass spectrometer (GC/C/IRMS) used to determine stable isotope ratios for individual PAH in a mixture (*Source:* O'Malley et al. 1996b).

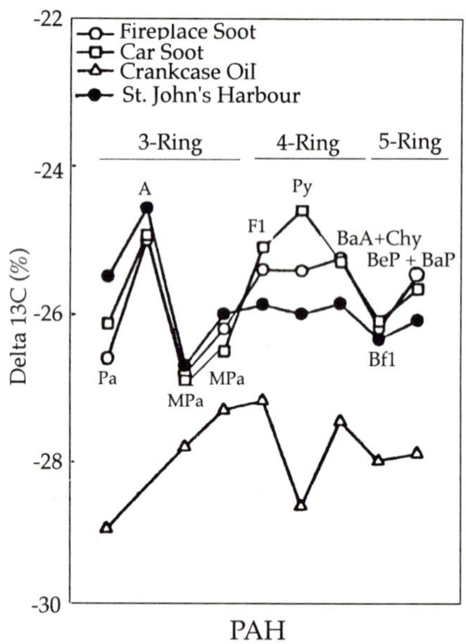

Figure 6. Comparison of isotopic signatures of PAH derived from carsoot and fireplace soot samples with those of PAH isolated from St. John's Harbor, Newfoundland and Conception Bay, Newfoundland (*Source*: O'Malley et al. 1996a).

Acknowledgments

We wish to thank V. O'Malley and R. Burke for useful discussions of isotope ratio techniques and the use of PAH as combustion indicators. We also acknowledge helpful information concerning PAH emissions from biomass burning that we obtained from M. Snook, W. Schlotzhauer, and D. Ward. Finally, we thank T. Kuhlbusch and P. Crutzen for providing pre-publication copies of their paper concerning black carbon in the environment and T. Kuhlbusch and V. O'Malley for their comments on the manuscript.

Literature References

Anderson RS, Davis RB, Miller NG, Stuckenrath R (1986) History of late- and post-glacial vegetation and disturbance around Upper South Branch Pond, northern Maine. *Can J Bot* 64:1977-1986

Andreae MO (1991) Biomass burning: Its history, use, and distribution and its impact on environmental quality and global climate. In Levine JS (ed) *Global Biomass Burning* pp 3-21 MIT Press Cambridge

Atkinson R, Arcy J, Zielinska B, Aschmann SM (1990) Kinetics and nitrarene product yields from the gas phas reactions of naphthalene, fluoranthene, and pyrene with N_2O_5 and OH radicals. *Atmos Environ* 22:999-1010

Backus DA, Gschwend PM (1990) Fluorescent polycyclic aromatic hydrocarbons as probes for studying the impact of colloids on pollutant transport in groundwater. *Environ Sci Technol* 24:1214-1223

Bauer JE, Capone DG (1985) Degradation and mineralization of the polycyclic aromatic hydrocarbons anthracene and naphthalene in intertidal marine sediments. *Appl Environ Microbiol* 50:81-90

Behymer TD, Hites RA (1985) Photolysis of polycyclic aromatic hydrocarbons adsorbed on simulated atmospheric particulates. *Environ Sci Technol* 19:1004-1009

Behymer TD, Hites RA (1988) Photolysis of polycyclic aromatic hydrocarbons adsorbed on fly ash. *Environ Sci Technol* 20:1311-1319

Bidleman TF, Foreman WT (1988) Vapor-particle partitioning of semi-volatile organic compounds. In Hites RA, Eisenreich SJ (eds) *Sources and Fates of Aquatic Pollutants Advances in Chemistry Series* American Chemical Society Washington DC 216:27-56

Burns LA, Cline D M, Lassiter R R (1982) *Exposure analysis modeling system (EXAMS): User manual and system documentation.* EPA-600/3-82-023 US Environmental Protection Agency Athens

Cachier H, Liousse C, Buat-Menard P, Gaudichet A (1996) Particulate content of savanna fire emissions. *J Atmos Chem* in press

Cerniglia CE, Heitkamp MA (1989) Microbial degradation of polycyclic aromatic hydrocarbons (PAH) in the aquatic environment. In Varanassi U (ed) *Metabolism of Polycyclic Aromatic Hydrocarbons in the Aquatic Environment* pp 41-68 CRC Press Inc Boca Raton

Charles MJ, Hites RA (1987) Sediments as archives of environmental pollution trends. In Hites RA, Eisenreich SJ (eds) *Sources and Fates of Aquatic Pollutants Advances in Chemistry Series* American Chemical Society Washington DC 216:365-389

Clark JS (1990) Fire and climate change during the last 750 years in northwestern Minnesota. *Ecol Mono* 60:135-159

Clark JS, Robinson J (1993) Paleoecology of fire. In Crutzen PJ, Goldammer JG (eds) *Fire in the Environment: The Ecological, Atmospheric, and Climatic Importance of Vegetation Fires* pp 193-214 Wiley Chichester

Chaloner, WG (1989) Fossil charcoal as an indicator of paleoatmospheric oxygen level. *J Geol Soc London* 146:171-174

Cofer WR III, Levine JS, Winstead EL, Stocks BJ (1990) Gaseous emissions from Canadian boreal forest fires. *Atmos Environ* 24A:1653-1659

Crutzen PJ, Andreae MO (1990) Biomass burning in the tropics: Impact on atmospheric chemistry and biogeochemical cycles. *Science* 250:1669-1678

Crutzen PJ, Carmichael GR (1993) Modeling the influence of fires on atmospheric chemistry. In Crutzen PJ, Goldammer JG (eds) *Fire in the Environment: The Ecological, Atmospheric, and Climatic Importance of Vegetation Fires* pp 89-106 Wiley Chichester

Daisey JM, Lewandow CG, Zorz M (1982) A photoreactor for investigations of the degradation of particle-bound polycyclic aromatic hydrocarbons under simulated atmospheric conditions. *Environ Sci Technol* 16:857-861

Eisenreich SJ, Baker JE, Franz T, Swanson M, Rapaport RA, Strachan WMJ, Hites RA (1992) Atmospheric deposition of hydrophobic organic contaminants to the Laurentian Great Lakes. In Schnoor JL (ed) *Fate of Pesticides and Chemicals in the Environment* pp 51-78 Wiley-Interscience New York

Gagosian RB, Zafiriou OC, Oektzer ET, Alford JB (1982) Lipids in aerosols from the tropical North Pacific: Temporal variability. *J Geophys Res* 87:11133-11144

Garstang M, Tyson P, Swap R, Ulanski S, Kallberg P, Macko S, Cachier H Atmospheric transport of particulate and gaseous aerosols (this volume)

Goldammer JG (1991) Tropical wild-land fires and global changes: Prehistoric evidence, present fire regimes, and future trends. In Levine JS (ed) *Global Biomass Burning* pp 83-91 MIT Press Cambridge

Goldberg ED (1985) *Black Carbon in the Environment.* Wiley New York

Griffin JJ, Goldberg ED (1981) Sphericity as a characteristic of solids from fossil fuel burning in a Lake Michigan sediment. *Geochim Cosmochim* Acta 45:763-769

Grimwalt J, Marfil C, Ablaiges J (1984) Analysis of hydrocarbons in aquatic sediments. I. Sampling, handling, and extraction. *Int J Anal Chem* 18:183-194

Gschwend PM, Hites RA (1981) Fluxes of polycyclic aromatic hydrocarbons to marine and lacustrine sediments in the northeastern United States. *Geochim Cosmochim* Acta 45:2359-2367

Halaby GA, Fagerson IS (1971) Polynuclear aromatic hydrocarbons in heat-treated foods. Pyrolysis of some lipids, ß-carotene, and cholesterol. In *Proceedings Third International Congress of Food Sciences and Technology* pp 820-829

Hites RA, LaFlamme RE, Windsor JG Jr, Farrington JW, Deuser WG (1980) *Geochim Cosmochim* Acta 44:873-878

Howard JB, Longwell JP (1983) Formation mechanisms of PAH and soot in flames. In Looke M, Dennis AJ (eds) *Polynuclear Aromatic Hydrocarbons: Formation, Metabolism, and Movement* pp 27-62 Battelle Press Columbus

Imboden DM, Schwarzenbach RP (1985) Spatial and temporal distribution of chemical substances in lakes: modeling concepts. In Stumm W (ed) *Chemical Processes in Lakes* pp 1-28 Wiley-Interscience New York

Junge CE (1977) Basic considerations about trace constituents in the atmosphere as related to the fate of global pollutants. In Suffet IH (ed) *Fate of Pollutants in the Air and Water Environment Part 1* 8:7-25 Wiley Interscience New York

Kamens R, Odum J, Fan Z-H (1995) Some observations on times to equilibrium for semivolatile polycyclic aromatic hydrocarbons. *Environ Sci Technol* 29:43-50

Karickhoff SW (1984) Organic pollutant sorption in aquatic systems. *J Hydraul Div* (ASCE) 110:707-735

Kasischke ES, Christensen NL Jr, Stocks BJ (1995) Fire, global warming, and the carbon balance of boreal forests. *Ecol Appls* in press

Killops SD, Massoud MS (1992) Polycyclic aromatic hydrocarbons of pyrolytic origin in ancient sediments: evidence for Jurassic vegetation fires. *Org Geochem* 18:1-7

Kuhlbusch TAJ, Crutzen PJ (1995) Toward a global estimate of black carbon in residues of vegetation fires representing a sink of atmospheric CO_2 and source of O_2. *Global Biogeochem Cycles* 9:491-501

Laflamme RE, Hites RA (1978) The global distribution of polycyclic aromatic hydrocarbons in recent sediments. *Geochim Cosmochim* Acta 42:289-303

Landrum PR, Bartell SM, Giesy JP, Leversee GJ, Bowling JW, Haddock J, LaGory K, Gerould S, Bruno M (1984) Fate of anthracene in an artificial stream: A case study. *Ecotoxicol Environ Saf* 8:183

Lee RF, Ryan C (1983) Microbial and photochemical degradation of polycyclic aromatic hydrocarbons in estuarine waters and sediments. *Can J Fish Aquatic Sci* 40:86-94

Liss P, Merlivat L (1986) Air-sea gas exchange rates: introduction and synthesis. In Buat-Menard P (ed) *The Role of Air-Sea Exchange in Geochemical Cycling*. NATO ASI Series C pp 113-127 D Reidel Publishing Co Dordrecht The Netherlands

Lim B, Renberg I Lake sediment records of fossil-fuel derived carbonaceous aerosols from combustion (this volume)

Lobert JM, Warnatz J (1993) Emissions for the combustion process in vegetation. In Crutzen PJ, Goldammer JG (eds) *Fire in the Environment: The Ecological, Atmospheric, and Climatic Importance of Vegetation Fires* pp 15-38 Wiley Chichester

Mackay D, Yuen ATK (1983) Mass transfer coefficient correlations for volatilization of organic solutes from water. *Environ Sci Technol* 17:211-217

Macko S Chemical and isotopic characterization of biomass burning materials (this volume)

Masclet P, Cachier H, Liousse C, Wortham H (1995) Emissions of polycyclic aromatic hydrocarbons by savanna fires. *J Atmos Chem* in press

Mazurek MA, Cofer III WR, Levine JS (1991) Carbonaceous aerosols from prescribed burning of a boreal forest ecosystem. In Levine JS (ed) *Global Biomass Burning* pp 258-263 MIT Press Cambridge

McElroy AE, Farrington JW, Teal JM Bioavailability of polycyclic aromatic hydrocarbons in the aquatic environment. In Varanasi V (ed) *Metabolism of Polycyclic Aromatic Hydrocarbons* pp 1-33 CRC Press Boca Raton

McMahon CK, Tsoukalas SN (1978) Polynuclear aromatic hydrocarbons in forest fire smoke. In Jones PW, Freudenthal RI (eds) *Carcinogenesis Vol 3: Polynuclear Aromatic Hydrocarbons* pp 61-73 Raven Press New York

McVeety BD, Hites RA (1988) Atmospheric deposition of polycyclic aromatic hydrocarbons to water surfaces: A mass balance approach. *Atmos Environ* 22:511-536

Mill T, Mabey WR, Lan BY, Baraze A (1981) Photolysis of polycyclic aromatic hydrocarbons in water. *Chemosphere* 10:1281-1290

Neff JM (1979) *Polycyclic Aromatic Hydrocarbons in the Aquatic Environment: Sources, Fates, and Biological Effects.* Applied Science London

Odum JR, McDow SR, Kamens RM (1994) Mechanistic and kinetic studies of the photodegradation of benz[a]anthracene in the presence of methoxyphenols. *Environ Sci Technol* 28:1285-1290

O'Malley VP, Abrajano TA Jr, Hellou J (1993) A novel approach to apportioning polycyclic aromatic hydrocarbon sources in estuarine environments. Joint CSCE-ASCE National Conference on Environmental Engineering July 1993

O'Malley VP (1994) *Compound-specific carbon isotope geochemistry of polycyclic aromatic hydrocarbons in Eastern Newfoundland estuaries.* PhD Thesis Memorial Univ of Newfoundland St. John's Newfoundland

O'Malley VP, Abrajano Jr TA, Hellou J (1994) Determination of the $^{13}C/^{12}C$ ratios of individual PAH from environmental samples: Can PAH sources be apportioned? *Org Geochem* 21:809-822

O'Malley VP, Abrajano Jr TA, Hellou J (1996a) Stable carbon isotopic apportionment of individual polycyclic aromatic hydrocarbons in St. John's Harbour, Newfoundland. *Environ Sci Technol* 30 in press

O'Malley VP, Stark A, Abrajano Jr TA, Hellou J, Windsor L (1996b) *Int J Polycyc Arom Comp* in press

Pankow JF (1987) Review and comparative analysis of the theories on partitioning between the gas and aerosol particulate phases in the atmosphere. *Atmos Environ* 21: 2275-2283

Patterson EM, McMahan CK, Ward DE (1986) Absorption properties and graphitic carbon emission factors of forest fire aerosols. *Geophys Res Lett* 13:129-132

Payette S (1992) Fire as a controlling process in the North American boreal forest. In Shugart HH, Leemans R, Bonan GB (eds) *A Systems Analysis of the Global Boreal Forest* pp 144-169 Cambridge University Press New York

Payette S, Gagnon R (1985) Late Halocene deforestation and tree regeneration in the forest-tundra of Québec. *Nature* 313:570-572

Readman JW, Mantoura RFC, Rhead MM (1984) The physico-chemical speciation of polycyclic aromatic hydrocarbons (PAH) in aquatic systems. *Fresenius Z Anal Chem* 319:126-131

Ramdahl T (1983) Retene, a molecular marker of wood combustion in ambient air. *Nature* 306:580-582

Robinson JM (1989) Phanerozoic O_2 variation, fire, and terrestrial ecology. *Global and Planetary Change* 1:223-240

Schlautman MA, Morgan JJ (1993) Effects of aqueous chemistry on the binding of polycyclic aromatic hydrocarbons by dissolved humic materials. *Environ Sci Technol* 27:961-969

Simoneit BRT, Chester R, Eglinton G (1977) Biogenic lipids in particles from the lower atmosphere over the eastern Atlantic, *Nature* 267:682-685

Simonich SL, Hites RA (1994a) Importance of vegetation in removing polycyclic aromatic hydrocarbons from the atmosphere. *Nature* 370:49-51

Simonich SL, Hites RA (1994b) Vegetation-atmosphere partitioning of polycyclic aromatic hydrocarbons. *Environ Sci Technol* 28:939-943

Simonich SL, Hites RA (1995) Organic pollutant accumulation in vegetation. *Environ Sci Technol* 29:2905-2914

Smith DM, Griffin JJ, Goldberg ED (1973) Elemental carbon in marine sediments: a baseline for burning. *Nature* 241:268-270

Smith TM, Shugart HH (1993) The transient response of carbon storage to a perturbed climate. *Nature* 361:523-526

Venkatesan MI, Dahl J (1989) Organic geochemical evidence for global fires at the Cretaceous/Tertiary boundary. *Nature* 338:57-60

Wakeham SG, Schaffner C, Giger W (1980) Polycyclic aromatic hydrocarbons in recent lake sediments-I. Compounds having anthropogenic origins. *Geochim Cosmochim* Acta 44:403-413

Wanninkhof R (1992) Relationship between wind-speed and gas-exchange over the ocean. *J Geophys Res* 97:7373-7382

Ward DE, Radke LF (1993) Emission measurements from vegetation fires: A comparative evaluation of methods and results. In Crutzen PJ, Goldammer JG (eds) Fire in the Environment: *The Ecological, Atmospheric, and Climatic Importance of Vegetation Fires* pp 53-76 Wiley, Chichester

Ward DE, Setzer AW, Kaufman YJ, Rasmussen RA (1991) Characteristics of smoke emissions from biomass fires of the Amazon region-BASE-A experiment. In Levine JS (ed) *Global Biomass Burning* pp 394-402 MIT Press Cambridge

Wehry EL, Mamantov G, (1993) Sorption and photochemical transformation of polycyclic aromatic compounds on coal stack ash particles. In Helz GR, Zepp RG, Crosby DG (eds) *Aquatic and Surface Photochemistry* pp 173-180 Lewis Publishers Boca Raton

Westall J (1987) Adsorption mechanisms in aquatic surface chemistry. In Stumm W (ed) *Aquatic Surface Chemistry: Chemical Processes at the Particle-Water Interface* Ch 1 pp 3-46 Wiley-Interscience New York

Whitlock C, Millspaugh S, Long C, Bradbury JP, Dean WE Representation of fire lake sediments from particles (this volume)

Wilber GG, Smith L, Malanchuk JL (1992) Emissions inventory of heavy metals and hydrophobic organics in the Great Lakes basin. In Schnoor JL (ed) *Fate of Pesticides and Chemicals in the Environment* pp 27-50 Wiley-Interscience New York

Yamasaki H, Kuwata K, Miyamoto H (1982) Effects of ambient temperature on aspects of airborne PAH. *Environ Sci Technol* 16:189-194

Youngblood WW, Blumer M (1975) Polycyclic aromatic hydrocarbons in the environment: Homologous series in soils and recent marine sediments. *Geochim Cosmochim* Acta 39:1303-1314

Zepp RG (1980) Assessing the photochemistry of organic pollutants in aquatic environments. In Haque R (ed) *Dynamics, Exposure, and Hazard Assessment* pp 69-110 Ann Arbor Science Ann Arbor

Zepp RG, Schlotzhauer PF (1979) Photoreactivity of selected aromatic hydrocarbons in water. In Jones PW, Leber P (eds) *Polynuclear Aromatic Hydrocarbons* pp 141-152 Ann Arbor Science Ann Arbor

Zepp RG, Wolfe NL (1987) Abiotic transformations of organic substances at the particle-water interface. In Stumm W (ed) *Aquatic Surface Chemistry: Chemical Processes at the Particle-Water Interface* pp 423-455 Wiley-Interscience New York

Part III:

Particulate emissions and transport to sediments

Biomass Consumption and Behavior of Wildland Fires in Boreal, Temperate, and Tropical Ecosystems: Parameters Necessary to Interpret Historic Fire Regimes and Future Fire Scenarios

B.J. Stocks and J.B. Kauffman[1]
Forest Fire Research
Canadian Forest Service - Ontario Region
Sault Ste. Marie, Ontario, Canada

Summary

The reconstruction and interpretation of historical fire regimes requires an understanding of fuel loads, fuel consumption, and fire behavior, and the effects these processes have on the nature and fate of combustion products. These parameters, as well as the areal extent and frequency of fire, vary between ecosystems. To provide that insight this chapter summarizes fuel consumption and fire behavior data from a variety of global ecosystems, including high-intensity wildfires, experimental fires, and prescribed fires in boreal, temperate, and tropical ecosystems.

While vast areas of tropical savanna burn with high combustion factors annually, fuel consumed per unit area is small, normally <1.0 kg m^{-2}. In contrast, fuel consumption in stand-replacing forest fires in the boreal and temperate zones averages 1.8 to 3.4 kg m^{-2}, while average fuel consumption in forest understory fires varies widely, from 0.6 to 8.8 kg m^{-2}. Land use activities associated with land conversion or timber harvest dramatically influence biomass consumption and hence charcoal production. For example, fires in temperate forest logging slash can have fuel consumption levels as high as 14 kg m^{-2}. The probability of natural fires in intact Amazon rainforest is negligible because precipitation patterns maintain fuel moisture conditions above combustion thresholds. However, modification of microclimate and fuel arrangement through forest cutting can result in slash fires consuming up to 23 kg m^{-2} of forest fuel.

We conclude that high-intensity wildfires are likely responsible for the majority of lake sediment charcoal found in boreal and temperate forest lake sediments. In the tropics soil and lake charcoal is likely the result of fires from climatic regimes that differ from today. Savanna fires have contributed to the global charcoal pool for millennia. Significant and intensified anthropogenic land use changes have dramatically altered natural fire patterns and hence charcoal deposition. Areas such as tropical forests and savannas burn much more frequently, while fire suppression has changed the pattern and behavior of fire in temperate forests and, to a lesser extent, in boreal forests.

Introduction

Wildland fire behavior and biomass consumption is highly variable both within and between ecosystems. This variability is primarily controlled by weather, fuel conditions, and the composition and structure of the ecosystem. Weather conditions during fires directly affect fire behavior. Antecedent weather has an effect through its influence on fuel moisture conditions, and therefore the

---

[1]Fisheries and Wildlife Department, Oregon State University, Corvallis, OR USA

NATO ASI Series, Vol. I 51
Sediment Records of Biomass Burning and Global Change
Edited by James S. Clark, Hélène Cachier, Johann G. Goldammer, and Brian Stocks

amount of fuel available for combustion, as well as the products of combustion. In addition to fuel moisture, other critical fuel properties that influence biomass consumption and fire behavior are fuel chemistry, fuel loading or quantity, fuel arrangement, fuel particle size distribution, and the ratio of air volume to fuel volume within the ecosystem (i.e., the packing ratio).

The amount of fuel consumed during the flaming stage of a wildland fire contributes directly to its overall intensity or energy-release-rate, thereby affecting convection column development and dynamics, which in turn determines the distribution and fate of combustion products such as aerosols and trace gases. Where possible in this chapter, the range of frontal fire intensities encountered in various ecosystems is also presented, in order to provide an indication of smoke and convection column dynamics. The amount and character of emissions from wildland fires can differ greatly depending on the combustion phase (flaming or smoldering) in which they are produced (e.g., Cofer et al. 1990, 1993), making the "rate" of fuel consumption an important parameter. However, almost all fuel consumption measurements made to date on wildland fires do not differentiate between combustion phases, relying instead on preburn and postburn sampling of all fuel strata (aerial, surface, and ground) to determine total fuel consumption only. However, it is clear that the majority of fuels in ecosystems with predominantly fine fuels (e.g., grasslands or savannas) burn during flaming combustion. In contrast, larger quantities of biomass are consumed during the smoldering phase of combustion in logging slash fires.

In this chapter we present examples of fuel loads and consumption during free-burning fires in boreal, temperate, and tropical ecosystems throughout the world. Additional data on logging slash burning in North America and rainforest slash-and-burn agricultural practices are also presented. Most of the fuel consumption data come from experimental fires conducted for research purposes. These fires, although generally smaller than wildfires, provide the opportunity for detailed preburn and postburn fuel sampling, and the thorough monitoring and documentation of fire behavior. In addition, they are good representations of fire behavior and biomass consumption of the ecosystems in which they burned.

Boreal and Temperate Ecosystems

Crown Fires

Crown fires are defined as fires that advance through the tree crown layer, spreading in conjunction with, and closely linked to, a surface fire (Merrill and Alexander 1987). A large dataset is available on fuel consumption in boreal and temperate ecosystem crown fires from a long-term experimental burning program in Canada designed to gather baseline data for fire behavior prediction purposes. The product of this program is the Canadian Forest Fire Behavior Prediction (FBP) System (Forestry Canada Fire Danger Group 1992), which quantitatively predicts a number of fire behavior parameters (including fuel consumption) for 16 major Canadian fuel types. Fuel consumption data from Canadian experimental crown fires in selected fuel types, given as ovendry weights (ODW), is presented in Table 1. Corresponding values for frontal fire intensity, or the rate of energy release along the flame front, are also included. Fireline intensity (I) is expressed in kW m^{-1}, and is calculated using the formula $I=Hwr$ (Byram 1959), where H is the fuel low heat of combustion (kJ kg^{-1}), w is the weight of fuel consumed (kg m^{-2}), and r is the linear rate of fire spread (m s^{-1}). Forest floor fuels consist of surface litter and organic horizons. Surface fuels comprise downed woody material and understory vegetation, and crown fuels consist of needles, lichens, and fine branches. Combustion factors (i.e. the percent of the total amount of fuel consumed) are also presented where available. More detailed information on these experimental fires can be found in: spruce-lichen woodland crown fires (Alexander et al. 1991), boreal spruce crown fires (Kiil 1975, M.E. Alexander (personal communication), Newstead and Alexander 1983), jack pine crown fires (Quintilio et al. 1977, Stocks 1987a, 1989), and budworm-killed balsam fir crown fires (Stocks 1987b).

The experimental crown fires summarized in Table 1 were deliberately conducted under a broad range of fire weather and fire danger conditions, and therefore covered a significant range in frontal fire intensity levels. Intermittent crown (torching) fires are possible when surface fires become highly vigorous and intensity levels exceed 2,000 kW m^{-1}. Allowing for some variability between fuel types, continuous crown fires usually develop when frontal fire intensities approach 8-10,000 kW m^{-1}. Total fuel loads and consumption were lowest in the spruce-lichen woodland as tree cover is sparse, and the forest floor is shallow. They were heaviest in budworm-killed balsam fir, where large quantities of dead material resulted in increases of both surface and aerial fuel loads and consumption. Throughout much of the boreal zone, however, where jack pine and black spruce dominate, crown fires generally result in fuel consumption levels of roughly 2.5 kg m^{-2}, with forest floor and crown fuels (needles and fine twigs) accounting for approximately 80% of that figure. Combustion factors represent the percentage of available fuel in each strata consumed during the fire. In the case of ground and surface fuels all fuel was considered to be available for consumption. In crown fires only needles and fine twigs within the tree crowns are consumed, and the combustion factors for crown fuels therefore express the percentage consumption of these fine fuels. Heavy branches and tree boles were not considered relevant and therefore were not measured.

Table 1. Average forest floor[1], surface fuel[2], crown fuel[3], and total fuel[4] consumption (ODW in kg/m^2), and associated rate of spread[5] (m min^{-1}) and frontal fire intensity ranges (kW m^{-1}), for experimental crown fires in various Canadian boreal fuel types. Numbers are the mean[a], standard deviation[b], and combustion factor[c] (the % of the total amount of fuel consumed).

Experimental Fire	FFC[1]	SFC[2]	CFC[3]	TFC[4]	RoS[5]	Intensity
Spruce-Lichen Woodland (n=6)	1.04[a] ± 0.10[b] 59-74%[c]	0.27[a] ± 0.13[b] 24-100%[c]	0.52[a] ± 0.17[b] 41-82%[c]	1.83[a] ± 0.22[b] 57-76%[c]	3.5-51.4	2,131-33,153
Boreal Spruce (n=9)	1.33[a] ± 0.44[b]	0.35[a] ± 0.15[b]	0.88[a] ± 0.36[b]	2.56[a] ± 0.59[b]	6.4-30.0	4,198-15,907
Jack Pine (n=15)	0.95[a] ± 0.39[b] 37-100%[c]	0.55[a] ± 0.36[b] 14-53%[c]	0.94[a] ± 0.29[b] 30-100%[c]	2.44[a] ± 0.76[b] 36-71%[c]	2.1-49.4	3,264-43,274
Dead Balsam Fir (n=6)	1.23[a] ± 0.54[b] 6-26%[c]	0.84[a] ± 0.34[b] 7-25%[c]	1.31[a] ± 0.36[b] 58-100%[c]	3.38[a] ±0.66[b] 19-33%[c]	11.5-82.0	11,730-68,880

For comparison purposes Table 2 presents fuel consumption, spread rate, and intensity data for six well-documented, high-intensity boreal/temperate wildfires in Canada, the United States, and Australia: the Mack Lake Fire in Michigan (Simard et al. 1980), the Bomb Range Fire in North Carolina (Wade and Ward 1973), the Sundance Fire in Idaho (Anderson 1968), Red Lake Fires # 31/74 and #7/86 in northwestern Ontario (Stocks 1975, Stocks and Flannigan 1987), the Gwatkin Lake Fire in eastern Ontario (Van Wagner 1965), the Bemm River Fire in Victoria, Australia (Buckley 1992), and the Manjimup Fire in Western Australia (McCaw et al. 1992). A typical high-intensity boreal forest crown fire is shown in Fig. 1.

Table 2. Total fuel consumption[1] (ODW in kg m^{-2}), forward rate of spread[2] (in m min^{-1}), and frontal fire intensity (in kW m^{-1}) data for eight selected boreal and temperate wildfires in Canada, the United States, and Australia.

Wildfire	TFC[1]	RoS[2]	Intensity
Mack Lake (1980 - Michigan)	2.34	56.3	39,523
Red Lake 31 (1974 - Ontario)	2.10	44.4	27,972
Red Lake 7 (1986 - Ontario)	5.00	36.7	55,050
Gwatkin Lake (1964 - Ontario)	2.44	24.4	17,858
Bomb Range Fire (1978 - North Carolina)	2.00	53.6	32,160
Sundance Fire (1967 - Idaho)	1.08	161.0	52,164
Bemm River Fire (1988 - Australia)	2.00	48.3	28,980
Manjimup Fire (1991 - Australia)	2.00	30.0	18,000

Figure 1: Typical high-intensity, stand-replacement boreal forest fire, Ontario, Canada.

High-intensity wildfires in boreal and temperate ecosystems, whether occurring in jack pine/black spruce-dominated ecosystems in Canada, subalpine larch/fir/spruce/hemlock or pine/pocosin forests in the United States, or in Australian eucalyptus ecosystems, can exhibit similar fire behavior characteristics. A comparison of Tables 1 and 2 clearly shows that experimental crown fires, although much smaller in scale, effectively simulated large-scale wildfires. Similar overall fuel consumption, spread rate, and intensity levels were observed in both wildfires and experimental fires. However, large-scale wildfires maintain this level of fire behavior over much longer periods of time, generating towering convection columns and dispersing smoke and other combustion by-products over large geographical areas.

Surface Fires

A surface fire is defined as a fire that burns in the surface fuel layer, excluding the crowns of the trees, as either a head fire, flank fire, or backfire (Merrill and Alexander 1982). Data from low-intensity surface fires include experimental fires in jack and lodgepole pine (Lawson 1973, Quintilio et al. 1977, and Stocks 1989), and prescribed underburn fires in red and white pine (Van Wagner 1965), slash and longleaf pine (Nelson and Ward 1980, Ward 1983), ponderosa pine (Shea 1993), mixed conifer (Kauffman and Martin 1989), and Australian eucalypts (Burrows 1985) (Table 3).

Table 3. Total fuel consumption[1] (ODW in kg m^{-2}), forward rate of spread[2] (m min^{-1}), and frontal fire intensity (kW m^{-1}) for prescribed underburning fires in Canada, the United States, and Australia. Numbers are the mean[a] and the range[b].

Surface Fires	TFC[1]	RoS[2]	Intensity
Jack/Lodgepole Pine (n=25)	1.01[a] 0.28-2.39[b]	1.40[a] 0.50-3.60[b]	62-1350[b]
Red & White Pine (n=10)	1.79[a] 0.40-3.07[b]	0.88[a] 0.50-1.60[b]	89-1474[b]
Slash/Longleaf Pine (n=17)	0.93[a] 0.25-2.22[b]	0.36[a] 0.12-0.89[b]	19-296[b] (backfires)
Slash/Longleaf Pine (n=8)	0.60[a] 0.10-0.86[b]	5.71[a] 1.64-9.14[b]	36-1724[b] (headfires)
Ponderosa Pine (n=6)	5.77[a] 3.60-7.90[b]	2.00[a] 1.30-3.07[b]	48-684[b]
Ponderosa Pine - Mixed Conifer (n=60)	8.80[a] 0.25-14.80[b]	1.3[a] 0.20-2.40[b]	16-273[b]
Australian Eucalyptus (n=18)	0.88[a] 0.42-14.0[b]	1.02[a] 0.20-3.68[b]	52-1497[b]

The understory surface fires represented in Table 3 generally exhibit lower total fuel consumption (~1 kg m^{-2}) than high-intensity crown fires. However, In some long-needled pine ecosystems (e.g. Ponderosa pine), heavy litter layers resulted in substantial fuel consumption (as high as 8-9 kg m^{-2}) and therefore charcoal production. Spread rates (and resulting frontal fire intensities) for surface fires in boreal and temperate forests are much lower than crown fires.

This is primarily because the in-stand, ground-level windspeeds affecting surface fires are much lower than the ambient above-crown winds that drive crowning fires. The resulting slower spread rates and lower intensity levels create smoke plumes that are generally weak and undefined. As a result, most of the charcoal particles produced by surface fires stay on-site rather than being transported long distances by convective processes.

Shrubland Fires

Quantitative data on shrubland fires are scarce, but Table 4 summarizes results from experimental fires conducted in sagebrush ecosystems in Oregon (Sapsis and Kauffman 1991, Kauffman and Cummings 1989, Young and Miller 1985), and a series of experimental fires in South African fynbos vegetation (van Wilgen et al. 1985).

Table 4. Total fuel consumption[1] (ODW in kg m^{-2}), forward rate of spread[2] (m min^{-1}), and frontal fire intensity data (kW m^{-1}) data for shrubland fires in the United States and South Africa. Numbers are the mean[a], the standard deviation[b], and the range[c].

Shrubland Fires	TFC[1]	RoS[2]	Intensity
Basin Big Sagebrush spring fires (n=5)	0.52[a] ± 0.09[b]	13.80[a] ± 1.80[b]	346[a] ± 42[b]
Basin Big Sagebrush fall fires (n=4)	0.98[a] ± 0.13[b]	94.20[a] ± 17.40[b]	591[a] ± 169[b]
Basin Big Sagebrush (n=5)	1.1-1.8[c]	no data	23,028-29,108[c]
Wyoming Big Sagebrush (n=?)	0.22[a] ± 0.11[b]	15-40[c]	1,046-2,790[c]
South African Fynbos (n=14)	1.292[a] ± 0.422[b]	28.16[a] ± 13.60[b]	12,128-36,032[c]

A large proportion of the fuels in shrubland ecosystems such as sagebrush, fynbos, or chaparral, are highly flammable fine fuels. Total fuel consumption in these fuel types was similar to understory surface fires, averaging ~1 kg m^{-2}. However, access to the ambient windfield results in very fast-moving shrubland fires that can generate high frontal fire intensity levels, even though fuel consumption levels are modest. Smoke/convection columns on this type of fire, although not as well-developed as on crowning fires in boreal/temperate ecosystems, can still be quite substantial.

Logging Slash Fires

Prescribed fires in post-harvest logging slash are routinely conducted by fire management agencies in Canada and the United States to prepare the site for regeneration purposes, as the removal of woody fuels and organic material through burning facilitates more efficient planting or seeding. Early in the program to develop quantitative fire behavior prediction guidelines for Canadian fire management agencies, a number of experimental fires were conducted in major logging slash fuel complexes. Unlike many prescribed burns conducted today for site preparation purposes, where ignition patterns are used to control fire behavior, these experimental fires were ignited as headfires, in order to simulate wildfire conditions. Table 5 summarizes data from three slash fuel types that are currently represented in the Canadian FBP System: jack/lodgepole pine slash in Ontario and Alberta (Stocks and Walker 1972, Quintilio 1972), and spruce/balsam (Lawson and Taylor 1986) and coastal cedar/hemlock/Douglas fir slash (Muraro 1971, Lafferty 1972) in British Columbia. Additional data from slash fires in Oregon and Washington hemlock/Douglas fir (Little et al. 1986), Oregon western juniper (Kauffman and Cummings 1989), and British Columbia coastal hemlock/sub-boreal spruce (Feller 1989) are also presented.

Table 5. Forest floor consumption[1], woody fuel consumption[2], and total fuel[3] consumption (ODW in kg m^{-2}) for logging slash fires in Canada and the United States. Numbers are the mean[a], the standard deviation or error[b], and the range[c].

Logging Slash Fires	FFC[1]	WFC[2]	TFC[3]
Jack/Lodgepole Pine (n=44)	2.18[a] ± 1.09[b]	.71[a] ± 1.56[b]	4.89[a] ± 2.19[b]
Spruce/Balsam (n=46)	6.53[a] ± 1.81[b]	5.92[a] ± 2.22[b]	12.44[a] ± 2.78[b]
Hemlock/Douglas fir (n=15)	1.1-7.1[c]	1.3-7.7[c]	1.5-12.2[c]
Western Juniper (n=5)	0.08-0.82[c]	0.75-2.51[c]	0.91-3.46[c]
Coastal Hemlock/Boreal Spruce (n=24)	no data	0.6-12.1[c]	0.7-17.3[c]
Coastal Cedar/Hemlock/ Douglas fir (n=8)	6.06[a] ± 3.31[b]	8.18[a] ± 4.03[b]	14.24[a] ± 5.97[b]

Logging slash fires are characterized by extremely high fuel consumption, ranging from 1-17 kg m^{-2}. Large quantities of dead fuel are distributed close to the ground in these fuel complexes, permitting a great deal of residual burning after flame front passage, resulting in higher levels of fuel consumption than wildfires in similar forest ecosystems. Forest floor and woody fuels generally contribute roughly equally to total fuel consumption amounts. Well-developed convection columns are common on logging slash fires, as ignition patterns used for control purposes often create strong indrafting.

Tropical Ecosystems

Evergreen Moist and Rain Forests

Given the importance of deforestation and fire in tropical evergreen forests, it is remarkable how little data exist concerning fuel loads, combustion factors, and fire behavior in these ecosystems. While little is known on biomass and fire behavior in slashed primary forests, even less is known about these parameters for slashed second-growth (or third-growth) forests and cattle pastures. Knowledge of these parameters is increasingly important, because in many areas of the tropics these are the primary land-use scenarios associated with biomass burning. Relevant data collected recently in Brazil (Kauffman et al. 1995, Hughes (in prep.), Fearnside et al. 1993, Kauffman et al. (in prep.-a), Kauffman et al. (in prep.-b), Kauffman et al. 1993) are summarized in Table 6.

Table 6. Ranges in fuel loads (ODW in kg m^{-2} , total fuel consumption (ODW in kg m^{--2}) and combustion factors (%) for Brazilian slashed tropical forests and cattle pastures converted from evergreen forest.

Site	Total Fuel Load	TFC	CF
Primary forests (Para, Brazil) (n=2)	29.2 - 43.5	15.2 - 22.8	52 - 52
Primary forests (Rondonia, Brazil) (n=3)	29.0 - 36.1	11.6 - 20.6	38 - 57
Primary forest (Amazonas, Brazil) (n=1)	26.5	7.7	29
Second-growth forests (Brazil) (n=3)	10.3 - 17.7	5.3 - 9.3	43 - 63
Third-growth forests (Brazil) (n=3)	6.4 - 12.0	4.5 - 5.8	60 - 88
Cattle pastures (Brazil) (n=4)	5.3 - 9.9	2.8 - 4.5	40 - 81
Tropical dry forest (Brazil) (n=3)	7.4 - 7.4	5.7 - 7.0	78 - 95

Total aboveground biomass of primary evergreen forests in the Brazilian Amazon ranges from 16.6 to 43.5 Kg m^{-2} (Brown and Lugo 1992, Fearnside 1992a, Kauffman et al. 1995). Total fuel loads of slashed primary forests in Amazonian regions undergoing rapid deforestation ranged from 29.2 to 43.5 kg m^{-2} Combustion factors for slash fires from these regions (Rondonia and Para States, Brazil) ranged from 38-57%. The quantity of biomass consumed during these fires ranged between 11.6 and 22.8 kg m^{-2}. Typical tropical slash fire behavior is shown in Fig. 2.

Figure 2: Slash fire in tropical dry forest, Northeastern Brazil.

Recent estimates of deforestation levels from remote sensing suggest that from 230,000 to 415,000 km^2 have been deforested in Brazil in the past few decades (Fearnside 1992b, Skole and Tucker 1993), which equates to ~8% of all Amazonian forests. Skole and Tucker (1993) also reported that the rate of deforestation averaged ~15,000 km^2 per year. From Table 6 we therefore estimate that 245 Tg of biomass is consumed annually in Brazil by primary slash burning alone (Kauffman et al. 1995b). As wildfires do not occur in this ecosystem under current climatic regimes, this represents a huge and unprecedented pulse of charcoal formation arising from this ecosystem.

Following deforestation and slash burning, land use normally centers around either shifting cultivation, pasture establishment, or some combination of the two. Given the nutrient-poor character of many Amazonian forests, shifting cultivation may only be feasible for two years. Kauffman and Cummings (in press) reported a land use scenario where active cultivation occurred for two years, followed by a 4-10 year fallow period. Biomass of second-growth Brazilian forests ranged from 10.3 to 17.7 kg m^{-2} (Table 6), with the majority of this biomass being composed of residual large woody debris carried-over from the primary forest. However, the fuel load of second-growth forests is characterized by a greater proportion of woody debris <20 cm in diameter than in primary forests (Kauffman et al. in prep.). Combustion factors of second-growth forest slash fires varied between 43 and 63%.

Land use following slashing and burning of second-growth forests is similar to that of recently burned primary forest sites; two years as shifting cultivation and then abandoned for ~4 years. Fuel loads for third growth forests ranged from 6.4-12.0 kg m^{-2} (Table 6). None of the woody fuels resulting from the regenerated forests exceeded 20 cm in diameter (Kauffman et al. in prep.). The relatively fine composition of the fuel load on these sites resulted in the highest combustion factors (60-88%) measured on any slash fires in the wet tropics.

Conversion of tropical forests to cattle pasture is the most prevalent land use in most areas of deforestation in the Brazilian Amazon. Fire is utilized as a tool for weed control and to maintain the productivity and palatability of the exotic pasture grasses (Uhl and Buschbacher 1985). Fires are normally set every two years in pastures, but accidental fires are also common. Total fuel loads in cattle pasture ranged from 5.3 to 9.9 kg m^{-2} (Table 6), with the vast majority of this being residual large woody debris from the primary forest (Kauffman et al. in prep.). Grasses are the most important fuel component with respect to fire propagation, but comprised <1.3 kg m^{-2} of the total fuel load. Combustion factors ranged from 40-88%. Higher combustion factors are related to older pastures containing low amounts of woody debris and/or pastures burned during exceptionally dry periods.

Globally, tropical dry forests are the most prevalent type of tropical forest, comprising ~42% of all areas occupied by tropical or subtropical forest (Murphy and Lugo 1986). In comparison to tropical evergreen forests, tropical dry forests are usually smaller in stature and diameter. Total aboveground biomass of this diverse forest type ranges from 6.8 to 27.5 kg m^{-2} (Murphy and Lugo 1986). Total fuel loads of slashed sites in the Brazilian Caatinga were ~ 7.4 kg m^{-2} (Table 6). As the climate of tropical dry forests are typified by a longer dry season, with lower precipitation and relative humidity levels, fuel moisture contents are usually quite low during slash burning. As a result, combustion factors in tropical dry forests are among the highest (78-95%) for slash burns in the world (Table 6). Kauffman et al. (1993) reported flame lengths of 8-10 m during tropical dry forest slash fires.

Tropical Savannas, Grasslands and Woodlands

Covering approximately 1530 million ha, tropical savannas, grasslands, and woodlands comprise an area greater than the area of tropical closed forest (1440 million ha) (Hao et al. 1990). Fuel load figures for savannas usually consist only of grasses and other associated fine fuels. Trees and palms present are rarely consumed as they are highly adapted to survive the frequent natural surface fires that are characteristic of this vegetation type. Fire return intervals are usually 1-5 years for tropical savannas. Savanna fuel loads globally range from 0.22 to 1.20 kg m^{-2} (Hao et al. 1990). Data collected recently from savanna fires in Brazil (Kauffman et al. 1994, de Castro 1995, Pivello and Coutinho 1992, and Kaufman et al. (in press)), and southern Africa savannas, woodlands, and grasslands (Shea et al. 1995, van Wilgen and Wills 1988, Everson et al. 1988) is summarized in Table 7.

Table 7. Ranges in fuel loads (ODW in kg m^{-2}), combustion factors (%), and frontal fire intensities (kW/m) for selected tropical savannas and woodlands.

Site	Fuel Load	CF	Fireline Intensity
Campo limpo, Brazil (n = 2)	0.55 - 0.71	92 -100	557 - 16,394
Campo sujo, Brazil (n = 2)	0.73 - 0.75	97 - 97	2,437
Campo cerrado, Brazil (n = 6)	0.49 - 0.77	88 - 94	no data
Campo cerrado, Brazil (n = 1)	0.86	72	2,842
Cerrado sensu stricto, Brazil (n = 5)	0.88 - 1.20	84 - 92	3,455 - 3,693
Lowveld savannas, South Africa (n = 10)	0.22 - 0.71	44 - 95	480 - 6,130
Natal savannas, South Africa (n=10)	0.16 - 1.02	no data	122 - 5,535
Dambo grasslands, Zambia (n = 1)	0.31	100	4,061
Natal grasslands, South Africa (n=16)	0.20 - 0.70	no data	1.203 - 17,437
Miombo woodland, Zambia (n = 3)	0.58 - 0.73	71-88	1,734 - 3,375

Hao et al. (1990) hypothesized that the total quantity of biomass consumed by fires in humid savannas was approximately ten times larger than that burned in closed nonfallow forests, even though the fuel loads are much lower in savannas. This is largely because Hao assumed that the area of savannas burned annually is 100 times greater than that of deforested tropical areas. Hao et al. (1990) assumed a mean combustion factor of 83%. From Table 7 it is evident that the grasslands of the Brazilian cerrado (campo limpo and campo sujo) have combustion factors ranging from 92-100%, while Brazilian woodland (campo cerrado and cerrado sensu stricto) combustion factors are slightly lower (72-94%) due to a higher proportion of less flammable fuels. Similar high levels of biomass consumption have been observed in the savannas of southern Africa. Grassland and savanna fires in the Natal Province of South Africa showed fuel consumption levels ranging from 0.16-1.02 kg m^{-2} (Everson et al. 1988, van Wilgen and Wills 1988). This is consistent with fuel consumption measurements by Shea et al. (1995), who also reported combustion factors for a variety of savannas and woodlands ranged from 44-100% (Table 7). Lower combustion factors were measured on sites with limited sustained ignition and spread due to the combination of heavy grazing and low productivity after a prolonged drought.

Because of the abundance of fine fuels and the long dry season typical of many tropical savannas, severe fire behavior, manifested in rapid spread rates, complete fuel consumption, and high fire intensity levels, is possible. The fireline intensity data available for the tropical savannas and woodlands shown in Table 7 ranges from 480-16,394 kW m^{-1}. Higher fire intensities were measured on productive sites with an abundance of fine fuels, with high winds or steep slopes, or during very dry weather conditions. Conversely, low fireline intensities were related to high fuel moisture contents or low levels of fuel biomass, high relative humidities, and windless, cloudy days.

Discussion

Crown fires account for a large proportion of the area burned in North American boreal forests, as fire-dependent species of pine and spruce predominate in this region and large expanses of contiguous even-aged forests promote the development of large fires. Fire return intervals in unprotected regions of the boreal zone vary from 50-150 years (Stocks et al. 1995). Under extreme fire weather conditions boreal fires can expand quickly, often growing by up to 50,000 ha in a single day. In the Canadian boreal zone fires >200 ha account for ~97% of the area burned annually (Stocks 1991). Although published information is lacking, it appears that crown fires account for less than 40% of the annual area burned in Russian boreal forests, perhaps due to forest stand structures that do not encourage crown fire development. In addition, large portions of the world's boreal zone remain uninfluenced by fire suppression activities, ensuring the development of large crown fires and continuing the natural, millennia-old role of fire in renewing and regenerating the fire-adapted boreal forest. Boreal crown fires demonstrate rapid spread rates and significant levels of fuel consumption in all fuel strata, developing towering convection columns that can transport combustion products long distances.

Crown fires in temperate forests are capable of exhibiting the same type of fire behavior as boreal crown fires, however crowning fires are less likely in temperate forests due to increased levels of protection and changes in forest composition. Surface fire is the dominant type of fire behavior in temperate forests, often accounting for ~75% of the area burned in these regions. Fire return intervals in temperate forests are highly variable, depending on the degree of human influence, and can range from 5-1000 years.

The surface fires summarized in this chapter are generally the result of understory prescribed burns carried out for fuel reduction purposes in pine and eucalyptus stands, removing forest floor and surface fuels in order to prevent fuel accumulation that could result in higher intensity wildfires. Such fires are routinely conducted under low to moderate fire danger conditions in order to remove understory fuels without damaging the overstory. Low in-stand windspeeds mitigate against low-intensity surface fires becoming crown fires, unless fire danger conditions become extreme. Such low-intensity fires spread slowly and generate weak and poorly-defined smoke plumes. Surface fires dominate in many of the temperate forests of the world. Many pine, eucalyptus, and oak forests evolved and developed with frequent natural understory fires that consumed surface fuels but resulted in low overstory mortality.

Shrubland fires in sagebrush and fynbos, despite modest fuel loadings in comparison to forests, exhibit fuel consumption levels similar to understory surface fires. This is related to the high ratio of consumption of aboveground biomass which is mostly composed of fine, readily-combustible fuels. Most fuel is consumed in the flaming phase, and high spread rates are common, resulting in well-defined convection column development. Large areas may be consumed in short periods of time. For example, 5000 ha day^{-1} is not an uncommon rate of fire growth in large sagebrush fires.

Slash fires, whether the result of logging operations in North America or deforestation/land conversion burning in tropical rainforest, exhibit high rates of fuel consumption, with more than 50% of the of this consumption occurring during the smoldering phase of combustion. Slash fires may smolder for more than a month following ignition. Tropical slash fires are normally ignited along the perimeter, creating high-intensity, convection-style fires that form large, well-defined convection columns. Ash biomass of these fires is 0.7-1.1 kg m^{-2} (Kauffman et al. 1995), indicating that large quantities of charcoal remain on site following fire.

Savannas and grasslands have much lower fuel loads than forest fuel complexes, but virtually all of this fuel is consumed during the flaming stage of combustion. High spread rates are common under strong winds, but low fuel consumption levels reduce overall energy release rates and convection columns are often weak and do not reach significant heights. Because of high levels of combustion efficiency charcoal productivity is low, consisting primarily of small (<1 mm) particles.

The fire behavior characteristics chart (Andrews and Rothermel 1982, Alexander et al. 1991) provides a means of visualizing the relationship of spread rate and fuel consumption to overall frontal fire intensity in a number of fuel complexes. Figure 3 illustrates the general range of data from fires in savanna, fynbos shrubland, boreal, and slash fuels plotted on such a chart. There is a strong contribution of fuel consumption to overall frontal fire intensity. Savanna fires, with low amounts of available fuel (generally 0.1-1.0 kg m^{-2}), require very high spread rates to reach intensity levels >10,000 kW m^{-1}, while fynbos shrubland fires can combine high spread rates and greater fuel consumption (1.0-2.0 kg m^{-2}) to achieve frontal fire intensities approaching 30,000 kW m^{-1}. Crowning wildfires in boreal forests exhibit a combination of substantial fuel consumption (1.5-5.0 kg m^{-2}) and high spread rates, resulting in fully-developed crown fires reaching intensity levels above 50,000 kW m^{-1}. Slash fires, although capable of attaining fireline intensities above 30,000 kW m^{-1}, usually achieve this primarily through a high degree of fuel consumption (2-20 kg m^{-2}), sometimes in combination with convection-style ignition patterns.

The large variation in intensity levels possible during wildland fires in various ecosystems has a direct impact on the character and impact of emissions from these fires, as well as greatly influencing biological diversity and ecosystem structure. Boreal fires have a much more significant smoldering phase than savanna fires, resulting in substantially different emissions, and these emissions are generally injected into the atmosphere at greater altitudes due to higher boreal fire intensities resulting in strong convection column development. In addition, combustion is less efficient in forest fires than in savanna or grassland fires, resulting in the production of larger quantities of charcoal.

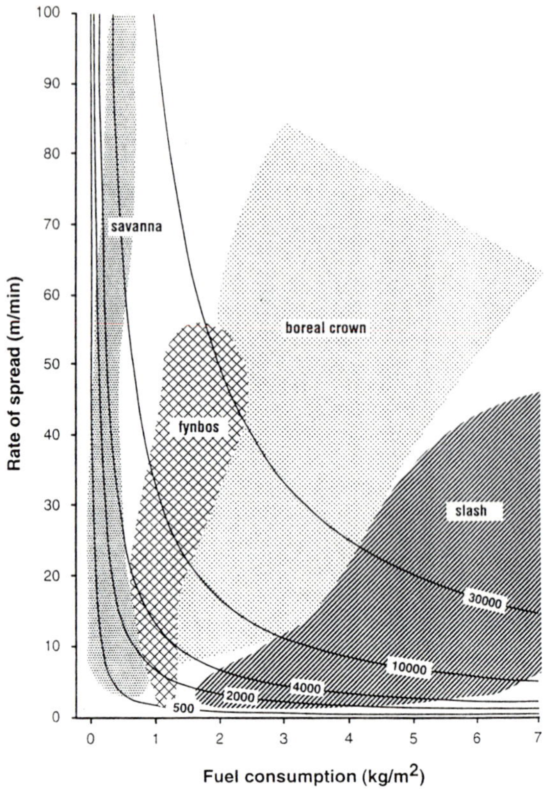

Figure 3: Fire behavior characteristics chart showing the range of rate of spread, fuel consumption, and frontal fire intensity (contours in kW m^{-2}) data for savanna, fynbos, boreal, and slash fires discussed in this paper.

Conclusions

High-intensity crowning wildfires in the boreal zone, where fire is a natural force, are arguably as common now as they have been for millennia, with an average of 5-10 million hectares burning annually. This trend is likely to continue as large portions of the boreal zone remain beyond the control of fire management agencies. Large crown fires account for more than 95% of the area burned in the North American boreal forest, consume a total of 2-3 kg m^{-2} on average (perhaps 50% in the flaming phase), reach frontal fire intensity levels often greater than 50,000 kW m^{-1}, and produce smoke/convection columns that can penetrate the upper troposphere. Fire return intervals in the unprotected boreal zone generally vary between 50 and 150 years, with this variation primarily due to differences in fire weather and fuels. In heavily protected boreal forests fire return intervals often exceed 500 years. While boreal fires release large quantities of carbonaceous aerosols and trace gases into the atmosphere, carbon sequestration begins promptly with the beginning of natural forest regeneration, and there is no net loss of carbon. This could change, however, if fires occur more frequently than at present. Shortening fire return intervals to the point where forest age-class distribution and net carbon sequestration are affected is a distinct possibility under current climate change scenarios that project significant warming at northern latitudes.

Fires were an integral ecosystem process in the evolutionary history and development of virtually all temperate forests. However, depending upon the forest composition, climate, and topographical features of the ecosystem, natural fires in temperate ecosystems will vary greatly. Fire return intervals in temperate systems will range from as little as 5-10 years to as long as 500-1000 years. Typically, natural understory burns occur in ecosystems with short fire return intervals, while severe, stand-replacement fires occur in ecosystems with fire return intervals longer than 80 years. Fire suppression policies, and land use activities such as livestocks grazing, logging, and agricultural conversion have greatly changed fire return intervals in many temperate ecosystems, particularly those with historically short fire return intervals. For example, fire suppression activities in ponderosa pine and mixed conifer forests of the western USA have resulted in a fire regime change to long-return severe stand-replacement fires. The fires that commonly occur today are likely dramatically different in terms of fire behavior, size, severity, and the charcoal signature they leave.

It is likely that more area of tropical savanna burns annually than any other vegetation type on earth, as fire return intervals in savannas are 2-10 years. The biota of savanna ecosystems is well adapted to frequent fires. Total biomass consumed per unit area is quite small in these ecosystems (generally < 1 kg m^{-2}) and carbon is quickly re-sequestered through rapid regrowth. In contrast, tropical deforestation and associated slash fires are largely anthropogenic in nature, and are a significant influence on the disruption of global carbon cycles. In addition, these land use activities are significant causes of unprecedented rates of species extinction and declines in biological diversity. Concomitant with the loss of forest structure, diversity, and carbon pools, anthropogenic slash fires in the tropical forests of Asia, Africa, and South America are resulting in an unprecedented pattern of charcoal production and deposition in these ecosystems. Prior to these contemporary land uses, charcoal in the soils and lake sediments of tropical ecosystems were likely deposited under different climatic regimes and/or cultural uses.

It is extremely important to recognize that fire behavior, effects, and charcoal production are highly variable among ecosystems. Rates of fuel consumption can vary 50-fold among between ecosystems, and fire return intervals range from 1-1000 years. A knowledge of contemporary fire behavior and ecology is of great value in interpreting historical fire records, as well as in predicting future fire regimes associated with climate change.

Literature References

Alexander ME, Stocks BJ, Lawson BD (1991) Predicting fire behavior in the black spruce-lichen woodland: the Porter Lake Project. *For Can North For Cent* Edmonton AB Inf Rep NOR-X-310

Anderson HE (1968) Sundance Fire: an analysis of fire phenomena. *USDA For Serv* Res Pap INT-56

Andrews PL, Rothermel RC (1982) Charts for interpreting wildland fire behavior characteristics. *USDA For Serv* Gen Tech Rep INT-131

Brown S, Lugo EE (1990) Aboveground biomass estimates for tropical moist forests in the Brazilian Amazon. *Interciencia* 17(1):8-18

Buckley AJ (1992) Fire behaviour and fuel reduction burning: Bemm River wildfire, October (1988). *Aust For* 55:135-147

Burrows ND (1985) Reducing the abundance of *Banksia grandis* in the jarrah forest by the use of controlled fire. *Aust For* 48(1):63-70

Byram, GM (1959) Combustion of forest fuels. In Davis KP (ed) *Forest fire: control and use* pp 61-89 McGraw-Hill New York

Cofer WR, Levine JS, Winstead EL, Stocks BJ (1990) Gaseous emissions from Canadian boreal forest fires. *Atmos Envir* 24A:1653-1659

Cofer WR, Levine JS, Winstead EL, Stocks BJ, Cahoon DR, Pinto JP (1993) Trace gas emissions from tropical biomass fires: Yucatan Peninsula, Mexico. *Atmos Environ* 27A:1903-1907

de Castro EA (1995) MS Thesis Oregon State University Corvallis OR (in prep)

Everson TM, van Wilgen BW, Everson, CS (1988) Adaptation of a model for fire danger rating in the Natal Drakensberg. *South Afr J Sci* 84:44-49

Fearnside PM (1992a) Forest biomass in Brazilian Amazonia: comments on the estimate by Brown and Lugo. *Interciencia* 17(1):19-27

Fearnside PM (1992b) Carbon emissions and sequestration in forests: case studies from seven developing countries. (Vol 2 Brazil) *US EPA Climate Change Div* Washington

Fearnside PM, Leal N, Fernandes, FM (1993) Rainforest burning and the global carbon budget: biomass, combustion efficiency, and charcoal formation in the Brazilian Amazon. *J Geophys Res* 98:16733-16743

Feller MC (1989) Estimation of nutrient loss to the atmosphere from slash burns in British Columbia. *Proc 10th Conf Fire and For Meteorol* pp 126-135 Ottawa

Forestry Canada Fire Danger Group (1992) Development and structure of the Canadian Forest Fire Behavior Prediction System. *For Can* Ottawa Inf Rep ST-X-3

Hao WM, Liu MH, Crutzen PJ (1990) Estimates of annual and regional releases of CO_2 and other trace gases to the atmosphere from fires based on FAO statistics for the period 1975-1980. In Goldammer JG (ed) *Fire in the Tropical Biota* pp 440-462 Springer-Verlag Berlin

Hughes RF (1996) PhD Thesis Oregon State University Corvallis (in prep)

Kauffman JB, Cummings DL (1989) Fuel loads and biomass consumption during spring and fall prescribed fires in central Oregon rangeland ecosystems. *USDA For Serv* Final report PNW Res Sta Seattle WA

Kauffman JB, Martin RE (1989) Fire behavior, fuel consumption, and forest floor changes following prescribed understory fires in Sierra Nevada mixed conifer forests. *Can J For Res* 19:455-462

Kauffman JB, Sanford RL, Cummings DL, Salcedo IH, Sampaio EVSB (1993) Biomass and nutrient dynamics associated with slash fires in neotropical dry forests. *Ecology* 74:140-151

Kauffman JB, Cummings DL, Ward DE (1994) Relationships of fire, biomass, and nutrient dynamics along a vegetation gradient in the Brazilian Cerrado. *J of Ecol* 82:519-521

Kauffman JB, Cummings DL, Hughes RF, de Castro EF, Shea RW, Ottmar RD (1995) Total aboveground biomass, fuel loads, and combustion factors of Brazilian tropical forests and savannas. A data and photographic summary. *USDA For Serv* Res Paper PNW Res Sta Portland OR

Kauffman JB, Cummings DL, Ward DE, Babbit RE (1995) Fire in the Brazilian Amazon: biomass, nutrient pools, and losses in slashed primary forests. *Oecologia* (in press)

Kaufmann JB, Cummings DL Biomass burning in the Brazilian Amazon. III. *Biomass and nutrient losses in secondary forests* Capoeira (in prep a)

Kauffman JB, Cummings DL Biomass burning in the Brazilian Amazon. II. *Biomass and nutrient losses in cattle pastures* (in prep b)

Kiil AD (1975) Fire spread in a black spruce stand. *Can For Serv* Bi-Monthly Res Notes 31:2-3

Lafferty RR (1972) Regeneration and plant succession as related to fire intensity on clearcut logged areas in coastal cedar-hemlock type: an interim report. *Can For Serv* Int Rep BC-33 Victoria BC

Lawson BD (1973) Fire behavior in lodgepole pine stands related to the Canadian Fire Weather Index. *Can For Serv* Inf Rep BC-X-76 Victoria BC

Lawson BD, Taylor SW (1986) Preliminary evaluation of prescribed fire impact relationships and predictors for spruce-balsam slash. In *Proc Fire Management Symp* 8-9 Apr 1986 pp 46-68 Prince George BC

Little SN, Ottmar RD, Ohmann JL (1986) Predicting duff consumption from prescribed burns on conifer clearcuts in western Oregon and western Washington. *USDA For Serv* Res Pap PNW-362

McCaw WL, Simpson G, Mair G (1992) Extreme wildfire behaviour in 3-year-old fuels in a Western Australian mixed Eucalyptus forest. *Aust For* 55:107-117

Merrill DF, Alexander ME (eds) (1987) Glossary of forest fire management terms. 4th ed *Natl Res Counc Can, Can Comm For Fire Mgt* Ottawa ON Publ NRCC No 26516

Muraro SJ (1971) Prescribed fire impact in cedar-hemlock logging slash. Env Can *Can For Serv* Ottawa ON Publ 1295

Murphy PG, Lugo AE (1986) Ecology of a tropical dry forest. *Annv Rev Ecol Syst* 17:67-88

Nelson RA, Ward DE (1980) Backfire particulate emissions and Byram's fire intensity. *USDA For Serv* Res Note SE-290

Newstead RG, Alexander ME (1983) Short-term fire retardant effectiveness in a lowland black spruce fuel complex. *Can For Serv* North For Cent Edmonton AB For Rep 28:3-4

Pivello VR, Coutinho LM (1992) Transfer of macro-nutrients to the atmosphere during experimental burnings in an open cerrado (Brazilian savanna). *J Tropical Ecol* 8:487-497

Quintilio D (1972) Fire spread and impact in lodgepole pine slash. MSc Thesis University of Montana Missoula

Quintilio D, Fahnestock GR, Dubé DE (1977) Fire behavior in upland jack pine: the Darwin Lake Project. *Can For Serv* Edmonton AB Inf Rep NOR-X-174

Saspis DB, Kauffman JB (1991) Fuel consumption and fire behavior associated with prescribed fires in sagebrush ecosystems. *Northwest Sci* 65(4):173-179

Shea RW (1993) *Effects of prescribed fire and silvicultural activities on fuel mass and nitrogen redistribution in* ***Pinus ponderosa*** *ecosystems of central Oregon.* MS Thesis Oregon State University Corvallis

Shea RW, Shea BW, Kauffman JB, Ward DE, Haskins CI, Scholes MC (1995) Fuel biomass and combustion factors associated with fires in savanna ecosystems of South Africa and Zambia. *J Geophys Res* (in press)

Simard AJ, Haines DA, Blank RW, Frost JS (1983) The Mack Lake Fire. *USDA For Serv* Gen Tech Rep NC-83

Skole DL, Tucker C (1993) Tropical deforestation and habitat fragmentation in the Amazon: satellite data from 1978 to 1988. *Science* 26:1905-1910

Stocks BJ, Walker JD (1972) Fire behavior and fuel consumption in jack pine slash in Ontario. *Can For Serv* Sault Ste Marie ON Inf Rep O-X-169

Stocks BJ (1975) The 1974 wildfire situation in northwestern Ontario. *Can For Serv* Great Lakes For Cent Sault Ste Marie ON Inf Rep O-X-232

Stocks BJ (1987a) Fire behavior in immature jack pine. *Can J For Res* 17:80-86

Stocks BJ (1987b) Fire potential in the spruce budworm-damaged forests of Ontario. *For Chron* 63(1):8-14

Stocks BJ (1989) Fire behavior in mature jack pine. *Can J For Res* 19:783-790

Stocks BJ (1991) The extent and impact of forest fires in northern circumpolar countries. In Levine JS (ed) *Chapman Conf on Global Biomass Burning: Atmospheric, Climatic, and Biospheric Implications* pp 197-202 MIT Press Cambridge MA

Stocks BJ, Flannigan MD (1987) Analysis of the behavior and associated weather for a northwestern Ontario wildfire: Red Lake #7. In *Proc Ninth Conf Fire and For Meteor* pp 94-100 Amer Met Soc Boston MA

Stocks BJ, Lee BS, Martell DL (1995) Some potential carbon budget implications of fire management in the boreal forest. In *The Role of Forest Ecosystems in the Global Carbon Cycle* NATO ARW Springer-Verlag (in press)

Uhl C, Buschbacher R (1985) A disturbing synergism between cattle ranching, burning practices, and selective tree harvesting in the eastern Amazon. *Biotropical* 17:265-268

Van Wagner CE (1965) Story of an intense crown fire at Petawawa. *Pulp and Paper Mag* Canada Woodlands Rev Sec 66:358-361

Van Wagner CE (1965) Prescribed burning experiments: red and white pine. *Can Dept For* Pub No 1020

van Wilgen BW, Le Maitre DC, Kruger FJ (1985) Fire behavior in South African fynbos (Macchia) vegetation and predictions from Rothermel's fire model. *J Appl Ecol* 22:207-216

van Wilgen BW, Wills AJ (1988) Fire behaviour prediction in savanna vegetation. *South Afr J Wildlife Res* 81:41-46

Wade DD, Ward DE (1973) An analysis of the Air Force Bomb Range Fire. *USDA For Serv* Res Pap SE-105

Ward DE (1983) Particulate matter emissions for fires in the palmetto-gallberry fuel type. *For Sci* 29(4):761-770

Young RP, Miller RF (1985) Response of *Sitanion hystix* to prescribed burning. *Amer Midl Natur* 113:182-187

Biomass Burning Emissions and the Atmosphere

W. R. Cofer III, K. P. Koutzenogii[1], A. Kokorin[2], and A. Ezcurra[3]
Atmospheric Sciences Division
NASA Langley Research Center
Hampton, VA 23681-0001 USA

Summary

Smoke produced by vegetation fires consists of a complex mixture of gaseous, liquid, and solid phases. Many particulates are generated during vegetation fires. Crutzen and Andreae (1990) have estimated that global emissions of pyrogenic smoke particles (50-150 Tg yr^{-1}) may currently exceed global sulfate aerosol produced from fossil fuel burning. These pyrogenic particulates are largely composed of carbonaceous materials (Crutzen and Andreae 1990, Cachier et al.,1991). It is clear that particulate organic carbon is the dominant (~2/3) aerosol product of biomass fires (Andreae et al. 1988). Carbonaceous smoke particles are usually divided into organic carbon (OC) and elemental carbon (EC) aerosols (Mazurek et al. 1991, Ward et al. 1991). The OC aerosol fraction usually predominates in the smoke plumes resulting from vegetation fires, and consists of almost anything, ranging from organic acids and aldehydes to waxes and tars. Results from biomass fires in the Canadian boreal system indicate average ratios of OC/EC of about 16:1 (Mazurek et al. 1991), from Africa of about 10:1 (Cachier et al. 1991), and from Brazil about 12:1 (Ward et al. 1992). Vegetation fires also produce charcoal carbon (Seiler and Crutzen 1980). Both EC and charcoal carbon show a relative lack of reactivity (both chemically and biochemically), and thereby, strongly resist decomposition. Thus, estimates of vegetation burning over geological times often have been based on deposits of charcoal and elemental carbon in marine, lacustrine, and terrestrial sediments (Herring 1985, Clark 1988, Anders et al. 1991). There are fundamental differences between charcoal particles and elemental carbon particles. Charcoal is formed during vegetation fires by means of low-temperature (relative to EC) pyrolysis. Charcoal carbon is produced from charring larger woody fragments under reducing conditions. This tends to produce large particles that are irregular in shape. While charcoal carbon fragments can be found at almost any size, they are usually large (>100 μm).

Elemental carbon is primarily produced during intense flaming combustion and the size distributions determined from vegetation fire smoke consist of a nucleation mode (<0.1 μm), an accumulation mode (0.1 - 2.0 μm), and a coarse mode (>2.0 μm). The largest number of smoke particles are found in the accumulation mode, centered around 0.3 μm, and they are typically spherical in shape. The large difference in the average particle sizes for EC and charcoal have profound implications regarding transport. The smaller EC particles would be expected to be much more broadly dispersed by atmospheric transport, since, once lofted by the fires, would have much longer atmospheric residence times. It is clear that the dispersion of small size smoke aerosols from large vegetation fires can be regional-to-global in scale. In contrast, the larger charcoal fragments would be expected to remain much closer to their source fires. The relationship of biomass burning emissions to the stratigraphic record is highly complex. It demands an understanding of emissions, transport, and deposition behavior, and global circulation/weather patterns at the time of the deposition.

---

[1] Institute of Chemical Kinetics and Combustion, Siberian Division of the Russian Academy of Science, Novosibirsk, Russia 630090

[2] Institute of Global Climate and Ecology, Glebovakaya 20B, Moscow, Russian Federation 107258

[3] Universidad del Pais Vasco, Nieves Cano 12, Vitoria, Spain 01006

NATO ASI Series, Vol. I 51
Sediment Records of Biomass Burning
and Global Change
Edited by James S. Clark, Hélène Cachier,
Johann G. Goldammer, and Brian Stocks

Introduction

While the burning of vegetation appears to currently be increasing globally (Levine et al. 1995), there exists no standard for assessing this over the time scale of human existence. Fire is a natural, periodic, and important occurrence in most major terrestrial ecosystems, often necessary for health and regeneration. It has been effectively argued that the increasing demands made by our increasing global population has changed land use patterns resulting in greater burning and deforestation. Although large amounts of burning occur naturally, most burning in the last century has been the result of human activities (Andreae 1991). Burning for agricultural purposes is extensively practiced worldwide, and fire is still commonly used in deforestation and land management. There are few reasons to believe that there will be any significant change in this in the near future. The release of gaseous and aerosol products into the atmosphere during vegetation fires exerts a global influence on both atmospheric chemistry and climate (Wofsy et al. 1994, Goldammer and Crutzen 1993, Andreae 1991, Crutzen and Andreae 1990). This is particularly clear in the tropics where fire emissions from vegetation can often dominate those from industrialization. Not only are large amounts of environmentally important gases and aerosols immediately released into the atmosphere during the active burning of vegetation, but fires may be modifiers of longer-term biogenic processes involved in the interchange of gases between the atmosphere and biosphere (Levine et al. 1990, Zepp 1988), may be shaping the present and future structure of vegetation (Richter et al. 1982), and may be one of the key factors in determining how terrestrial vegetation responds to future climate change (Walker 1991, Clark and Reed 1995). Indeed, global warming may cause more frequent and extended drought, increasing fire frequency, which in turn would generate additional emissions and further surface albedo changes with accompanying warming, forming an escalating spiral (Dixon and Turner 1991).

Since Crutzen et al. (1979) first suggested that biomass fires could be a significant contributor to the budgets of several important atmospheric trace gases, substantial amounts of research have been conducted to understand trace gas and aerosol production from vegetation fires and their subsequent interaction with the environment. Many combustion products of significant climatic and photochemical importance are released into the atmosphere during biomass burning (see Table 1). Three of the major trace gas emissions from biomass burning are greenhouse gases (CO_2, CH_4, and N_2O) and are known to have been steadily increasing in the modern atmosphere until quite recently (Khalil and Rasmussen 1993, 1992, Keeling et al. 1984, Sarmiento 1993). While the causes of this increasing tendency have not been defined quantitatively, global vegetation burning has most certainly been a significant factor (Seiler and Conrad 1987). Many of the vegetation fire-produced gases are chemically and photochemically active in the atmosphere leading to enhanced ozone production in the troposphere (Delaney et al. 1985, Fishman et al. 1990), ozone depletion in the stratosphere (Rasmussen et al. 1980, Khalil et al. 1993, Mano and Andreae 1994, Cofer et al. 1991), acid rain (Lacaux et al. 1991, Sanhueza et al. 1989), and various other environmental consequences.

Table 1. Biomass combustion produced emissions of atmospheric significance.

Species	Emissions (Tg/yr)	Atmospheric Impact
CO_2	9.0 - 15 x 10^3	greenhouse gas
CO	4.0 - 9.0 x 10^2	photochemistry
CH_4	1.7 - 3.8 x 10^1	greenhouse
NMHC[a]	1.5 - 3.5 x 10^1	photochemistry
Tot. Part.	4.0 - 17 x 10^1	radiation
OC	3.0 - 9.0 x 10^1	
EC	0.5 - 2.0 x 10^1	
NO_x	0.5 - 3.8 x 10^1	photochemistry
H_2	3.0 - 9.0	photochemistry, stratosphere
N_2O	0.4 - 2.0	greenhouse, stratosphere
NH_3	0.5 - 6.0	tropospheric chemistry
RCN	0.5 - 2.0	photochemistry
SO_x	1.0 - 7.0	tropospheric chemistry
CH_3Cl	0.2 - 0.9	stratosphere
CH_3Br	?	stratosphere
COS	0.7 - 3.1 x 10^{-1}	stratosphere

In addition to the gaseous emissions from biomass fires, smoke aerosols and particulates are also influencing global climate and atmospheric chemistry (Radke et al. 1991, Penner et al. 1994). Smoke aerosols can interact directly with solar radiation (Lenoble 1991, Kaufman 1987), serve as a source of condensation nuclei (Penner et al. 1992, Rogers et al. 1991) and perhaps lead to global cooling

(Penner et al. 1994), or form substrates for a variety of heterogeneous atmospheric chemical reactions (Andreae 1991). There are also complex biogeochemical interactions between biomass burning and ecosystems, often involving the intricate exchange of aerosol-dispersed nutrients over large distances (Lacaux et al. 1991, Andreae 1991).

Most current attention to the environmental aspects of biomass burning has been directed toward the tropics, specifically to the tropical rain forests of Brazil and to the savannas of Africa and South America, where most of the world's biomass burning occurs (Andreae 1991). Interest is also peaking on Southeast Asia (Crutzen and Andreae 1990) where several major vegetation fire experiments are planned for the near future (Goldammer, SEAFIRE, IUFRO News, January 1994). In terms of biomass burned and emissions released into the atmosphere, tropical fires are estimated to account for 85% of all biomass burning emissions, with savanna fires clearly dominating (~65%) other tropical biomass fires (Goldammer 1991, Hao and Liu 1994). Tropical burning can be expected to continue since it is used extensively for agriculture, clearing, and land management. In fact, the very persistence of savanna ecosystems depends largely upon recurring fires.

Although the bulk of studies dealing with biomass burning have been focused on the tropics, atmospheric impacts from fires in temperate and higher latitude forests have not been neglected (Wofsy et al. 1994, Laursen et al. 1992, Dixon and Krankina 1993, Susott et al. 1991, Cofer et al 1990a), though global impacts are much harder to discern due to larger amounts of industrial emissions present at northern latitudes.

Combustion

The principal components of all vegetation are carbon, oxygen, and hydrogen, in a blend of about 45, 49, and 6% by mass, respectively. Combustion of vegetation, thus, leads primarily to water and carbon dioxide as reaction products according to the following scheme based on cellulose.

$$(C_6H_{10}O_5)_n + (6O_2)_n = (6CO_2)_n + (5H_2O)_n$$

Depending upon the biomass fuel nitrogen, sulfur, halogen, and mineral contents, gaseous and particulate forms of these will also be released in small, but often very important, amounts. But combustion of vegetation is never completely efficient and in addition to water and carbon dioxide, varying amounts of CO, CH_4, H_2, hydrocarbons, particulates, etc., are produced. These are the products of incomplete combustion (oxidation) and are produced at the expense of conversion of fuel carbon and hydrogen to CO_2 and H_2O. Thus, an inverse relationship exists between the fully typically range from about 5-20% of the CO_2 product. The amount is strongly tied to combustion influencing factors such as winds, temperature, fuel moisture, fuel size, humidity, etc. The combustion efficiency (CE) for biomass fires is usually expressed as the ratio of the CO_2 produced by the fire to the total carbon product in terms of the mass of carbon, as expressed in the below.

$$CE = CO_2 / (CO_2 + CO + CH_4 + \text{carbonaceous particles} + \text{hydrocarbons})$$

Since quantitative measurements of all the carbon products produced during biomass burning can be very difficult, if not impossible to obtain, the measured ratio of CO/CO_2 in smoke plumes is often used as a measure of combustion efficiency (Hegg et al. 1990, Ward et al. 1991). The larger the CO/CO_2 ratio, the less efficient the combustion. The use of ratios to compare biomass burning emissions, first used by Crutzen (1979), is still commonly practiced. Emission ratios (ERs) reveal the relationship of one fire product to another. Generally, trace products are normalized to a major product, usually carbon dioxide. Emission ratio is defined below in terms of carbon dioxide.

$$[ER's = (\text{trace gas X - background X}) / (CO_2 - \text{background } CO_2)]$$

The ambient background levels of selected gases (or particles) are subtracted from their smoke plume concentrations and ratioed to the background corrected CO_2. The assumption is that the difference between the background and plume measurements represents the production by the fire. The resulting ratios paint a picture of the relationship between products. Thus, an ER for CO normalized to CO_2 of 0.1 indicates that CO is produced at about 10% of the amount of CO_2. Likewise, an ER for CH_4 of 0.01 would indicate CH_4 production at the 1% of CO_2 level. Sometimes carbon monoxide is used as the normalizing parameter. The use of CO as a normalizing parameter is advantageous when smoke plume CO_2 concentrations are not sufficiently elevated above background concentrations to provide an accurate determination of excess CO_2. This typically occurs in aged or diluted smoke plumes since in-plume CO_2 concentrations converge with their higher ambient background concentrations more rapidly than CO.

Vegetation

Since there is a strong coupling between types of vegetation and their combustion, indigenous properties of an ecosystem can frequently be used to predict combustion behavior, and thus, to estimate emission product stoichiometries. For example, savannas tend to have high proportions of grassy fuels. Grass fuels have very high surface areas per unit of mass (excellent for heat transfer), a relatively consistent size and mass, and are generally well arrayed (allowing easy access to oxygen during combustion). These characteristics lead to easy ignition, very consistent and efficient combustion, and minimal amounts of smoldering. In contrast, the boreal forest system is composed of a very large mixture of fuel sizes and types, ranging from trees and needles to a forest floor consisting of a mixture of litter resting on a partially decomposed and compacted organic surface layer. Combustion in this diverse mixture of fuels would not be expected to proceed similarly to that of the grassy savannas. These generalizations are illustrated by the data shown in Table 2. An examination of the emission ratios presented in Table 2 reveal the relatively low ERs (reflecting efficient combustion) determined for gaseous emissions in both South African savannas (Cofer et al. 1996) and graminoid wetlands in Florida (Cofer et al. 1990b). It also suggests that either the composition of the smoldering emissions are the same as for flaming, or as we believe, that the smoldering phase is nearly absent and what we sampled was dominated by residual flaming combustion products. Examination of boreal forest fire ERs, in Table 2, however, quickly reveal a dramatic difference. An obviously different set of ER's is seen for the smoldering emissions. Thus, the ecosystem characteristics of the boreal forest dictates a different type of combustion and resulting emissions chemistry. These types of indigenous ecosystem dictated properties can often be used to effectively predict emissions.

Smoke Composition

Smoke consists of a complex and changing mixture of gaseous, liquid, and solid phases. It changes because the nature of the combustion itself changes at the source (e.g., flaming vs. smoldering emissions) producing a different emissions composition, and it changes temporally in the atmosphere because it is intrinsically a chemically and photochemically reactive mixture. Many particulates are generated during biomass fires. Crutzen and Andreae (1990) have estimated that emissions of pyrogenic smoke particles (50-150 Tg yr^{-1}) may exceed sulfate aerosol produced from fossil fuel burning. These pyrogenic particulates are largely composed of carbonaceous materials (Crutzen and Andreae 1990, Cachier et al. 1991), although a significant portion of smoke particles consist of or contain inorganic material (Cofer et al. 1988, Lacauex et al. 1993). Andreae et al. (1988) have reported results from the Amazon that conclude about 35% by mass of biomass fire smoke particles are soluble in water as ionic species. This may help explain why smoke particles often act as very efficient cloud condensation nuclei (Penner et al. 1992, Radke et al. 1991, Rogers et al. 1991). What fraction of the inorganic particulates are produced chemically during combustion, or later in the aging smoke plume via gas-to-particle reactions, or simply advected soil particles suspended in the smoke during vigorous updrafts is unclear. Nevertheless, it is clear that particulate organic carbon is the dominant (~2/3) aerosol product of biomass fires (Andreae et al. 1988). Carbonaceous smoke particles are usually sub-divided into organic carbon (OC) and elemental carbon (EC) aerosols (Mazurek et al. 1991, Ward et al. 1991). The OC aerosol fraction predominates in smoke plumes, and consists of almost anything, ranging from organic acids and aldehydes to waxes and tars. Results from biomass fires in the Canadian boreal system indicate average ratios of OC/EC of about 16:1 (Mazurek et al. 1991), from Africa of about 10:1 (Cachier et al. 1991), and from Brazil about 12:1 (Ward et al. 1992). In all cases, it is apparent that production of organic carbon dominates production of black carbon in vegetation fires. It has been found that black carbon is primarily produced during intensely flaming combustion (Andreae 1991, Enfield et al. 1991, Patterson et al. 1986). The actual amount of black carbon that is produced and suspended in the atmosphere in smoke is significant because black carbon has an absorption component that can exert a warming influence on the atmosphere-surface system and redirect warming to different altitudes (Lenoble 1991, Penner et al. 1994). The OC particles impact radiation primarily through scattering, and their potential role as CCN may add significantly (through cloud formation) to this impact.

Table 2. CO_2-normalized emission ratios (in %) for biomass fires in several different ecosystems.

	CO	H_2	CH_4	TNMHC	N_2O
Wetlands					
F(13)*	4.7 ± 1.1	1.2 ± 0.5	0.30 ± 0.11	0.45 ± 0.21	0.019 ± 0.004
S(19)*	5.3 ± 1.2	1.4 ± 0.5	0.36 ± 0.13	0.47 ± 0.15	0.025 ± 0.008
Savanna					
F(21)*	4.8 ± 0.8	0.9 ± 0.3	0.30 ± 0.10	0.47 ± 0.22	0.008 ± 0.003
S(11)*	4.6 ± 1.9	1.5 ± 0.4	0.55 ± 0.22	0.59 ± 0.25	0.013 ± 0.003
Mediterranean					
F(11)*	5.7 ± 1.5	2.2 ± 0.6	0.3 ± 0.10	0.43 ± 0.07	0.016 ± 0.003
S(9)*	7.2 ± 1.7	2.5 ± 0.6	0.4 ± 0.10	0.73 ± 0.16	0.020 ± 0.005
Boreal					
F(28)*	6.7 ± 1.2	2.1 ±0.5	0.6 ± 0.20	0.61 ± 0.13	0.016 ± 0.003
S(22)*	12.1 ± 1.9	3.1 ± 0.6	1.2 ± 0.28	1.15 ± 0.25	0.017 ± 0.002

* = combustion phase and number of samples.

Particle size distributions determined from vegetation fire smoke appear to show little variation from fire to fire, although particle concentrations vary substantially (Radke et al. 1988). A nucleation mode(<0.1 μm), an accumulation mode (0.1 - 2.0 μm), and a coarse mode (> 2.0 μm) are all observed (Radke et al. 1991, Cachier 1989). The largest number of smoke particles are found in the accumulation mode, centered around 0.3 μm. Radke et al. (1991) found particles about this size to consist mostly of tarry condensed hydrocarbons of spherical shape. A typical particle size distribution of smoke aerosol, determined by Radke et al. (1991), is shown in Fig. 1. Figure 1 is a composite developed from 4 fires. No attempt has been made by the authors of this work to make any distinctions between the 4 fires, only to show how similar the resulting distributions were. Accompanying the carbonaceous component of the particles are numerous water soluble inorganic compounds such as sulfates (SO_4^{-2}), nitrates (NO_3^-), phosphates (PO_4^{-3}) and chlorides (Cl^-). This combination of carbon and inorganic material can produce hydrophillic particles that can act as cloud condensation nuclei and ultimately contribute to the formation of acidic rain. Acidic rain in the tropics is usually linked to nitric, formic, and acetic acids (Lacaux et al. 1993). Water soluble cations (+ ions) typically associated with the inorganic constituents consist primarily of potassium (K^+), sodium, (Na^+), ammonium (NH_4^+), and calcium (Ca^{2+}). Note that these inorganic compounds are the primary soluble inorganic particulate emissions, and that other trace minerals exists in vegetation fire emissions. Some of these probably result from advection of soil particles. It should also be noted that some of these may not begin as particles, but are transformed chemically with time in the smoke plumes (e.g., NO_x/NO_3^{--}). Large amounts of potassium and phosphates are always found in vegetation fire aerosols, and the large enrichment of potassium in vegetation fire produced aerosols has permitted its use as a tracer for these aerosols (Andreae 1983).

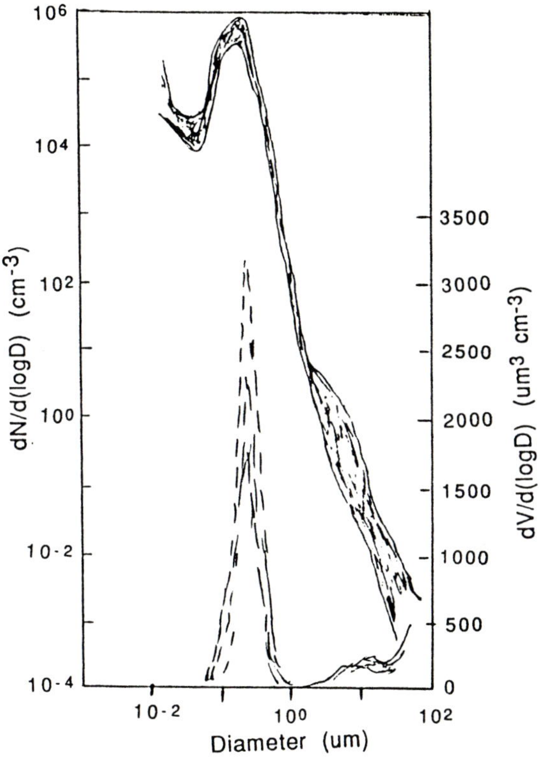

Figure 1. Number and volume distribution of smoke particles measured near (<5 km) four fires (from Radke et al. 1991).

Of vegetation fire aerosol, the largest particles (>100 μm) are the least well characterized. Few large particles are transported far enough in the atmosphere to be important with regard to atmospheric chemistry, climate, or air quality (odors, visibility, etc.). Thus, little attention has been devoted to their characterization. Conversely, the gasses and small particles (≤1 μm) produced by vegetation burning dominate atmospheric chemistry and climatic processes because of their facility for dispersal, transport, and residence time in the atmosphere. Vegetation fires produce elemental and charcoal carbon (Seiler and Crutzen 1980). Both show a relative lack of reactivity (both chemically and biochemically), and thereby, strongly resist decomposition. Thus, estimates of vegtation burning over geological times are based on deposits of charcoal and black carbon in marine, lacustrine, and terrestrial sediments (Herring 1985, Clark 1988, Anders et al. 1991). The large particles tend to be irregular in shape and are usually identified as charcoal, while the small particles are typically spherical and classified as elemental carbon. Both types are used to examine past fire regimes. Seiler and Crutzen have estimated that about 20% of the carbon residue left on the ground after biomass fires is in the charcoal state. However, Fearnside et al. (1990) estimated much lower levels (~5%) of charcoal formation. Since black carbon production in fires is environmentally important and appears to vary substantially (Ward 1986, Patterson and McMahon 1984, Cachier et al. 1989), largely a function of combustion intensity, the quantities of black carbon produced in vegetation fires needs to be better understood. One of the most important aspects of the formation of black carbon in fires is that it may provide one of the most effective pathways for removing carbon dioxide from the atmosphere while freeing oxygen (Crutzen and Andreae 1990, Andreae 1991).

While the greenhouse gasses carbon dioxide and water are the main products of vegetation fires, the atmospheric impacts associated with the release of CO_2 are complicated to assess. For example, CO_2 released during savanna burning in the tropics (about 65% of all biomass burning CO_2) is reincorporated into new savanna growth rapidly. Thus this CO_2 can be considered to be recycled by means of photosynthesis in short time scales (Hao et al. 1990). Recycling of CO_2 released during boreal forest fires, however, is quite another thing. Boreal forests represent one of the world's largest terrestrial organic carbon pools (Kauppi et al. 1992). Once burned, they cannot quickly reincorporate equivalent amounts of carbon into new growth and into a new rich organic surface layer (forest floor). A significant part (~33 %) of the fuel carbon stored in the boreal biome consists of a partially decomposed and compacted organic surface layer formed from years of accumulation of forest litter. This requires, in some cases, at least a century to reestablish preburn levels (Auclair 1985). So, although the recycling of carbon occurs in all vegetation systems, the time scales for recycling can differ substantially.

Many of the trace gasses produced during vegetation fires have potentially dramatic impacts on atmospheric chemistry. The number and types of hydrocarbons emitted by vegetation fires are too numerous to list. Alkanes, olefins, aromatic and polycyclic compounds are all represented (see Blake et al. 1994, Lobert et al. 1991, Greenberg et al. 1984). Nitrogen oxides (e.g., NO_x) and hydrocarbons emitted from biomass fires are likely to be very important in tropospheric chemistry, serving as precursors for ozone production in the troposphere (Delany et al. 1985, Crutzen et al. 1985). Satellite data have suggested that enhanced tropospheric ozone concentrations over the tropical Atlantic Ocean may result from NO_x/HC emissions from fires in South America and Africa (Fishman et al. 1990). Andreae et al. (1994) have reported observing this off the coast of Brazil. Large biomass fires also appear to be an important source of other nitrogen gases. About 7% of the atmospheric nitrous oxide (N_2O) budget is attributable to biomass fires (Cofer et al. 1991). N_2O is not only important as a greenhouse gas, but also as a large contributor to startospheric ozone destruction. Other nitrogen species released during vegetation fires include ammonia (NH_3), nitric acid (HNO_3), hydrogen cyanide and methyl cyanide (Hurst et al. 1994, Hurst and Griffin 1994, LeBel et al. 1988). Biomass fires may be the main terrestrial source of atmospheric methyl chloride (Rasmussen et al. 1980), and methyl bromide (Khalil et al. 1993, Mano and Andreae 1994), which likely contribute to ozone depletion in the stratosphere (Andreae 1991). Biomass burning has also been credited with generating large amounts of CO observed in the free troposphere (Reichle et al. 1990). CO is one of the main regulators of atmospheric oxidations through its reactions with hydroxyl (OH) radical. Biomass burning constitutes a source of sulfur dioxide to the troposphere and sulfur to the stratosphere via formation of carbonyl sulfide (Nguyen et al. 1994). Thus, biomass burning can be seen to contribute in a significant way to the trace gas composition of the atmosphere.

Emission Factors

Emission factors (EFs) are used to relate the mass of a particular species released into the atmosphere during burning to the mass of fuel combusted (Ward et al. 1979). EFs are usually expressed as the ratio of grams of product to kg of dry fuel burned. EFs are developed from fires where fuel types and loadings before a fire, and fuel consumptions and fire behavior after a fire, have been measured or characterized. Usually this is accomplished through research on prescribed burns (intentional fires). The development of EFs require the precise correlation of fuel consumption, vegetation type, and fire conditions, with the resulting emissions.

Once developed, EFs can have broad application to the estimation of emissions for generic types of vegetation fires. For example, a set of EFs developed for boreal fires has been used to estimate emissions from major fires in China and Siberia in 1987 (see Cahoon et al. 1994). In this work, satellite measurements of burned areas (14.4 million ha) were combined with fuel consumption figures estimated for southeastern Siberian taiga (boreal forest) and EFs developed for boreal fires to estimate the resulting atmospheric emissions. The choice and application of EFs, however, must involve a high degree of prudence. That is, EFs developed from logging slash fires in the boreal system would not necessarily be expected to accurately reflect emissions from boreal wildfires. There are to many potential dissimilarities between prescribed fires in logging slash and wildfires in standing live timber. However, one might reasonably expect EFs determined from prescribed savanna fires to accurately reflect savanna wildfires, since the state of the fuels would be the same.

Ward et al. (1979) and Nelson (1982) have developed a procedure referred to as the carbon mass balance (CMB) technique for calculating EFs, which can be used to develop EFs for wildfires and fires where no prefire or postfire fuel characterizations have been done. The following synopsizes the CMB approach. Total excess carbon (TC_{ex}) is calculated from all measured excesses of carbon products, i.e., CO_2 + CO + CH_4 + NMHC + Particles. This calculation is done on a mass of carbon/volume basis. The TC_{ex} is related to the original mass of fuel through multiplication by 2 since the C content in woody fuels is about 50% of the dry mass. By convention, it is expressed in g kg^{-1}. Thus the EF for species X can be determined by:

$$EF(X) = EX's(X)/(2)TC_{ex}$$

where EX's(X) is excess of species X. Since more than 97% of the carbon released during biomass combustion is in the form of CO_2, CO, CH_4, and particulates, EFs determined by the CMB technique would be expected to be reasonably good. This technique has been applied to aircraft obtained smoke plume measurements by Radke et al. (1988). The establishment of the CMB method for determining EFs should greatly assist in the development of global biomass burning emission budgets.

Emission factors determined for North American boreal and temperate forest fires by fixed-wing, helicopter, and ground-based measurement techniques are presented in Table 3. The EFs in Table 3 represent the averaged emissions from flaming, transitional, and smoldering combustion. The agreement among the EFs determined by different techniques and by the different research groups can be seen to be exceptionally good. The higher varibilities shown in the ground measurements should be expected due to less mixing, which would tend to dampen oscillations.

Table 3. Average emission factors (g/kg) for major combustion products determined from fires in North American boreal and temperate forests. (Source 1, 2 = fixed wing aircraft, 3 =heliocopter, 4,5 = tower measurements)

CO_2	CO	CH_4	TNMHs	Particles	N_2O	NO_x	H_2	Source
1650±35	93±16	3.8±1.0	1.8±0.4	20±15	0.23±0.05	4±6	1.8±0.7	(1,2)
1595±45	105±20	3.5±1.1	3.7±1.1	15±10	0.15±0.05	NM	1.4±0.6	(3)
1625±85	107±45	4.1±1.7	3.2±2/1	20±15	NM	2±2*	1.9±0.7	(4,5)

(1) Radke et al. 1991. (2) Laursen et al. 1992. (3) Cofer et al. 1990a. (4) Susott et al. 1991. (5) Ward et al. 1992. NM=not measured, *= NO not NO_x

Geographical Distribution

While fire has been utilized throughout history for agriculture, heat, hunting, pest control, land management, etc., the heightened level of concern stems largely from the increasing world population's demands for energy and space. In the industrialized regions, this substantially translates into fossil fuel usage and some burning for clearing to support an expanding and mobile population. In less developed regions, vegetation is a more prominent source of energy, with large amounts of clearing and burning for cultivation, for heating and cooking, and land management. Since many of the emissions from fossil fuel combustion and vegetation fires are the same, they are potentially acting on the atmosphere in tandem. Most burning occurs in the tropics (Goldammer 1991, Hao and Liu 1994). Table 4 summarized the amount of tropical burning estimated by Seiler and Crutzen (1980), Crutzen and Andreae (1990), and Hao and Liu (1994). It can be determined from Table 4 that most vegetation fires occur in tropical savanna. The second largest catageory is shifting cultivation. All in all, about 6 x 10^{12} kg of biomass is burned yearly in the tropics. This compares to about 7 x 10^{12} kg of biomass burned yearly worldwide. Most other burning is located in the regions of temperate and boreal forests. Fires in these systems account for about 5 -15% of global burning. However, unlike in the tropics, the degree of burning in the temperate and boreal forest varies greatly from year to year. For example, in 1987 more than 22 million ha burned in the world's boreal forest. Assuming an average fuel consumption of 2.5 kg m^{-2} (Stocks 1989), this would translate into more than 10% of global biomass burning emisions. By comparison, in 1992, it has been estimated that no more than 2-3 million ha of boreal forest burned (Donald Cahoon, personal communication), thus translating into only about 1% of global emissions.

Table 4. Estimates of teragrams/yr of biomass burned from various sources in the tropics.

	Seiler and Crutzen (1980)	Crutzen and Andreae (1990)	Hao and Liu (1994)
Deforestation	715	1000	510
Shifting cultivation	1700	1665	1310
Savanna fires	1190	2115	2670
Fuel wood	620	1000	620
Agricultural residue	710	1445	280

Conclusions

There are no reasons to believe that the bulk of biomass burning emissions will leave any fingerprints in the sediments. There are, however, a few notable exceptions to this. Some carbon is always reduced during vegetation fires to nearly its elemental state. Small (0.01 - 1.0 µm) black carbon particulates are formed by the condensation of carbon in oxygen deficient flame zones (reducing flames). These spherical condensates, though individually very small, typically agglomerate into chains or aggregates producing particles with much larger (1-20 µm) effective diameters (Cachier et al. 1991). This type of elemental carbon particle is referred to as black carbon or soot carbon. Charcoal is also formed during vegetation fires by means of a lower-temperature pyrolysis process. Charcoal carbon is produced from charring larger woody fragments under reducing conditions. Both the soot and charcoal produced in fires are extremely resistant to chemical and biochemical decomposition, and thereby, in principle, permit evaluation of past fire activity based upon their accumulations in the sediment record. The relationship of biomass burning emissions to the stratigraphic record, however, is complex, demanding an understanding of emissions, transport, and deposition behavior. Charcoal carbon fragments can be found at almost any size, but are usually much larger (>100 µm) than soot carbon aggregates. Thus, the soot aggregates would be expected to be much more broadly dispersed by atmospheric transport, since the smaller particles, once lofted by the fires, would have much longer atmospheric residence times. It is clear that the dispersion of smoke aerosols from large vegetation fires can be regional-to-global in scale. In contrast, the large charcoal fragments should remain much closer to their source fires. This has been confirmed by Herring (1985) who observed from marine cores that the charcoal particles found closer to land were larger and retained more of their original plant morphology.

A consequence of this may be that large concentrations of small soot carbon in the sediments would indicate periods of very extensive (spatially) fire activity, since the dispersion of these small particles is anticipated to be so efficient. Enhancements of charcoal carbon in sediments, however, might only indicate high levels of localized fire activity. Certain other hydrocarbons, such as polynuclear aromatic hydrocarbons (PAH's) are known products of combustion, and may also be resistant enough to chemical and biochemical decomposition to indicate biomass fire activity, even to the extent of identifying vegetation types.

Literature References

Anders E, Wolbach WS, Gilmour I (1991) Major wildfires at the cretaceous-tertiary boundary. In Levine JS (ed) *Global Biomass Burning: Atmospheric, Climatic, and Biospheric Implications* pp 485-492 MIT Press Cambridge

Andreae MO, Anderson BE, Blake DR, Bradshaw JD, Collins JE, Gregory GL, Sachse GW, Shipman MC (1994) Influence of plumes from biomass burning on atmospheric chemistry over the equitorial and tropical South Atlantic during CITE 3. *J Geophys Res* 99:12793-12808

Andreae MO et al. (1988) Biomass-burning emissions and associated haze layers over Amozonia. *J Geophys Res* 93:1509-1527

Andreae MO (1983) Soot carbon and excess fine potassium: Long-term transport of combustion derived aerosols. *Science* 220:1148-1151

Andreae MO (1991) Biomass burning: Its history, use, and distribution and impact on ennvironmental quality and global climate. In Levine JS (ed) *Global Biomass Burning: Atmospheric, Climatic, and Biospheric Implications* pp 3-21 MIT Press Cambridge

Auclair AND (1985) Postfire regeneration of plant and soil organic pools in a *Picea mariana-Cladonia stellaris* ecosystem. *Can J For Res* 15:279-291

Blake DR, Smith Jr TW, Chen TY, Whipple WJ, Rowland FS, (1994) Effects of biomass burning on summertime nonmethane hydrocarbon concentrations in the Canadian wetlands. *J Geophys Res* 99:1699-1719

Cachier H, Ducret J, Bremond M, Yoboue V, Lacaux J, Gaudichet A, Baudet J (1991) Biomass burning aerosols in a savanna region of the Ivory Coast. In Levine JS (ed) *Global Biomass Burning: Atmospheric, Climatic, and Biospheric Implications* pp 174-180 MIT Press Cambridge

Cachier H, Bremond MP, Baut-Menard P (1989) Carbonaceous aerosols from different tropical biomass burning sources. *Nature* 340:371-373

Cahoon Jr DR, Stocks BJ, Levine JS, Cofer III WR, Pierson JM (1994) Satellite analysis of the severe 1987 forest fires in northern China and Siberia. *J Geophys Res* 99(18):627-638

Clark JS (1988) Particle motion and the theory of charcoal analysis: Source area, transport, deposition, and sampling. *Quat Res* 30:67-80

Clark JS, Reid CD (1994) Sensitivity of North American ecosystems to global change. In Towmmann MA (ed) *Non-Linear Responses to Climate Change* pp 53-89 Resources for the Future Washington DC

Cofer III WR, Levine JS, Winstead EL, Cahoon DR, Sebacher DI, Pinto JP, BJ Stocks (1996) Source compositions of trace gases released during African savanna fires. *J Geophys Res* (in press)

Cofer III WR, Levine JS, Winstead EL, Stocks BJ (1990a) Gaseous emissions from Canadian boreal forest fires. *Atmos Environ* 24A:1653-1659

Cofer III WR, Levine JS, Winstead EL, LeBel PJ Koller AM, Hinkle CR (1990b) Trace gas emissions from burning Florida wetlands. *J Geophys Res* 95:1865-1870

Cofer III WR, Levine JS, Winstead EL, Stocks BJ (1991) Nitrous oxide emissions from biomass burning. *Nature* 349:689-691

Cofer III WR, Levine JS, Sebacher DI, Winstead EL, Riggan PJ, Brass JA, Ambrosia VG (1988) Particulate emissions from a mid-latitude prescribed chaparral fire. *J Geophys Res* 93:5207-5212

Crutzen PJ, Heidt LE, Krasnec JP, Pollock WH, Seiler W (1979) Biomass burning as a source of atmospheric gases CO, H_2, N_2O, NO, CH_3Cl, and COS. *Nature* 282:253-256

Crutzen PJ, Delany AC, Greenberg J, Haagenson P, Heidt L, Lueb R, Pollock W, Seiler W, Wartburg A, Zimmerman P (1985) Tropospheric chemical composition measurements in Brazil during the dry season. *J Atmos Chem* 2:233-256

Crutzen PJ, Andreae MO (1990) Biomass burning in the tropics: Impact on atmospheric chemistry and biogeochemical cycles. *Science* 250:1669-1678

Delany AC, Haagensen P, Walters S, Wartburg AF, Crutzen PJ (1985) Photochemically produced ozone in the emissions from large-scale tropical vegetation fires. *J Geophys Res* 90(2):425-429

Dixon RK, Krankina ON (1993) Forest fires in Russia: Carbon dioxide emissions to the atmosphere. *Can J Res* 23:700-705

Dixon RK, Turner DP (1991) The global carbon cycle and climate change: Responses and feedbacks from below-ground systems. *Environ Pollut* 73:245-262

Einfeld W, Ward DE, Hardy CC (1991) Effects of fire behavior on prescribed fire smoke characteristics: A case study. In Levine JS (ed) *Global Biomass Burning: Atmospheric, Climatic, and Biospheric Implications* pp 412-419 MIT Press Cambridge

Fearnside PM (1990) Fire in the tropical rain forest of the Amazon basin. In *Fire in the Tropical and Subtropical Biota* Goldammer JG (ed) pp 106-116 Springer-Verlag Heidelberg

Fishman JC, Watson E, Larsen JC, Logan JA (1990) Distribution of tropospheric ozone determined from satellite data. *J Geophys Res* 95(3):599-617

Goldammer JG (1991) Tropical wild-land fires and global changes: Prehistoric evidence, present fire regimes, and future trends. In Levine JS (ed) *Global Biomass Burning: Atmospheric, Climatic, and Biospheric Implications* pp 83-91 MIT Press Cambridge

Goldammer JG, Crutzen PJ (1993) Fire in the environment: Scientific rationale and summary of results of the Dahlem Workshop. In Crutzen PJ, Goldammer JG (eds) *Fire in the Environment: The Ecological, Atmospheric, and Climatic Importance of Vegetation Fires* pp 1-14 John Wiley & Sons London

Greenberg JP, Zimmerman PR, Heidt L, Pollock W (1984) Hydrocarbon and carbon monoxide emissions from biomass burning in Brazil. *J Geophys Res* 89:1350-1354

Hao WM, Liu MH, Crutzen PJ (1990) Estimates of the annual and regional release of CO_2 and other trace gases to the atmosphere from fires in the tropics, based on FAO statistics for the period 1975-1980. In Goldammer JG (ed) *Fire in the Tropical Biota: Ecosystem Processes and Global Challanges* pp 106-116 Springer-Verlag Berlin-Heidelberg

Hao WM, Liu MH (1994) Spatial and temporal distribution of tropical biomass burning. *Global Biogeochem Cycles* 8:495-503

Hegg DA, Radke LF, Hobbs PV, Rasmussen RA, Riggan PJ (1990) Emissions of some trace gases from biomass fires. *J Geophys Res* 95(5):669-675

Herring JR (1985) Charcoal fluxes into sediments of the north Pacific ocean: The cenozic record of burning. In Sundquist ET, Broecker WS (eds) *The Carbon Cycle and Atmospheric CO_2: Natural Variations Archean to Present* American Geophysical Union Wasjington DC Geophysical Monograph 32

Hurst DF, Griffith DT, Cook GD (1994) Trace gas emissions from biomass burning in tropical Australian savannas. *J Geophys Res* 99(16):441-456

Hurst DF, Griffith DT, Carras JN, Williams DJ, Fraser PJ (1994) Measurements of trace gases emitted by Australian savanna fires during the 1990 dry season. *J Atmos Chem* 18:33-56

Kauppi P, Mielikainen K, Kuusela K (1992) Biomass and the carbon budget of European forests: 1971-1990. *Science* 256:70-74

Kaufman YJ (1987) Satellite sensing of aerosol absorption. *J Geophys Res* 92:4307-4317

Keeling CD, Carter AF, Mook WG (1984) Seasonal, latitudinal, and secular variations in the abundance and isotopic ratios of atmospheric CO_2. *J Geophys Res* 89:4615-4628

Khalil MAK, Rasmussen RA (1993) Decreasing trend of methane: Unpredictability of future concentrations. *Chemosphere* 26:803-814

Khalil MAK, Rasmussen RA, Gunawardena R (1993) Atmospheric methyl bromide: trends and global mass balance. *J Geophys Res* 98:2887-2896

Kurz WA, Apps MJ, Stocks BJ, Volney JA (1994) Global climate change: Disturbance regimes and biospheric feedbacks of temperature and boreal forests. In Woodwell G (ed) *Biotic Feedbacks in the Global Climate System: Will the Warming Speed the Warming?* Oxford University Press Oxford UK

Lacaux JP, Delmas RA, Cros B, Lefeivre B, Andreae MO (1991) Influence of biomass burning emissions on precipitation chemistry in the equatorial forests of Africa. In Levine JS (ed) *Global Biomass Burning: Atmospheric, Climatic, and Biospheric Implications* pp 167-173 MIT Press Cambridge

Lacaux JP, Cachier H, Delmas R (1993) Biomass burning in Africa: An overview of its impact on atmospheric chemistry. In Crutzen PJ, Goldammer JG (eds) *Fire in the Environment: The ecological, atmospheric, and climatic importance of vegetation fires* John Wiley and Sons London

Laursen KK, Hobbs PV, Radke LF, Rasmussen RA (1992) *J Geophys Res* 97:20687-20701

LeBel PJ, Cofer III WR, Levine JS, Vay SA, Roberts PA (1988) Nitric Acid and Ammonia Emissions from a Mid-Latitude Prescribed Wetlands Fire. *Geophys Res Lett* 15:792-795

Lenoble J (1991) The particulate matter from biomass burning: A tutorial and critical review of its radiative impact. In Levine JS (ed) *Global Biomass Burning: Atmospheric, Climatic, and Biospheric Implications* pp 381-386 MIT Press Cambridge

Levine JS, Cofer III WR, Cahoon Jr DR, Winstead EL (1995) Biomass burning: A driver for global change. *Environ Sci and Technol* 29:120A-126A

Levine JS, Cofer III WR, Sebacher DI, Rhinehart RP, Winstead EL, Sebacher SS, Hinkle CR, Schmalzer PA, Koller AM (1990) The effects of fire on biogenic emissions of methane and nitric oxide from wetlands. *J Geophys Res* 95:1853-1864

Lobert JM, Scharffe DH, Hao WM, Kuhlbush TA, Seuwen R, Warneck P, Crutzen PJ (1991) Experimental evaluation of biomass burning emissions: Nitrogen and carbon containing compounds. In Levine JS (ed) *Global Biomass Burning: Atmospheric, Climatic, and Biospheric Implications* pp 289-304 MIT Press Cambridge

Mano S, Andreae MO (1994) Emission of methyl bromide from biomass burning. *Science* 263:1255-1258

Mazurek MA, Cofer III WR, Levine JS (1991) Carbonaceous aerosols from prescribed burning of a boreal forest ecosystem. In Levine JS (ed) *Global Biomass Burning: Atmospheric, Climatic, and Biospheric Implications* pp 258-263 MIT Press Cambridge

Nelson Jr RM (1982) An evaluation of the carbon mass balance technique for estimating emission factors and fuel consumption in forest fires. *USDA Forest Service* Southeastern Forest Experimental Station Asheville NC Res Paper SE-231

Nguyen BC, Mihalopous N, Putaud JP (1994) Rice straw burning in Southeast Asia as a source of CO and COS to the atmosphere. *J Geophys Res* 99(16):435-439

Patterson EM, McMahon CK (1984) Absorption characteristics of forest fire particulate matter. *Atmos Environ* 18:2541-2551

Patterson EM, McMahon CK, Ward DE (1986) Absorption properties and graphitic carbon emission factors of forest fire aerosols. *Geophys Res Lett* 13:129-132

Penner JE, Dickinson RE, O'Neil CA (1992) Effects of aerosol from biomass burning on the global radiation budget. *Science* 256:1432-1433

Penner JE, Charlson R, Hales J, Laulainen N, Leifer R, Novakov T, Ogren J, Radke L, Schwartz S, Travis L (1994) Quantifying and minimizing uncertainty of climate forcing by anthropogenic aerosols. *Bull Amer Meteor Soc* 75:375-400

Radke LF, Hegg DA, Hobbs PV, Nance JD, Lyons JH, Laursen KK, Weiss RE, Riggan PJ, Ward DE (1991) Particulate and trace gas emissions from large biomass fires in North America. In Levine JS (ed) *Global Biomass Burning: Atmospheric, Climatic, and Biospheric Implications* pp 209-224 MIT Press Cambridge

Radke LF, Hegg DA, Lyons JH, Brock CA, Hobbs PV, Weiss R, Rasmussen R (1988) Airborne measurements on smokes from biomass burning. In Hobbs PV, McCormick MP (eds) *Aerosols and Climate* pp 411-422 A Deepak Publishing Hampton VA

Rasmussen RA, Rasmussen LE, Khalil MAK, Dalluge RW (1980) Concentration distribution of methyl chloride in the atmosphere. *J Geophys Res* 85:7350-7356

Reichle Jr HG, Connors VS, Holland JA, Sherrill RT, Wallio HA, Casas JC, Condon EP, Gormsen BB, Seiler W (1990) The distribution of middle tropospheric carbon monoxide during early October 1984. *J Geophys Res* 95:9845-9856

Richter DD, Raalston CW, Harms WR (1982) Prescribed fire: Effects on water quality and forest nutrient cycling. *Science* 215:661-663

Rogers FR, Hudson JG, Zielinska B, Tanner RL, Hallett J, Watson JG (1991) Cloud condensation nuclei from biomass burning. In Levine JS (ed) *Global Biomass Burning: Atmospheric, Climatic, and Biospheric Implications* pp 431-438 MIT Press Cambridge

Sanhueza E, Elbert W, Rondon A, Corina Arias M, Hermoso M (1989) Organic and nonorganic acids in rain from a remote site of the Venezuelan savanna. *Tellus* 41B:170-176

Sarmiento JL (1993) Atmospheric CO_2 stalled. *Nature* 365:697-698

Seiler W, Conrad R (1987) Contributions of tropical ecosystems to the global budgets of trace gases, especially CH_4, H_2, CO, and N_2O. In Dixon RE (ed) *The Geophysiology of Amazonia* ch 9 John Wiley and Sons New York

Seiler W Crutzen PJ (1980) Estimates of gross and net fluxes of carbon between the biosphere and the atmosphere from biomass burning. *Climate Change* 2:207-247

Stocks BJ (1989) Fire behavior in mature jack pine. *Can J For Res* 19:783-790

Susott RA, Ward DE, Babbitt RE, Latham DJ (1991) The measurement of trace emissions and combustion characteristics for a mass fire. In Levine JS (ed) *Global Biomass Burning: Atmospheric, Climatic, and Biospheric Implications* pp 245-257 MIT Press Cambridge

Walker BH (1991) Ecological consequences of atmospheric and climate change. *Climatic Change* 18:301-316

Ward DE, Nelson RM, Adams DF (1979) Forest fire smoke plume documentation. Paper presented at the 77th Annual Meeting *Air Pollution Assoc, Air and Waste Mgt Assoc* Pittsburg PA

Ward DE, Setzer AW, Kaufman, YJ, Rasmussen RA (1991) Characteristics of smoke emissions from biomass fires of the Amazon region-Base-A experiment. In Levine JS (ed) *Global Biomass Burning: Atmospheric, Climatic, and Biospheric Implications* pp 394-402 MIT Press Cambridge

Ward DE, Susott RA, Kauffman JB, Babbitt RE, Cummings DL, Dias B, Holben BN, Kaufman YJ, Rasmussen RA, Setzer AW (1992) Smoke and fire characteristics for cerrado and deforestation burns in Brazil: BASE-B experiment. *J Geophys Res* 97(14):601-619

Wofsy SC, Fan SM, Blake DR, Bradshaw JD, Sandholm ST, Sing HB, Sachse GW, Harriss RC (1994) Factors influencing atmospheric composition over subartic North America during summer. *J Geophys Res* 99:1887-1897

Zepp RG (1988) Environmental photoprocesses involving natural organic matter. In Frimmel FH, Christman RF (eds) *Humic Substances and their Role in the Environment* pp 193-214 John Wiley and Sons New York

Atmospheric Transports of Particulate and Gaseous Products by Fires

M. Garstang, P.D. Tyson[1], H. Cachier[2], and L. Radke[3]
Department of Environmental Sciences
University of Virginia
Charlottesville, VA 22903 USA

Summary

Atmospheric transports of fire-generated products are scale-dependent phenomena. The occurrence of fires in time and space is also related to atmospheric scales of motion. Large scales of motion (>1000 km), such as the semi-permanent subtropical anticyclones, may control rainfall and rainfall in turn influences the water balance and the probability of fires. Long term drought conditions influence biomass and hence the production of any fire. This complex role of the atmosphere in generating and transporting products from fires is the focus of this chapter.

Atmospheric transports are treated in terms of scales of motion which occur at preferred frequencies. Four scales of atmospheric motion are dealt with: the large-scale, long term scale typical of features such as the large semi-permanent subtropical anticyclones which operate on space scales of >1000 km and time scales of >10's of days; the synoptic or storm scales typical of a frontal cyclone with space scales between 100 - 1000 km and time scales of tens of minutes to less than 24 hours, and finally, the turbulent and smaller scales of motion which operate on space scales of much less than 1 km and time scales of minutes to microseconds.

The occurrence of fires in space and time and the transports of the fire products are dealt with in this chapter in each of the above scales of motion. Interaction between scales occurs and is often non-linear. Such interaction is referred to in each of the scales of motion treated.

Fundamental transport pathways are seen to occur as a function of the large scale of motion. These scales similarly influence the occurrence of fires. Zonal motions in the atmosphere predominate over meridional motion but there is a relationship between zonal and meridional motion, dry and wet conditions and the occurrence of fires.

Convection and convective systems like squall lines operate to both transport fire-generated products vertically in the atmosphere and involve the fire particles in droplet, precipitation and deposition processes. Direct fire-generated convection can be intense although typically occurring under conditions (dry) which do not support natural convection. Interaction between the convective motions of the fire and smaller scale motions in the atmosphere, however, may result in velocity fields capable of elevating and transporting particles >100 μm in size to distances in excess of 10 km.

Turbulent and smaller scale motions can couple with mesoscale and larger scale conditions under circumstances which, for example, produce high velocity (>20 m s^{-1}) low level nocturnal jets which can transport fire-generated particles distances of 1000 km from the fire in 10 hours.

Sedimentation of fire-generated products in a given location such as a lake can be due as much to changes in atmospheric circulation fields as in changes of any fire regime. Perhaps more importantly, a given fire regime may be reflected quite differently in different fixed locations due to variations in transport pathways which can be discrete and defined in space and time.

---

[1] Climatology Research Group, University of the Witwatersrand, Johannesburg 2050, South Africa
[2] Centre des Faibles Radioactivités, Laboratoire mixte CNRS–CEA, avenue de la Terrasse, 91198 – GIF sur Yvette CEDEX, France
[3] NCAR/Research Aviation Facility, P.O. Box 3000, Boulder, CO 80307–3000. USA

NATO ASI Series, Vol. I 51
Sediment Records of Biomass Burning
and Global Change
Edited by James S. Clark, Hélène Cachier,
Johann G. Goldammer, and Brian Stocks

Introduction

The complex non–linear interactions between fire and the biological, chemical, and physical systems comprising the environment of the fire are becoming increasingly recognized (Crutzen and Goldammer 1993). Non–linear effects of water balance on past fire frequency and intensity in deciduous forests has been determined by geographic trends in charcoal abundance (Patterson and Backman 1988, Clark and Royall 1994). In fragile arctic and alpine environments subtle shifts in climate translate into broad–scale and long lasting changes in physiognomy of the dominant vegetation (Agee and Smith 1984, Payette and Gagnon 1985). The Holocene record indicates that rather small changes in climate in the boreal forests are converted by fire into broad transitions in vegetation type with consequent impact on carbon storage (Clark and Robinson 1993).

The role of wildfires in atmospheric chemistry and in the budgets of elements such as carbon on geologic time scales remains uncertain. Carbon burial may result in a net increase in atmospheric O_2 but a decline in CO_2. Clark (1988), in particular, has shown that fires contribute to the carbon budget stored in sediments (lakes) due to atmospheric and surface water transport and deposition. Paleoecologists, generally assume that large fragments (10^2–10^5 mm) in the sediments derive from local (catchment) fires, while small particles (10^1–10^2 mm) may derive from geographically wide and distant sources (Clark and Robinson 1993). Biomass burning produces large amounts of particulates in a wide size range. The largest particles are referred to as charcoal and show recognizable cellular structure. They are considered to be primary particles originating from burned plant debris. Submicron (fine) particles are of secondary origin from gas–to–particle oxidative conversion processes. Two main components are distinguished for these fine carbonaceous particles: an organic carbon (OC) fraction which is always more abundant (~ 80% of particulate carbon) and the black carbon (BC) fraction. The production rate of charcoal and aerosol particles is dependent on the quality of the combustion and fire efficiency. Charcoal re–entrainment by wind erosion of post–burn soils could also be an important source of atmospheric charcoal. Some preliminary estimates of aerosol production indicate that biomass burning could account for 7% of global total particulate matter (TPM) (Andreae 1991). However, its contribution to atmospheric particulate carbon burdens is more important, of the order of 65% of total OC and 45% of BC, according to most recent estimates (Liousse et al. 1995).

Whereas OC aerosol particles are likely to undergo chemical degradation with time, charcoal and BC particles are chemically resistant. These latter particles are produced by combustion only and thus have the potential to trace fires in sediments unambiguously, where they are often referred to as "charcoal", whether of primary or secondary origin.

Suitable sedimentary environments, such as lakes, are not randomly distributed in space or with respect to changing climatic conditions over time. Adequate lake sites are rare or absent in dry environments. The paleofire record is poorest for arid and semi–arid environments. In more favorable locations it is not clear why lakes do not show a charcoal signature where fires are known to occur in their catchments and why periods characterized by no local fires often display as much charcoal as times when fires have been known to be frequent (Clark and others, this volume).

Fires in ecosystems clearly do not occur in isolation. They are influenced by and, in turn, influence the environment, including its biology, chemistry, and physics. This chapter focuses on the role of the atmosphere in generating and transporting products from fires. Atmospheric conditions and transports are considered within a complex mosaic of biological, chemical, and manmade conditions. Thus, changes in biome composition, shifting biomes, and attendant changes in fuel characteristics and loadings, which may both affect and be affected by the atmosphere, are recognized as important but are left to other chapters of this volume.

Atmospheric transport is seen as a scale–dependent phenomenon. Large–scale, long–term scales of motion in the atmosphere interact with turbulent processes within the space and time scales of a given fire. The fire itself is a product of atmospheric scales of motion (e.g., dry conditions). Transport of material from the fire starts with the initial elevation of a particle by turbulent motion. Where the particle is finally deposited depends upon a series of interacting scales of motion, which range from the convective– (dry and moist), meso-synoptic– (cloud ensemble to storm) to the planetary– (subtropical anticyclone) scales. These scales are not independent but interact non–linearly. A relatively modest elevation of a particle into a narrow jet stream will transport that particle two orders of magnitude further from the source than a comparable particle not entering the jet.

The main purpose of this chapter is to review the scale–dependent processes in the atmosphere that might influence the sediment record in a given lake. Such atmospheric processes might affect the production, transport, and deposition of fire generated material. A receptor, such as a lake, must be seen as responding to time and space changes in atmospheric influences on fires as well as on transport and deposition. While major atmospheric transport pathways may always have been present on the planet, the relationship between these transport pathways and any given receptor(s) is continually changing. Changes in deposition may be as much a function of changes in space and time of the transport pathway as any change in the rate of production by fires such as fire frequency or intensity. Differences in deposition rates in varied sediments in lakes in boreal forests may be, in part, due to changes in transport patterns (Clark and Richards 1995).

The discussion below starts with the large scale and progresses scale–wise to the small scale. By definition, however, scales of motion in the atmosphere are interactive. Because this interaction is non–linear, effects of one scale on another are often large and unpredictable. Treatment by scale is adopted as a necessary measure by which complex processes can be ordered. No scale, however, can be treated in isolation so that cross–referencing to the roles played by other scales will be necessary in each section.

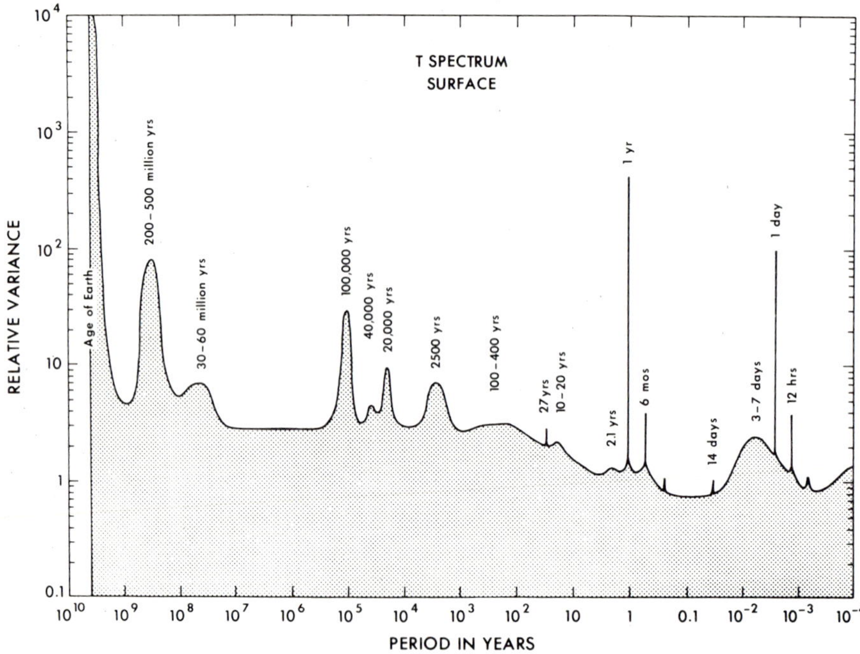

Figure 1. Idealized power spectrum of atmospheric temperature between 10^{-4} and 10^{10} yr. Peaks in variance are related to weather and climate processes discussed in the text. (after Peixoto and Oort 1992).

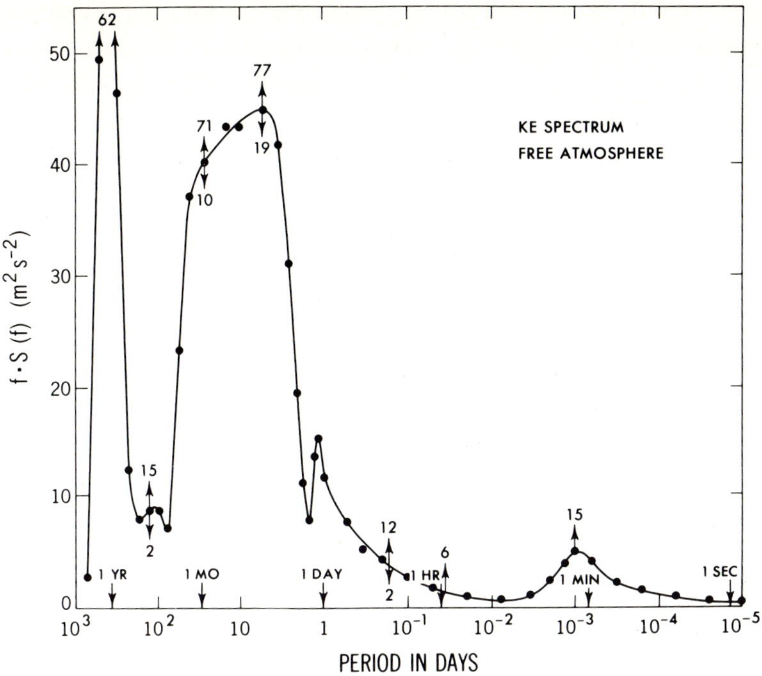

Figure 2. Spectrum of atmospheric kinetic energy between 10^{-5} and 10^{3} days adapted from Vinnichenko (1970). The kinetic energy spectrum is tentative and somewhat schematic, being patched together from limited data at few stations. The abscissa is Period = 1/f and the ordinate is $f * S^2(f)$, where f is frequency and $S^2(f)$ is variance associated with frequency f. This representation is used if the frequency range is large (necessitating a log 1/f scale). The area under the curve is normalized to total variance. Peaks in the variance are related to weather events in the text. (After Peixoto and Oort 1992.)

Atmospheric scales

A schematic variance spectrum of surface air temperature (Fig. 1) illustrates that internal energy in the atmosphere near the surface is not uniformly distributed over time but exhibits a range of distinct peaks (Mitchell 1976, Peixoto and Oort 1992). An approximate kinetic energy spectrum derived from wind velocities representative of only a few stations (Fig. 2) provides greater detail for periods below < 1 yr (frequencies above 10^{-3} days). The kinetic energy in the atmosphere is concentrated at specific frequencies. Most of the energy resides between 1 and 30 d, periods typical of the planetary short– (cyclones) and long–waves (large ridges/anticyclones and troughs) in the atmosphere. The sharp peaks on either side of the synoptic–planetary–scale maximum are the annual and diurnal cycles. The relative maximum with a period around 15 min is due to local wind systems (Tyson 1968) and small–

scale turbulent motions (Pasquill 1974). Cloud– or convective–scale motions and the ensemble of convection occurring in mesoscale squall lines do not appear in either spectrum. The convective– and meso–scale motions, which are important in atmospheric processes involving deep vertical transfers of energy as well as matter, are inadequately sampled by conventional observational systems and are consequently missing in these spectra.

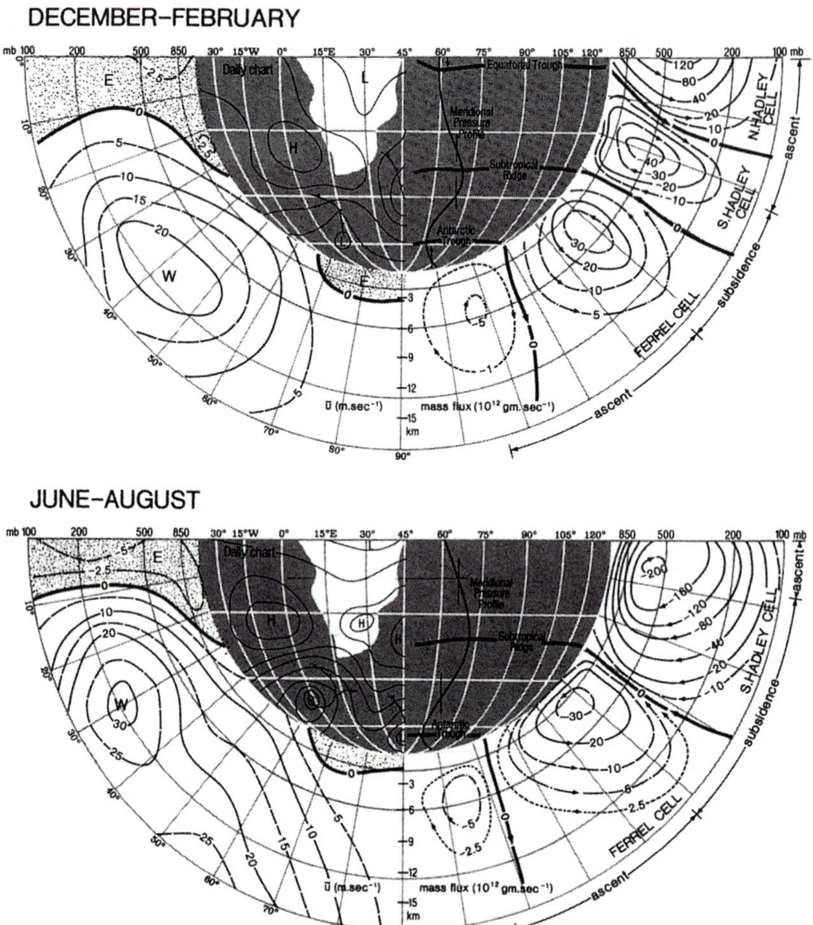

Figure 3. Mean December–February and June–August zonal and meridional wind and mass flux in the southern hemisphere. The shaded region of the atmosphere on the left hand side shows easterly zonal flow, the clear region westerly zonal flow. The right hand side shows the meridional Hadley, Walker, Ferrel and Polar cells (data taken from Newell et al. 1972, after Tyson 1986.)

Figure 1 contains peaks that suggest longer periodicities than those discussed earlier. Periodic phenomena appear as peaks in the spectrum, such as the quasi–biennial oscillation with a period of 2.1 years (Reed et al. 1961, Veryard and Ebdon 1961, Maujokat 1986, Lau 1988) and sunspot cycles near 22 years (Burroughs 1992). Increased variance can be associated with aperiodic phenomena such as the Little Ice Age at 100–400 yr, which began with rapid expansion of mountain glaciers in Europe (Grove 1988). The peak near 2500 yr has been ascribed to cooling after the "climate optimum" about 5000 yr BP (Imbrie 1985). The three peaks at 20,000, 40,000 and 100,000 yr have been related to periodic astronomical variations of the orbital parameters of the earth used by Milankovitch (1941) to explain the ice ages (Imbrie 1985). Each major shift in the energy spectrum of the atmosphere implies a change in the circulation fields. Because these circulation fields are scale–wise interactive, changes in circulation and atmospheric structure can occur on all of the time and space scales identified in the above spectra. Changes in the sediment record must take into account these changes in the atmosphere if the sediment record is to be adequately interpreted. The spectrum of changes in atmospheric motion and

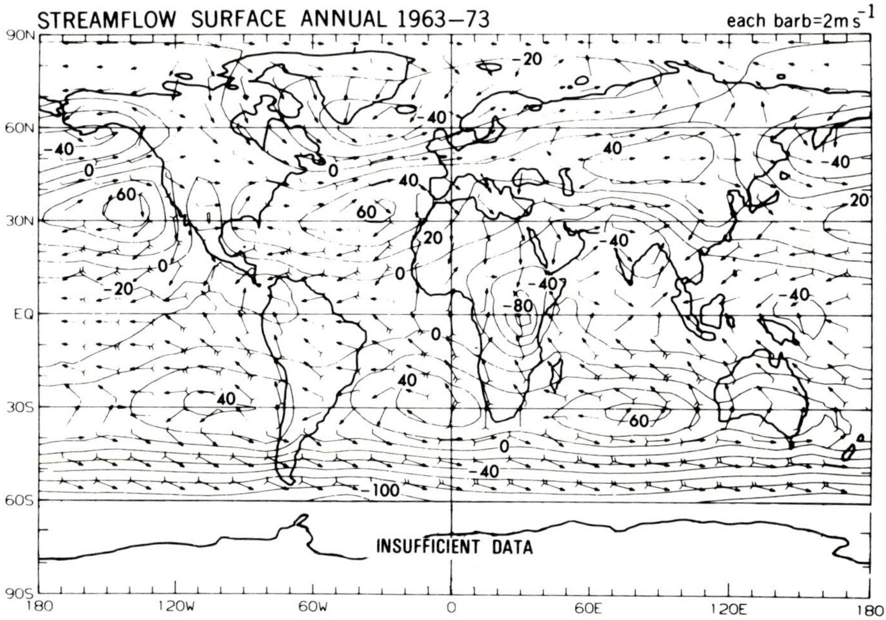

Figure 4. Global distributions of the height anomalies of the 1000 hPa pressure field, $z_{1000} - z^{SA}{}_{1000}$, in gpm for annual–mean, northern winter. The quantity $z^{SA}{}_{1000}$ (= 113 gpm) represents the height of the 1000–mb level. Vector surface winds are shown by arrows. Each barb on the tail of an arrow represents a wind speed of 2 m s^{-1}. (After Peixoto and Oort, 1992.)

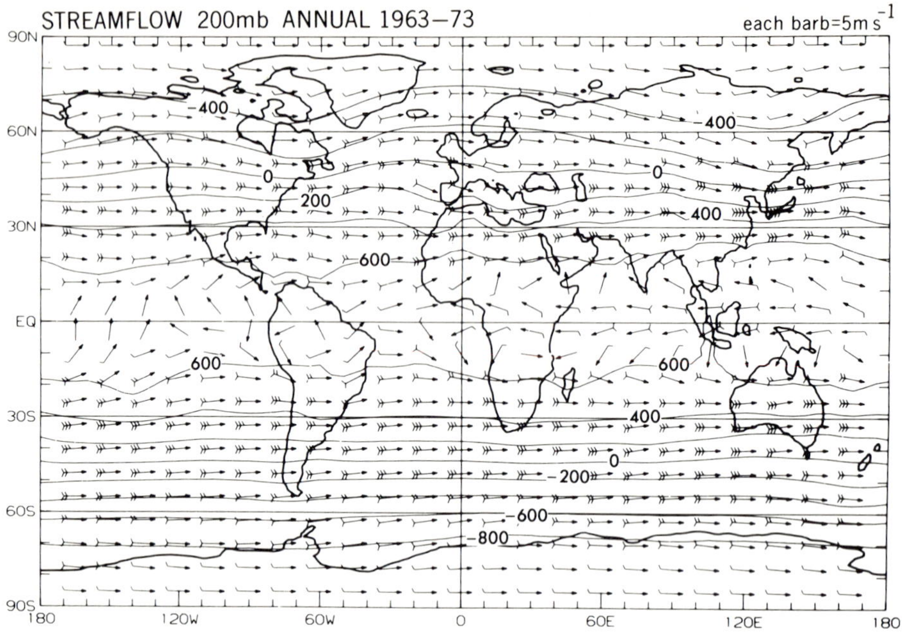

Figure 5. Global distribution of the height of the 200–mb pressure field, z_{200} – 11784 gpm, in gpm for annual mean, northern winter. Also shown are vector winds at 200 hPa. Each barb on the tail of an arrow represents a wind speed of 5 m s^{-1}. (After Peixoto and Oort, 1992.)

structure that impact the transport and deposition of fire generated material will be treated in terms of three fundamental scales of motion in the atmosphere.

Planetary and synoptic scales

Global radiative balances producing warm tropical and cold polar surface temperatures combine with the angular rotation of the earth to produce a three–cell general circulation shown under present conditions in Fig. 3. Figure 4 shows the horizontal wind and pressure fields that make up the trades or easterlies of the tropics, the westerlies of middle latitudes, and the polar easterlies. At higher elevations (200 hPa, approximately 10 km) a more nearly zonal flow dominated by the westerlies is seen in Fig. 5. The longitudinal profile of the zonal–mean sea level pressure in Fig. 6 shows that these large scale wind fields are associated with meridional gradients of pressure, with two dominant high pressure systems in the subtropics near latitudes 30° N and S. The zonal–mean wind components in Fig. 7 reflect the surface, upper air, and longitudinal pressure gradients of Figs. 4 to 6 in producing broad zones of east–to–west flow (easterlies) in the tropics and west–to–east flow (westerlies) in midlatitudes. Westerly flow reaches a maximum of 25 m s^{-1}

in both hemispheres near 200 mb (~ 10 km), but westerly flow, consistent with the pole to equator pressure gradients of Fig. 6, is higher at all levels (by about 5 m s^{-1}) in the southern hemisphere. Seasonal migration of the zonal wind fields in both hemispheres is about 5° latitude exhibiting a broadening of the westerlies in the northern hemisphere winter.

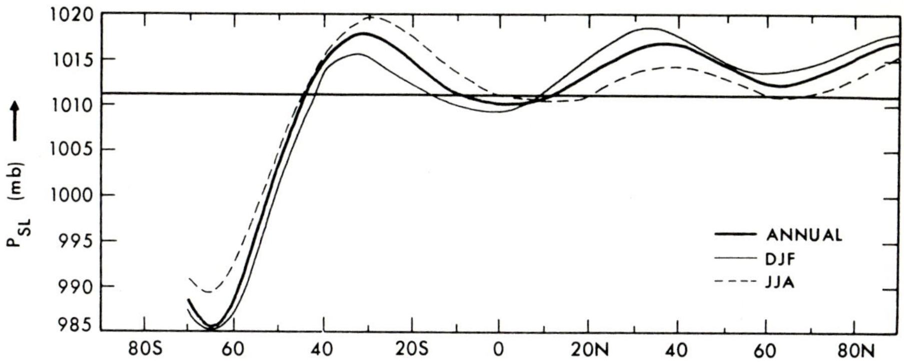

Figure 6. Meridional profiles of the zonal–mean sea level pressure for annual–mean conditions after Trenberth (1981) and for the DJF and JJA seasons after Oort (1983) in mb. (After Peixoto and Oort 1992.)

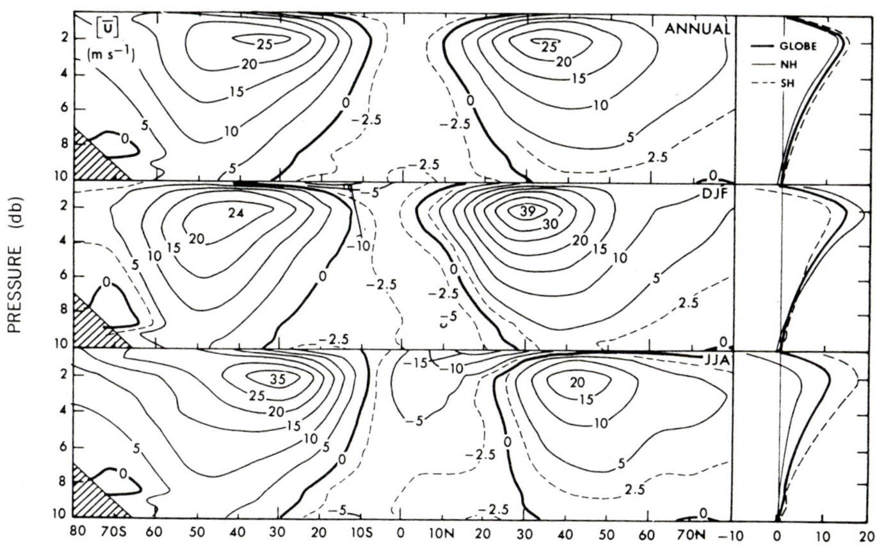

Figure 7. Zonal–mean cross sections of the zonal wind component in m s^{-1} for annual, DJF, and JJA mean conditions. Vertical profiles of the hemispheric and global mean values are shown on the right. (After Peixoto and Oort 1992.)

Global distributions of mean annual precipitation (Fig. 8) show that regions of maximum precipitation coincide with the equatorial trough and that the midlatitude cyclone tracks on either side of the subtropical anticyclones. Precipitation minima are located in the large anticyclones of the subtropics and in polar regions. Meridional profiles of zonal–mean precipitation over land (Fig. 9) emphasize the equatorial maximum but also show a much less pronounced maximum (mainly in winter) in precipitation at higher latitudes. The peak in rainfall over land in the southern hemisphere near 50° S is an artifact of the few isolated areas of land at that latitude. In the subtropics (35° to 40°) annual rainfall is not only low (< 50 cm), but is also strongly seasonal. Mean zonal cloudiness (Fig. 10) corresponds with the general rainfall distributions of Figs. 8 and 9, but shows the distinction between tropical convective rain (cumuliform and lower percent sky cover) and midlatitude stratiform rain (stratiform clouds with high percent sky cover).

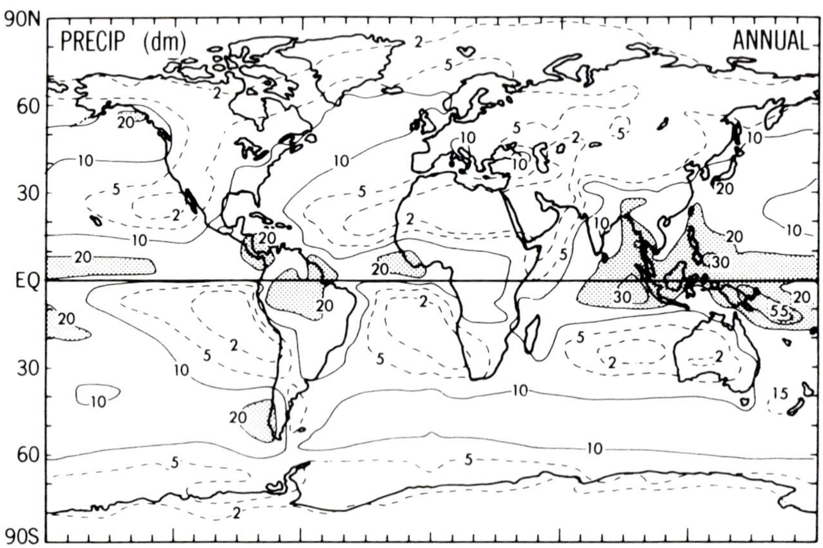

Figure 8. Global distributions of precipitation for annual–mean conditions in dm yr^{-1} based on data from Jaeger (1976). (After Peixoto and Oort 1992.)

Subtropical anticyclones transport burning emissions on the planetary scale. Figure 11 shows the relationship between transport and precipitation over land. The anticyclones are semi–permanent planetary scale features occupying a significant fraction of the earth's lower atmosphere. They are maintained by sinking air between Hadley–Walker circulations along the equatorial trough and midlatitude Ferrel cells (Fig. 3). Adiabatic warming of descending anticyclonic air produces near cloudless skies and, over land, low humidities, low rainfall, and high surface temperatures. Large anticyclonic gyres extend outwards from their arid continental centers to produce major transport pathways over adjacent areas of higher rainfall and vegetation cover (Fig. 11). The subtropical highs migrate seasonally 5° latitude into the summer hemisphere, but are most intense over continents in the winter hemisphere, where both fires and transport pathways are most prominent. Cachier et al. (1995) and Moody et al. (1991) demonstrate that transport of continental species (e.g. radon, BC) to Amsterdam Island in the central southern Indian Ocean increases during the winter dry season. In the summer hemisphere in the tropics the near steady state condition that prevails under anticyclones is replaced with synoptic scale traveling or stationary easterly waves. Rainfall associated with these synoptic scale systems inhibits both natural and manmade fires (Fig. 11).

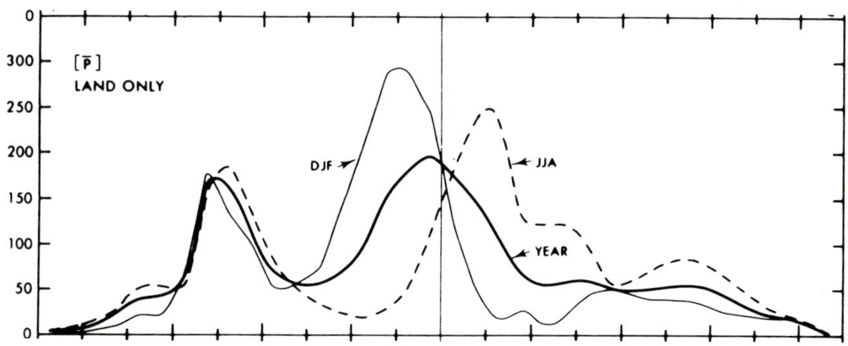

Figure 9. Meridional profiles of the zonal–mean precipitation rate in cm yr–1 for the land area for annual, DJF and JJA mean. (After Peixoto and Oort 1992.)

In midlatitudes the boundary between the westerlies and polar easterlies (Fig. 3) defines the location of the polar front. Polar frontal cyclones, in the form of short waves traveling from W to E along the long waves, produce rainfall (Figs. 8, 9) over a relatively broad, midlatitude band. While frontal cyclones are followed by slow–moving high pressure systems, which share some characteristics of subtropical highs, the frontal precipitation and cyclones that precede the highs disrupt flow and inhibit fires. The region of preferred westerly transport of emissions, as well as fires themselves, in midlatitudes therefore lies between the poleward margins of the semi–permanent anticyclones and the equatorward extent of the polar front. Both fire occurrence and transport pathways are modulated seasonally by seasonal migrations of semi–permanent anticyclones and the polar front.

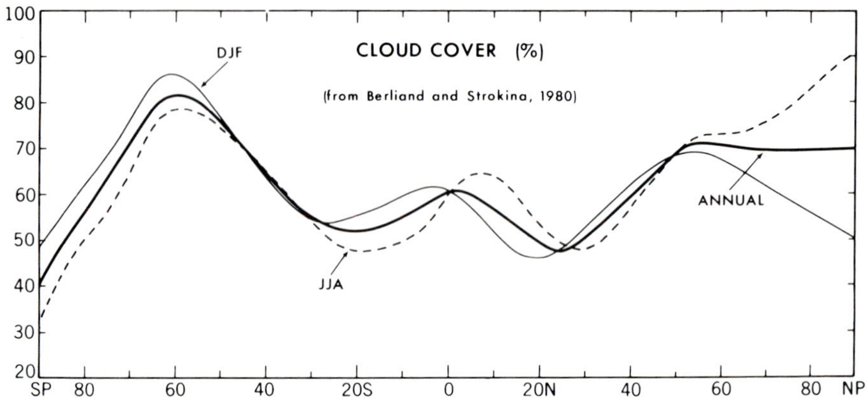

Figure 10. Meridional profiles of the zonal–mean cloud cover (in %) for annual, DJF, and JJA mean conditions based on data from Berlyand and Strokina (1980). (After Peixoto and Oort 1992.)

Structure and transport of the subtropical anticyclones. The role played by the semi–permanent anticyclones in governing the structure and transports of the atmosphere is illustrated below for a given location. Tyson et al. (1995) and Garstang and Tyson (1996) examined atmospheric structure and transports over southern Africa. Figure 12 shows the seasonal march over southern Africa of anticyclonic circulations, westerly, and easterly disturbances. During a large fraction of the year (March–September) anticyclones occur on 50% or more of all days reaching a peak of 80% in July.

On the larger scale, multiple persistent stable layers trap and concentrate products into altitude bands and inhibit both their subsequent upward as well as downward displacements. Horizontal transport distances are greater by as much as two orders of magnitude under these stratified versus unstratified conditions. Dispersal of fire generated particulates must occur over large areas. Rates of sedimentation in any given location will be reduced under stratified compared to unstratified conditions. The size spectrum of fire generated particles transported by the atmosphere is likely to be significantly different than in a well mixed atmosphere. Thus, local as well as distant deposition is likely to be affected with local and distant deposition rates being small but total mass transport being large. Current evidence (Andreae 1991) suggests that only 3% of the total charcoal produced by a fire is transported by the atmosphere; the much larger fraction is transported by surface water. It is likely that the partitioning between the transport by the atmosphere and by surface water runoff is significantly different from the above figures in the dry environments of the anticyclones and their margins.

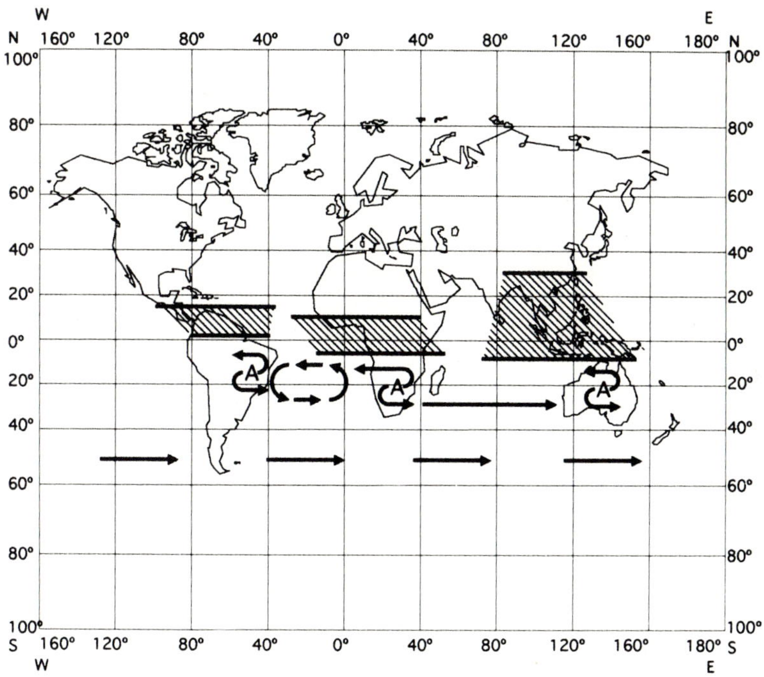

Figure 11a. Zonal near surface mean air transport and rainfall in the southern hemisphere winter months of JJA over the continents. Relative centers of rainfall maxima are hatched. Bold arrows represent westward and eastward transport pathways out of the continental subtropical anticyclones labelled A. Transport in the southern hemispheric westerlies is shown for SH winter only.

Cachier et al. (1991, 1995) made detailed studies of particulates in close proximity to fires. In the Ivory Coast, atmospheric background burning emissions increased in concentration during the dry season, when extensive savanna fires occur. In the dense humid savannas, these fires develop primarily during the day when atmospheric humidity is at a minimum. Observed values both in savanna (Liousse 1993) and in equatorial forest (Artaxo et al. 1995) regions show that the seasonal pattern of tropical biomass burning production of particles may display important interannual variability. This variability appears both in the timing of fires and in their intensity, and may be related to year–to–year climate variability.

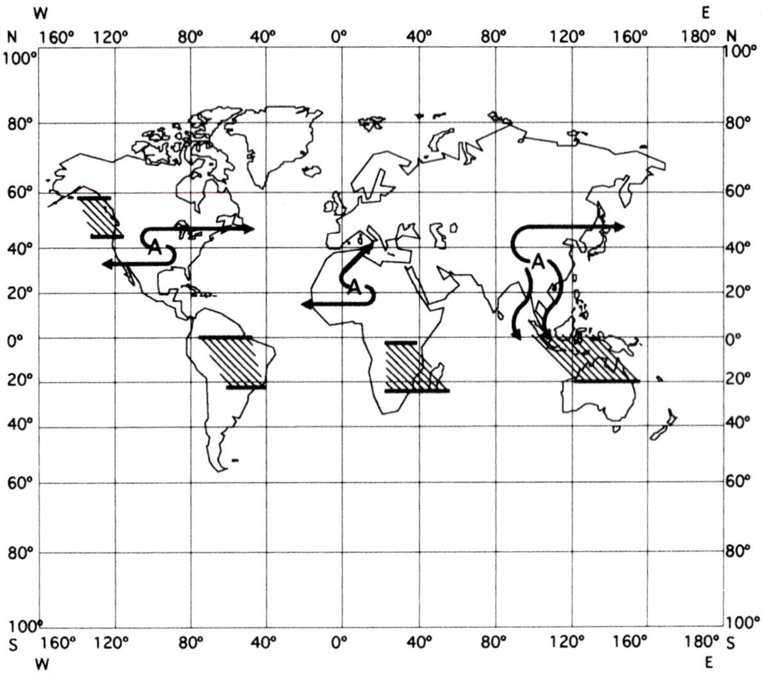

Figure 11b. Zonal near surface mean air transport and rainfall in the northern hemisphere winter months of DJF over the continents. Remainder as for (a) above.

The large–scale sinking within the subtropical highs results in near cloudless skies in a dry atmosphere. Radiative balances at the surface are negative for as much as 16 hours of the 24 hours of the day resulting in cold surface temperatures at night. Inversions of the vertical lapse rate of temperature or absolutely stable layers form and persist in such an environment. Absolutely stable layers, in which the lapse rate of the atmosphere is less than that of a moist adiabatic parcel, as well as inversions of temperature occur at multiple levels in the atmospheric column. The temporal persistence, multiple–level structure and spatial extent of the inverted and stable layers is shown in Fig. 13. Marenco et al. (1990) show, in Fig. 14, pronounced layering of CO, CO_2, NO_y and O_3 over central Africa during TROPOZ.

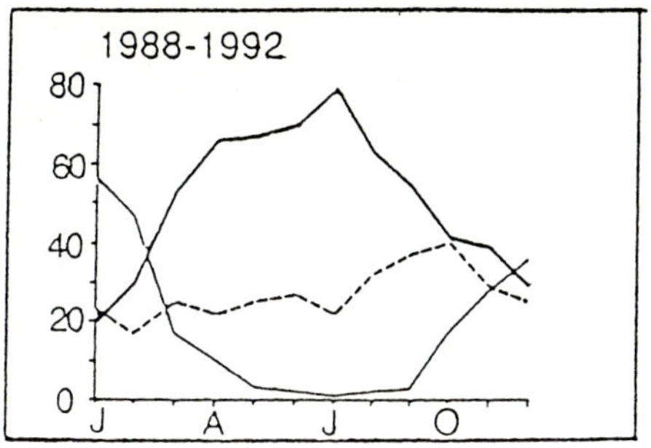

Figure 12. Frequency of occurrence over the period 1988–1992 of the combined continental and ridging anticyclones (heavy solid line), the westerly waves/troughs and cut off lows (light solid line), and tropical easterly waves (dashed lines). (After Tyson 1986).

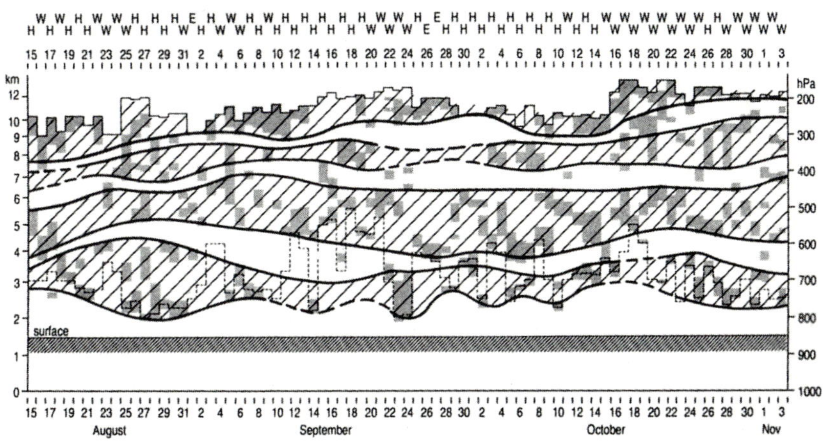

Figure 13a. Daily variation of absolutely stable layers over southern Africa at Pretoria during SAFARI (August 15–November 4, 1992). The duration, height and depth of each stable layer on each day is shown by rectilinear stippling. Continuous stable layers are cross hatched and bounded by heavy solid lines, discontinuous stable layers by heavy dashed lines. The height of the 1200 UT mixing depth is shown by light dashed lines. The circulation class for each day is shown by H – Continental high, W – Westerly disturbance, and E – Easterly disturbance.

The temporal and spatial persistence of inversions and stable layers has major consequences for the vertical and horizontal transport of fire generated products. Interaction at convective and turbulent scales occurs in the immediate vicinity of the fire. These interactions are very different in this anticyclonic environment from other weather regimes and will be dealt with on the smaller scales in sections that follow.

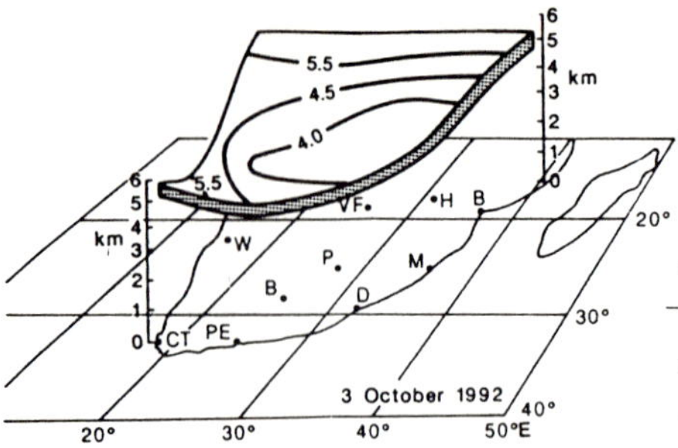

Figure 13b. Spatial extent of the 5 km absolutely stable layers over southern Africa on October 3, 1992. Thickness of layers is diagrammatic. The layer is about 50 hPa thick.

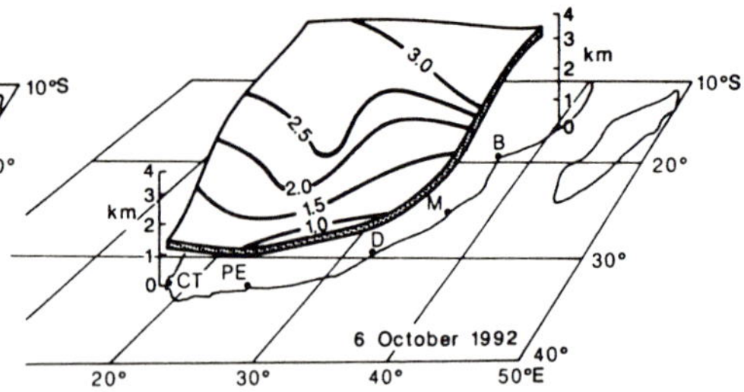

13c. As for (b) except at 3 km on October 6, 1992. (after Garstang et al. 1996.)

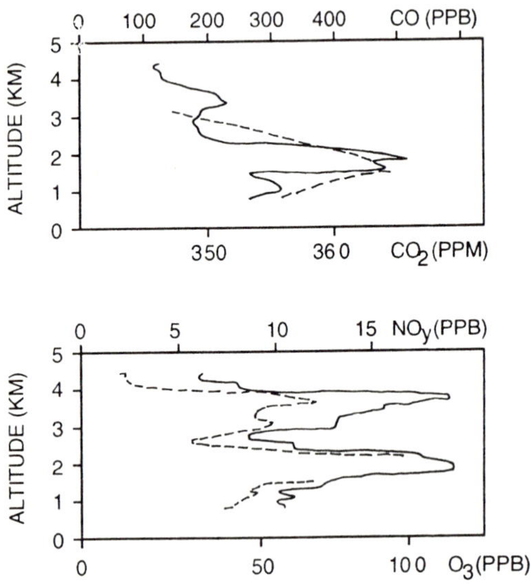

Figure 14. Profiles of CO and CO_2 (upper) and O_3 and NO_4 (lower) over central Africa during TROPOZ (after Marenco et al. 1990).

The biomass burning particulate emissions are primarily carbonaceous (~ 50–80% C by mass). As in other combustion particles, two main fractions may be distinguished, black (BC) and organic carbon (OC). BC is always the less abundant component of the carbonaceous aerosol, but, due to the various modes of combustion, its relative abundance ranges widely, from 4–8% during smoldering to

12–20% for flaming fires. During fires in mature forests (with large trees) it is now assumed that more than half of the fuel is burned under smoldering conditions, after the fire front has passed (Susott et al. 1991). Flaming conditions yield higher total particulate emissions, with emission factors (EF) an order of magnitude higher than observed for smoldering. Atmospheric BC particle emissions range from EF = 1.0 gC gC^{-1} dry matter to EF = 0.4 gC gC^{-1} dry matter in flaming and smoldering, respectively. Biomass burning aerosols may be distinguished from industrial combustion aerosols by their enrichment in potassium (K:BC ~ 0.2) which may serve as a tracer of biomass burning plumes far from source regions (Andreae 1983).

The combined effects of horizontal and vertical transport in and on the margins of subtropical anticyclones is to produce transport pathways that extend over planetary scale distances. An important part of the anticyclonic transport is that of recirculation, which retains fire generated products over the continental source region as well as injecting recirculated material into major transport pathways which exit the continent.

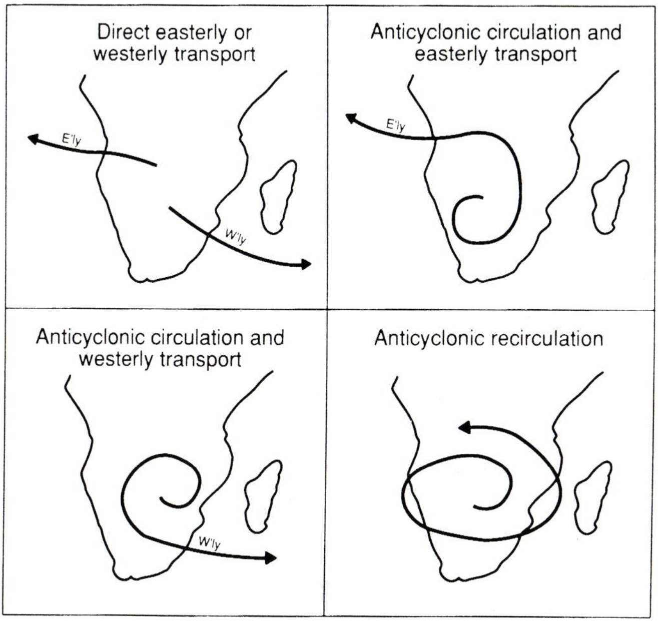

Figure 15. Major transport trajectory modes likely to result in easterly or westerly exiting of material from southern Africa or in recirculation over the subcontinent. (After Garstang and Tyson, 1996.)

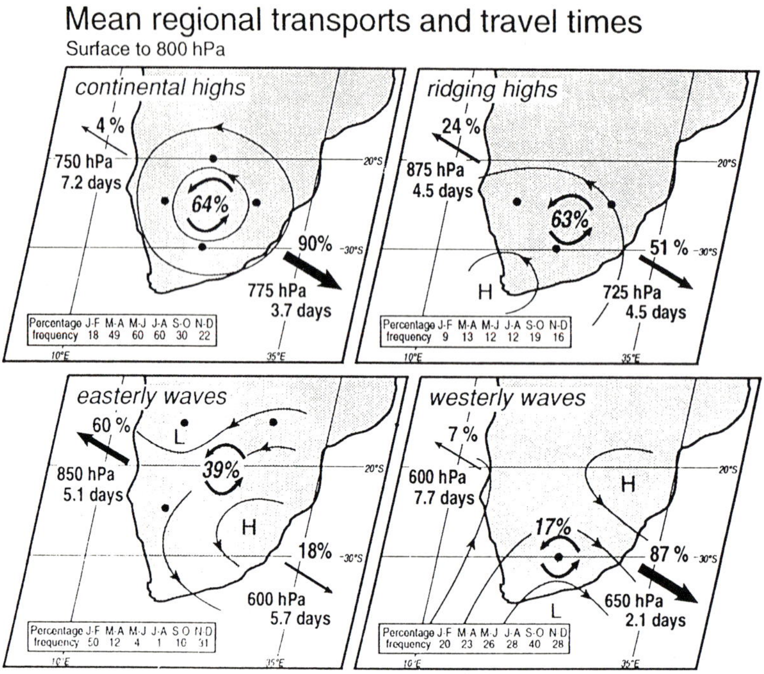

Figure 16. Mean percentage surface–850 hPa transport and recirculation by circulation type over and out of southern Africa. Mean heights of transport and transit times to reach beyond the west coast at 10° E and beyond the east coast at 35° E are given. (After Garstang and Tyson 1996.)

Garstang and Tyson (1996) show that transport is circulation–type dependent. The direction of transport, establishment of any dominant transport pathway, and the resultant frequency and total mass transport depends upon the location of the source within the spatial domain of the circulation type. Figure 15 shows five transport modes over southern Africa associated with the dominant circulation types contained in the frequency distribution of Fig. 12.

Air transport has been traditionally inferred from trajectory calculations (Danielson 1959, 1961). All trajectory methods are limited by the ability to observe or determine three–dimensional velocity fields in the atmosphere. Trajectories best approximate actual atmospheric transports under large scale steady state conditions. Under these circumstances vertical motions due to storm and convective scale systems are small or absent. Such conditions are approximated in and near the semi–permanent subtropical anticyclones and are typical of southern Africa in austral winter and spring. Garstang et al. (1996) used a three–D trajectory scheme developed by Kållberg (1984), which is based on the European Center for Medium

Range Weather Forecasting (ECMWF) global operational analysis (Bengtsson 1985). Pickering et al. (1994), in comparing trajectories calculated by different techniques over southern Africa, concluded the ECMWF data base to be the best available observations for that location and purpose.

Tyson et al. (1995), however, used an ensemble approach in which a large number of trajectories were calculated by circulation type. All trajectories for each circulation type were then followed through meridional and latitudinal planes. These planes were spaced at 5° intervals over the continent of Africa and at 10° intervals over surrounding oceans. Planes extended from the surface to 500 hPa (6 km). A record was kept of the direction of the trajectory passing through the plane such that recirculated air could be distinguished from air being transported directly to the plane from the source. The distance and time taken from the source to the plane was recorded.

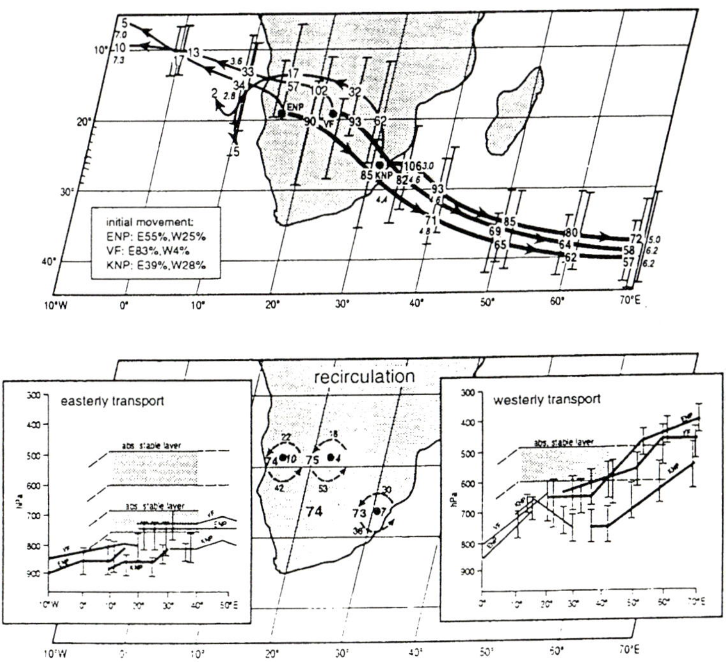

Figure 17. Vertically–integrated surface to 850 hPa transport from Etosha National Park, Victoria Falls and Kruger National Park during SAFARI. Above are maximum frequency pathways. Below are recirculated transport and transport variations with pressure level, latitude, and absolute stability of the atmosphere (occurring when observed lapse rate is less than the saturated adiabatic lapse rate). Confidence limits on horizontal and vertical dispersion of transports are given. (After Tyson et al. 1995.)

Figure 16 shows the mean regional low level transports over southern Africa by circulation class based upon a 7–year climatology (Tyson et al. 1996). A much shorter record (80 days) of the transport picture is shown from three locations in southern Africa for Southern African Fire–Atmosphere Research Initiative (SAFARI) in 1992 in Fig. 17. Both figures support the general transport picture presented in Fig. 11 and both figures are regarded as typical of the transport pathways which exist over the continents in both hemispheres.

Table 1. Mean annual zonal volume flux as a function of circulation class through 10° E and 35° E in $m^3\ y^{-1} x\ 10^{16}$ and in percent.

Circulation Type	Mean Annual Frequency of Occurrence (%)	Annual Zonal Volume Flux ($m^3\ yr^{-1}$ ¥ 10^{16})	Percent Volume Flux Eastwards and Westwards	Percent of Total Zonal Volume Flux
continental – 10° E	40	31.9	67.7	26.6
anticyclone – 35° E		46.2	63.5	38.5
ridging – 10° E	13.6	2.7	5.7	2.2
anticyclone – 35° E		8.5	11.7	7.1
westerly – 10° E	27.6	6.1	13.0	5.1
wave – 35° E		7.7	10.6	6.4
easterly – 10° E	18	6.4	13.6	5.3
wave – 35° E		10.4	14.3	8.7
Totals		47.1		39.2
		72.8		60.7

The total zonal volume flux of air over southern Africa is substantially greater than the meridional volume flux. The volume fluxes of air depend on size (horizontal and vertical dimensions) of the transport plume and on plume velocity. These factors are related, in turn, to circulation class and to frequency of occurrence of the circulation class (Garstang and Tyson 1996). Figure 17 shows locations of the pathways of maximum frequency zonal transport originating from three locations across southern Africa during SAFARI. Frequency distributions of air trajectories crossing meridional planes located at intervals of 5° longitude over the continent and 10° longitude over the adjacent oceans are shown. The limits from the central core transport contain 98% of all trajectories originating from each location. The percent eastward and westward transport is given for each location. Table 1 gives the long–

term mean annual zonal volume flux of air out of southern Africa across longitudes 10° E and 35° E into the Atlantic and Indian Oceans respectively (Garstang and Tyson 1996). Approximately 40% of the air is transported westwards, 60% eastwards. Nearly three quarters of the westward and eastward transports are achieved by the anticyclonic circulations. More than half of the total elevated mass is recirculated and contained over the subcontinent of Africa. Variability of these numbers, both on an intra-annual and longer term basis, is likely to be high. Extremely dry conditions, for example, prevailed during the 1992 SAFARI experiment (Andreae et al. 1994, Alexander and van Heerden 1992). Transport to the west was about 27% (compared with a 40% long-term mean) and to the east 73% (60% long-term mean). Transports by anticyclonic circulations in SAFARI were 55%, lower than the 75% long-term mean. Correspondingly, transports during SAFARI by westerly waves were substantially greater than the long-term mean.

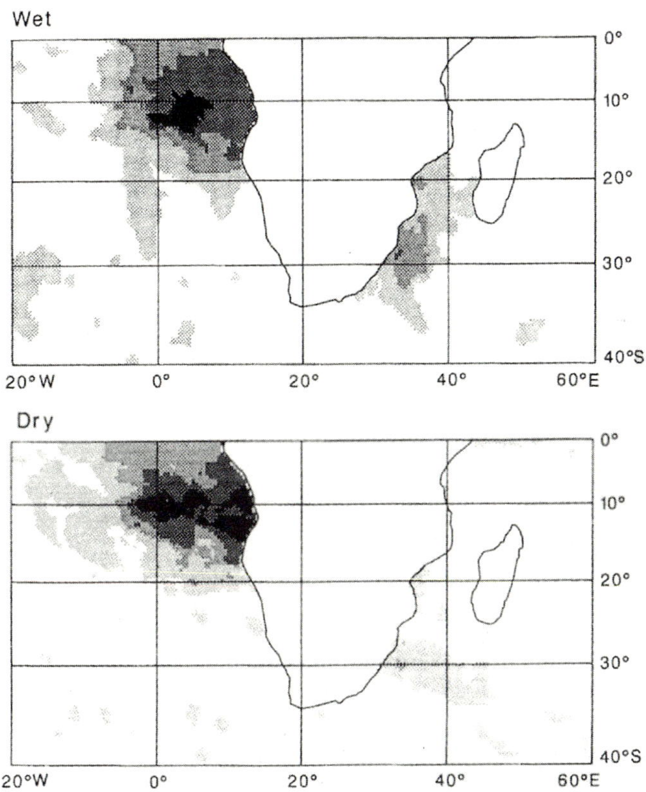

Figure 18. Aerosol optical density contours show the percentage occurrence of weekly aerosol loadings exceeding 0.3 during August to October in the wet year of 1987 and dry year of 1992. Heaviest shading indicates 100%; lightest 25%. (After Garstang et al. 1996.)

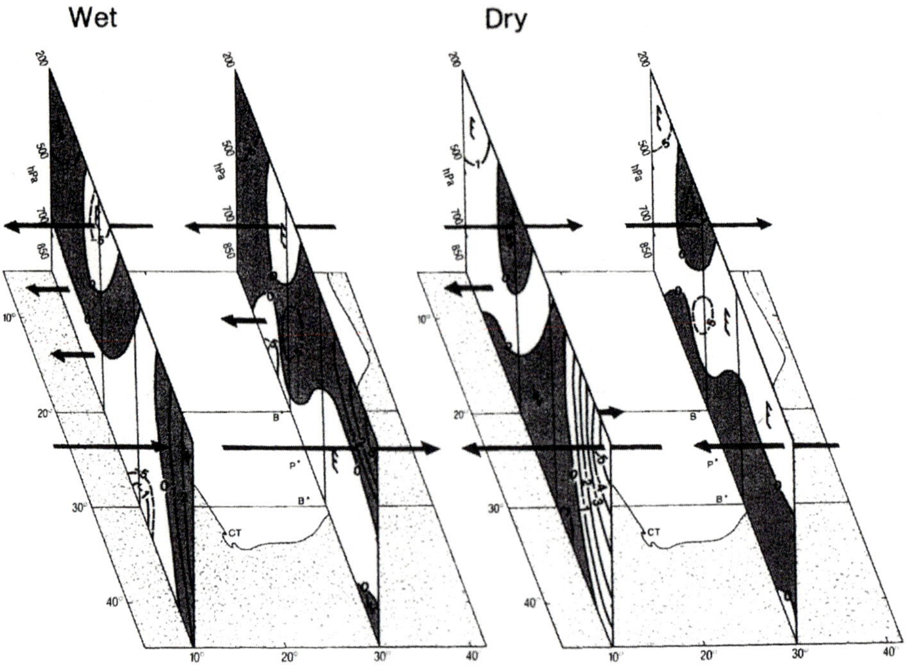

Figure 19. Summer (October–March) zonal wind anomalies (m s^{-1}) during the wet spell of 1971/1972 to 1980/1981 and dry spell of 1962/1963 to 1970/1971. Positive anomalies indicate westerly (W) flow; negative anomalies indicate easterly (E) flow. (After Tyson 1986).

Aerosol optical thickness, derived from AVHRR measurements (Swap et al. 1996b) for the dry conditions of SAFARI, are compared in Fig. 18 with a corresponding three-month wet period. Little difference is observed off the west coast. During the dry period the east coast of southern Africa shows aerosols extending over the Indian Ocean. During the wet period the plume is replaced by a shaded region that extends southward along the coast. Changes in rainfall regimes over southern Africa (Tyson 1986) are directly related to changes in the large scale circulation fields (Harrison 1988, Lindesay 1988, Mason 1992). Figure 19, spanning the summer rainfall period of October to March, shows zonal wind anomalies for wet and dry years. During wet years tropical easterlies strengthen as midlatitude westerlies weaken. Anticyclonic recirculation disperses transports towards low latitudes along the coast and into these strengthened easterlies. In dry years the situation is reversed; diminished transports to the west in weakened easterlies of dry years correspond to the findings in SAFARI. These changes in circulation fields and rainfall are reflected in recently documented changes in strength of the subtropical anticyclones (Davis et al. 1996) discussed below. Swap et al. (1996a), using similar analyses, show aerosol plumes off northwest Africa following the same spatial and temporal pathways as depicted for the southern hemisphere (Fig. 20).

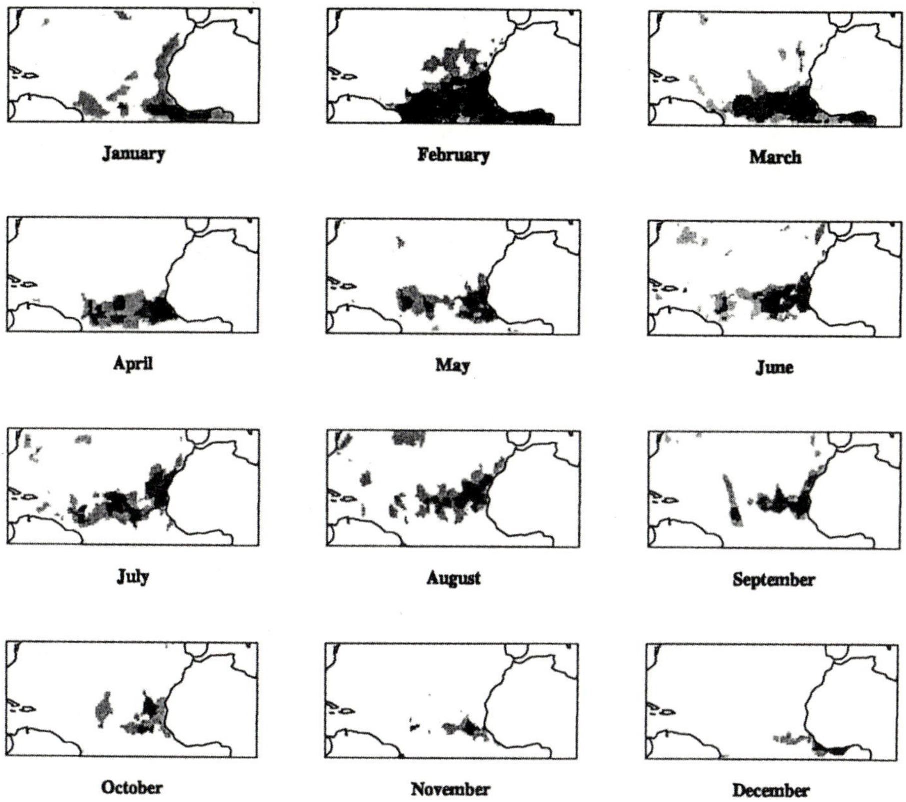

Figure 20. Monthly composites for 1989. Moderate to heavy aerosol events are indicated by NOAA AVHRR aerosol optical thickness values of ≥ 0.30. Temporal characteristics are indicated by gray shading with the lightest shade representing $\leq 25\%$ frequency and each successively darker shade signifying an increase of an additional 25% (after Swap et al. 1996b).

Long–term changes in atmospheric circulations. Foregoing discussion establishes that planetary scale circulations play a dominant role in determining transport pathways from localized sources. Location of source in the prevailing atmospheric circulation controls the fate of products from that source. Changes in the circulation fields that persist over time will change the transports from fixed geographic source regions.

Similarly, if receptors such as lakes are fixed in space changes in transport pathways induce changes at the receptor.

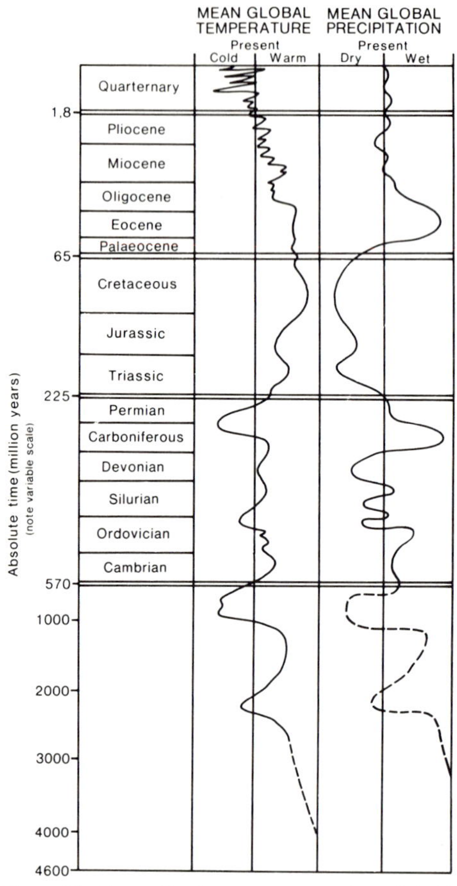

Figure 21. Generalized temperature and precipitation history of the earth (after Frakes 1979).

The temperature and precipitation record for the earth (Frakes 1979) (Fig. 21) includes a number of severe cold periods followed by extended warm periods. Although warm periods tend to have corresponded with wet conditions, this is not consistently the case. Inferences, however, can be drawn from such a record to suggest that a) surface winds, in general, are likely to be weak during wet and strong during dry periods, b) zonal transports in the middle latitudes are greater during dry than during wet periods, c) zonal transports in the tropics are greater during wet than during dry periods, d) zonal wind transport pathways in the westerlies are likely to migrate poleward during warm and equatorward during cold periods, e) vertical stratification of the atmosphere, in terms of inversions and stable layers, is likely to be greater during dry/cold than during warm/wet periods, and f) changes are likely to be greatest in mid and high latitudes and smallest in low latitudes.

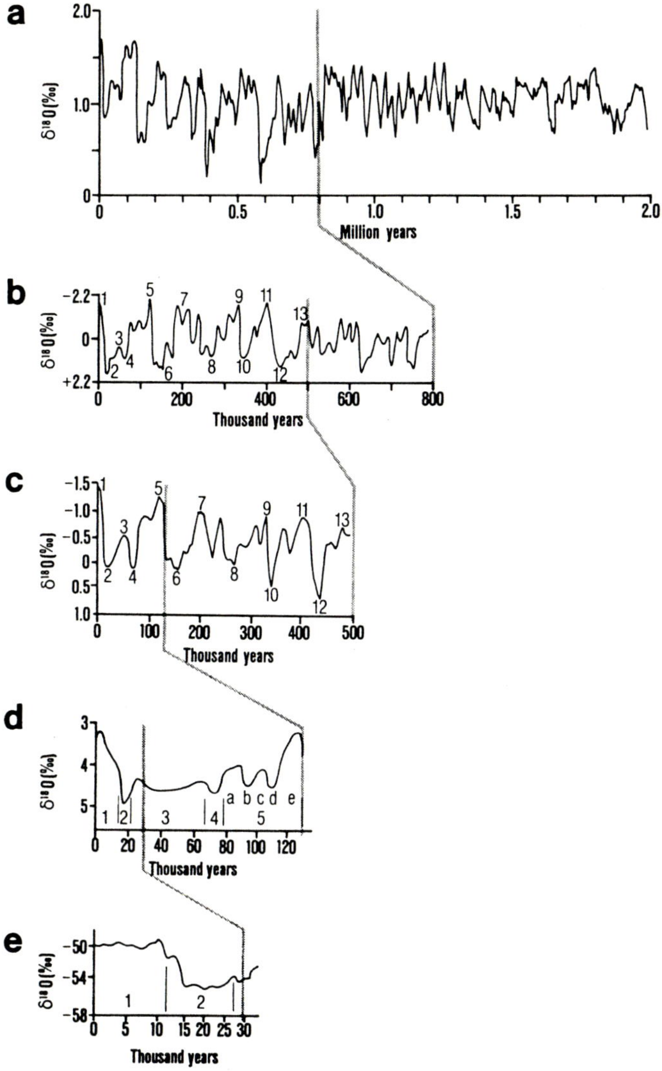

Figure 22. Oxygen isotope chronologies for (a) ocean core V28–239 located in the western equatorial Pacific (after Shackleton and Opdyke, 1973), (b) the averaged records for cores V28–239, DSDP502, V22–174, RC11–120 and V30–40 (after Imbrie et al. 1984), (c) RC17–69, located off the Natal coast (after Prell et al. 1979), (d) smoothed version of the benthic record for core MD73–025, situated south of Madagascar at 44° S, 51° E (after Shackleton, 1977), and (e) an ice core from Dome C, eastern Antarctica (after Tyson 1986).

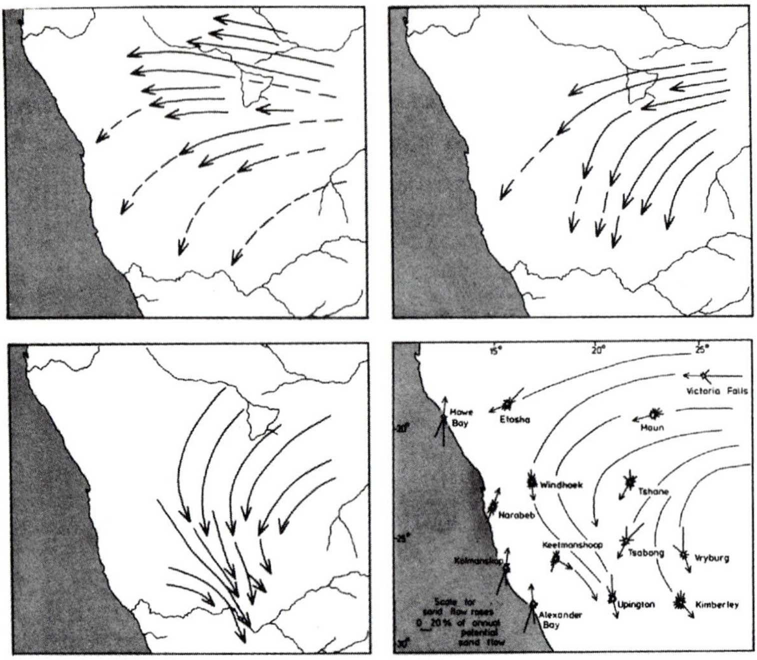

Figure 23. Patterns of palaeo–sand flows inferred from the alignment of presently fixed dunes before 30,000 yr BP (upper left), during the Last Glacial Maximum around 18,000 yr BP (upper right), after 12,000 yr BP (lower left), and annual potential sand flows based on the present wind field (lower right) (after Lancaster 1981).

Relatively subtle shifts of a few degrees (< 5°) in the large scale circulation fields may result in significant changes in transport pathways, particularly when referenced to fixed sources or potential receptors (such as lakes). A possible example is the transport pathways illustrated in Fig. 11 for Australia. Sediment records contain confirming evidence of the two major exit pathways from Australia as illustrated in Fig. 11. Analysis of the current seven-yr record of aerosol optical thickness (AOT) observed from AVHRR show AOT values off Australia to be much lower than AOT values seen off Africa. This observation is particularly true for the expected pathway into the Tasman Sea. A possible explanation is that under current conditions the westerly maximum lies to the south of Australia.

Changes in oxygen isotope values in deep ocean cores document changes in the earth's climate over the past 2 million yr (Fig. 22). Higher ^{18}O values suggest warmer climates. For the globe as a whole, alternating glacial and interglacial conditions have occurred frequently, with a quasi–periodicity of 100,000 yr (Pestiaux and Berger 1984, Imbrie 1985, Kukla 1985). All of the essential features of the global curve are evident in the record from the core off the Natal coast of South Africa. Changes over the last 130,000 yr in the same locality at about 44°S, 51°E (Fig. 22) illustrate the warm Last Interglacial and the cooling that culminated in the Last Glacial Maximum about 15,000 yr BP. At that time the oceanic Antarctic Convergence in the Indian Ocean moved 5° to 8° north of its present position, and the Subtropical Convergence shifted 3° to 5° equatorward (Prell et al. 1979, Howard 1985). These movements would have been accompanied by a commensurate expansion of the circumpolar vortex and equatorward migration of the semi–permanent anticyclones and transport pathways.

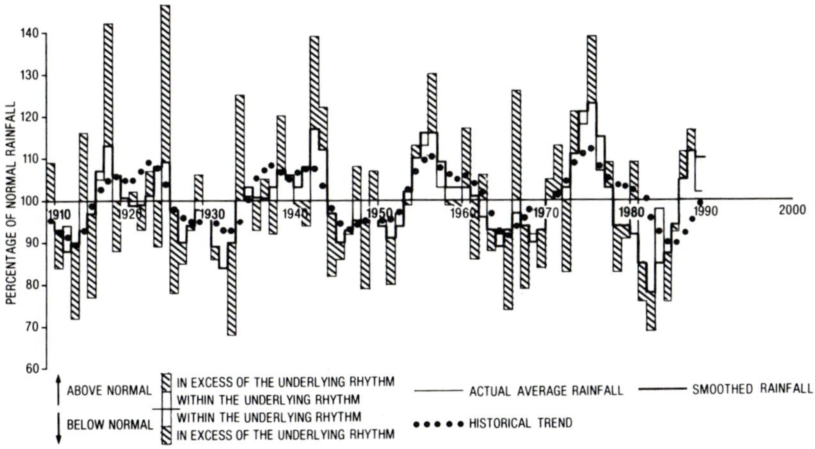

Figure 24. Spatially–averaged rainfall as a percentage of normal for the summer rainfall region of South Africa as a whole. The October–September rainfall series is for 33 stations. Dates on the horizontal axis give the year in which October rains began. The series is smoothed using a five–term binomial filter. Extrapolation is based on analysis of 1910/11 to 1989/90 data (updated after Tyson 1986).

Alignment of patterns of Kalahari dunes (Lancaster 1981, 1984) suggest that anticyclonic flow has existed over this part of the southern Africa for more than 30,000 yr (Fig. 23). There is remarkable correspondence in alignment between the paleo–dunes and modern day wind fields. The long–term existence of dominant anticyclonic flow over southern Africa is supported by the paleo–dune evidence. At around 15,000 yr BP, during the cool, dry glacial maximum in southern Africa, the continental anticyclone was up to 10° north of its present position and stronger than today. The long term dominance of the subtropical anticyclone in the southern hemisphere is probably reflected in the persistence of a northern hemisphere counterpart. The main transport pathways would have undergone commensurate migrations and would have been associated with stronger winds during the colder period.

The model of an expanding, strengthened circumpolar vortex and weakening tropical easterlies producing cool, dry conditions over the summer rainfall region of southern Africa and strengthened easterlies and a contracted westerly wind vortex associated with warm, wet conditions appears to hold for the last two millennia (Tyson and Lindesay 1992, Cohen and Tyson 1995). The same explanation applies to extended wet and dry spells that have occurred this century in a near 18–yr oscillation in the summer rainfall region of southern Africa (Fig. 24) (Tyson 1986). Such periodicities (18 to 20 yr) have been seen in Australia (Cornish 1954), eastern Brazil (King 1975), the southern United States (Currie 1976, Mitchell et al. 1979) and elsewhere. Changes in the strengths of the easterly and westerly zonal flows shown in Fig. 19 accompany wetter and drier years (Lindesay 1988, Mason 1992).

Davis et al. (1996) examined changes in the frequencies of Atlantic basin subtropical anticyclones ($\geq$ 1020 hPa) particularly as these systems influence weather over North America. They found a statistically significant decline in strength of these systems over the past 100 yr. Dominant anticyclones have occupied roughly the first half of this century with a declining trend beginning in mid–century. High frequency of occurrence of strong subtropical anticyclonic activity corresponds to strong zonal (E–W, W–E) flow illustrated in Fig. 11. Strong zonal flow corresponds with wet conditions in the tropics and subtropics and dry conditions in middle latitudes. Low frequency of occurrence of strong subtropical anticyclone activity corresponds to weak zonal and increased meridional flow (N–S, S–N). Strong meridional flow corresponds with diminished zonal transports, dry conditions in the tropics and subtropics, and wet conditions in middle latitudes. While zonal transports may weaken in the tropics, fires may increase in number under dry conditions. Under meridional flow, transport in midlatitudes shifts from W–E to S–N, and fires are reduced in numbers under wet conditions. The implications for transport are clear: During wet spells the degree of transport by non–anticyclonic systems will increase, while in dry spells the intensified anticyclonic conditions over the subcontinent will produce increased trapping of material awaiting transport, increased recirculation, and an increase in the anticyclonic transportation mode. While these changes impact transports, feedback effects from these planetary scale motions to smaller scales can be expected.

Meso– and convective–scales

Meso– and convective scales of motion occur on horizontal scales of a few to 100 km and on time scales of tens of minutes to a few tens of hours. Typical mesoscale phenomena are squall lines and land–sea breeze fronts. Deep convective clouds, such as individual cumulus congestus or cumulonimbus embedded in mesoscale cloud clusters, are typical of convective scale motions. Dry convection occurring in the absence of condensation will be discussed with turbulent or small scale motions below.

Meso– and convective–scales of motion are subject to the large scale fields of motion that create the necessary if not sufficient conditions for the formation of meso–convective–scale systems. Adequate moisture in a convectively unstable atmosphere is a necessary precursor for deep moist convective clouds to form. Although deep moist convection will form in isolation in a moist unstable atmosphere, most deep moist convection will form in response to organized low level convergence, such as that created on a sea–land boundary, over a topographic feature, or as part of a larger synoptic scale system, such as a cold front. Such meteorological conditions are not typical of biomass burning conditions. Fires and burning more typically occur under the fair, dry conditions of anticyclonic circulations. Neither mesoscale cloud complexes nor deep convection form within high pressure systems. Two situations in which meso– to convective–scale circulations form are, however, important to the transport of products from fires.

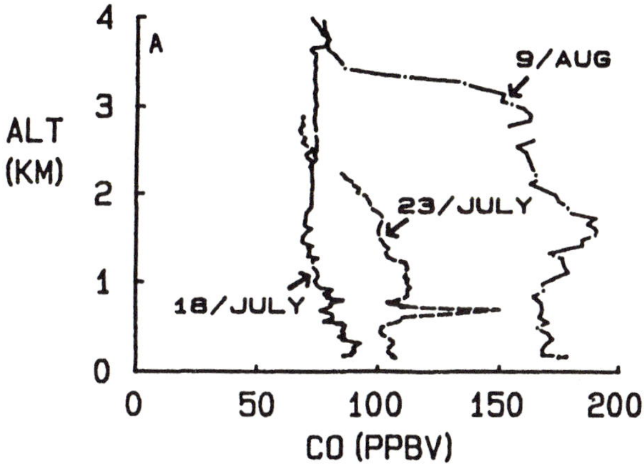

Figure 25. CO profiles illustrating CO buildup during the early phases of the 1985 dry season at Reserva Ducke (2° 56' S, 59° 58' W) (after Sachse et al. 1988).

Entrainment into deep moist convection. Fire–generated products may be transported rapidly over distances of up to 1000 km in 10 hours. Whereas such transport is most likely to be zonal and to remain in the transport plume of the large scale circulation regime, transport could be into regions of active convection. An example of the latter process is encountered in Brazil, where burning products in the cerrado are advected into the convectively unstable atmosphere of the Amazon Basin (Sachse et al. 1988). Fig. 25 illustrates the build–up of carbon monoxide in the Central Amazon Basin (2° 56' S, 59° 58' W) over the course of the dry season of 1985. As the dry season progresses, CO from fires more than 1000 km to the south is advected northwards into the Basin and elevated by dry convective mixing in the mixing layer (surface to ~ 1.2 km) and then upwards by moist convection to just below a persistent capping inversion at 3.0 to 3.5 km. Connors et al. (1991), using shuttle–borne Measurements of Air Pollution from Space (MAPS) showed that regional transports of CO are related to deep convective clouds over the eastern region of southern Africa. Garstang et al. (1988) and Scala et al. (1990) provide further evidence of moist convective transports. They show that in a dry unstable environment cumulus clouds transport air to cloud top heights as suggested by Chatfield and Crutzen (1984). In a moist more stable environment a significant portion of the air transported by cumulus congestus and cumulonimbus clouds detrains in the middle troposphere. Under such conditions, much of the air that reaches anvil heights originates at mid–tropospheric levels (Scala et al. 1990). Pickering et al. (1991) and Thompson et al. (1994) made extensive use of the Goddard Cumulus Ensemble (GCE) Model (Simpson and Tao 1993) as well as photochemical models to evaluate moist convective transports and interactions of trace gases.

Biomass burning particles display cloud condensation nuclei (CCN) capabilities even at very low supersaturation levels (Stith et al. 1981, Rogers et al. 1991, Dinh et al. 1992). These properties are likely to be due to the presence of a hydrophilic coating of organic substances at the carbonaceous particle surface rapidly acquired after emissions. Thus, such particles may be conveyed in cloud structures and are found in rains far from source regions. As an example, Cachier and Ducret (1991) found that acidic rains over the equatorial forest of the Congo were heavily polluted by fires especially during the savanna burning seasons.

The mechanisms by which the particulate smokes from biomass fires are removed from the atmosphere can be stated in order of decreasing importance:

- Wet scavenging by cloud and precipitation.
- Gravitational removal by sedimentation assisted by coagulation.
- Dry removal by wind–driven impaction near the earth "s surface.

The efficiencies and relative importance of these mechanisms are modified by interrelated characteristics of the fires, concurrent and subsequent meteorology, and the size and nature of the smoke particles themselves.

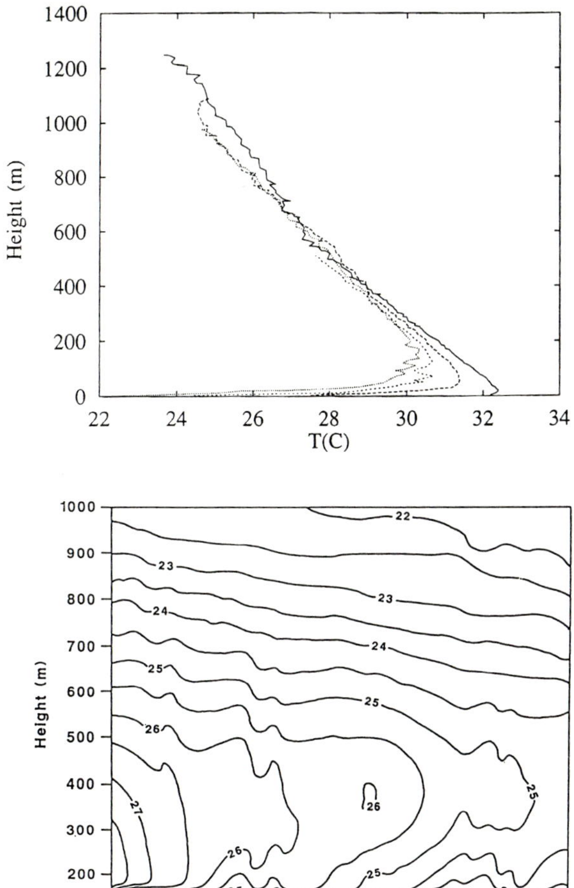

Figure 26. Above: Tethered balloon soundings of temperature at Okaukuejo, Etosha National Park, Namibia, starting at 1700 LST (——), 1800 sunset (----), 1900 (- - -) and 2000 (· · · ·) on a clear, calm evening (18 September 1992) (After Garstang et al. 1996). Below: Time–height contours of temperature measured by tethersonde at Ducke, Central Amazon Basin (2° S, 60° W) 30–31 July 1985. Height is meters above mean sea level (Ducke is 78 m above MSL) (after Greco et al. 1992).

Smoke particles from biomass fires have a curious and modestly unexpected property that affects their rates of removal from the atmosphere: they are meteorologically active (Twomey and Warner 1967, Radke et al. 1978). Their primary activity is as nuclei for cloud water condensation. Aerosols that serve as cloud condensation nuclei (CCN) vary remarkably in both atmospheric concentration and in the relative ease or efficiency with which they act. One CCN is needed for each cloud droplet formed in the atmosphere. The yield of CCN is about 5 ¥ 10^{12} per gram of biomass burned, a prolific production. The CCN concentration at cloud base directly influences the cloud droplet concentration and the cloud albedo. It less directly influences the probability of precipitation formation.

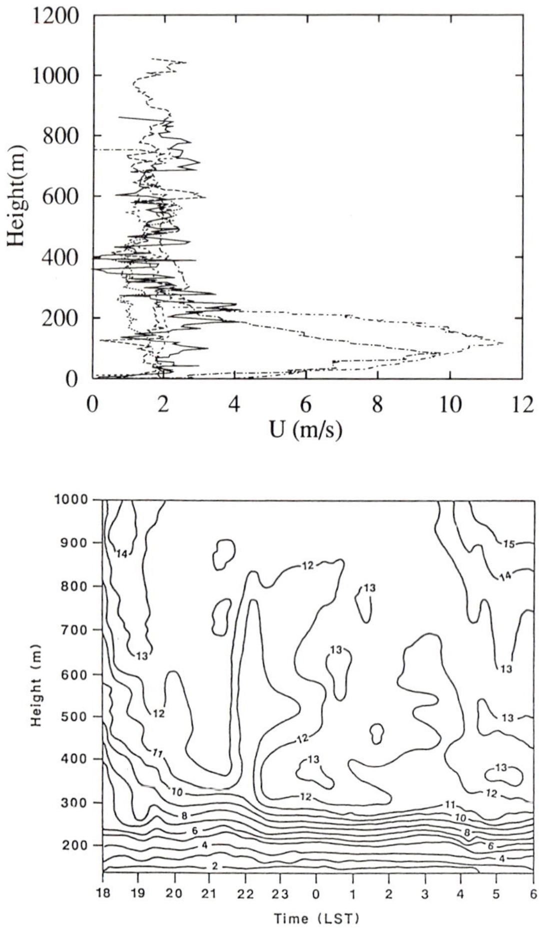

Figure 27. Above: Evolution of nocturnal jets in a dry environment at Okaukuejo, Etosha National Park, Namibia, September 19, 1992. Below) Time–height contours of wind speed measured by tethersonde at Ducke, Central Amazon Basin (2° S, 60° W) 30–31 July 1985. Height is meters above mean sea level (Ducke is 78 m above MSL) (after Greco et al. 1992). Symbolism follows Figure 26.

Recent calculations by both Radke et al. (1990a, b) and Rogers et al. (1990) indicate that the CCN produced by biomass burning is a significant fraction of the total global production of CCN. Thus, in the pre–industrial era biomass fire CCN would have had an even larger role in both hydrologic cycle and global heat balance than they do today. While smoke particulate size distribution is event dependent, a remarkable consistency is often observed (Radke et al. 1988). Usually about 80% of the smoke mass emitted is smaller than 2 mm, a size range that is ineffectively removed from the atmosphere by sedimentation. These small particles are aerodynamically mobile. Beyond 2 km from the fire the remaining mass fraction is mostly smaller than ~ 1 cm and subject to efficient removal by gravitational sedimentation. About 2% of the biomass burned becomes particulate smoke particles smaller than 2 mm (Radke et al. 1990a). Estimates of the total biomass burned annually are ~ 10^4 Tg yr^{-1} (Radke 1989, Levine 1990a, 1990b).

Little is known about how biomass fire smoke particles are partitioned between cellulosic fragments, condensed hydrocarbons, and black carbon (BC). Optical properties of the biomass smokes suggest that about 50% of the smoke less than 2 mm is BC or charcoal (Radke et al. 1990a). The fraction of smoke larger than 2 mm that is charcoal is not known, but might be similar to that of small particles. Thus, a reasonable estimate is that about 1% of total biomass consumed becomes airborne charcoal or about 100 Tg yr^{-1}.

Fire–generated convection. Vertical stratification of the atmosphere, often present in widespread and pervasive multiple layers (Fig. 13), is typically present in regions where burning occurs. Marenco et al. (1990) show in Fig. 14 pronounced layering of CO, CO_2, NO_y, and O_3 over central Africa during TROPOZ. These stable layers or inversions are products of the large–scale circulation, which produces seasonal and regional dry conditions conducive to natural and manmade fires. Thus, fires typically occur in stably stratified atmospheres that do not support natural convective motions. There is substantial feedback between the large scales, small scales and fires under these circumstances. Under dry conditions, skies are typically clear, moisture content in the atmosphere is low, and longwave radiation losses from the surface of the earth intense. Large scale sinking motions are enhanced, inversions strengthened, and, in particular, strong low level nocturnal inversions form. Vertical shear of horizontal winds is intensified. Convective motions generated by the fire under these circumstances are discussed in this section. Feedback between fires and smaller scale turbulent motions are dealt with in the section that follows.

Figure 26 shows the evolution of a nocturnal inversion in the dry environment of northern Namibia (above) and in the moist Central Amazon Basin (below). Such intense low–level inversions result in a decoupling of the lower atmosphere from the rough surface of the earth. Frictional retardation of the flow is greatly reduced. Other forces such as gravity, which acts where slopes exist, and inertial forces at latitudes > 10° to 15° from the equator, accelerate air above the surface (Greco et al. 1992, Zunckel et al. 1996, Preston–Whyte et al. 1995). Figure 27 shows examples of nocturnal jets in the corresponding inversion environments.

A fire embedded in such stratified environments initiates thermodynamic and kinematic feedback effects. The thermodynamic effect may be seen as a concentration of buoyant energy below the inversion analogous to that described by Garstang et al (1987). Buoyant accelerations due to the strong temperature gradient between the fire and its environment are discussed below. Kinematic effects are twofold, a) large horizontal convergence at the boundary between the low level jet and the fire plume, and b) downward (and upward) exchanges of horizontal momentum embedded the jet. The difference in horizontal wind velocity at the boundary between the fire plume and the jet can exceed 10 m s^{-1}. If this change in horizontal velocity takes place over distances as small as 0.25 to 0.5 km mass convergence across the plume boundary of 2 to 4 ¥ 10^{-2} s^{-1} will result. This convergence implies upward velocities caused by the interaction between the jet and the fire plume of 20 and 40 m s^{-1} in the kilometer of air above the inversion.

Intense turbulent mixing driven by buoyant forces within and across the boundary of the fire plume causes high velocities in the core of the jet to be brought down to the surface fire. Velocities brought to the surface (both the magnitude and the direction) are likely to be close to those observed in the core of the jet. These responses suggest a diurnal cycle in fire intensity predicting a maximum during the early evening and at night (Stocks, personal communication).

Numerical simulations using cloud models applied to fires (Penner et al. 1986, 1991) predict simulated plume velocities > 20 m s^{-1}. Such models likely underestimate magnitudes of vertical velocity fields, especially in strongly shearing environments such as those occurring in the presence of jet–like structures. Upward motions of the magnitude described above are capable of transporting a large spectrum of particle sizes and trace gases generated in the fire to considerable heights. A 20 m s^{-1} upward velocity operating over 10 min will elevate air 12 km. Entrainment of outside air into the plume in the presence of stratified layers in the atmosphere will substantially reduce vertical transports. In the first few kilometers of the atmosphere, however, upward forces are large and capable of lifting particles of millimeter to centimeter size. Vertical transport of particles will also be strongly shape–dependent.

If airborne elemental carbon or charcoal is to be used as a geological or paleo–climatic marker the relationship between source and sink locations needs to be known. The relationship includes complex scale interactions and mechanisms that make precision difficult to attain. For example, it is evident that airborne charcoal will be transported downwind. Average wind speed at most locations has the useful property that the zonal (east–west) winds are roughly an order of magnitude stronger than the meridional (north–south) component. Thus, a rule of thumb can be applied without recourse to meteorological data or analysis. However, the magnitude of the zonal transport may change by an order of magnitude depending upon altitude.

Downwind of a fire several signatures can be expected. First, a surface debris trail starting at the fire location showing sedimentational sorting by particle size and terminal velocity as a function of distance (time) downwind. Second, precipitation prompts removal of charcoal. Large fires can be capped with convective clouds that often grow to cumulonimbus proportions accompanied by significant precipitation. Data concerning smoke scavenging efficiencies, however, are limited (Radke et al. 1990a, Penner et al. 1990). Between 50 and 100% of the smoke the cloud ingested during its lifetime will be deposited with the precipitation creating another event–driven "trail" downwind. However, while this prompt charcoal deposition mechanism is regularly observed, a variety of meteorological processes included here argue against local fire–spawned cloud and precipitation scavenging playing a major role in removing the total smoke emitted from all biomass fires. Most of the airborne charcoal particulate material is between 0.1 and 2 mm. These particles first incorporated in cloud droplets, and cloud droplets grow to form precipitation size drops. Most cloud drops do not rain out of the cloud but re–evaporate. Evaporation of the coagulated cloud droplets substantially reduces the total number of particles while increasing both their average size and enhancing their suitability for subsequent scavenging on their next encounter with a cloud.

Knowledge of how deposition recurs does not specify where the charcoal is deposited. The atmospheric residence time of particulate pollution in this same size range as biomass burning material is known to vary strongly with season and latitude (Charlson et al. 1987). In mid–latitudes in the troposphere an estimate of about 7 days has been used (Prospero et al. 1983). Aerosol residence time in Arctic air pollution and long–range transport at high latitudes can be scaled with relative annual precipitation amounts. For example, mean annual precipitation in the arctic is about an order of magnitude less than at mid–latitudes. Zonal distributions of precipitation and changes over time in these distributions provides a first approximation to the locations of deposition from fires occurring within or on the margins of the precipitation bands. A global climatology of precipitation such as Selles (1965) may provide this initial guide.

Horizontal transport distances and fallout patterns from fires in non–precipitating environments will be strongly dependent upon the thermodynamic and kinematic structure of the atmosphere in which the fire is occurring. Nocturnal jets in environments that are dominated by strong low– and mid–level inversions as illustrated above, will be capable, in the course of a single night, of transporting fire products considerable distances from the source (Garstang and Tyson 1996, Zunckel et al. 1996). Jets with core velocities of 15 to 25 m s^{-1} lasting for 10 hours will transport fire products over distances of 500 to 1000 km. Fire products will be stratified in the vertical column by both jet and inversion structures. In southern Africa, maximum concentrations are likely to occur between the low level nocturnal inversion and the main subsidence inversion near 5 km above the surface. Deposition of even the larger particles will not occur in the immediate vicinity of the fire but particles will be conveyed downstream in the layer between the low and main inversions. Larger particles, once clear of strong upward velocities of the fire plume, must fall but could be transported some tens of kilometers from the source before reaching the ground. Smaller particles (< 100 mm) will remain suspended in the atmosphere over distances of 100s to 1000s of kilometers and most likely removed by wet deposition in rainfall.

Patterns of deposition in the immediate vicinity of a fixed source, such as a catchment fire, are a product of the above complex set of interactions. The magnitude of the vertical velocity field may be greatest in the immediate vicinity of the fire. Horizontal transport velocities, however, may reach this maximum value at some distance from the fire. Neither velocity component is likely to be symmetrically distributed about the fire. "Dead zones" of deposition, which are particle size dependent, could occur around the fire. Variations in large scale circulation fields over time will result in changes in direction of the mean transport pathways away from any fixed source and relative to fixed receptors such as lakes. Because the transport pathway for the larger particle may be limited in horizontal dimensions (both width and length) by the processes described above, marked changes in sediment signals in the receptor may result.

Turbulent and smaller–scales

Turbulent to molecular scales of motion occur on horizontal scales of meters down to micrometers and time scales of a few minutes to microseconds. Turbulent scales of motion in the fire plume are driven by buoyant and shearing stress forces in complex interactions that cannot be explicitly resolved. Statistical mean results of turbulent mixing can be estimated. Once beyond the strong turbulent motions of the fire, fire products will eventually be subject to molecular diffusion.

Conventional physics suggest that particles smaller than 100 mm are not easily entrained from the surface and particles larger than 100 mm are too massive to entrain (Clark and Richard 1996). Conditions within and on the boundary of a fire are likely to depart substantially from any such steady state view. Two conditions must be considered in a fire: a) generation of particles above ground, and b) elevation of particles at the surface. The first condition is largely confined to the active fire, the second to the smoldering and post–fire stage. Turbulent motions in the active portion of the fire are likely to be most violent at the leading edge where gradients of velocity, temperature and density are greatest. Upward motions in the fire will be due to buoyancy and kinematic forces. Kinematic forces were discussed above.

Buoyancy driven forces are primarily a result of the large horizontal temperature gradient between the fire and its surroundings modified only weakly by moisture (positive) and entrainment (negative). Vertical accelerations can be expressed in terms of the virtual temperatures of the fire plume and the environment, where virtual temperature is the temperature of moist air with the equivalent density (and pressure) of the mixture,

$$\frac{dw}{dt} = g\left[\frac{T_p - T}{T}\right] - \frac{w}{M_p}\frac{dM_p}{dt} \tag{1}$$

where w is the vertical velocity of a unit mass (momentum), M_p is mass of the plume, T_p is the virtual temperature of the plume, and T is the virtual temperature of the environment outside the plume. The entrainment term on the righthand side of (1) diminishes in importance as the fire size increases from small (< 1 km^2) to large (> 1 km^2). Although temperatures in the core of the fire reach 800° C, the temperature of the fire plume decays rapidly over the first kilometer above the heat source due to infrared radiative cooling. Buoyant motions driven by large temperature gradients below 1 km are difficult to estimate. Above 1 km it is reasonable to assume that temperature differences between the fire plume and the environment are in the order of 10° C. Such temperature differences produce vertical velocities in the order of 30 m s^{-1}. Upward velocities of these magnitudes produce horizontal inflow into the fire plume below 1 km which, in turn, add to the upward motion in the fire column.

Feedback effects between these turbulent motions and the convectively produced effects of convergence, upward motions, and momentum exchanges, produce strong turbulent motions within and on the edges of the fire at the surface. These turbulent motions will suspend and elevate particles ranging in size from microns to centimeters. Flaming debris has been encountered at elevations of as much as 3 km and distances as great as 10 kilometers from the fire by research aircraft (Radke, personal observation). Particles in the active fire are at the surface and above the surface in the burning vegetation. In the smoldering and post fire environment the number of particles subject to vertical removal are increasingly found at the surface. Turbulent forces decay rapidly outside of the active fire so that it is likely that the limits of size (< 100 mm) in vertical transport apply in the post fire regime. Surface wind speeds are low in the high pressure systems in which a large fraction of the burning occurs. Conventional mechanical shearing forces dependent on the vertical gradient of the horizontal wind at the surface will not be sufficient to elevate particles into the atmosphere in the post fire environment. Strong daytime heating of a low albedo surface on the other hand will create small vortices in the form of dust devils that are capable of elevating material to heights of 500 m or more in the mixing layer. Swap (personal communication) observing from a 100 ft tower noted as many as 30 dust devils operating at one time at Okaukuejo in the Etosha National Park of Namibia during the end of the dry season. Fitzjarrald (1976) found dust devils in a California desert to be frequent and pervasive. Golden (1968, 1973, 1974a, 1974b, 1977) found a phenomena analogous to dust devils, waterspouts, occur at a frequency of hundreds per summer month over the Florida Keys. It is likely that large burn areas are particularly favorable locations for the formation of dust devils and that this is a primary mechanism for elevating surface fire produced particles in the post burn phase.

Concluding remarks

The atmosphere operates on a range of scales associated with processes that affect transport of fire– generated products. Scales of motion are not independent of the occurrence of natural and manmade fires. Interactions between scales of motion effect the production and distribution of fire-generated products. Fire occurrence is controlled on a wide range of scales (especially temporal) by atmospheric circulations.

We argue here that large and persistent subtropical anticyclones are fundamental features of the atmosphere that control fire occurrence and distribution. Transport pathways emanating from these large, semi–permanent high pressure systems tend to be zonal and relatively persistent in their location in the tropical easterlies and the midlatitude westerlies. In the middle and high latitude boreal forests anticyclonic conditions will still exert a controlling influence on long term periodicities, inter– and intra–annual periodicities and the day–to–day behavior of fires. Anticyclone behavior at the boundary between low and high latitude air masses is transient and tied to the traveling wave cyclones of these latitudes. Transport pathways of fire generated products will reflect the W to E fields of motion of the poleward margins of high pressure cells, but these transports will contain significant meridional components. Fluctuations in zonal and meridional transports will be much greater polewards from the semi–permanent anticyclones than the persistent zonal transport on the equatorward margin.

Convective and turbulent scale processes interact with each other and with larger scales of atmospheric motion to produce horizontal and vertical transport conditions peculiar to a given fire. Vertical transports can elevate a wide spectrum of particles from microns to centimeters. Horizontal transport of large particles is limited to 10's of kilometers while transports of the smaller particles are measured in 100's and 1000's of kilometers. Vertical stratification of the atmosphere in the form of multiple inversions or stable layers couples with the velocity fields to influence the horizontal transports. Distribution of fire products in the atmosphere is confined horizontally to major transport pathways and vertically to often thin layers of the atmosphere.

Turbulent motions generated in the active fire are sufficient to elevate large (cm) particles from sources already above the ground. In the smouldering and post fire region mean motions of the atmosphere are not able to elevate significant amounts of material. Instead it is postulated that intermittent vortices, such as dust devils, may play an important but unknown role.

Altered sedimentation of fire–generated products in a given lake can be due to changes in fire regime and to changes in atmospheric circulation fields that affect both fire regime and transport. Rainfall, a function of the large scale circulation fields, combines with the fields of motion to dictate fire frequency and location. Frequencies of fires and of transport pathways emanating from fire regions undergo modest latitudinal shifts with changes in the general circulation. Changes of fire regimes over time are also a function of changes in terrestrial biomass. Products such as particulate carbon from fires are then a combined response to the number of fires, current biomass levels, and dispersal mechanisms. Fixed measurement sites, such as lakes where sediments are monitored, are thus subject to a complex set of factors.

Long range and relatively widespread transport and ultimate deposition of particulate and elemental carbon may disperse carbon over wide although reasonably prescribed areas. Many of these areas will be in the deep oceans, but ocean cores may still record fire histories from relatively well prescribed exit regions off the continents. Sequestration of elemental carbon as part of the global carbon budget must, in part, be contained in these sediments.

Literature References

Agee JK, Smith L (1984) Subalpine tree reestablishment after fire in the Olympic Mountains, Washington. *Ecology* 65:810–819

Alexander WJR, van Heerden J (1992) An assessment of the severity of the current drought, October 1991 to March 1992. *Water Resources Study* University of Pretoria

Andreae MO (1983) Soot carbon and excess fine–potassium: long range transport of combustion–derived aerosols. *Science* 220:1148–1151

Andreae MO (1991) Biomass burning: Its history, use, and distribution and its impact on environmental quality and global climate. In Levine JS (ed) *Global Biomass Burning: Atmospheric, Climatic and Biospheric Implications* pp 3–21 MIT Press Cambridge

Andreae MO, Fishman J, Garstang M, Goldammer JG, Justice CO, Levine JS, Scholes RJ, Stocks BJ, Thompson AM, van Wilgen B, the STARE/TRACE–A/SAFARI–92 Science Team (1994) Biomass burning in the global environment: first results from the IGAC/BIBEX field campaign STARE/TRACE–A/SAFARI–92. In Prinn RG (ed) *Global Atmosphere–Biosphere Chemistry* pp 83–101 Plenum Press New York

Artaxo P, Gerab F, Yamasoe MA, Martins JV (1996) Fine mode aerosol composition at three long term atmospheric monitoring sites in the Amazon basin. *J Geophys Res* (in press)

Bengtsson L (1985) Medium–range forecasting at ECMWF. *Adv Geophys* 238:3-56

Burroughs WJ (1992) *Winter Cycles: Real or Imaginary?* Cambridge University Press Cambridge

Cachier H, Ducret J (1991) Influence of biomass burning on equatorial African rains. *Nature* 352:228–230

Cachier H, Ducret J, Brémond M-P, Yoboué V, Lacaux J-P, Gaudichet A, Baudit J (1991) Biomass burning aerosols in a savanna region of the Ivory Coast. In Levine JS (ed) *Global Biomass Burning: Atmospheric, Climatic and Biospheric Implications* pp 174–180 MIT Press Cambridge

Cachier H, Liousse C, Buat–Menard P, Gaudichet A (1995) Particulate content of savanna fire emissions. *J Atmos Chem* 22:123–148

Charlson RJ, Lovelock JE, Andreae MO, Warren SG (1987) Oceanic phytoplankton, atmospheric sulphur, cloud albedo and climate. *Nature* 32:655–661

Chatfield RB, Crutzen PJ (1984) Sulfur dioxide in remote ocean air cloud transport of reactive precursors. *J Geophys Res* 89:7111–7132

Clark JS (1988) Charcoal–stratigraphic analysis on petrographic thin sections: recent fire history in northwest Minnesota. *Quat Res* 30:67–80

Clark JS, Robinson J (1993) Paleoecology of fire. In Crutzen P, Goldammer JG (eds) *Fire in the Environment: Its Ecological, Climatic and Atmospheric Chemical Importance* pp193–214 Wiley New York

Clark JS, Royall PD (1994) Pre–industrial particulate emissions and carbon sequestration from biomass burning in North America. *Biogeochem* 23:35–51

Clark JS, Richard PJH (1995) The role of paleofire in boreal and other cool–coniferous forests. In Goldammer JG (ed) *Fire in Ecosystems of Northern Eurasia* Wiley New York

Clark JS, Stocks BJ, Richard PJH (1996) Climate implications of biomass burning since the 19th century in eastern North America. *Global Change Biol* (in press)

Cohen A, Tyson PD (1996) Sea surface temperatures during the Holocene on the south coast of Africa: Implications for terrestrial climate and rainfall. *The Holocene* (in press)

Connors V, Cahoon Jr D, Reiche H, Brunke E-G, Garstang M, Seiler W, Scheel HE (1991) Savanna burning and convective mixing in southern Africa: Implications for CO emissions and transport. In Levine JS (ed) *Global Biomass Burning: Atmospheric, Climatic and Biospheric Implications* pp 147–154 MIT Press Cambridge MA

Cornish EA (1954) On the secular variation of rainfall at Adelaide. *Austr J Phys* 7:334–346

Crutzen P, Goldammer JG (1993) *Fire in the Environment: Its Ecological, Climatic and Atmospheric Chemical Importance* Wiley New York

Currie RG (1976) The spectrum of sea level from 4 to 40 years. *Geophys J Roy Astron Soc* 46:513–520

Danielson EF (1959) The laminar structure of the atmosphere and its relation to the concept of a tropopause. *Arch Met Geoph* 293–332

Danielson EF (1961) Trajectories: isobaric, isentropic and actual. *J Meteor* 18:479–486

Davis RE, Hayden BP, Gay DA, Phillips WL (1996) The Atlantic subtropical anticyclone. *J Clim* (in press)

Dinh PV, Lacaux JP, Serpolay R (1992) Cloud forming properties of biomass burning aerosols. In Fukuta N, Wagner PE (eds)*Nucleation and Atmospheric Aerosols* pp 173–176 Elsevier Science Publishers Amsterdam, The Netherlands

Fitzjarrald DE (1976) A field investigation of dust devils in a California desert. *J Appl Meteor* 12:808–813

Frakes LA (1979) *Climates Throughout Geologic Time*. Elsevier Press Amsterdam

Garstang M, Kelbe B, Emmitt G, London W (1987) Generation of convective storms over the escarpment of northeastern South Africa. *Mon Wea Rev* 115:429–443

Garstang M, Scala J, Greco S, Harriss R, Beck S, Browell E, Sachse G, Gregory G, Hill G, Simpson J, Tao W-K, Torres A (1988) Trace gas exchanges and convective transports over the Amazon rain forest. *J Geophys Res* 93:1528–1550

Garstang M, Tyson PD (1996) Atmospheric circulation, vertical structure and transport. In van Wilgen BW, Andreae MO, Goldammer JG, Lindesay JA (eds) *Fire in Southern African Savannas: Ecological and Atmospheric Perspectives* Witwatersrand University Press Johannesburg South Africa (in press)

Garstang M, Tyson PD, Swap R, Edwards M, Kållberg P, Lindesay JA (1996) Horizontal and vertical transport of air over southern Africa. *J Geophys Res* (in press)

Golden J (1968) Waterspouts at Lower Matecumbe Key, Florida, September 2, 1967. *Weatherwise* 23:103–114

Golden J (1973) Some statistical aspects of waterspout formation. *Weatherwise* 26:108–117

Golden J (1974a) The life cycle of Florida Keys' waterspouts. *J Appl Meteor* 13:676–692

Golden J (1974b) Scale–interaction implications for the waterspout life cycle II. *J Appl Meteor* 13:693–709

Golden J (1977) An assessment of waterspout frequencies along the U.S. east and gulf coasts. *J Appl Meteor* 16:231–236

Greco S, Ulanski S, Garstang M, Houston S (1992) Low–level nocturnal wind maximum over the central Amazon basin. *Bound–Layer Meteor* 58:91–115

Grove J (1988) *The Little Ice Age*. Methuen London

Harrison MSJ (1988) *A synoptic climatology of South African rainfall variations.* PhD thesis University of the Witwatersrand Johannesburg South Africa

Howard WR (1985) Late Quaternary southern Indian Ocean circulation. *South African J Sci* 81:253–254

Imbrie J (1985) A theoretical framework for the Pleistocene ice ages. *J Geol Soc London* 142:417–432

Imbrie J, Hays JD, Martinson DC, McIntyre A, Mix AC, Morley JJ, Pisias NC, Prell WL, Shackleton NJ (1984) The orbital theory of Pleistocene climate: support from a revised chronology of the marine

$\partial^{18}O$ record. In Berger AL, Imbrie J, Hays J, Kukla G, Saltzman B (eds) *Milankovitch and Climate* Part 1 pp 269-305 D Reidel Dordrecht

Jaeger L (1976) Monatskarten des Nuderschlages fur die ganzen Erden *Ber.Deutsch Wetterd* Ofenback/ Main NR. 139 v.18

Kållberg P (1984) Air parcel trajectories from analysed or forecast windfields. *Swedish Meteorol Hydrol Inst Res Developm Note* 37 Noorköping Sweden

King JW (1975) Sun–weather relationships. *Astron Aeron* 13:10–19

Kukla G (1985) Quaternary palaeoclimates and evolution on land. *South African J Sci* 81:255

Lancaster IN (1981) Palaeoenvironmental implications of fixed dune systems in southern Africa. *Palaeo Palaeo Palaeo* 33:327–346

Lancaster IN (1984) Aridity in southern Africa: age, origins and expression in landforms and sediments. In Vogel JC (eds) *Late Cainozoic Palaeoclimates of the Southern Hemisphere* pp 433-444 AA Balkema Publishers Rotterdam

Lau K–M (1988) Annual cycle, quasi–biennal oscillation and the southern oscillation in global precipitation. *J Geophys Res* 93:10975–10988

Levine J (1990a) Global biomass burning: Atmospheric, climatic and biospheric implications. *EOS Trans* 71:1075–1077

Levine J (1990b) Atmospheric trace gases: burning trees and bridges. *Nature* 346:511–512

Lindesay JA (1988) South African rainfall, the southern oscillation and a southern hemisphere semi–annual cycle. *J Clim* 8:17–30

Liousse C (1993) *Biomass burning particulate emissions from savanna fires in Africa: smoke plume ground experiments and remote sensing*. PhD dissertation Paris 7 University

Liousse C, Devaux C, Dulac F, Cachier H (1995) Aging of savanna biomass burning aerosols – consequences on their optical properties. *J Atmos Chem* 22:1–17

Marenco A, Dale JCM, Prieur S (1990) Study of tropospheric ozone in the tropical belt from STRATOZ and TROPOZ campaigns. *Atmos Environ* 24A:2823–2834

Mason SJ (1992) *Sea surface temperatures and southern Africa rainfall variability*. PhD thesis University of the Witwatersrand Johannesburg South Africa

Maujokat B (1986) An update of the observed quasi–biennal oscillation of the stratospheric winds over the tropics. *J Atmos Sci* 44:1873–1877

Milankovitch M (1941) History of radiation on the earth and its use for the problem of the ice ages. *K Serb Akad Beogr Spec Publ* 132 (Translated by the Israel Prog For Scientific Translations Jerusalem 1969)

Mitchell JM (1976) An overview of climatic variability and its causal mechanisms. *Quat Res* 6:481–493

Mitchell JM, Stockton CW, Meko DM (1979) Evidence of a 22–year rhythm of drought in the western United States related to the Hale cycle since the turn of the century. In McCormac BM, Seliga TA (eds) *Solar–Terrestrial Influences on Weather and Climate* pp 125–144 D Reidel Hingham MA

Moody JL, Pszenny AAP, Grandry A, Keene WC, Galloway JN, Polian G (1991) Precipitation composition and its variability in the southern Indian Ocean: Amsterdam Island, 1980–1986. *J Geophys Res* 96(20):769–786

Newell RE, Kidson JW, Vincent DG, Boer GJ (1972) *The General Circulation of the Tropical Atmosphere and Interactions with Extratropical Latitudes* v.1 The MIT Press Cambridge MA

Oort AH (1983) Global atmospheric circulation statistics, 1958-1973. *NOAA Prof Paper* No 14

Pasquill F (1974) *Atmospheric Diffusion*. Ellis Horwood Chichester England

Patterson WA, Backman AE (1988) Fire and disease history of forests. In Huntley B, Webb III T (eds) *Vegetation History* pp 603–632 Kluwer Academic Press Dordrecht

Payette S, Gagnon R (1985) Late holocene deforestation and tree regeneration in the forest–tundra of Québec. *Nature* 313:570–572

Peixoto JP, Oort AH (1992) *Physics of Climate*. American Institute of Physics New York

Penner J, Haselmen Jr LC, Edwards L (1986) Smoke–plume distributions above large–scale fires: implications for simulations of nuclear winter. *J Clim Appl Meteor* 25:1434–1444

Penner JE, Atherton CS, Dignon J, Ghan SJ, Walton JJ, Hameed S (1991a) Tropospheric nitrogen: A three–dimensional study of sources, distribution and deposition. *J Geophys Res* 96:959-990

Penner J, Ghan S, Walton J (1991b) The role of biomass burning in the budget and cycle of carbonaceous soot aerosols and their climate impact. In Levine JS (ed) *Global Biomass Burning: Atmospheric, Climatic and Biospheric Implications* pp 387–393 MIT Press Cambridge MA

Pestiaux P, Berger A (1984) An optimal approach to the spectral analysis of deep–sea climatic records. In Berger AL, Imbie J, Hays J, Kukla G, Saltzman B (eds)*Milankovitch and Climate* pp 417–445 D Reidel Dordrecht

Pickering K, Thompson A, Scala J, Tao W-K, Simpson J, Garstang M (1991) Photochemical ozone production in tropical squall line convection during NASA Global Tropospheric Experiment/Amazon Boundary Layer Experiment 2A. *J Geophys Res* 96:3099–3144

Pickering K, Thompson A, McNamara D, Schoeberl M (1994) An intercomparison of isentropic trajectories over the south Atlantic. *Mon Wea Rev* 122:864–879

Prell WL, Hutson WH, Williams DF (1979) The subtropical convergence and late Quaternary circulation in the southern Indian Ocean. *Mar Micropalaeontol* 4:225–234

Preston–Whyte RA, Diab R, Sokolic F (1996) Wind regimes at Okaukuejo during the SAFARI–92 experiment. *South African Geograph J* (in press)

Prospero C, Charlson RJ, Mohnen V, Jaenicke R, Delany AC, Moyers J, Zoller W, Rahn K (1983) The atmospheric aerosol system: An overview. *Rev Geophys Space Phys* 21:1607–1629

Radke LF (1989) Airborne observations of cloud microphysics modified by anthropogenic forcing. *Proc AMS Symp Role of Clouds Atmos Chem Glob Clim* 3:310–315 Anaheim CA

Radke LF, Stith JL, Hegg DA, Hobbs PV (1978) Airborne studies of particulates and gases from forest fires. *J Air Poll Control Assoc* 28:30–33

Radke LF, Hegg DA, Lyons JH, Brock CA, Hobbs PV (1988) Airborne measurements on smokes from biomass burning. In Hobbs PV, McCormick MP (eds) *Aerosols and Climate* pp 411–422 Deepak Publishing Hampton VA

Radke LF, Lyons JH, Hobbs PV, Hegg DA, Sandberg DV, Ward DE (1990a) Airborne monitoring and smoke characterization of prescribed fires on forest lands in western Washington and Oregon. *USDA For Serv Gen Tech Rep* PNW–GTR–21

Radke LF, Lyons JH, Hobbs PV, Weiss RE (1990b) Smokes from the burning of aviation fuels and their self–lofting by solar heating. *J Geophys Res* 95:14,071

Reed RJ, Campbell WJ, Rasmussen LA, Rogers DG (1961) Evidence of a downward propagating, annual wind reversal in the equatorial stratosphere. *J Geophys Res* 66:813–818

Rogers GF, Hudson JG, Hallet J, Penner JE (1990) Cloud droplet nucleation by crude oil smoke and coagulated crude oil/wood smoke particles. *Atmos Environ* 25A:2571–2580

Rogers CF, Hudson JG, Zielinska B, Tanner RL, Hallett J, Watson JG (1991) Cloud condensation nuclei from biomass burning. In Levine JS (ed)*Global Biomass Burning: Atmospheric, Climatic and Biospheric Implications* pp 431–440 MIT Press Cambridge MA

Sachse G, Harriss R, Fishman J, Hill G, Cahoon D (1988) Carbon monoxide over the Amazon Basin during the 1985 dry season. *J Geophys Res* 93:1422–1430

Scala J, Garstang M, Tao W-K, Pickering K, Thompson A, Simpson J, Kirchhoff VWJH, Browell E, Sachse G, Torres A, Gregory G, Rasmussen R, Khalil MAK (1990) Cloud draft structure and trace gas transport. *J Geophys Res* 95(17):015–030

Selles WD (1965) *Physical Climatology*. University of Chicago Press Chicago

Shackleton NJ (1977) Carbon-13 in Uvigerina: tropical forest history and the equatorial Pacific carbonate dissolution cycles. In Anderson NR, Malakoff A (eds) The Fate of Fossil Fuel CO_2 in the oceans. *Mar ineSci* 6:401-427

Shackleton NJ, Opdyke ND (1973) Oxgen isotope and palaeomagnetic stratigraphy of Equatorial Pacific Core V 28-239: oxygen isotope temperatures and ice volumes on a 10^5 year and a 10^6 year scale. *Quat Res* 3:39-55

Simpson J, Tao W-K (1993) The Goddard Cumulus Ensemble Model. Part II: Applications for studying cloud precipitating processes and for NASA TRMM. *Terres Atmos Oceanic Sci* 4:55–96

Stith JF, Radke LF, Hobbs PV (1981) Particle emissions and the production of ozone and nitrogen oxides from the burning of forest slash. *Atmos Environ* 15:73–82

Susott RA, Ward DR, Babbitt RE, Latham DJ (1991) The measurement of trace gas emissions and combustion characteristics for a mass fire. In Levine JS (ed) *Global Biomass Burning: Atmospheric, Climatic and Biosphere Implications* pp 245–257 MIT Press Cambridge MA

Swap R, Garstang M, Macko S, Tyson P, Maenhaut W, Artaxo P, Kållberg P, Talbot R (1996a) The long–range transport of southern African aerosols to the tropical south Atlantic. *J Geophys Res* (in press)

Swap R, Ulanski S, Cobbett M, Garstang M (1996b) Temporal and spatial characteristics of Saharan dust outbreaks. *J Geophys Res* (in press)

Thompson A, Pickering K, Dickerson R, Ellis W, Jacobs D, Scala J, Tao W-K, McNamara D (1994) Convective transport over the Central United States and its role in regional CO and ozone budgets. *J Geophys Res* 99(18):703–711

Trenberth KE (1981) Seasonal variations in global sea level pressure and the total mass of the atmosphere. *J Geophys Res* 86:5238-5246

Twomey S, Warner J (1967) Comparison of measurements of cloud droplets and cloud nuclei. *J Atmos Sci* 24:702–703

Tyson PD (1968) Velocity fluctuations in the mountain wind. *J Atmos Sci* 25:381–384

Tyson PD (1986) *Climatic Changes and Variability in Southern Africa*. Oxford University Press Cape Town South Africa

Tyson PD, Lindesay JA (1992) The climate of the past 2000 years in southern Africa. *Holocene* 2:271–278

Tyson PD, Garstang M, Swap R, Edwards M, Kållberg P (1996) An air transport climatology for subtropical southern Africa. *Internl J Clim* 15 (in press)

Veryard RG, Ebdon RA (1961) Fluctuations in tropical stratospheric winds. *Meteor Mag* 90:125–143

Vinnichenko ML (1970) The kinetic energy spectrum in the free atmosphere - 1 second to 5 years. *Tellus* 22:158-166

Zunckel M, Hong Y, Brassel K, O'Beirne S (1996) Characteristics of the nocturnal boundary layer: Okaukuejo, Namibia during SAFARI–92. *J Geophys Res* (in press)

PART IV:

STRATIGRAPHIC RECORDS

Pre-Quaternary Records of Wildfire

Jennifer M. Robinson, William G. Chaloner[1], and Timothy P. Jones[2]
College of Biological and Environmental Sciences
Murdoch University
Perth, Western Australia 6150

Summary

The fossil record of charcoal extends back to the early history of the plant kingdom. This vast record has been approached in three ways: (1) Studies of specific events or deposits and attempting reconstructions; (2) Study of changes in fossil charcoal abundance and attributes across time; and (3) Experimental work that relates chemical and physical characteristics of the fossil charcoal (fusain) to the formative event. Before the Devonian, we have no evidence, only speculation about the possibility of fire. The Devonian record consists of a few specimens, whose identity and whose credentials as charcoal are in question. In the Carboniferous, unambiguous fossil charcoal is not only common, but often far too abundant to be explained by modern analogues. The Permian record has been little studied and is obscured by poor explanation of the peculiar features of Permian coals. Early work verifying the authenticity of fossil charcoal was done with Mesozoic charcoals, and charcoal is abundant in some Mesozoic sediments. However, almost nothing has been deduced about the fire ecology of Mesozoic environments. By the standards of more ancient sediments, fossil charcoal is present, but scarce in Late Tertiary sediments. The apparent variation of charcoal abundance over geologic time may reflect fundamental changes in the Earth system, including changes in climate, changes in atmospheric chemistry, and changes in the terrestrial carbon cycle. Alternately, it may be an artifact of selective preservation or selective sampling. The large amount of uncertainty surrounding the fossil charcoal record can be explained by the vastness of the material and the small number of people who have worked it. Many of the unknowns are open to experimental research, field study, and interpretive work.

Introduction

Because charcoal is inert under a wide range of pH, redox, and temperature conditions (Goldberg 1985, Retallack 1984), it has extremely high fossilization potential. The fossil record of charcoal has been traced for hundreds of millions of years to the early history of the plant kingdom. The interpretation of this record is in its infancy. The first step, that of recognizing that fossil charcoal in older sediments is evidence of fire, has only been taken with any confidence in the last few decades. The handful of researchers who have seriously studied pre-Quaternary charcoal records as evidence of past fire have used different methods and have focused on

[1] Royal Halloway, Bedford New College, University of London, Eggham, Surrey, TW20 0EX, UK
[2] Eberhard-Karls-Universität Tübingen, Institut und Museum für Geologie und Palaeontologie, Sigwartstrasse 10, Tübingen D-72076, Germany

NATO ASI Series, Vol. I 51
Sediment Records of Biomass Burning
and Global Change
Edited by James S. Clark, Hélène Cachier,
Johann G. Goldammer, and Brian Stocks

different issues and different sorts of evidence from different segments of the geologic record. The result is a fragmentary understanding of fire history and numerous unanswered questions. This paper aims to illustrate the nature of the evidence available for studying pre-Quaternary fire and to review the methodological issues, interpretive problems, and knowledge gaps that surround the use of that evidence.

In this Introduction we point out some basic differences between Quaternary and pre-Quaternary records, review the history of observation of fossil charcoal, discusses terminology, and describe three approaches to studying fossil charcoal. The second section takes some forays into fire-history reconstruction. These are arranged chronologically and represent methodological approaches ranging from informed speculation, in the case of consideration of the first fire, to studies of charcoal fossil in specific locations, to attempts to draw inference from industrial coal-characterization data. The third section reflects on the way in which atmospheric, evolutionary, and geochemical factors may have impinged on fire regimes and char formation over the last several hundred million years.

The study of the pre-Quaternary fire differs from the study of Quaternary fire in many ways, the most basic of which relate to time. 'Pre-Quaternary' potentially refers to the full four billion plus years of Earth history. As described below, it is unclear when fire first appeared on Earth, and thus how far back the charcoal record might extend. The length of the record exacerbates taphonomic problems. Few of the fire events studied in the Quaternary record will survive into the long-term fossil record, because sediments accumulated in most small lakes and bogs will be eroded; their records are strongly biased. Upland and dryland systems are unlikely to survive and are poorly represented; large lowland depositional basins have higher survival rates, and are relatively well represented.

Moreover, the clock, as read in older sediments, is difficult to relate to fire regimes and ecosystems. A pre-Quaternary geologist's attention is likely to be drawn to the million-year (plus) fire event, and events 'only' thousands of years apart may be difficult, if not impossible, to distinguish. In most cases, the calibration of the sedimentary clock is vague. Variable charcoal abundance in the horizons of a coal seam presumably represent periods with more and less charcoal formation. Generalized interpretation of the chronology of such periods requires knowledge of factors such as biomass production and decomposition rates, peat and mineral sediment accumulation rates, and the effects of differential preservation, diagenesis, and sedimentary compaction on the organic portion of the sediment. Paleoecological understanding of such factors is seldom sufficient to determine whether a given millimeter of sediment represents 10 or 1000 years. Short time-span changes associated with particular events are sometimes tractable. For example, Ahrens (1991) undertook detailed reconstruction of an ecological succession after one Carboniferous fire event, and Nichols and Jones (1992, see below) deduce that an enormous amount of charcoal was deposited over a period a month or less based on apparent tidal laminations of sediments. Such studies, however, are few and far between.

Early observations of fossil charcoal

Charcoal-like substances have long been noted by geologists in coals, sedimentary rocks, and paleosols. Grand 'Eury (1882) , writing in French of such material in coal, simply called it 'fusain', which is French for 'charcoal'. English geologists were more cautious, preferring terms such as 'mother of coal' or 'mineral charcoal'. According to Francis (1961) , Stevenson adopted the French word 'fusain' in 1911 to avoid cumbersome terminology. When Stopes (1919) proposed what has since become the standard system of coal taxonomy, she distinguished between 'fusain', the visually identifiable components of coal units and hand specimens, and 'fusinite' the microscopically or chemically identifiable component of coal (maceral). The International Commission on Coal Petrology (ICCP 1966) adopted Stopes' taxonomy, and 'fusain' and 'fusinite' became accepted terms. Here we use 'fusain' to refer to substances in pre-Quaternary sediments that are morphologically, optically, and chemically indistinguishable from modern charcoal.

Fusain is the 'dirty' constituent of coal. It readily turns to coal dust, and its friability is a major contributor to collapses of mine pillars and walls. It is macroscopically notable for leaving a black mark on anything it touches—i.e., it has a 'black streak' in the usage of mineralogists.

Fusain typically shows well-preserved, open, cell wall structures, which may either be empty or infilled with minerals. Seen through a microscope under transmitted light, fusain is opaque (black to reddish) in thin section. A polished section or block under oil immersion is highly reflective (light gray to white). Chemically, fusain is more oxidized (hydrogen-depleted and oxygen-enriched) than most coal constituents, and less chemically reactive. Crystallographically it shows (usually weak) indications of graphitic hexagonal lattice structure. Fusain is classed, along with other oxidized, unreactive, and reflective coal constituents in the group of macerals known as inertinites. It is distinguishable from other intertinites by the presence of well-preserved cell walls.

Despite the similarities between fusain and charcoal, many prominent geologists, including White (1925), Stopes (1919), and Schopf (1975), did not accept the hypothesis that fusain was of fire origin. White (1925), for example, wrote:

> "The assumption that [fusain] results from fires on the surface of exposed peat is inconsistent with the great extent of the fusain layers, the evenness of the bedding, the absence of corrosion pits or ash accumulations with supposed cinders, the normal state of the coal just beneath the fusain, the regularity and parallelism of the distribution of the fusain, generally in thin layers, the sometimes incredible closeness of the layers which may form a large part of the deposit in some beds, the purity of some of the layers, and the delicacy and orientation of some of the carbonized debris. Fortuitous forest fires could hardly have been so frequent and so regularly recurrent over the same great areas, and so widely destructive as these should have been, without leaving traces of ash accumulation or residue."

Other geologists such as Crickmay (1935) and Skolnick (1958) advocated a fire origin and noted the ubiquity of fusain in the rock record of the last few hundred million years. Velikovsky (1955) enlivened, if not enlightened, the debate by proposing that fires were evidence of cataclysmic events.

Some objections to the theory of a fire origin of fusain seem to be based on misunderstanding of the taphonomy of fire traces. Ash tends to be water soluble and is not generally found in association with fossil charcoal. Wildfires do not leave 'cinders'; thus the absence of ash and cinders presents no problem for the fire origin hypothesis. Harris (1981) demonstrated experimentally that seemingly delicate carbonized debris can be robust enough to permit fossilization. On the other hand, noted in the case examples below, the bedding characteristics and abundance of fusain in some sediments is perplexing or problematic when compared to modern analogies.

Methodological approaches

The link between fusain and ancient fire regimes has been studied in three basic ways, i) by looking at specific events or deposits and attempting reconstructions (for example, Nichols and Jones (1992)), ii) by attempting to gather information about changes in fusain abundance and attributes across time (Robinson 1991), and iii) by experimental work that relates chemical and physical characteristics of the fusain to the formative event (for example, Harris 1981).

Before the 1950s, fusain was mostly studied using the first approach. With the emergence of the ICCP after World War II, industrial requirements resulted in standardized coal characterization (e.g. ICCP 1966) with resultant sacrifice of attention to specific events and their biological and geological context. Polished thin sections fell into disuse in favor of pellets made from the macerated material, sampled in such a way as to represent the mean characteristics a coal unit. This lead to systematic loss of information on bedding horizons and the organic morphology of coals, and much information crucial to reconstruction of individual fire events.

Coal petrographers tend, after Diessel (1986) , to use the proportions of inertinite vs. other maceral categories as indices of paleoenvironmental conditions on the assumption that inertinite indicates a xeric, oxidizing environment. Such studies, however, rarely make direct mention of fire and tend to be superficial in their consideration of biological evidence. Meanwhile, coal thin sections, which preserve anatomical and stratigraphic detail, have gone out of fashion, and thin section collections made earlier in the century are now in deep storage in many coal-petrographic institutions.

Detailed experimental studies of fusain and modern charcoal have demonstrated that i) fusain fragments have morphologies and width/length ratios consistent with experimentally created charcoals (Harris 1981) , ii) 'fragile' structures, such as charred leaves are sufficiently robust to be preserved (Harris,1981), and iii) resins can survive combustion (Jones 1991) . The latter is significant because the presence of resin in association with fusain had been used to argue against a fire origin on the assumption that resin is too combustible to survive burning (Schopf 1975). Semifusinite grades into fusinite just as less charred wood grades into more charred wood on a partially-charred log (Jones et al. 1993)

Case Examples

The hypothetical first fire

At some point in the Earth's history there must have been a first wildfire–a first time that biomass, formed as a product of autotrophic carbon fixation, accumulated and dried on a land surface and was ignited in an atmosphere with sufficient oxygen to support combustion. When might this have occurred?

From the perspective of fuel availability there is the possibility of wildfire well before the Cambrian. Some organic-rich shales readily ignite and burn–for example, oil shale in the marine Jurassic Kimmeridge Shale Formation of southern Britain is "used as a fuel in cottages locally ... and has long been known as 'Kimmeridge coal'" (Chatwin 1948). The same author reports that, where the shale outcrops near Kimmeridge in Dorset, "spontaneous ignition of the shale gave rise in 1826 to a fire that lasted four years, and caused the place to be known as the "Burning Cliff". Flammable shales could presumably have formed through marine deposition of planktonic algae or photosynthetic bacteria. The latter have a geologic record tracing back about three billion years, so atmospheric oxygen permitting, fire fueled by biomass could be quite ancient.

A more probable scenario for wildfire before vascular plants colonized the land in the late Silurian (420 million years before present) involves the dead remains of macrophytic algae of marine or freshwater origin. Algal debris is commonly aggregated by wind action along the lee shore, accumulating well above water level, as in the familiar kelp and wrack that shows up on beaches of the temperate northern hemisphere. In theory, such dried-out accumulations of plant material could be ignited by lightning strike, to produce a localized wildfire derived from 'aquatic biomass' production. Authenticated finds of fossil algae, for example Fry's (1983) illustration of the non-calcareous macrophyte alga, Winnipegia, date at least to the late Ordovician. Earlier reports of macrophytic green algae date to the late Pre-Cambrian (see review by Edwards et al. 1993). If 'the first wildfire' was in fuel derived from aquatic-based photosynthesis, it would seem that this could have occurred as early as the late Pre-Cambrian and certainly well before land colonization by vascular plants. However, algae lack thickened, lignified cell walls, and the putative algal combustion may have produced something quite unlike wood charcoal.

The first fire in terrestrial biomass is likely to have occurred in situations comparable to a moorland fire, from upright reed-like plants forming a sward and then drying out. This scenario could be envisaged for early Devonian land vegetation. Alternatively, peat might have accumulated beneath such vegetation, as evidently occurred in the early Devonian Rhynie Chert setting (Stewart and Rothwell 1993). Such peat, formed on a land surface at or just below a fluctuating water table, might have periodically dried out and been ignited by lightning. This would closely parallel the natural burning of peat that occurs in the North American tundra, and is well documented (Racine et al. 1985).

Wildfire in the Paleozoic

The earliest record of fusain. The record of fusinized land plant remains begins in the Devonian with two problematic occurrences. The earliest candidate is associated with the enigmatic fossil, Nematasketum, from the basal Devonian (Gedinian) of Shropshire and other localities in South Wales (Burgess and Edwards 1988) . This is know from three-dimensional fragments of fusinised tissue. These are anatomically similar to the genus Prototaxites, which occurs widely in the Devonian from Europe and North America. Although the habit and habitat of Nematasketum is unknown, Prototaxites is widely regarded as a terrestrial plant, with a main axis (trunk) up to a meter in diameter. The affinities of Nematasketum and Prototaxites remain enigmatic, but the present consensus is that both were autotrophic land plants.

What is more relevant in the present context is that these Nematasketum fragments share with fossil charcoal the typical characteristics of open cell lumina (indicating a robust rigidity at the time of burial) and apparent homogeneous wall ultrastructure. These two characteristics have been invoked as evidence of fusinization by pyrolysis in secondary wood (Jones and Chaloner 1991) . Poor understanding of Nematasketum biology (for example, ambiguity over whether its cell walls were originally stratified during life) and ecology, however, limits application of these criteria in this particular case. Burgess and Edwards (1988) comment that their fossil shows "complete fusion (homogenization) of adjacent cell walls. This may result from charring, but more probably from burial diagenesis". From all available palaeobotanical evidence, we favor the former explanation.

Another candidate for the earliest fusain record comes from a coal sequence from Melville Island in the Canadian Arctic (Goodarzi et al. 1989) . Based on palynological evidence, this is cited as Givetian and Frasnian (late to middle Devonian). Fusain is said to be rare in the coal lithology and observed only as isolated fragments. Unfortunately, no illustration is published. If the dating and fire origin of this substance can be confirmed, this specimen will be of interest, not only as the first record of a wildfire, but also the first record in a coal-forming environment.

The next possible candidate is the seemingly charred remains of the fossil wood, Callixylon, reported by Beck et al. (1982) . Some regard this as a reasonably secure 'earliest record' of wood charred by combustion (see discussion in Cope and Chaloner 1985, Jones and Chaloner 1991) . Beck et al. (1982), however, favor 'fusinization' by some unspecified non-pyrolysis process within the sediments in which Callixylon was buried.

Large plants (trees) affect wildfires in several ways. They generally sustain more standing biomass, and may also produce large amounts of litter, as compared with herbaceous plants. When dry, this provides high fuel loadings. The height and bulk of arborescent plants elevates fuel above the ground level, which may improve ventilation, allow drying from the bottom, and enhance flammability. Furthermore, trees attract lightning strikes to areas in which there is fuel. Thus, it is reasonable to presume that wildfire became more intense with the evolution of the arborescent habit.

Moderate-sized plants existed in the Upper Devonian. Cyclostigma kiltorkense, for example, achieved a height of 8 m and a trunk diameter of 30 cm (Chaloner and Boureau 1967). By the Late Devonian–Early Carboniferous, the progymnosperm Callixylon had trunks up to 8.4 m length and 1.5 m in diameter (Stewart and Rothwell 1993). Fusinized (probably charcoalified) fragments of Callixylon secondary wood have been recovered from the Lower Mississippian, Berea Sandstone near Amhurst Ohio, USA (Beck et al. 1982).

Lower Carboniferous. A case study. The shallow marine clastic deposits at Shalwy Point exposed in a sea-cliff near Donegal, Ireland provide fossil evidence for a Lower Carboniferous 'catastrophic' wildfire [or series of wildfires] (George and Oswald 1957, Nichols and Jones 1992). At this site, a continuous series of virtually non-fossiliferous sandstone beds and thin mudstones is abruptly interrupted by a single (possibly extensive) undulating, fusain-rich horizon ('fusain unit') which clearly indicates an abrupt disruption of an established sedimentary regime. Above the 'fusain unit' the sediments revert to virtually non-fossiliferous sandstone beds similar to those seen below the horizon. The 'fusain unit' itself consists of approximately 50 thin laminations which have been emphasized by a combination of varying charcoal content and cliff-face weathering. The laminations have been interpreted as tidally-controlled deposits, in which each lamination was laid down during slack water, i.e., each lamination represents a half a day's deposition. With a maximum of 50 laminations, the whole unit must have been deposited in a little less than 1 month (Nichols and Jones 1992).

The charcoal fragments in the 'fusain unit' consist of roughly equal proportions of lycopod fragments and secondary wood fragments of putative gymnosperms. The fragments have been moderately size-sorted by winnowing, typically to around 0.5 cm^3 and are concentrated on the lamination's top surfaces. Charcoal fragments are interspersed with angular quartz grains as is typically seen in soils, along with ostracods and marine gastropods, all in a muddy limestone matrix.

The excellent three-dimensional delicate anatomical preservation of the charcoal and the lack of evidence of extensive transportation (such as rounded corners) suggest that fragments were deposited near the wildfire site. The stable carbon isotopic composition of the charcoal is anomalously heavy (δ13C of -16‰ to -17‰) when compared with Recent land plants (C3 photosynthetic pathway) and similar plant assemblages from around the same age (Jones 1994). Although the cause(s) of this anomaly are not known, photosynthesis under water stressed conditions can result in relative enrichment in 13C (Beardsall et al. 1982), and the isotopic evidence may be testimony to fires arising from/after severe and prolonged drought, as opposed, for example, to fires caused by a meteorite impact.

Nichols and Jones (1992) note that the single event horizon within the sediments does not preclude the 'forests' of that time experiencing regular wildfires, but does indicate that this particular event was exceptionally severe. The debris from smaller fires would have been insufficient to disrupt the established sedimentary system and create similar deposits, and most of any charcoal produced would have washed out to sea and been deposited in a lower energy environment. The abundance of charcoal in this deposit is extraordinary in comparison to modern analogs. For example, the 'Great Fire of Borneo, 1982-3', which affected around 37,000 square kilometers of Indonesia (Johnson 1984) apparently left no charcoal in the adjacent sediments (pers. comm. R. Gastaldo 1995).

Estimating the area burned by this Lower Carboniferous fire is problematic. Using modern analogs, Nichols and Jones (1992) estimated an extent of 95,000 square km, an area equivalent to the present area of Ireland. Following the fire and loss of the vegetation's cover and binding root systems, the environment would have been especially vulnerable to soil erosion and slope failures. With rainstorms, the river system would have become full of excess run-off water heavily loaded with eroded sediment and fire-debris. It is probably the dumping of this load into the estuarine/shallow marine environment over a short time-span that so drastically changed the sedimentation regime, and caused the deposition of the 'fusain unit' event horizon.

Pennsylvanian. Fusain generally detracts from the industrial value of a coal; hence fusain content is regularly evaluated for industrial purposes. In theory, composites of industrially mined coal profiles provide evidence on the relative fusain-richness of coals of different ages. In practice, there are problems. Despite the ICCP standards, different laboratories use different techniques and criteria. For example, different standards are employed in separating fusinite from semifusinite, and in counting smaller fragments, which some laboratories classify as inertodetrinite. Furthermore, much of the information is in unpublished reports, and much of the published literature reports only values for total inertinite, thus lumping fusinite, semifusinite, sclerotinite (coalified fungal remains), resinite (coalified resin), and other highly-reflective macerals.

The Pennsylvania State University Coal Data Base (Anonymous 1988), which contains data on around 1300 US coal samples analyzed according to a common standard, offers a glimpse of what this approach might yield if common standards were rigorously applied. Pennsylvania State's data base is particularly rich on regional coals, most of which are of Pennsylvanian age. Stratigraphic series information was used to convert seam names and formations given in the Data Base into millions of years on the geologic time scale. The resulting "history" of fusain-richness for the ~35 Ma period from the Morrowian to the Virgilianis is shown in Fig. 1, in context of fusain-richness figures for the Mississippian and the Miocene and Pliocene extracted from other data bases. To avoid anomalous results arising from the arbitrary distinction between fusinite and semifusinite, standard deviations are presented for the sum of fusinite plus semifusinite, rather than for the two separately.

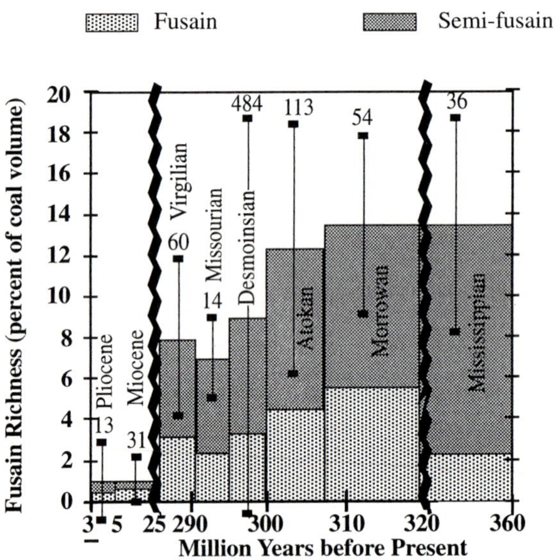

Figure 1. Fusain and semi-fusain richness for various Pennsylvanian series in context of Mississippian and Late Tertiary values. Error bars represent one standard deviation around the sum of fusain and semi-fusain. Number of observations above error bars. Miocene, Pliocene, and Mississippian not to scale.

The statistical attributes of these data deserve note. First, observations are unevenly distributed in time. The Desmoinsian, with a length of 5 Ma is represented by 484 observations, almost 100 observations per My, while the 4 Ma of the Missourian are represented by only 14 observations, or 3.5 observations per Ma. Second, standard deviations are large, especially for the relatively-heavily sampled Desmoinsian. Third, because the locations at which samples were taken came from an opportunistic, commercially-oriented sampling frame, observations are likely to be biased. Coals whose commercial value is lessened due to high fusinite content, high mineral content, or other 'defects' are probably under-represented. Thus the samples may not accurately represent the composition of coal of the ages they represent. On the other hand, coal-preserving sediments are poorly representative of past biospheres as a whole, and it is unclear how much a 'representative' sample of coals would reveal about past global fire regimes.

Caveats aside, these statistics suggest that the Pennsylvanian coal-forming ecosystems, like modern ecosystems, had highly varied propensities to accumulate charcoal, and perhaps also varied propensities to burn. This, in turn, implies that attempts at historical reconstruction based on small and possibly unrepresentative sets of case studies must be viewed skeptically.

Taken at face value, the coal record show a decline in fusinite-richness over time. All Pennsylvanian coals are extremely fusinite rich compared to late Tertiary coals, lignites, and modern peats. In comparison to the few older (Mississippian) coals for which data could be found, Pennsylvanian coals are unexceptional. Cretaceous coals, as represented in the PSU data base, are equally fusinite-rich (Robinson 1989). The long-term trends shown in Figure 1 are generally consistent with verbal accounts (c.f. Teichmüller (1982) and this paper, below, on Miocene lignites).

Within the Pennsylvanian, the data suggest declining fusain-richness. This is curious, as climatic reconstructions for the period based on detailed study of taxonomic information and fusinite content, support the conclusion that the Atokan and Missourian were the driest series of the Pennsylvanian, while the Desmoinsian was the wettest (Phillips et al. 1985). Meanwhile, geochemical reconstructions suggest a monotonic rise in atmospheric oxygen content (Berner and Canfield 1989), which, if anything, should have lead to increased fire activity.

Note on the Permian. Most of the early work on coals, and thus scientific understanding of fusain, was done in Europe and North America. Euramerican coals are predominantly Pennsylvanian, Cretaceous and Tertiary. In general, for these coals there is no terminological difficulty for fusain, because authorities in all countries agree on its macroscopic and microscopic appearance, if not on its origin (Francis 1961). Difficulty arises however, in characterizing the inertinites of Permian coals of India and Australia (probably also China and Siberia). Published data indicate these coals are extremely fusain-rich. It is common, for example, to find Permian coals classed as 20% or more fusain (Cook 1975). 'Fusain' in these coals, however, includes a variety of opaque and inert substances, most of which are morphologically unlike charcoal. Beeston (1987) proposes that for such coals it is necessary to distinguish between oxyfusinite, degradofusinite, and pyrofusinite. Only the last of these is relevant to present discussion, and it is typically rare in fusain-rich Permian coals. Pending further work on the historical biology of Permian coals little can be said about fire in the Permian.

Wildfire in the Mesozoic

Middle to Late Jurassic fluviatile, coastal, and fully marine clastic deposits of the UK and North Sea contain abundant fossil charcoal (Cope 1980, Harris 1958). Often charcoal is the most common form of fossil plant preservation found in these sediments. This reflects both charcoal's inert nature (non-charcoalified plant remains biodegrading away leaving a charcoal-rich refractory component), as well as the vast amount of charcoal that was produced in the Jurassic wildfires. Typically there is a trend of the fluviatile and coastal deposits containing larger fragments, whereas the deeper marine sediments contain smaller fragments often in laminations. Scanning electron microscope studies (Cope 1984, Jones and Chaloner 1991) show that gymnospermous secondary wood charcoal usually dominates the assemblages. As well as identifying the plant types and parts, electron microscopy has revealed anatomical structures in the charcoal that indicate the fuel supply consisted of both desiccated material (probably floor litter or attached dead branches) as well as 'green' wood (living trees).

Wildfire in the Cenozoic

Charcoal comprises at most a few percent, and in many cases a fraction of a percent, of the German Miocene brown coals of the Lower Rhine Embayment—a trivial amount compared to several percent, sometimes over ten percent, in Carboniferous coals. This is, in part an artefact. Brown coals are only partially coalified. In the Lower Rhine Embayment Miocene, one volume unit of coal represents an estimated

2.7 units of peat (Hager 1993) . Were they subjected to the deep burial and high pressures that convert the peat, lignite, and brown coal to bitumenous and anthricite coals, the German Miocene coals would loose mass through volitalization of cellulose and more mobile or more readily degraded components. This would leave a higher fraction of inert components, including charcoal and lignins. Coalification, however, does not reduce mass enough to account for the differences in charcoal richness in German Miocene coals—or other coals of similar age (Robinson 1989). This leaves us to wonder if the difference in charcoal richness are real—if Miocene peat-forming environments were significantly less fire-prone than the Carboniferous peat-forming environments.

If the Miocene was truly less fire-prone, then why? Of the three requirements for wildfires—fuel, atmospheric oxygen, and ignition—there was no shortage of potential fuel. Mosbrugger et al. (1994), in a somewhat unusual reconstruction of a forest responsible for lignite, proposed an above-ground biomass (living and dead) of 750 Mg ha^{-1}, most of it conifer trees. Ignition is difficult to reconstruct. Mosbrugger et al. (1994) calculated a mean tree height of 9.9 m, which would result in excellent inflammable targets for lightning strikes. Nonetheless, the paucity of charcoal in these deposits might reflect an environment where lightning was infrequent or regularly associated with wetness adequate to stifle fire. This leaves the possibility that fire was inhibited by lower-than-present levels of atmospheric oxygen. As discussed below, this possibility cannot be excluded.

Fire and the Changing Earth System

Terrestrial systems have changed radically since the plant kingdom became abundant on land about 400 Ma ago. To the extent that changes in the Earth system have affected the underlying determinants of fire behavior, they have changed fire regimes and patterns of charcoal formation. To the extent that the past departs from modern examples, modern fire and charcoal deposition regimes are imperfect guides to the past. Below, we note that fire regimes may have been affected by changes in atmospheric oxygen concentration as affecting propensity to burn, climatic change as affecting sources of ignition and fuels, and evolutionary change as affecting fuels.

Oxygen

Oxygen enrichment greatly enhances burning; oxygen starvation stifles it. There is little basis for predicting the effect of oxygen enhancement or starvation on charcoal formation in a wildfire environment. Thus, if the oxygen content of Earth's atmosphere has varied, fire regimes have changed in step, but we are poorly equipped to interpret the record.

Oxygen history is highly uncertain, but it is not unlikely that ecologically-significant changes have occurred in the last 400 Ma. A number of reconstruction methods have been tried, ranging from mass balance models parameterized with estimates of the quantities of reduced carbon buried in sediments over time, to the mineralogy of ancient soil profiles, to the analysis of ancient air trapped in amber. Different approaches have yielded different scenarios for the variation of atmospheric oxygen over Phanerozoic (approximately the last 600 Ma) time. The reconstructions of Tappan (1968) , Berner and Canfield (1989) , and Budyko et al. (1987) all suggest intervals in the last 600 Ma when there has been significantly more, along with intervals with significantly less, oxygen in the atmosphere than at present.

The fusain record suggests constraints on oxygen variation. Cope and Chaloner (1980) and Chaloner (1989) argue that relatively uninterrupted fusain from the early Carboniferous onwards implies that wildfire has persisted in at least some terrestrial ecosystems since the time when large woody land plants evolved. Based on Watson's (Watson 1978, Watson et al. 1978) experimental evidence of the relationship between oxygen and flammability, they suggest that the charcoal fossil record means that oxygen must have remained between 13% and 35% over that period. Below the lower figure, there would have been too little oxygen for plant material to burn; above the higher figure, by extrapolation of Watson's experimental results, the flammability of plant tissue would have been such as to make sustained forest growth virtually impossible. The validity of Watson's experimental analog for vegetation-atmosphere interaction, especially the use of shredded paper rather than living plants in establishing these thresholds, has been questioned (Robinson 1989) , although the importance of his use of experimental methods for studying fire-atmosphere interaction is not disputed.

Constraining oxygen concentration to between 13 and 35% leaves latitude for large variations in global pyricity. Robinson (1989) attempted to test Berner and Canfield's (1989) oxygen reconstruction against evidence from fusain abundance in coals and paleobotanic reconstructions of forest structure in different ages. She found the record was consistent with general features of the oxygen reconstruction, but uncertainties, both in fire behavior and char formation as a function of oxygen content and in interpretation of the fusain record, preclude rigorous testing.

Likewise, in the above case studies, oxygen variation is a possible explaination for relative scarcity of fusain in Miocene brown coals compared to Pennsylvanian coals.

Climate

The last few million years are exceptionally cold by Earth history standards (Frakes 1979). Many past climates, for example megathermal mesic high latitude zones or the megamonsoons predicted in association with the assembly of Pangaea (Kutzbach and Gallimore 1989), are outside the ranges of modern experience. This makes reconstruction of past fire regimes highly problematic. Furthermore, declining atmospheric CO_2, one suggested cause of planetary cooling, should have decreased plant productivity, and thus altered fuel accumulation rates and carbon fluxes in general.

Biological Evolution

Plant evolution has probably affected fuel abundance, geometry, chemistry, and distribution on the landscape. Lignin has a higher ignition temperature, a greater propensity to smolder and char, and a lower propensity to flame than do polysaccharides. Based on comparisons of the bulk chemistries of older and more modern taxa, Robinson (1990) proposed that the ratios of lignin to cellulose and other polysaccharides in bulk plant material were high in the Paleozoic, and have declined over the Mesozoic and Cenozoic. If this is true, the earliest forests would have been less flammable than modern forests, and would have had a high propensity to form charcoal when they did burn.

If, however, as reconstructed by Berner et al. (1989), the late Paleozoic atmosphere contained more oxygen than the modern atmosphere, high oxygen would have counteracted low-flammability. Thus, conceivably, fusain-richness in Pennsylvanian coals, reflects lignin-rich forests growing under high atmospheric oxygen.

Fuel loadings and geometries, likewise, may have varied greatly over geologic time. Attempts to quantitatively evaluate biomass (or volume) of past forests by measurement of stump floras (DiMichele and DeMaris 1987, Mosbrugger et al. 1994) have found stem densities much higher than those found in modern ecosystems. Many modern vegetative formations and associated fire regimes are geologically young features of Earth's landscape. In particular, grasses and herbs went though major bursts of evolution and range expansion in the last 20 to 25 Ma (Singh 1988).

Such landscapes tend to be fire-prone, if not fire–dependent, and their fire regimes are intimately associated with features of non-woody angiosperms, such as rapid growth and post-fire recovery, finely-divided fuel, and ability to form highly continuous fuel beds in arid and semi-arid climates (Daubenmire 1968) . It is unclear, for example, whether anything like modern savanna or prairie fire regimes could have occurred in the arid and semi-arid lands of the Triassic or Jurassic because it is unclear whether a flora of gymnosperms and pteridophytes would have developed sufficiently continuous and resilient cover to support extensive and frequent burning.

Herbivore evolution may also have affected fuel geometry and loading. The community of browsing animals, be they sauropods, horses, camels and antelope, or elephants and mammoths, can have a major impact on the three dimensional structure of forest, and hence on its propensity to dry out and ability to carry fire. This community has changed dramatically over time, undoubtedly affecting fire regimes as it changed.

Lastly, the evolution of detritivores cannot be ignored. The abundance of fungal remains in coals suggests that the extent to which wood is subject to fungal degradation has increased greatly in the Tertiary (Robinson 1990) . Likewise, insect consumption of plant material appears to have increased over time, with notable increases occurring in angiosperm material in the Cretaceous and thereafter (Scott et al. 1992). If, in the distant past, the decomposers of plant tissue were significantly less active and decay slower, it would have major ramifications for accumulation of fuels and the propensity of systems to burn and char.

Conclusions

Fire is an important ecological variable whose traces survive well in the geologic record. The present understanding of fire as seen in the fossil record turns up more questions than answers. While some aspects of the history of fire may remain forever uncertain, the subject as a whole has received insufficient systematic research, and many important questions are suitable for experimental work and structured observational research. Improved understanding of ancient fires would fill in a missing dimension to paleoecological reconstructions and bring in additional lines of evidence in reconstructions of the cycles of carbon and oxygen.

Literature References

Arens NC (1991) Wildfire in the Paleozoic: preliminary results of a case study on the fire ecology of a Pennsylvanian floodplain forest, Joggins, Nova Scotia, Canada. In Nodvin SC, Waldrop TA (eds) *Fire and Environment: Ecological and Cultural Perspectives* pp 279-288 SE For Exp Sta Ashville NC

Anon (1988) The Penn State Coal Sample Bank and Data Base. Energy and Fuels Research Center, the Penn State Univ

Beardsall J, Griffiths H, Raven JA (1982) Carbon Isotope Discrimination and the CO_2 Accumulating Mechanism in *Chlorella emersonii*. *J Exp Bot* 33:729-737

Beck CB, Coy K, Schmid R (1982) Observations on the fine structure of Callixylon wood. *Amer J Bot* 69:54-76

Beeston JW (1987) Aspects of inertinite formation and deposition in the Denison Trough, Queensland. *Austral J Coal Geol* 7:33-45

Berner RA, Canfield DE (1989) A new model for atmospheric oxygen over Phanerozoic time. *Amer J Sci* 289:333-361

Burgess ND, Edwards D (1988) A new Palaeozoic plant closely allied to Prototaxites Dawson. *Bot J Linnean Soc* 97:189-203

Budyko MI, Ronov AB, Yannshin AL (1987) *History of the Earth's Atmosphere*. Springer-Verlag Berlin

Chaloner WG (1989) Fossil charcoal as an indicator of palaeoatmospheric oxygen level. *J Geol Soc Lond* 146:171-174

Chaloner WG, Boureau E (1967) Lycophyta. In Boureau E (ed) *Traité de Paléotanique* Masson Paris

Chatwin CP (1948) *The Hampshire Basin and Adjoining Areas*. Geological Survey and Museum HMSO London

Cook AC (1975) The spatial and temporal variation of the type and rank of Australian coals. In Cook AC (ed) *Australian Black Coal: Its Occurrence, Mining, Preparation and Use* pp 63-84 Australasian Inst Mining Metall, Illawarra, NSW

Cope MJ (1980) Physical and chemical properties of coalified and charcoalified phytoclasts from some British Mesozoic sediments; an organic geochemical approach to palaeobotany. In Douglas AG, Maxwell JR (eds) *Advances in Organic Geochemistry* pp 663-677 Pergamon Press Oxford

Cope MJ (1984) *Some Studies of the Origin, Nature and Occurrence of Charcoalified Plant Fossils*. PhD Thesis Univ London UK

Cope MJ, Chaloner WG (1980) Fossil charcoal as evidence of past atmospheric composition. *Nature* 283:647-649

Cope MJ, Chaloner WG (1985) Wildfire: An interaction of biological and physical processes. In Tiffney BH (ed) *Geological Factors and the Evolution of Plants* pp 257-278 Yale University Press New Haven

Crickmay CH (1935) The nature and origin of fusain. *Amer Midl Natur* 16:94-98

Daubenmire R (1968) The ecology of fire in grasslands. *Adv Ecol Res* 5:209-266

Diessel CF (1986) The correlation between coal facies and depositional environments. In *Advances in the Study of the Sydney Basin, Proc. 20th Newcastle Symposium* pp 19-22 Univ of Newcastle Newcastle NSW

DiMichele WA, DeMaris PJ (1987) Structure and dynamics of a Pennsylvanian-Age Lepidendron forest: Colonizers of a disturbed swamp habitat in the Herrin (No. 6) Coal of Illinois. *Palaios* 2:146-157

Edwards D et al (1994) Algae. In MJ Benton (ed) *The Fossil Record*. Chapman Hall

Frakes LA (1979) *Climates throughout Geologic Time*. Elsevier/North-Holland Inc Amsterdam

Francis W (1961) *Coal: Its Formation and Composition*. Edward Arnold Ltd London

Fry WL (1983) An algal flora from the Upper Ordovician of the Lake Winnipeg region, Manitoba, Canada. *Rev Palaeobot Palynol* 39:313-341

George TN, Oswald DH (1957) The Carboniferous Rocks of the Donegal Syncline. *Quart J Geol Soc London* CXIII 137-179

Goldberg ED (1985) *Black Carbon in the Environment: Properties and Distribution*. Wiley Interscience New York

Goodarzi F, Gentzis T, Embry AF (1989) Organic petrology of two coal-bearing sequences from the Middle to Upper Devonian of Melville Island, Arctic, Canada. *Cont Canadian Coal Geosci* 89-8:120-130

Grand'Eury (1882) Memoire sur la formation de la houille. *Ann Mines* 8:99-290
Hager H (1993) The origin of the Tertiary lignite deposits in the Lower Rhine region, Germany. *Intern J Coal Geol* 23:251-262
Harris TM (1958) Forest fire in the Mesozoic. *J Ecol* 46:447-453
Harris TM (1981) Burnt ferns from the English Wealden. *Proc Geol Assoc* 92:47-58
ICCP (1966) *International Handbook of Coal Petrography*. Centre National de la Recherche Scientifique Paris
Johnson B (1984) *The Great Fire of Borneo, Report of a Visit to Kalimantan-Timur a Year Later.* World Wildlife Fund
Jones TP, Chaloner WG (1991) Fossil charcoal, its recognition and palaeoatmospheric significance. *Palaeo Palaeo Palaeo* 97:30-50
Jones TP, Scott AC, Mattey DP (1993) Investigations of "fusain transition fossils" from the Lower Carboniferous: comparisons with modern partially charred wood. *Intern J Coal Geol* 22:37-59
Jones TP (1991) *The Nature, Origin and Recognition of Fusain.* PhD thesis University of London UK
Jones TP (1994) ^{13}C enriched lower Carboniferous fossil plants from Donegal, Ireland: carbon isotope constraints on taphonomy, diagenesis and palaeoenvironment. *Rev Palaeobot Palynol* 81:53-64
Kutzbach JE, Gallimore, RG (1989) Pangean climates: Megamonsoons of the megacontinent. *J Geophys Res* 94:3341-3357
Mosbrugger V, Gee CT, Belz G, Ashraf AR (1994) Three-dimensional reconstruction of an *in-situ* Miocene peat forest from the Lower Rhine Embayment, northwestern Germany: new methods in palaeovegetation analysis.*Palaeobot Palaeo Palaeo* 110:295-317
Nichols G, Jones TP (1992) Fusain in Carboniferous shallow marine sediments, Donegal, Ireland: the sedimentological effects of wildfire. *Sedimentology* 39:487-502
Phillips TJ, Peppers RA, DiMichele WA, (1985) Stratigraphic and interregional changes in Pennsylvanian coal-swamp vegetation: environmental inferences. *Intern J Coal Geol* 5:43-109
Racine CH, Dennis JG, Patterson WA (1985) Tundra fire regimes in the Noatak River watershed, Alaska: 1956-83. *Arctic* 38
Retallack G (1984) Completeness of the rock and fossil record: some estimates using fossil soils. *Paleobiol* 10:59-78
Robinson JM (1989) Phanerozoic O_2 variation, fire, and terrestrial ecology. *Global Planetary Change* 1:223-240
Robinson JM (1990) Lignin, land plants, and fungi: Biological evolution as affecting Phanerozoic oxygen balance. *Geology* 15:607-610
Robinson JM (1991) Phanerozoic atmospheric reconstructions: A terrestrial perspective. *Global Planetary Change* 5:51-62
Schopf JM (1975) Modes of fossil preservation. *Rev Paleobot Palynol* 20:27-53
Scott AC, Stephenson J, Chaloner WG (1992) Interaction and coevolution of plants and arthropods during the Palaeozoic and Mesozoic. *Phil Trans Royal Soc London* B 335:129-165
Singh G (1988) History of aridland vegetation and climate: A global perspective. *Biol Rev* 63:159-195
Skolnick H (1958) Observations on fusain. *Bull Amer Assoc Petrol Geol* 42:2223-2236
Stewart WN, Rothwell GW (1993) *Paleobotany and the Evolution of Plants.* Cambridge Univ Press Cambridge
Stopes MC (1919) On the four visible ingredients of banded bituminous coals. *Proc Roy Soc London* 90:470-487
Tappan H (1968) Primary production, isotopes, extinctions and the atmosphere. *Palaeo Palaeo Palaeo* 4:187-210
Teichmüller M (1982) Origin of the petrographic constituents of coal. In Stach E (ed) *Stach's Textbook of Coal Petrology* pp 219-283 Gebrüder Borntraeger Berlin FRG
Velikowsky I (1955) *Earth in Upheaval.* Doubleday Garden City NY
Watson AJ, Lovelock JE, Margulis L (1978) Methanogenesis, fires, and the regulation of atmospheric oxygen. *BioSystems* 10:293-298
Watson AJ (1978) *Consequences for the Biosphere of Grassland and Forest Fires.* Thesis, Reading Univ UK
White D (1925) Environmental conditions of deposition of coal. *Am Inst Mining Metall Eng Trans* 71:3-34

Marine Sediments: A Reservoir for Black Carbon and their Use as Spatial and Temporal Records of Combustion

Daniel O. Suman, T.A.J. Kuhlbusch[1], B. Lim[2]
Division of Marine Affairs & Policy
Rosenstiel School of Marine & Atmospheric Science
University of Miami
4600 Rickenbacker Causeway
Miami, Florida 33149, USA

Summary

Black carbon (BC) found in ocean sediments provides regional and global information regarding biomass burning activities and transport mechanisms. BC fluxes to surface sediments, largely from the Pacific Ocean, range between 0.002-3.6 µg BC cm^{-2} yr^{-1} for deep sea sediments and 26-354 µg BC cm^{-2} yr^{-1} for continental margin sediments. The BC flux may be a function of the magnitude of biomass burning, the distance to the source region, and riverine and eolian transport mechanisms.

This manuscript presents some initial calculations of the global BC cycle. Using BC sediment flux data, we calculate global BC deposition to sediments in the deep ocean and continental margins, and compare these values to atmospheric deposition of BC to the global open and coastal ocean surface based on atmospheriç BC concentration fields. About 10.4 Tg BC yr^{-1} are deposited in ocean sediments, largely in the coastal ocean. BC deposition to the ocean surface is much more homogeneous throughout the ocean, accounting for deposition of 6.9 Tg BC yr^{-1}. We hypothesize that rivers may transport 12.2 Tg BC yr^{-1} to the ocean and that these particulates are largely deposited in coastal ocean sediments. The manuscript discusses gaps in our knowledge of sedimentary BC and suggests future research initiatives that might further understanding of the global cycle and temporal distribution in ocean sediments of this product of biomass burning.

Introduction

Research of the past 20 years has demonstrated that the black carbon fraction of marine sediments can provide a regional signature, as well as a global background, of global biomass burning. The term "black carbon" is used here to define the carbon fraction that has been thermally altered during vegetation fires and is thought to persist in the natural environment over geological timescales. As a result of its stability, black carbon can be used as a tracer of past fire activity and represents a short term sink in the global bio-atmospheric carbon cycle. We do not use the term "charcoal" because it includes degradable organic matter, as well as black carbon.

---

[1] Atmospheric Chemistry Department, Max Planck Institut für Chemie, Postfach 3060, D-55020, Mainz, Germany
[2] OECD Environmental Directorate, 2 Rue André-Pascal, 75016 Paris, France

NATO ASI Series, Vol. I 51
Sediment Records of Biomass Burning and Global Change
Edited by James S. Clark, Hélène Cachier, Johann G. Goldammer, and Brian Stocks

Researchers have studied particulate morphologies and black carbon fluxes to the sediments and interpreted these data in terms of the temporal and spatial variabilities of biomass burning, as well as eolian or atmospheric transport processes. Nevertheless, marine sediment black carbon data are few, and existing sediment data largely represent the North Pacific Ocean. Black carbon records in marine sediments can greatly improve our knowledge of spatial and temporal trends in global and regional biomass burning and, thus, enhance our understanding of biomass burning's importance for the atmosphere (greenhouse gases, photochemically active gases, etc.) and the biogeochemical carbon cycle (carbon sequestration).

This manuscript first reports the current knowledge of black carbon concentrations in marine sediments and then develops a model for a global black carbon cycle. We relate the sediment black carbon dataset to the modeled global atmospheric black carbon concentrations by estimating deposition to the world ocean sediments and comparing these values to the atmospheric deposition of black carbon to the ocean surface. We suggest that river transport of black carbon to the ocean must be of the same order of magnitude as the eolian flux of black carbon to the ocean. The manuscript also discusses how marine sediment cores can hold temporal records of past biomass burning.

Spatial distributions of black carbon in marine sediments.

Global Distributions of Black Carbon in Marine Sediments

Smith et al. (1973) reported the first black carbon (BC) concentrations for pelagic (deep sea) sediments from 27 sites in the Pacific and Atlantic Oceans (Fig. 1). All sediment samples were extracted from no deeper than 15 cm in sediment cores. Black carbon concentrations ranged from 0.001-0.099% of bulk dry sediment. Assuming sediment densities of 2 g cm^{-3} and pelagic sedimentation rates of 10^{-3} mm yr^{-1}, black carbon fluxes to pelagic sediments ranged from 0.002 to 0.2 μg BC cm^{-2} yr^{-1}. These values are average fluxes that represent approximately the last 150,000 yrs.

Herring (1977, 1985) reported black carbon fluxes to eight additional North Pacific pelagic sites. His results were consistent with those of Smith et al. (1973) and ranged from $\leq$ 0.0014 to 3.6 μg BC cm^{-2} yr^{-1}. However, few of Herring's samples originated from the top of the sediment cores, and the dates of many of these sediment strata approximated 10^6 years.

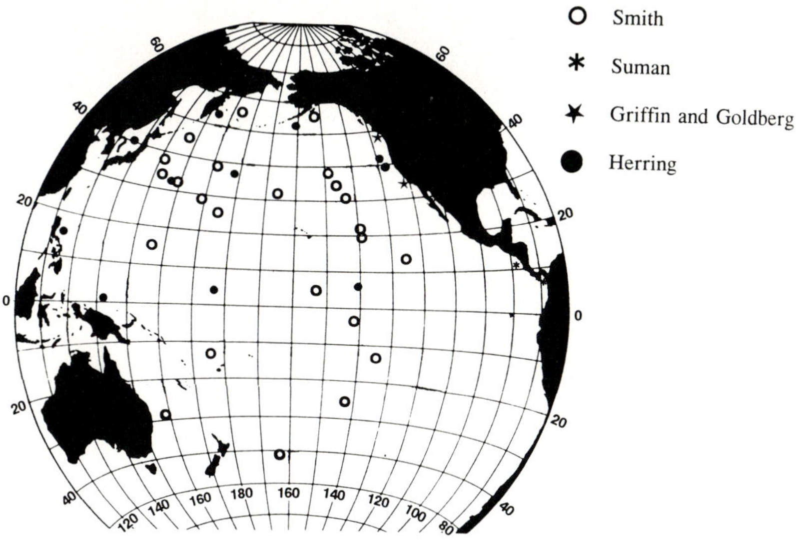

Figure 1. Map of the Pacific Ocean showing sites of sediments that have been analyzed for black carbon.

Black carbon flux data are also available for several other sites on the continental margins (Fig. 1). Griffin and Goldberg (1975) calculated BC fluxes to sediments in Saanich Inlet (British Columbia, Canada) and suggested an average of 74 μg BC cm^{-2} yr^{-1}. Average fluxes to the Santa Barbara Basin (California, USA) sediments were 26 μg BC cm^{-2} yr^{-1}. Suman (1983, 1991) determined that BC fluxes to the continental rise adjacent to Nicaragua (Central America) ranged between 23 - 135 μg BC cm^{-2} yr^{-1}. The same study (1983, 1986) reported higher BC fluxes to three shallow Gulf of Panama (Central America) sites that averaged from 109 to 354 μg BC cm^{-2} yr^{-1}. Lim and Cachier (accepted) report fluxes of 120 μg BC cm^{-2} yr^{-1} to a deep water site (2300 m) in the Mediterranean Sea. The high BC fluxes to all these sediments are most likely a function of biomass burning on the adjacent continents.

All these researchers, except Lim and Cachier, used a similar infrared spectrometric analytical technique that facilitates comparison of their flux data. Fluxes were determined over different temporal scales, however. Because of the low sedimentation rates, the strata used to determine pelagic BC fluxes represent approximately 10^4 - 10^5 years of deposition. On the other hand, continental margin strata contain on the order of 10^1 years of deposition.

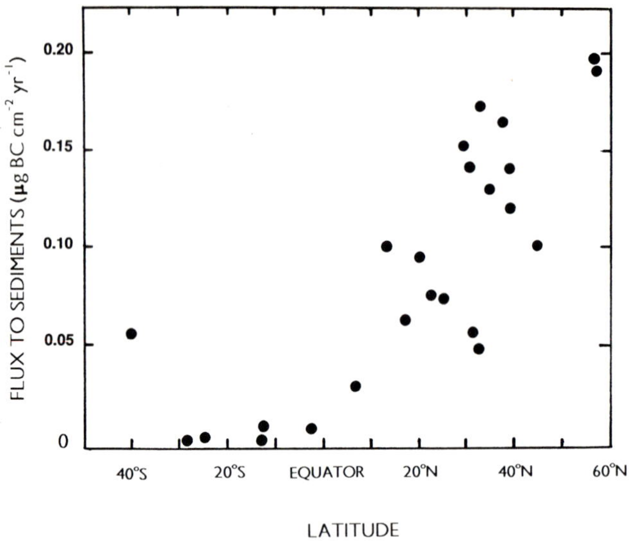

Figure 2. Black carbon flux to pelagic sediments in the Pacific Ocean as a function of latitude. Flux data are based on Smith et al. (1973).

Significance of the Spatial Distribution of Fluxes to Sediments

Smith et al. (1973) reported percent dry weight of BC in sediments as a function of latitude. They noted that BC concentrations in Pacific pelagic sediments increased from the equator to higher northern latitudes. Estimated BC fluxes to deep sea sediments exceeded 0.10 μg BC cm^{-2} yr^{-1} north of 30° N latitude (Fig. 2). Herring (1977) reported that in the Bering Sea and the Gulf of Alaska fluxes surpassed 1 μg BC cm^{-2} yr^{-1}. Black carbon fluxes to the equatorial Pacific sampling sites registered the lowest values of the series. Data from the South Pacific are essentially absent.

Smith et al. (1973) suggested that the spatial distribution of BC is a product of forest fires and transport by zonal wind systems. These authors hypothesized that forest fires in temperate coniferous and non-coniferous forests have produced large amounts of BC that has been delivered to Pacific sediments by the mid-latitude westerlies. They insisted that fires were uncommon in tropical forests because of the humid conditions in these areas.

These speculations, however, may be at odds with much evidence that suggests that recent tropical biomass burning has an extremely important impact on global biogeochemical cycles (Andreae 1991, Crutzen and Andreae 1990).

Moreover, the greatest measured fluxes to coastal marine sediments were reported for the equatorial Gulf of Panama sediments by Suman (1986). These fluxes were at least four times greater than those to coastal marine sediments of California and British Columbia, Canada.

Moreover, Smith et al.'s (1973) six data points are all from the South Pacific Ocean. It is possible that BC fluxes to the southeast Pacific Ocean may approach values of the mid-latitudes of the North Pacific. Fluxes to the southeast may be lower than those of the high latitudes of the North Pacific, because few major river systems flow from the Pacific watershed of South America, and the present-day major biomass burning region in South America, the Amazon Basin, drains into the Atlantic Ocean. The Andes act as a meteorological barrier to east winds emanating over the Amazon Basin. Both these factors may limit atmospheric and fluvial transport to the southeast Pacific Ocean.

Perhaps observations and hypotheses can be reconciled. Widespread tropical biomass burning is probably a recent human-induced phenomenon that is not recorded in buried pelagic sediments. Moreover, BC fluxes to marine sediments should depend on several factors, including magnitudes of the biomass burning sources, strength of the wind and river transport, distance from sources, and transport mechanisms through the water column to marine sediments. We plotted Smith et al.'s (1973) data for BC fluxes to pelagic Pacific sediments against distance from the continental land mass (Fig. 3). BC fluxes to sediments are highest in pelagic sediments that are adjacent to the continental margins and diminish toward the center of the Pacific Ocean. It is therefore simplistic to conclude from these data that temperate burning in Asia and North America produces more BC than tropical biomass burning. Source strength is only one of many factors that determine BC fluxes to marine sediments. Continental land masses surround the high latitude North Pacific, so the BC source is relatively close to the sediment deposition sites. The central equatorial Pacific is thousands of kilometers from sources and, therefore, its sediments have lower fluxes of continentally-derived particulates.

Assessment of the spatial distribution of biomass burning based on BC fluxes depends on both eolian and fluvial transport. Decreasing BC fluxes to marine sediments with increasing distance from river mouths suggests fluvial transport. We suspect that sedimentation of BC particles that are fluvially transported to the coastal oceans occurs largely on the continental shelf, and, in many cases, fluvial delivery may greatly exceed eolian sources to these sediments. Long-range eolian transport of BC will become increasingly dominant at greater distances from the coast to the point where this flux may represent the sole transport mechanism for BC deposition to marine sediments. Measurements of eolian and riverine fluxes of BC to the ocean do not exist. However, in the following sections, we compile some first order estimates of atmospheric BC deposition and riverine BC influx.

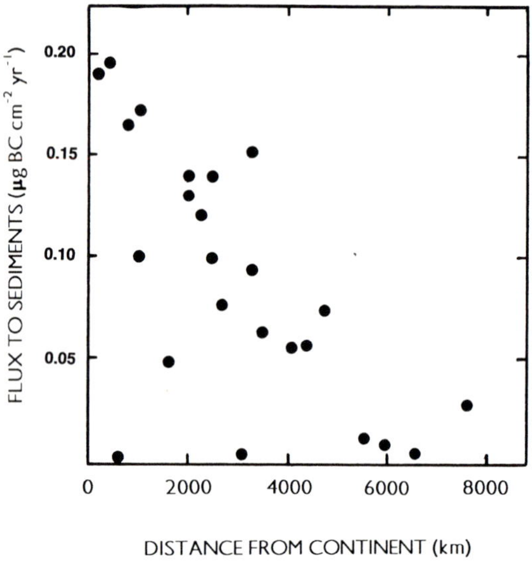

Figure 3. Black carbon flux to pelagic sediments in the Pacific Ocean as a function of distance from the nearest continent. Flux data are based on research by Smith et al. (1973).

Modeling Atmospheric Deposition of Black Carbon to the Oceans

Estimates of eolian fluxes of BC to the oceans are essential to understand the geochemical cycling of BC and the coupling between oceans and atmosphere. Despite debates concerning calibration of optical methods for BC (Liousse et al. 1993), the BC aerosol database for remote sites is of relatively high quality. In this section we show global distributions of BC in the atmosphere, and we estimate deposition to open and coastal oceans. We then compare atmospheric fluxes of BC to the ocean surface with fluxes to marine sediments.

Figures 4 a and b illustrate BC concentration fields for winter and summer months (Cooke and Wilson, submitted). Highest concentrations occur over continental emission sources, notably Europe, South and North America, and Sub-Saharan Africa. Seasonality of anthropogenic fossil fuel emissions from North America and Europe and biomass burning from the African and South

American continents are reflected in high BC concentrations during winter months and dry seasons, respectively.

Direct atmospheric deposition of BC particulates to the oceans occurs by wet and dry deposition. Dry deposition is estimated as

$$F_d = V_d \cdot C_d \cdot \text{constant},$$

where F_d is dry deposition flux (g BC m^{-2} yr^{-1}), V_d is dry deposition velocity (cm s^1) and C_d is aerosol concentration (µg BC m^{-3}). Wet deposition is estimated as

$$F_w = P \cdot S \cdot C_d \cdot Y^{-1} \cdot \text{constant},$$

where F_w is the wet deposition flux (g BC m^{-2} yr^{-1}), P is annual precipitation (mm yr^{-1}), S is the scavenging ratio, C_d is the BC aerosol concentration (µg BC cm^{-3}), and Y is the density of air (g m^{-3}) at standard temperature and pressure. S is defined as the amount of air cleansed of all BC by 1 kg of rain. Total deposition then is

$$F_T = F_d + F_w.$$

A variety of sources were to used to parameterize the model. For aerosol BC we used January and July concentrations (Fig. 4) and assumed that averages were representative of annual concentrations. Seasonal precipitation fields for the winter and summer months were taken from Jaeger (1983). For further details on the atmospheric deposition model, see GESAMP (1990). Based on its hydrophilic nature (Ducret and Cachier 1992) and its submicron size (< 1 µm), we believe that BC aerosols are removed from the atmosphere by wet and dry deposition in much the same manner as pollutant-type particles, such as those containing lead. Wet deposition is thought to be the dominant removal mechanism. The efficiency of wet removal will vary as a function of both distance from the source and the season. Field studies (Ducret 1994) suggest scavenging ratios (S) over open oceans average 280, with lower values of 120 expected close to continental sources. For the purposes of these calculations, we chose a rather high S of 500 (derived from marine sites) as an extreme upper limit and recognize that the top range of our estimates errs on the high side.

Dry deposition velocities are difficult to determine experimentally, because dry deposition collectors do not mimic the ocean surface. Based on lead data for particles of a similar size range, our best estimate for deposition velocity is 0.1 cm s^{-1} (Patterson and Settle 1987) with an upper limit of 0.3 cm s^{-1} (GESAMP 1990). Unlike particles > 1 µm, dry deposition velocity is not expected to vary significantly with distance from source, and, thus, we use the same velocities in calculations for open and coastal oceans. Uncertainties of wet and dry deposition fluxes presented here result principally from uncertainties of scavenging ratios and deposition velocities. We estimate that calculated wet and dry deposition values have associated uncertainty factors of ± 4 and ± 3, respectively.

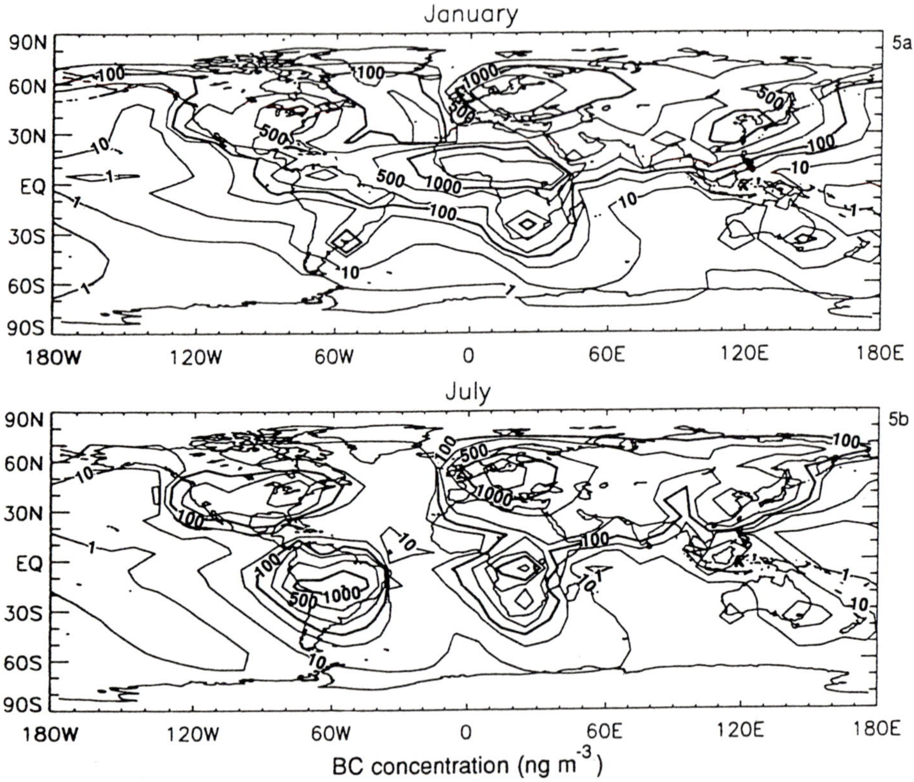

Figures 4 a & b. Tropospheric black carbon concentration fields (ng m^{-3}) during January and July modeled by Cooke and Wilson (reprinted with permission).

Using upper and lower limits of the parameters discussed above, we estimated atmospheric BC fluxes and converted them to annual deposition to the coastal and open Atlantic, Pacific, and Indian Oceans (Table 1). Values are reported in teragrams (Tg = 10^{12} g) BC yr^{-1}. The surface area of the oceans and the respective coastal margins used in our calculations are based on bathymetric data compiled by Sverdrup et al. (1946). We define coastal regions as waters above the continental margins (0 - 2000 m water depth) and open ocean regions as areas with greater water depths. Continental margins account for 12.7%, 25.7%, and 10.7% of the area of the Pacific, Atlantic, and Indian Oceans, respectively.

Table 1. Estimated atmospheric (wet plus dry) deposition of black carbon to the open and coastal oceans (Tg BC yr^{-1})

	Open Oceans		Coastal Oceans	
	1*	2*	1*	2*
Atlantic				
Wet	2.53	0.61	1.65	0.92
Dry	0.12	0.04	0.60	0.20
Total	2.65	0.65	2.25	1.12
Pacific				
Wet	1.24	0.29	1.39	0.77
Dry	0.27	0.09	0.39	0.13
Total	1.51	0.38	1.76	0.90
Indian				
Wet	0.98	0.24	0.61	0.34
Dry	0.24	0.08	0.12	0.04
Total	1.22	0.32	0.73	0.38
Global Total	5.38	1.35	4.74	2.40

*Two scenarios are presented for open and coastal oceans. Scenario "1" assumes wet deposition scavenging ratios of 500 (GESAMP 1990) and dry deposition velocities of 0.3 cm s^{-1}. Scenario "2" assumes wet deposition scavenging ratios of 280 for coastal oceans and 120 for open oceans (Ducret 1994) and dry deposition velocities of 0.1 cm s^{-1} (GESAMP 1990).

We calculate the total estimated annual BC deposition to open and coastal areas of the Pacific Ocean in terms of BC mass per unit area in order to facilitate comparison with sediment core data in the Pacific Ocean. For the Pacific, the atmospheric deposition model estimates BC fluxes to open ocean waters that range from 0.24 to 0.96 μg BC cm^{-2} yr^{-1}. We estimate eolian BC fluxes to Pacific coastal waters from 3.9 to 7.7 μg BC cm^{-2} yr^{-1}.

Table 1 suggests global BC deposition to the oceans of 7 ± 3 Tg yr^{-1}. However, because of the recycled sea-salt component, gross BC deposition may overestimate net deposition (possibly by as much as 50%, as is the case with lead). Until more data become available, estimation of the net deposition of BC is impossible. Comparison of these deposition values (median 7 μg BC yr^{-1}) to BC atmospheric emissions estimated by Crutzen and Andreae (1990) [20-30 Tg BC yr^{-1} or median 25 Tg BC yr^{-1}] and Cooke and Wilson (submitted) [median 13.5 Tg BC yr^{-1}] implies that 28 to 52% of the BC emitted is subjected to long-range transport from continental source regions and deposited to the ocean surface. Liousse and Cachier (1992), however, have estimated global BC emissions to be on the order to 4 Tg yr^{-1}, a value lower than the above estimates. Our results also indicate that wet deposition is the dominant atmospheric removal mechanism in oceanic sites with wet deposition accounting for 73 to 95% of total BC deposition (Table 1).

Carbon sequestered in ocean sediments.

Because BC is assumed to be microbially and chemically inert under normal environmental and sedimentary conditions, its deposition in marine sediments removes carbon from the short-term bio-atmospheric to the long-term geological carbon cycle. Thus, its sequestration represents a sink of atmospheric CO_2.

Kuhlbusch and Crutzen (in press) have estimated the global BC formation by vegetation fires to be 70 to 260 Tg yr^{-1} with more than 90% of the BC produced remaining after the fires as residues on the ground. Little is known of the fate of the BC remaining on the site. It may be incorporated into soils, reburned, or later remobilized by water and deposited in river, lacustrine, or marine sediments. Whether some of the BC that remains on site later becomes airborne transported to the open ocean is unknown. Nevertheless, we believe that the primary pathway of BC to pelagic marine sediments is eolian transport of BC particles that are emitted during burns. These particles are most likely submicron and may be as large as a few microns.

Current knowledge permits rough estimates of black carbon sequestered in marine sediments. A simplistic model divides ocean sediments into three water depth categories: 0 - 200 m, 200 - 2000 m, and > 2000 m (Table 2). BC fluxes to each of these three categories are assumed to have average values of 35, 1, and 0.1 μg BC cm^{-2} yr^{-1}, respectively. These fluxes are all within the ranges reported previously for the appropriate depth categories (Griffin and Goldberg 1975, Herring 1977, Smith et al. 1973, Suman 1983). Calculations were based on bathymetric data compiled by Sverdrup et al. (1946), which include all adjacent seas. The product of the appropriate flux and the area is the deposition of BC in sediments of each depth category (Table 3).

Table 2. Depth Categories of the World Oceans used to estimate BC deposition in Table 3.

	Pacific	Atlantic	Indian
Area (km^2)	179,679,000	106,463,000	74,917,000
		% Area in Depth Categories	
0 - 200 m	5.7%	13.3%	4.2%
200 - 2,000 m	7.0%	12.4%	6.5%
>2,000 m	87.3%	74.3%	89.3%

Table 3. Estimated black carbon deposition in marine sediments (Tg BC yr^{-1}).

Depth (m)	Pacific	Atlantic	Indian	Total World Oceans
0 - 200	3.58	4.96	1.10	9.64
200 - 2,000	0.13	0.13	0.05	0.31
>2,000	0.16	0.09	0.07	0.32
Total	3.87	5.18	1.22	10.27

The limitations of these data are many. They ignore the regional (latitudinal) variations within each depth class of marine sediments. The model also assumes global fluxes when only several flux values have been reported in the literature. Fluxes to nearshore sediments were averaged over decades while pelagic fluxes were averaged over 10^5 years.

Despite these limitations, the estimate clearly shows that the main depositional zone in marine sediments is the depth category of 0 - 200 m, which is predominantly adjacent to the coast. More than 90% of marine sedimentary BC is buried in this depth category, which comprises less than 10% of the area under the world ocean.

There is good agreement between the estimated global BC flux to marine sediments of 10 Tg BC yr^{-1} and estimated worldwide annual BC emissions to the atmosphere by vegetation fires (6 - 28 Tg BC yr^{-1}) or from all sources (20 - 30 Tg BC

yr^{-1}) (Crutzen and Andreae 1990). An upper limit for BC emissions might be 50 Tg BC yr^{-1} if we consider Penner et al.'s (1993) estimate of 12 - 24 Tg BC yr^{-1} from industrial sources. More recent modeling (Cooke and Wilson, submitted) suggests lower global BC emissions to the atmosphere of the order of 13.5 Tg BC yr^{-1}. Given this range of estimates, we accept Crutzen and Andreae's intermediate value. We suggest that marine sediments, especially those of the continental shelves, may be important depositories of BC. The mass of BC sequestered in marine sediments may approximate the mass that is emitted to the atmosphere during a similar time frame. We do not hypothesize, however, that the BC emitted to the atmosphere during biomass burning has a sink itself in continental shelf sediments.

Comparison of atmospheric deposition and sedimentation of black carbon

One objective of this chapter is to evaluate the importance of riverine inputs and surface runoff relative to atmospheric inputs. Suman (1986) noted that eolian transport accounts for a minor percentage (3%) of the BC flux to nearshore sediments close to agricultural burning. This observation has not been verified in other coastal regions, however; we are unaware of other comparisons between atmospheric BC deposition to the ocean surface and deposition in ocean sediments. Although a number of uncertainties are associated with such an exercise, we believe that synthesis of these BC data is appropriate and will provide first steps toward identification of the constraints of current knowledge.

Comparing atmospheric fluxes with sedimentation rates is frustrated by different analytical techniques used for BC determination. Aerosols are commonly measured thermally (Cadle and Groblicki 1982, Cachier et al. 1989), optically (Hansen et al. 1984) or in combination (Chow et al. 1993). BC sedimentation estimates mostly come from a chemical oxidation followed by infrared spectroscopy (Smith et al. 1973, 1975). The adoption of a standard analytical technique might help matters as no chemical definition of BC exists (Cachier and Liousse 1994). Until recently, there have been no systematic attempts to intercalibrate BC in aerosol and sediment datasets. Preliminary results (Lim and Renberg, unpublished data) suggest thermal treatment may be more efficient than chemical oxidation in separating BC from charcoal in sediments. For example, BC concentrations could be two orders of magnitude higher in chemically than in thermally-treated sediments. Analytical differences, therefore, may prove to be of fundamental importance.

There is a difference in time scales describing controls on and measurement of aerosol deposition from air vs. its sedimentation in oceans that further frustrates comparison. Sediment fluxes are integrated over geological timescales of 10^1 to 10^5 years, while the timescale for the atmospheric data is 10^0 to 10^1 years. The difference is rather important, because BC in sediments and in aerosols represent different emission sources. BC in ancient sediments was produced by biomass burning, whereas recent aerosols contain BC from both fossil fuel and biomass combustion. Recently deposited marine sediments might be a sink for anthropogenic fossil fuel CO_2 expressed as an increase in deposition of BC. Other factors, such as climate, vegetation type and distribution, and, thus, BC emissions, have also varied during the Holocene and earlier. Lateral advective transport of BC from coastal to open oceans, including turbidite flows, introduces an additional time scale for sedimentation. Acknowledging these limitations, we compare the atmospheric and sediment deposition for the Atlantic, Pacific, and Indian Oceans in Table 4.

Table 4. Comparison of atmospheric black carbon deposition to the ocean surface and black carbon deposition to ocean sediments (Tg BC yr^{-1})

	Open Ocean	Coastal Ocean	Coastal Contribution (%)
Atlantic			
Atmosphere[1]	1.65	1.69	51
Sediment[2]	0.19	5.09	96
Pacific			
Atmosphere	0.95	1.33	58
Sediment	0.16	3.71	96
Indian			
Atmosphere	0.77	0.55	42
Sediment	0.07	1.15	94

[1]Atmospheric deposition is the median of the upper and lower estimates given in Table 1.
[2]Sediment data are from Table 3. Sediment deposition parameters are based on Sverdrup et al. (1946) (Table 2). We define coastal ocean as waters above the continental margin (0 - 2000 m).

A number of significant patterns emerge from this exercise. Almost all the black carbon deposition (94-96%) to the ocean sediments occurs in nearshore sediments rather than in open ocean sediments, whereas only about half of the total atmospheric deposition to the ocean surface occurs over the continental margins. Black carbon concentration gradients in the sediments are much steeper than those in the atmosphere. These results suggest that black carbon distribution in the atmosphere is much more homogeneous than in marine sediments. This picture is entirely consistent with our knowledge of the atmosphere, which is dynamic and responds rapidly (within days) to changes in particulate emissions. The model also suggests that the absent factor, riverine input, makes a much more significant black carbon contribution to nearshore sediments than to open ocean sediments.

Our results from these models yield several puzzling results. In the open ocean, atmospheric inputs appear to be up to an order of magnitude larger than the BC deposition to sediments. However, atmospheric fluxes are based on current aerosol measurements and include a significant anthropogenic (fossil fuel and biomass) source of BC. This is not the case for ancient sediments in which BC inputs are dominated by emissions from natural wildfires and, thus, may be lower than present day atmospheric inputs of BC to the ocean surface. The atmospheric model also predicts a more significant atmospheric BC flux to the coastal ocean than our sediment model suggests and Suman's (1986) work estimated.

The model also demonstrates that the BC deposition to the continental shelf sediments are 2 to 3 times more important than the eolian pathway to these waters. This again implies that riverine fluxes and surface runoff are significant contributors to the BC found in the sediments of marginal seas and the continental shelf, while the eolian contribution dominates in the open ocean. A caveat to conclusions regarding shelf sediments is that the BC measured in coastal sediments or estimated in river waters may contain a significant charcoal component and, therefore, overestimate BC.

For the reasons we have already discussed, we interpret these crude deposition fluxes with caution. Nevertheless, these estimates and comparisons of atmospheric BC deposition and marine BC sedimentation are an important first step in the compilation of an atmosphere:ocean BC budget.

Comparison of Riverine Black Carbon Inputs and Marine Sedimentation.

We are unaware of BC measurements in river water. Nevertheless, for the reasons outlined above, rivers are likely to be the most significant transport mechanism of BC to marine sediments, and their influence on nearshore sediment composition is of great importance. A crude calculation demonstrates the importance of the riverine BC flux. Milliman and Meade (1983) estimated that rivers carry 122 x 10^{14} g of suspended sediments to the oceans each year. Assuming an average BC (really "charcoal") concentration in soils of 0.1 % (Goldberg 1985) and that the composition of suspended solids mirrors soil composition, we suggest that rivers may transport 12.2 Tg BC to the oceans annually. This estimate is the same order of magnitude as the annual eolian flux to the ocean surface of 7 Tg BC. It also compares favorably with the global BC deposition to ocean sediments of 10.1 Tg yr^{-1}. Assuming the validity of the global eolian and riverine BC fluxes, then 8 - 28% of the BC that is annually produced in biomass burning (70 - 260 Tg yr^{-1}) (Kuhlbusch and Crutzen, in press) may be transported to the oceans. This percentage is an upper limit, because industrial combustion processes (fossil fuel, wood and bagasse fuel burning) also produce BC.

Temporal Distribution: Fire Histories Recorded in Marine Sediment Cores

Stability of Black Carbon in Marine Sediments

Before discussing some results obtained by analysis of marine sediment cores, we consider whether BC is inert to degradative processes in marine sediments over timescales of millions of years. Any interpretation of BC fluxes to marine sediments and consequent conclusions about past fire activity depend on the inertness of BC. A correct assessment of the sediment sink of BC from the biosphere further demands understanding of potential oxidation.

No evidence suggests that BC is degraded in the natural environment, but this does not prove that degradation does not occur. Herring (1977) analyzed 11 marine sediment cores, six indicating decreasing and two increasing BC fluxes since 65 mY BP. These latter two cores suggest that BC was not degraded over this timescale. Thus, the substance appears to be quite stable in marine sediments.

Sedimentary Records

The literature reports relatively few analyses of fire histories from marine sediment cores, and research efforts have largely focused in the North Pacific Ocean. The restricted geographical range of samples limits the global significance of the results.

Holocene records. High sedimentation rates on continental margins may record BC deposition with sufficient detail to permit meaningful historical analyses. All analyses to date have used lead-210 to date sediment cores. Researchers isolated BC from sediment samples by means of chemical digestion first reported in detail by Griffin and Goldberg (1975). Quantitative determination of the BC involved application of an infrared spectrometric technique developed by Smith et al. (1975). Griffin and Goldberg (1975) used these techniques to obtain BC fluxes during the past 150 years to two anoxic basins adjacent to the North American continent. The varved strata of these anoxic sediments are ideal, because bioturbation and sediment resuspension are minimal. BC fluxes to the Santa Barbara Basin ranged from 9 to 38 μg BC cm^{-2} yr^{-1}, while those of Saanich Inlet ranged from 43 to 122 μg BC cm^{-2} yr^{-1}. The data show much scatter with minima and maxima in adjacent sections. Linear regression suggested that BC fluxes to both sites may have remained constant or increased slightly during the past century. These authors were uncertain about the precise cause of three–fold variations in flux. They proposed variation in some combination of 1) biomass burning sources, 2) fossil fuel combustion sources, and 3) eolian and riverine transport.

Suman (1983, 1986) employed similar methodologies to analyze biomass burning histories in coastal marine sediments of Panama. Gulf of Panama sediment cores recorded several centuries of depositional history. Fluxes in one core varied between 94 and 139 μg BC cm^{-2} yr^{-1}, while in another they ranged from 84 to 165 μg BC cm^{-2} yr^{-1}. During this time, BC fluxes remained rather constant within a factor of two. Trends in flux were not evident, being within uncertainties of BC flux estimates to the sediments. As in California and British Columbia, biomass burning was significant during the 19th and 20th Centuries. Flux variations again may be explained by changing sources, transport mechanisms, or remobilization of the sediments.

A core analyzed by Suman (1983, 1991) from a low oxygen zone of the Nicaraguan continental margin revealed a relatively undisturbed sedimentary environment and five centuries of depositional history. BC fluxes were relatively constant and varied only by a factor of three (44 to 135 μg BC cm^{-2} yr^{-1}), again, close to the uncertainties of BC flux estimates. This core displayed unexpected evidence for decreasing BC fluxes toward the present. The author hypothesized that large pre-Columbian populations in central and northern

Central America made ample use of fire for agriculture. As a result of the Spanish Conquest, the native population probably declined in the 16th and 17th Centuries and caused a reduction in BC source.

Once again, we emphasize that BC fluxes to coastal sites (9 to 165 μg BC cm^{-2} yr^{-1}) are larger than the estimated BC eolian fluxes to Pacific coastal waters (3.9 to 7.7 μg BC cm^{-2} yr^{-1}). This discrepancy suggests that riverine transport of BC to coastal sediments is important and that coastal watersheds adjacent to analyzed sediment cores experienced relatively large amounts of biomass burning.

Glacial/Interglacial records. Global climatic changes marked by glacial and interglacial stages of the Pleistocene altered the spatial distribution of vegetation and, most likely, the occurrence of biomass burning. Wind velocities may also have increased during glacial stages (Kennett 1982). Recent preliminary analysis of a tropical equatorial Atlantic core (0° 21'S, 11° 55'W) by Verardo et al. (1994) determined BC concentrations and fluxes at approximately thousand-year intervals during the past 100,000 yrs. This research has the potential of linking terrestrial particulates and the pelagic Pleistocene sedimentary record. Results should indicate marked variations in biomass burning between glacial and interglacial cycles. Under glacial conditions, more than 50% of the total organic carbon component was "particulate charcoal". "Charcoal" accumulation rates during the glacials reach 40 μg cm^{-2} yr^{-1} and are orders of magnitude lower during the interglacials (Verardo, pers. comm.). Verardo attributes the high flux to stronger wind stresses and drier land conditions. We hesitate to compare these preliminary data to other data sets because these researchers employed a different analytical technique to measure "charcoal". Moreover, previous work on this core noted the equatorial upwelling over this pelagic site and the core's relatively high organic carbon concentrations (Verardo and McIntyre 1994). These unusual circumstances might impair comparisons with data from other pelagic sites.

Cenozoic records. Herring (1977, 1985) calculated BC concentrations and fluxes to 11 North Pacific Deep Sea Drilling project sediment cores that extend back to the early Cenozoic (65 million years BP). Recent sediment fluxes were much greater than those in the earlier Cenozoic. Herring determined that BC fluxes were generally low (0.001 - 0.01 μg BC cm^{-2} yr^{-1}) in the Palaeogene period (65 to 26 mY BP), and that, with the beginning of the Neogene, fluxes increased significantly. By the end of the late Neogene, BC fluxes were two or three orders of magnitude greater than those of the Palaeogene (0.1 - 10 μg BC cm^{-2} yr^{-1}). Herring attributed the low fluxes in the early Tertiary to 1) the warm, humid climate and more abundant tropical vegetation that was not prone to biomass burning and 2) low zonal wind strengths. As the climate deteriorated in the Cenozoic, drier temperate forests and grasslands emerged, which may have provided more fuel. Herring also observed the same latitudinal variation in BC fluxes to open ocean sites (i.e. higher in the North Pacific and lower in the equatorial Pacific) in Quaternary sediments, as well as in sediments that are 10 million years old.

Future Research Directions

In order to improve our knowledge of the spatial distribution and fluxes of BC in pelagic sediments and to correctly assess the sedimentary reservoir of BC, we must obtain samples that are representative of each ocean basin. Significant data exist in the North Pacific Ocean. Future samples, therefore, should represent the southern ocean, particularly areas adjacent to South America and Australia. Atlantic Ocean samples should represent the zonal wind systems and assess BC deposited from Northern Hemisphere industrial areas and Southern Hemisphere biomass burning areas in Brazil and tropical Africa. No BC data exist from Indian Ocean sediments, and a sampling strategy in that ocean should follow the same considerations as we discuss above.

The prominence of the continental shelf as a BC depository suggests the need for additional research here. We need to clarify whether BC fluxes to continental shelf sediments throughout the world ocean are similar to those of Saanich Inlet, the Santa Barbara Basin, and the shelf sediments of Nicaragua and Panama. We also need a sense of how this BC flux attenuates with distance from source, specifically across the continental margin. Transects of sediment cores perpendicular to the coastline are needed. Comparison of results from transects located in different zonal wind bands, as well as in areas with varying riverine input, will begin to shed light on the relative importance of eolian and riverine transport of BC to the coastal ocean.

The relative importance of BC transport mechanisms needs to be evaluated. We must determine the contributions of wet and dry deposition of BC to the ocean surface through field data that may or may not confirm estimates from the atmospheric model we discuss here. Parameters for wet deposition need to be better constrained, because deposition occurs predominantly by the wet mode. Although it is most likely to be the dominant BC pathway, riverine transport of BC remains a black box. Studies of transport mechanisms are beset by experimental difficulties. Nevertheless, they are essential to our understanding of the global BC budget.

Improved assessment and identification of present BC source regions is critical for understanding the BC signal in marine sediments. This will require detailed information about the frequency of biomass burning activities, the areal extent of burning, the type of vegetation burned, and the production of BC from burning. Eventually, we hope to be able to compare BC deposition throughout the world's oceans with BC production in continental regions. Plausible correlations between these two datasets will permit us to meaningfully address past burning activity as recorded in sediment cores.

An integrated research project should focus on a specific region and obtain detailed information on a) vegetation biomass and type, b) biomass burning activity and strength of the BC emissions, c) the relative importance of eolian and fluvial transport of black carbon from the burning site to nearshore marine sediments and pelagic sediments that are adjacent to the region, d) fluxes of BC to nearshore and pelagic sediments, and e) past records of burning from BC concentrations in sediment cores and their possible verification through land use records or written and oral histories. We propose that alluvial fan sediments of the Ganges River in the Bay of Bengal might provide rich results.

Future analyses of well-sited sediment cores will provide qualitative and quantitative information about past fire activity that could greatly help assess the role that fire has played in the history of the earth and the atmosphere. We will obtain fruitful results from analysis of fire and land-use practices in human societies during the Holocene. Glacial/interglacial comparisons will be important in addressing climate and, therefore, vegetational variations.

To estimate the bio-atmospheric sink and to insure that fire histories are meaningful, we must investigate the stability of BC in marine sediments. One possible methodology to address this question will be to analyze cores from zones of low and high oxygen concentrations but similar combustion source regions. Degradative processes might be responsible for the flux differences. Additionally, an evaluation of microbial processes in marine sediments that have the capacity to oxidize BC would provide more conclusive evidence for degradation.

In order to meaningfully assess the relative importance of the eolian and riverine BC fluxes to the ocean surface and compare these estimates with BC deposition to marine sediments, we need to adopt an experimental definition of BC. We must establish a operational distinction between charcoal and BC for all analytical methods. Intercalibration of aerosol and sediment analytical techniques and datasets is essential and should be in first place on the agenda of future research needs.

Conclusions

Ubiquitous in ocean sediments, BC provides both regional and global information about biomass burning activities. Its distribution in surface sediments of the Pacific Ocean suggests that BC fluxes to the ocean floor depend on numerous factors, including magnitudes of biomass burning sources, distances between core locations and sources, eolian and river transport mechanisms, transport through the water column, and reworking in the marine sediments. Presently, we lack a sense of the relative importance of these factors. Fluxes to pelagic sites are generally less than 0.20 μg BC cm^{-2} yr^{-1}, while fluxes to sites in nearshore sediments are two to three orders of magnitude greater (23 to 354 μg BC cm^{-2} yr^{-1}). Current sediment data are not representative of the world oceans, however, as researchers have largely ignored the Atlantic and Indian Oceans.

Analysis of sediment cores provides information about past burning activities, although uncertainty about the relative importance of burning and transport clouds interpretation. BC analyses from North and Central American nearshore sites suggest that burning in nearby source regions have not changed radically during the past century or two. Analyses of deep sea cores detects an increasing BC flux with the beginning of the Neogene (26 mY BP).

We have created several simple models to evaluate the BC global budget. Of the 70 - 260 Tg BC yr^{-1} produced during biomass burning, perhaps 6 to 28 Tg BC yr^{-1} are emitted to the atmosphere. Emissions from all combustion sources may average 25 Tg yr^{-1}. About a third of the BC emitted to the atmosphere from continental sources is transported and deposited on the ocean surface. We estimate that atmospheric deposition to the ocean surface may reach 7 ± 3 Tg BC yr^{-1}, of which approximately half may be deposited on the surface of the coastal ocean. Rivers may transport another 12 Tg BC yr^{-1} to the oceans. Through sediment research, we estimate the BC flux to ocean sediments to be approximately 10 Tg BC yr^{-1}, of which 95% may be deposited on the continental margins. This may represent 4 to 14% of the total BC (aerosol and residual) produced by biomass burning. Thus, significant BC is sequestered in marine sediments.

BC fluxes to the ocean estimated by us may be placed in perspective of the ocean carbon budget. The riverine flux of BC (0.012 Pg BC yr^{-1}) is about 3% of the total riverine flux of organic carbon (0.40 Pg C yr^{-1}) (Schlesinger and Melack 1981, Sarmiento and Sundquist 1992). Burial of BC in ocean sediments (0.01 Pg BC yr^{-1}) is equivalent to 11% of the estimated burial of organic carbon in marine sediments (0.09 Pg C yr^{-1}) (Sarmiento and Sundquist 1992). Considering BC as if it were completely distinguishable from organic carbon, the flux of carbon (black carbon plus organic carbon) from rivers to the ocean (0.41 Pg C yr^{-1}) minus the burial of carbon in marine sediments (0.10 Pg C yr^{-1}) still implies a CO_2 flux of about 0.3 Pg C yr^{-1} that must be transferred to land from the oceans (Sarmiento and Sundquist 1992). BC fluxes do not alter the conclusions of Sarmiento and Sundquist's (1992) global CO_2 budget.

The uncertainties in these models and the different analytical techniques used for BC determination in aerosols and sediments could partially explain the discrepancy between atmospheric deposition and sedimentation estimates. Although our models have great uncertainties and limitations, they do produce estimates fairly consistent with the global BC budget.

Future research directions in this field must focus on development of an improved sampling strategy that will be representative of the world ocean and allow us to assess the relative importances of transport mechanisms and sources. BC analysis has often not been the *raison d'être* of our past sampling strategies. Most importantly, we need to develop an experimentally acceptable definition of BC and intercalibrate various analytical techniques if we wish to build a meaningful dataset that represents reservoirs and pathways.

Acknowledgments

We are grateful to W.F. Cooke and J.N. Wilson for making their submitted data available to us and to H. Cachier for helpful comments. This work was partially funded by the Human Capital and Mobility Programme of the European Community

Literature References

Andreae MO (1991) Biomass burning: Its history, use, and distribution and its impact on environmental quality and global climate. In Levine JS (ed) *Global Biomass Burning* pp 3-21 MIT Press Cambridge

Cachier H, Brémond MP, Buat-Ménard P (1989) Determination of atmospheric soot carbon with a simple thermal method. *Tellus* 41B:379-390

Cachier H, Liousse C (1994) Combustion aerosols: facts and vocabulary. *Proc Fifth Conf Carbonaceous Particles Atmosph* p 1

Cadle SH, Groblicki PJ (1982) An evaluation of methods for the determination of organic and elemental carbon in particulate samples. In Wolff GT, Klimisch RL (eds) *Particulate Carbon: Atmospheric Life Cycle* pp 89-109 Plenum Press New York

Chow JC, Watson JG, Pritchett LC, Pierson WR, Frazier CA, Purcell RG (1993) The DRI thermal/optical reflectance carbon analysis system: description, evaluation and applications in US air quality studies. *Atmosp Envir* 27A:1185-1201

Cooke WF, Wilson JN A global black carbon aerosol model. *J Geophys Res* (submitted)

Crutzen PJ, Andreae MO (1990) Biomass burning in the tropics: impact on atmospheric chemistry and biogeochemical cycles. *Science* 250:1669-1678

Ducret J (1994) *Incorporation du carbone particulaire aux retombées humides: Implications geochimiques* PhD thesis Université de Paris 7 Paris

Ducret J, Cachier H (1992) Particulate carbon content in rain at various temperate and tropical locations. *J Atmosph Chem* 15:55-67

GESAMP (Joint Group of Experts on the Scientific Aspects of Marine Pollution) (1990) *The Atmospheric Input of Trace Species to the World Oceans.* UNEP Regional Seas Reports and Studies No 119

Goldberg ED (1985) *Black Carbon in the Environment.* John Wiley New York

Griffin JJ, Goldberg ED (1975) The fluxes of elemental carbon in coastal marine sediments. *Limnol Oceanogr* 20:456-463

Hansen ADA, Rosen H, Novakov T (1984) The aethalometer - an instrument for the real time measurement of optical absorption by aerosol particles. *Sci Total Envir*36:191-196

Herring JR (1977) *Charcoal fluxes into Cenozoic sediments of the North Pacific* PhD Thesis University of California San Diego

Herring JR (1985) Charcoal fluxes into sediments of the North Pacific Ocean: the Cenozoic record of burning. In Sundquist ET, Broecker WS (eds) *The Carbon Cycle and Atmospheric CO_2: Natural variations, Archean to Present* pp 419-442 Amer Geophys Union Washington DC

Jaeger L (1983) Monthly and areal patterns of mean global precipitation. In Street-Perrott A, Beran M, Ratcliffe R (eds) *Variations in the Global Water Budget* pp 129-140 D Reidel: Dordrecht The Netherlands

Kennett J (1982) *Marine Geology.* Prentice-Hall Englewood Cliffs NJ

Kuhlbusch TAJ, Crutzen PJ Toward a global estimate of black carbon in residues of vegetation fires representing a sink of atmospheric CO_2 and a source of O_2. *Global BiogeochemCycles* (in press)

Lim B, Cachier H Determination of black carbon by chemical oxidation and thermal treatment in recent marine and lake sediments and Cretaceous-Tertiary clays. *Chem Geology* (accepted)

Liousse C, Cachier H (1992) Measurement of black carbon aerosols in the atmosphere of two different source regions: real-time data for the Paris region and a savanna site of the Ivory Coast. *EnvirTechnol* 13:959-967

Liousse C, Cachier H, Jennings SG (1993) Optical and thermal measurements of black carbon aerosol content in different environments: variation of the specific attenuation cross-section, sigma. *Atmos Envir* 27A:1203-1211

Milliman JD, Meade RH (1983) World-wide delivery of river sediment to the oceans. *J Geol* 91:1-21

Patterson CC, Settle DM (1987) Review of data on eolian fluxes of industrial and natural lead to the lands and seas in remote regions on a global scale. *Marine Chem* 22:137-162

Penner JE, Eddleman H, Novakov T (1993) Towards the development of a global inventory for black carbon emission. *Atmos Envir* 27A:1277-1295

Sarmiento JL, Sundquist ET(1992) Revised budget for the oceanic uptake of anthropogenic carbon dioxide. *Nature* 356:589-593

Schlesinger WH, Melack JM (1981) Transport of organic carbon in the world's rivers. *Tellus* 33:172-187

Smith DM, Griffin JJ, Goldberg ED (1973) Elemental carbon in marine sediments: a baseline for burning. *Nature* 241:268-270

_____ (1975) A spectroscopic method for the quantitative determination of elemental carbon. *Anal Chem* 47:233-238

Suman DO (1983) *Agricultural burning in Panama and Central America: Burning parameters and the coastal sedimentary record.* PhD thesis University of California San Diego

_____ (1986) Charcoal production from agricultural burning in Central Panama and its deposition in the sediments of the Gulf of Panama. *Envir Conserv* 13:51-60

_____ (1991) A five-century geochronology of biomass burning in Nicaragua and Central America. In Levine JS (ed) *Global Biomass Burning* pp 512-518 MIT Press Cambridge

Sverdrup HU, Johnson MW, Fleming RH (1946) *The Oceans: Their Physics, Chemistry, and General Biology.* Prentice-Hall Englewood Cliffs NJ

Verardo DJ, Ruddiman WF, Freedland C (1994) High resolution record of charcoal deposition in late Pleistocene marine sediments of the tropical Atlantic. *EOS, Trans Amer Geophys Union* Supplement to 19 April 1994 issue

Verardo DJ, McIntyre A (1994) Production and destruction: control of biogenous sedimentation in the tropical Atlantic 0-300,000 years B.P. *Paleoceanography* 9:63-86

Reconstructing Environmental Impacts of Fire from the Holocene Sedimentary Record

H.J.B. Birks
Botanical Institute
University of Bergen
Allegaten 41
N-5007 Bergen, Norway
and
Environmental Change Research Centre
University College London
26 Bedford Way
London WC1H 0AP, UK

Summary

Assuming that the charcoal record preserved in lake sediments is a record, albeit an incomplete and biased record, of past fire occurrence and extent, I outline quantitative approaches to the reconstruction of the impacts that fires may have had on environmental variables during the Holocene. An important distinction in considering the environmental impacts of fires is between natural fires and fires initiated by humans for forest destruction and land-use management. The palaeoecological record of fire-related environmental impacts is strongest when considering fire as a management tool. Such fires can be shown from palaeolimnological studies to raise the nutrient level and pH of lakes and to increase soil erosion and inwashing of minerogenic matter. The environmental impact of natural fires is more difficult to demonstrate unambiguously. Such demonstrations require extremely fine-resolution stratigraphical data and quantitative palaeoenvironmental reconstructions with a high precision. As is so often the case in palaeoecology, alternative hypotheses may be possible to explain observed stratigraphical patterns. Statistical techniques such as constrained ordinations, variance partitioning, and Monte Carlo permutation tests to derive empirical probability values provide useful means of testing and falsifying alternative hypotheses involving complex stratigraphical data. Recent developments in quantitative environmental reconstructions and in the statistical analysis of stratigraphical data that are relevant to the assessment of the environmental impacts of fire during the Holocene are briefly outlined and illustrated with three recently published examples. Current research techniques are briefly discussed and some difficulties and limitations are outlined.

Introduction

Fires represent a major disturbance to a landscape. Such a catastrophic disturbance can have impacts on the environment of that landscape, such as the water chemistry of streams and lakes and the hydrology and sediment yield of the catchment, and hence on the composition of stream and lake biota (Wright 1981). This chapter outlines the use of numerical techniques in assessing the possible impacts of fires on

NATO ASI Series, Vol. I 51
Sediment Records of Biomass Burning
and Global Change
Edited by James S. Clark, Hélène Cachier,
Johann G. Goldammer, and Brian Stocks

the environment during the Holocene, as recorded by the biological and chemical composition of lake sediments.

The structure of the chapter is as follows. After outlining the assumptions of the charcoal and fossil biological records preserved in lake sediments, I discuss the critical distinction between natural fires and human–initiated fires used as a management tool. I present three basic questions that can be asked of sedimentary records and indicate what the basic research requirements are before any attempt is made to address these questions. I then outline current quantitative procedures for reconstructing past environmental variables from fossil assemblages and discuss statistical techniques for testing competing causal hypotheses. I summarise three published studies to illustrate the use of numerical procedures for the analysis of stratigraphical charcoal data and the reconstruction and assessment of the environmental impacts of fires. I conclude by briefly discussing current research techniques and their potentialities and limitations.

Assumptions

There are three major assumptions in the use of the stratigraphical record of charcoal, pollen, and diatoms (and other limnic biota) to assess the environmental impacts of fire in the past. First, the stratigraphical record of charcoal preserved in the sediments of a small– or medium–sized lake, particularly of large charcoal pieces, is a reflection, admittedly an incomplete and potentially biased record, of past fires within the catchment of that lake. Charcoal is well preserved in lake sediments and large (> 50 μm length) particles are thought to be primarily local in origin and to record what Clark (1988a) terms the "catchment fire regime". Large charcoal particles provide the most unambiguous evidence for local, within–catchment fires, whereas small charcoal particles predominantly indicate non–local regional or extra–regional fires (Clark 1988a, 1988b, Clark and Royall 1995).

Second, the pollen–stratigraphical record preserved in lake sediments is a reflection, also no doubt an incomplete and biased record, of past plant populations and vegetation in the catchment of the lake (the "basin watershed model" of Jackson 1994). Sugita (1994) provides a detailed analysis of the likely size of representative pollen source–areas of lakes of different sizes.

Third, and least contentious, the diatom–stratigraphical record (or of other limnic biota such as chrysophytes, chironomids, and cladocera) is a record of the past diatom populations and assemblages occurring in the lake. In other words, interpretations of the charcoal, pollen, and palaeolimnological record preserved in sediments of a small– or medium–size lake are based on the assumptions that large charcoal particles and the pollen assemblages reflect the fire and plant population history of the catchment of the lake and that the palaeolimnological record is a reflection of past lake biota.

Natural and Anthropogenic Fires

Unlike pollen, which is produced annually in fairly consistent amounts from one year to the next, charcoal is produced in large quantities for a short time and at irregular intervals in natural fires, depending on the fire frequency (e.g. 60–250 years between fires; Swain 1973, 1978, Cwynar 1978). Fire frequency is a complex function of climate, moisture, vegetation, fuels, landscape type, land–use, etc. (e.g. Green 1981, Clark 1988c). The magnitude of the environmental impacts of any fire will depend not only on the frequency and extent of the fires but also on the vegetation and fuel types, site type, spatial and temporal scales, ecological setting, topography, bedrock geology, soils, and interactions between these factors.

Natural fire frequencies of 60–250 years contrast with a fire frequency of 10–20 years when fire is used as a management tool, for example in controlled burning to maintain *Calluna vulgaris*–dominated heathland and to produce new pasture (e.g. Kaland 1986, Gimingham 1994, Odgaard 1994). A burn frequency of 10–20 years is required, for example, to stimulate the growth of *Calluna* and to provide suitable grazing for sheep and for red grouse. As fires can influence the nutrient status of terrestrial ecosystems and of lakes and streams within the burned area (e.g. R.F. Wright 1976, H.E. Wright 1981, Gimingham 1994), the environmental impacts from natural and anthropogenic fires for management purposes may differ.

Basic Questions

Three basic questions should be addressed in considering the possible environmental impacts of fires in the past. First, what are the impacts of fire on the environment in terms of, for example, changes in lake nutrients, lake–water pH, and lake sediment accumulation rates? Second, as there may be often two or more competing hypotheses to explain some observed stratigraphical pattern, is it possible to test competing hypotheses statistically? Observed changes in the pollen record may be a result of climate, human disturbance, natural fires, soil changes, pathogens, and complex interactions between these factors (e.g. Peglar and Birks 1993). Is it possible to test between competing hypotheses statistically, given the statistical complications of (a) non–random data, (b) stratigraphically ordered sampling units (pollen spectra), (c) percentage data, and (d) an undesigned, uncontrolled impact experiment in the past (a palaeoecological equivalent of a before–after–control–impact (BACI) study but with no control and few, if any, replicates)? Third, what are the impacts of fire on vegetation and vegetational dynamics over the time–scale of the Holocene? This question is not considered in this chapter, as it has been intensively discussed by, for example, Green (1981, 1982, 1987), Walker (1982), Swain (1973), Clark et al. (1989), Bradshaw (1993a, 1993b), Huntley (1993), and Bradshaw and Hannon (1992).

Research Requirements

In any attempt to address questions one and two above, there are three research requirements. First, detailed fine–resolution stratigraphical data (e.g. charcoal, pollen, diatoms, etc.) are required with a reliable temporal chronology and a sample resolution of about 5–10 years. Second, statistically robust, numerically accurate, and ecologically realistic methods are needed for reconstructing past environmental variables (e.g. lake–water pH) from fossil assemblages. Third, statistical methods are required for testing competing causal hypotheses to explain the observed stratigraphical patterns, given the statistical complexities of stratigraphical data. The first requirement is not discussed here. There are well–established techniques for fine–resolution pollen and diatom analysis (e.g. Peglar 1993, Renberg 1990b, Green and Dolman 1988) and less well–established techniques for fine–resolution charcoal analysis (e.g. Patterson et al. 1987, Clark 1988b, Clark et al. 1989, MacDonald et al. 1991, Clark and Royall 1995). I do, however, discuss the second and third requirements.

Quantitative Palaeoenvironmental Reconstruction Methods

In the last decade the development of quantitative "transfer functions" has revolutionised palaeolimnological interpretations by providing quantitative reconstructions of important environmental variables such as lake–water pH and total phosphorus (e.g. Battarbee 1991, Dixit et al. 1992, Anderson 1993, 1995). The mathematical basis for such transfer functions is reviewed by ter Braak (1995), and Birks (1995a) provides a review of existing methods and their application in Quaternary palaeoecology. In contrast to palaeolimnology, transfer functions have been used in palynology for over 25 years and have been derived using mathematical techniques related, but not identical, to those developed in palaeolimnology. Bartlein and Whitlock (1993) provide an excellent example of the critical use of pollen–climate transfer functions.

The primary aim of all palaeoecological transfer functions is to express the value of an environmental variable (e.g. lake–water pH) as a function of biological data (e.g., diatom assemblages). This function is termed a transfer function or biotic index and its construction is called calibration (ter Braak and Prentice 1988). Calibration is the opposite of regression where the aim is to model the response of a single taxon as a function of one or more environmental variables. In practice nearly all quantitative palaeoenvironmental reconstructions involve two stages (Birks et al. 1990). First, the responses of modern taxa to the contemporary environment are modeled. This is a regression problem and involves a modern training set of biological assemblages from surface samples with associated environmental data. Second, these modeled responses are used to infer the past environmental variables from the composition of the fossil assemblages (ter Braak and van Dam 1989, Birks 1994). Biological data, both in modern training sets and as fossil assemblages preserved in lake sediments, usually contain many (e.g. 50–200) taxa. The data are usually quantitative, are commonly expressed as percentages, and invariably contain many zero values for sites where a taxon is absent. Taxon abundance is often a non–linear unimodal function of the environmental variables (ter Braak 1987a).

There are at least six major requirements in any quantitative palaeoenvironmental reconstruction (Birks 1994, 1995a). These are as follows. (1) A large high–quality training set of modern surface–samples and associated environmental data is essential. This should be representative of the likely range of environmental variables in the past, be of consistent taxonomy and nomenclature, be of comparable quality, and be from the same sedimentary environment. (2) The fossil data used for reconstruction should be of comparable taxonomy and nomenclature, quality, and sedimentary environment as the modern training set. (3) Good independent chronology is available for the fossil data to allow an assessment of the timings and rates of change. (4) Robust statistical methods are required that can adequately model the complex, non–linear, unimodal relationships between modern taxa and their environment. (5) Reliable and realistic statistical estimates of standard errors of prediction for each reconstructed value are essential. (6) Critical ecological and statistical evaluations of all reconstructions are needed, as any numerical calibration procedure is designed to generate a result. It is essential, however, to evaluate if the result obtained is ecologically sensible and statistically reliable.

There are now several powerful and robust reconstruction procedures in palaeoecology (see Birks (1995a) and ter Braak (1995) for reviews). The choice between them depends on whether a linear response model of taxa to the environmental variable of interest or a unimodal response model is appropriate, and what estimation procedure (least squares, partial least squares, maximum likelihood) is adopted (Birks 1995a). Some techniques used for quantitative palaeoenvironmental reconstruction have a clear underlying mathematical theory and rationale, some are computationally simpler than others, and some are more ecologically realistic than others (Birks 1995a). At present weighted–averaging (WA) (ter Braak and van Dam 1989) and its extension weighted–averaging partial least squares regression (WA–PLS) (ter Braak and Juggins 1993) are recommended for species–rich data over long environmental gradients (ter Braak 1995), whereas inverse linear regression and its extension partial least squares (PLS) regression (Martens and Næs 1989) are recommended for data over shorter gradients (Birks 1995a). PLS and its unimodal–based equivalent WA–PLS combine the desirable features of inverse linear regression and WA regression (high correlation) and of principal components regression or correspondence analysis regression (stable predictors and high variance). PLS and WA–PLS are attractive and robust reconstruction techniques for a wide range of palaeoecological data, although care is needed to select the optimal number of components using statistical cross–validation (ter Braak 1995). Prediction errors can be estimated for individual fossil samples by computer–intensive procedures such as bootstrapping (Birks et al. 1990) or Monte Carlo simulation (Birks 1995a).

ter Braak and Juggins (1993) and ter Braak et al. (1993) have shown, using artificial simulated data, the superiority of WA and WA–PLS for reconstruction of environmental variables along medium or long gradients, with WA–PLS out–performing WA (in terms of prediction errors) on data containing low statistical noise. The root mean square error of prediction (as assessed by cross–validation) for inferring lake–water pH from diatom assemblages using different training sets lies between 0.28 – 0.38 pH units, based on WA–PLS (ter Braak and Juggins 1993, Birks et al. unpublished results). Other diatom–based training sets exist for the quantitative reconstruction of total P, salinity, total Al, dissolved organic carbon, and alkalinity, whereas other modern training sets include chrysophytes and lake–water pH, chironomids and lake–water temperature, regional pollen and regional climate, and local pollen and local soil and land–use variables (Birks 1995a).

Statistical methods for testing competing hypotheses

In any environmental impact study there is always one or more response variables **Y** (e.g. lake–water pH, sediment organic content, sediment accumulation rate, sediment K, N, or P composition, etc.). What is a response variable clearly depends on the nature of the study. In addition there is always one or more predictor or explanatory variables **X** (e.g. charcoal content, land–use indicator taxa, independent climate records) that are thought, on ecological grounds, to influence, to some extent, the response variable(s). There may also be covariables – predictor variables of secondary interest whose effects on the response variables are adjusted for and partialled out statistically.

The basic statistical model for modeling the effects of the explanatory variables **X** on the response variables **Y** is of the general regression type

$$\mathbf{Y} = \mathbf{XB}$$

where **B** is a matrix of regression coefficients. If **Y** contains 1 variable and **X** consists of 1 variable, the relevant technique is simple linear regression. If **Y** consists of 1 variable and **X** contrains 2 or more variables, appropriate techniques include multiple linear regression or biased procedures that guard against colinearity between the predictor variables such as principal components regression and partial least squares regression. If **Y** contains 2 or more response variables and **X** contains 1 or more predictor variables, the most useful technique is redundancy analysis (ter Braak 1994) (also known as constrained principal components analysis, reduced–rank regression, PCA of y with respect to x, etc. – ter Braak 1987b).

The basic idea in all these regression–based techniques is to model and hence to "explain", in a statistical sense, the observed variation in **Y** in terms of a linear combination of the **X** predictor variables (including product and cross–product terms to allow for non–linear relationships and interactions between variables) that gives the highest amount of variance of explained by **X**. With one response variable, the use of linear and multiple linear regression is relatively straightforward, so long as due regard is paid to regression diagnostics (e.g. Montgomery and Peck 1992). The major problem is how to evaluate the significance of the fitted model, given the complex statistical properties of stratigraphical data. It is not possible to rely on conventional statistical tests and tables of F statistics used to evaluate an analysis of variance of a regression model. A more appropriate statistical evaluation procedure is to use distribution–free Monte Carlo permutation tests (Manly 1991, ter Braak 1992, Verdonschot and ter Braak 1994). In these an empirical distribution of the test statistic of interest is derived by repeated permutations of the data, and a comparison made between the observed test statistic and, say, 999 values of the same statistic based on permuted data to derive a Monte Carlo test of significance and an exact probability for the observed statistic (ter Braak 1992). Such Monte Carlo tests are distribution–free, as they do not assume or require normality of the error distribution. They do, however, require independence or exchangeability of samples (ter Braak and Wiertz 1994, Verdonschot and ter Braak 1994). The validity of a permutation test clearly rests on the appropriateness of the type of permutation used. Stratigraphical and other time–series data require a special permutation type (ter Braak 1990). Such restricted permutation tests provide a non–parametric way of overcoming the difficulty of performing statistical tests in the presence of temporal autocorrelation in stratigraphical data (Besag and Clifford 1989, ter Braak 1990).

If there are two or more response variables, redundancy analysis (RDA) is a powerful means of modeling response variables in relation to one or more predictor variables (ter Braak 1994). It finds axes of variation within the response variables that are linear combinations of the predictor variables and that give the lowest possible total residual sum of squares. In other words, RDA finds a weighted sum of predictor variables that fits or "explains" the response variables best and gives the maximum total regression sum of squares. It is intermediate between principal components analysis and separate multiple regressions for each response variable (ter Braak and Prentice 1988). However, because of the dimensional reduction of the response variables, RDA requires fewer regression parameters than separate multiple regressions. The statistical significance of the various RDA axes can be assessed by Monte Carlo permutation tests, as for multiple regression (ter Braak 1990, 1992).

In all these regression models, it is possible to allow statistically for the effects of covariables by performing partial linear regression, partial multiple linear regression, or partial RDA (ter Braak 1988, ter Braak and Prentice 1988). These involve analysis of the residual variation in the response variables that remains after the fitting of the effects of the covariables. The resulting analysis shows the effect that is "uniquely" attributable to a particular predictor variable(s). The statistical significance of the resulting model can also be assessed by Monte Carlo permutation tests (ter Braak 1990, 1992). Partial techniques provide a powerful basis for testing competing hypotheses, as the effects of particular predictor variables can be partialled out and the relative importance of other predictors assessed statistically (e.g. Lotter and Birks 1993, Birks and Lotter 1994, Gaillard et al. 1992, Walker et al. 1991).

In simple or multiple regression analysis it is possible to partition or decompose the total variance into a series of independent components to form a ANOVA table for the regression model (Whittaker 1984). Borcard et al. (1992) extend this decomposition procedure to include constrained ordination analyses such as RDA. Given two different groups of predictor variables $\mathbf{X_a}$, $\mathbf{X_b}$, it is possible by means of a series of RDA and partial RDA to decompose the variance in **Y** into four independent components, (1) variance due to $\mathbf{X_a}$ independent of $\mathbf{X_b}$, (2) variance due to $\mathbf{X_b}$ independent of $\mathbf{X_a}$, (3) variance due to the covariance or interaction of $\mathbf{X_a}$ with $\mathbf{X_b}$, and (4) variance not explained by $\mathbf{X_a}$ and $\mathbf{X_b}$. The statistical importance of these components can be assessed by Monte Carlo permutation tests (Legendre 1993, ter Braak and Wiertz 1994), thereby permitting the relative importance of competing hypotheses to be evaluated and quantified (e.g. Birks 1993). Examples of such variance partitionings in palaeoecological contexts include Lotter and Birks (1993) and Birks and Lotter (1994).

Illustrative Examples

I outline three examples to illustrate the application of constrained ordinations, Monte Carlo permutation tests, variance partitionings, and quantitative palaeoenvironmental reconstructions in the analysis of stratigraphical charcoal data and in the assessment of the environmental impacts of fire during the Holocene. The first concerns the statistical modeling of the observed stratigraphical patterns of charcoal in lake sediments in relation to pollen taxa and is not specifically involved with assessing environmental impacts of fire. The second and third examples concern changes in soil and lake–water chemistry as a result of burning and associated land–uses. They involve the quantitative reconstruction of environmental variables from palaeoecological data and illustrate the impacts of anthropogenic fires on the environment.

Odgaard (1992) developed a statistical model using RDA for the concentrations of charcoal particles in lake sediments at Solsø in north–west Jutland, Denmark in relation to several predictor variables including the concentrations of 20 pollen taxa and of dark fungal hyphae. Statistically significant predictors were found by means of a forward selection procedure in RDA (ter Braak 1990) and associated Monte Carlo permutation tests. The resulting model involves *Calluna vulgaris* pollen as the only significant predictor variable and explains 66% of the variance in the charcoal sequence. No systematic trends are apparent in the regression residuals. Odgaard (1992) then applied the Solsø model to another lake, Skansø, to predict, from the observed pollen sequence of *Calluna vulgaris*, the charcoal stratigraphy. When this is done, a strong systematic trend in the residuals between observed and predicted charcoal concentrations emerges. This trend is strongly related to sediment depth. Odgaard (1992) proposes that there is differential settling of charcoal particles relative to pollen, because the settling rate of charcoal is lower than pollen due to the trapping of air in the highly porous charcoal particles. Charcoal accumulation, in the absence of any resuspension and focusing, would therefore be concentrated in the calmest parts of the basin, whereas pollen sedimentation would be more even. In the presence of resuspension and focusing, over–representation of charcoal relative to pollen is likely in the focal, deep parts of the lake. Odgaard argues that the systematic bias in the residuals is the result of such differential focusing. A partial RDA with the effects of depth partialled out statistically as a covariable gives predicted charcoal values that correspond more closely to the observed values ($r = 0.81$) and the resulting residuals show no systematic trend. From these analyses, Odgaard concluded that the source area and transport mechanism for charcoal particles and pollen are probably similar but that pollen and charcoal may be focused differentially within a lake. He also showed that high fire intensities controlled by humans triggered the regional expansion of *Calluna* heathland and that these heaths were maintained by fire over millennia (see also Odgaard 1994).

The second example illustrates the local environmental impact of repeated anthropogenic fire, as recorded in a local pollen sequence from a soil humus profile covering the last 2–3000 years in south–east Sweden (Gaillard et al. 1992). Using a modern pollen–land–use–soil training set for southern Sweden, Gaillard et al. (1992) reconstructed local changes in soil pH, water content, total N, total K, and total P, and grazing pressure from the fossil pollen stratigraphy by means of weighted averaging regression and calibration. The pollen stratigraphy (Gaillard et al. 1992) indicates that the local vegetation was formerly a *Tilia*–dominated forest that was transformed by clearance and burning into open *Betula* woodland. This, in turn, developed into heath with *Juniperus communis* as a result of grazing and other land–use changes. At the time of forest clearance and burning of the *Tilia* forest, soluble elements like K may have been lost from the ash by leaching and effectively released from the litter, resulting in a decrease in K at the time of maximum local fire. With subsequent but less intense burning and increased grazing, soil moisture decreased, reducing the water–holding capacity of the soil, total N declined, possibly because of volatilisation, and total P increased, possibly because of an effective mineralisation of P. These soil chemical reconstructions suggest that there were considerable environmental impacts of repeated fire and associated land–use factors on some local soil conditions (e.g. soil N, P, K and moisture) but not on soil pH (see Gaillard et al. 1992 for further details).

The third example (Renberg et al. 1993b) is also from southern Sweden, but from south–west Sweden, the area of former *Calluna* heathland (Blennow and Hammerlund 1993). It concerns the impacts of burning and other land–uses on lake–water pH, as reconstructed by weighted averaging regression and calibration of fossil diatom assemblages (Birks et al. 1990). Renberg (1990a) showed at Lilla Öresjön that the diatom–inferred pH record decreased from about pH 7 12000 years ago to pH 6 about 8000 years and to pH 5 by 2500 years ago. About 2300 BP lake–water pH increased to 6 until about 1900 AD, when it began to fall to the former level of pH 5. In the last 30 years lake pH dropped below pH 5 to about pH 4.5 as a result of the input of strong acids as 'acid rain' following combustion of fossil fuels. In an attempt to understand the cause of the rise in pH about 2000 years ago, Renberg et al. (1993a, 1993b) reconstructed lake–water pH at a further 13 lakes on acid granitic bedrock in or near the former *Calluna* heath landscape in south–west Sweden. Eleven lakes show a rise of pH at different times between 1000 and 2300 BP. Partial multiple regression analyses were done using sample depth as a covariable to eliminate statistically any effects of autocorrelation, with diatom–inferred pH as the response variable, and a forward selection procedure (Miller 1990, ter Braak 1990) and associated restricted Monte Carlo permutation tests to assess statistical significance. These analyses show that the most effective predictors of the observed changes in lake–water pH are apophyte (native plants favored by human impact, e.g. Gramineae and *Urtica*) and other cultural (e.g. cereal) pollen types. Between 60% and 95% of the variance in the reconstructed lake–water pH values is explained by 2 or 3 groups of pollen types such as apophytes (see Renberg et al. 1993b for further details).

The phase of increased lake–water pH is interpreted as resulting from the land–use activities of the expanding agrarian culture during the Iron Age involving animal husbandry, some cultivation, and repeated burning. The charcoal record indicates that fire was used as a management tool. Regular burning (every 5–10 years in south–west Sweden) destroyed the acid humus, released base cations and nutrients, and increased the base saturation and pH of the soils. Leaching of ash resulted in increases in surface–water pH and nutrient status. In addition during fires, large amounts of particulates were emitted including ash and nutrients that may have enhanced the nutrient status of the lakes within the heath landscape. With a fire frequency of 5–10 years, there would have been a recurrent regeneration of soil pH, higher pH in surface waters, and a constant supply of base cations and nutrients to the lakes (Renberg et al. 1993b). This 'anthropogenic alkalization period" (Renberg et al. 1993a) continued until the cessation of these land–use practices about 100 years ago.

Partitioning of the variance in the diatom–inferred pH at Stora Skärsjön and Lille Holmevatten into four independent components (Table 1) shows that the sedimentary charcoal record independent of anthropochorous and apophyte pollen types that are indicative of land–use only explains 0.2–1.9%, whereas charcoal interacting and covarying with anthropochorous and apophyte pollen explains 25.9–34.0% of the variance. These results illustrate the potential value of variance partitioning in considering alternative hypotheses to explain observed stratigraphical patterns and highlight the strong interactions between potential causal factors, in this case fire and agricultural land–use.

Discussion and Conclusions

There are six general discussion topics and conclusions to this chapter. First, numerical methods exist for deriving fairly precise reconstructions for palaeoenvironmental variables such as lake–water pH from fossil assemblages. At present the major limitation in such reconstructions is the small number of high–quality, taxonomically detailed and consistent, modern training sets (Birks 1994).

Table 1. Results of partitioning the variance in the diatom–inferred pH sequence at Stora Skärsjön and Lilla Holmevatten. Entries are sum of squares (= variances) expressed as percentages of the total variance in the pH record.

Variance component	Stora Skärsjön	Lilla Holmsvatten
Unexplained variance	12.8[a]	19.8[a]
Related to charcoal independent of anthropochorous[c] and apophyte[c] pollen types	1.9[b]	0.2[b]
Related to anthropochorous and apophyte pollen types independent of charcoal	51.3[a]	55.1[a]
Charcoal covarying with anthropochorous and apophytepollen types	34.0[a]	25.9[a]

[a]Significance level (as assessed by restricted Monte Carlo permutation tests) = $p \leq 0.05$.
[b]Significance level (as assessed by restricted Monte Carlo permutation tests) = $p > 0.05$.
[c]Anthropochorous taxa include *Secale, Hordeum*, and introduced weeds; apophyte taxa are mainly Gramineae, *Rumex acetosa/R. acetosella, Plantago major/media*, and *Urtica*. See Renberg et al. (1993b) for further details.

Second, statistical techniques exist for testing hypotheses about the impacts of fires on the environment in the past. Such methods have recently been used to assess the impacts of volcanic ash on aquatic and terrestrial systems (Lotter and Birks 1993, Birks and Lotter 1994) and the impact of the arrival and expansion of *Picea abies* on lake–water acidity (Korsman et al. 1994). Techniques such as constrained ordinations, variance partitioning, and restricted Monte Carlo permutation tests discussed above provide a means of assessing environmental impacts in the past from stratigraphical palaeoecological data and of testing alternative causal hypotheses to explain the observed stratigraphical patterns (Birks 1995b).

Third, the environmental impacts of fire as a management tool with a fire frequency of 10–25 years can be demonstrated statistically at some sites, resulting in increased lake–water pH and nutrient status (e.g. Renberg et al. 1993b), and also increased erosion and sedimentation rates (e.g. Birks 1972, Odgaard 1994).

Fourth, the environmental impacts of natural fires are much more difficult to discern from the palaeoecological record. MacDonald et al. (1991) concluded that "the inability of evidence from lake sediments to provide detailed histories of past fire activity is not surprising given the wide range of variation in the spatial extent, proximity, intensity and impact of individual fires". Effects such as fire–induced erosion and increased Al, V, and silt content of sediments (e.g. Cwynar 1978) immediately following a fire are difficult to detect without very fine–resolution sampling on an annual or near–annual basis. This requires sites with annually laminated sediments and with a large catchment to lake–area ratio to maximise the catchment–derived "signal" in the sedimentary record (Hall and Smol 1993, Rhodes and Davis 1995). Environmental responses to recent fires in Minnesota, Ontario, and Washington occur for 1–5 years only and appear to have little or no impact on the biotic assemblages in lakes (e.g. H.E. Wright 1981, Tarapchak and Wright 1986, Bradbury 1986). Korhola et al. (1996) have, however, recently interpreted a rise in lake–water pH of 0.9 units (from pH 4.6) between 1890 and 1930 in a naturally acid hill–top lake in southern Finland (Korhola and Tikkanen 1991) as a result of presumed natural catchment forest–fires. These fires may also have resulted in changes in the lake's stratification and may have increased lake turbidity. Rhodes and Davis (1995) have also demonstrated rises in lake–water pH from about 5 to 6.1 at times of natural forest fire and associated soil erosion at an oligotrophic site in Maine, presumably due to the input of base cations as ash and inwashed soil immediately after fire. The pH peaks are short–lived and are followed by a return of lake–water pH to pre–fire values within a few decades.

Fifth, there is considerable scope for statistical testing of competing hypotheses (e.g. Huntley 1993, Bradshaw 1993b) about the role of fire in influencing long–term vegetation dynamics in the Holocene using numerical techniques such as constrained ordinations, variance partitioning, and restricted Monte Carlo permutation tests. Such techniques would permit the relative importance of climatic change, human impact, and fire and of interactions between these factors to be assessed and tested statistically.

Sixth, important conceptual and methodological problems remain in palaeoecological studies of past fire impacts, in particular the different source areas of large and small charcoal particles and of pollen assemblages preserved in lake sediments are difficult to quantify. A further problem is the apparent lack of standardisation amongst palaeoecologists in techniques for estimating the charcoal content of sediments (e.g. Patterson et al. 1987, Clark 1988b, MacDonald et al. 1991). No matter how detailed the statistical analyses are, the most important factor in quantitative palaeoecological studies is the quality and consistency of the palaeoecological data used in the statistical analyses.

Acknowledgments

I am grateful to Jim Clark for the invitation to participate in the NATO Advanced Research Workshop (ARW 931616) on Sediment records of biomass burning and global change in Portimâo, Portugal, to Brigitta Ammann, Richard Bradshaw, Bill Chaloner, Jim Clark, Glen MacDonald, Shinya Sugita, and Cathy Whitlock for helpful discussions at the workshop, to Atte Korhola for permission to mention his (then) unpublished work in southern Finland, and to Anne Birgit Ruud Hage for her invaluable help in word–processing the manuscript.

Literature References

Anderson NJ (1993) Natural versus anthropogenic change in lakes: The role of the sediment record. *Trends Ecol Evol* 8:356–361

Anderson NJ (1995) Using the past to predict the future: lake sediments and the modelling of limnological disturbance. *Ecol Modelling* 78:149–172

Bartlein PJ, Whitlock C (1993) Paleoclimatic interpretation of the Elk Lake pollen record. In Bradbury JP, Dean WE (eds) *Elk Lake, Minnesota: Evidence for Rapid Climate Change in the North–Central United States Geol Soc Amer* Spec Pap 276:275–293

Battarbee RW (1991) Recent paleolimnology and diatom–based environmental reconstruction. In Shane LCK, Cushing EJ (eds) *Quaternary Landscapes* pp 129–174 University of Minnesota Press Minneapolis

Besag J, Clifford P (1989) Generalized Monte Carlo significance tests. *Biometrika* 76:633–642

Birks HH (1972) Studies in the vegetational history of Scotland. II. Two pollen diagrams from the Galloway Hills, Kirkcudbrightshire. *J Ecol* 60:183–217

Birks HJB (1993) Is the hypothesis of survival on glacial nunataks necessary to explain the present-day distributions of Norwegian mountain plants? *Phytocoenologia* 23:399–426

Birks HJB (1994) The importance of pollen and diatom taxonomic precision in quantitative palaeoenvironmental reconstruction. *Rev Palaeobot Palynol* 83:107–117

Birks HJB (1995a) Quantitative palaeoenvironmental reconstructions. In Maddy D, Brew JS (eds) *Statistical Modelling of Quaternary Science Data* (in press)

Birks HJB (1995b) Did Icelandic volcanic eruptions influence the post–glacial history of the British Isles? *Trends Ecol Evol* 9:312–314

Birks HJB, Line JM, Juggins S, Stevenson AC, ter Braak CJF (1990) Diatoms and pH reconstruction. *Phil Trans Roy Soc London* B 327:263–278

Birks HJB, Lotter AF (1994) The impact of the Laacher See Volcano (11000 yr B.P.) on terrestrial vegetation and diatoms. *J Paleolimnol* 11:313–322

Blennow K, Hammarlund K (1993) From heath to forest: land–use transformation in Halland, Sweden. *Ambio* 22:561–567

Borcard D, Legendre P, Drapeau P (1992) Partialling out the spatial component of ecological variation. *Ecology* 73:1045–1055

Bradbury JP (1986) Effects of forest fire and other disturbances on wilderness lakes in northeastern Minnesota II. Paleolimnology. *Arch Hydrobiol* 106:203–217

Bradshaw RHW (1993a) Tree species dynamics and disturbance in three Swedish boreal forest stands during the last two thousand years. *J Vegetation Sci* 4:759–764

Bradshaw RHW (1993b) Forest response to Holocene climatic change: equilibrium or non–equilibrium. In Chambers FM (ed) *Climate Change and Human Impact on the Landscape* pp 57–65 Chapman and Hall London

Bradshaw RHW, Hannon GE (1992) Climatic change, human influence and disturbance regime in the control of vegetation dynamics within Fiby Forest, Sweden. *J Ecol* 80:625–632

Clark JS (1988a) Particle motion and the theory of charcoal analysis: source area, transport, deposition, and sampling. *Quat Res* 30:67–80

Clark JS (1988b) Stratigraphic charcoal analysis on petrographic thin sections: application to fire history in northwestern Minnesota. *Quat Res* 30:81–91

Clark JS (1988c) Effect of climate change on fire regimes in northwestern Minnesota. *Nature* 334:233–235

Clark JS, Merkt J, Müller H (1989) Post–glacial fire, vegetation, and human history on the northern Alpine Foreland, south–western Germany. *J Ecol* 77:897–925

Clark JS, Royall PD (1995) Particle–size evidence for source areas of charcoal accumulation in late Holocene sediments of eastern North American lakes. *Quat Res* 43:80–89

Cwynar LC (1978) Recent history of fire and vegetation from laminated sediment of Greenleaf Lake, Algonquin Park, Ontario. *Can J Bot* 56:10–21

Dixit SS, Smol JP, Kingston JC, Charles DF (1992) Diatoms: powerful indicators of environmental change. *Environm Sci Technol* 26:23–33

Gaillard M–J, Birks HJB, Emanualsson U, Berglund BE (1992) Modern pollen/land–uses and cultural landscapes: an example from south Sweden. *Veg Hist Archaeobot* 1:3–17

Gimingham CH (1994) Lowland heaths of west Europe: management for conservation. *Phytocoenologia* 24:615–626

Green DG (1981) Time series and postglacial forest ecology. *Quat Res* 15:265–277

Green DG (1982) Fire and stability in the postglacial forests of southwest Nova Scotia. *J Biogeogr* 9:29–40

Green DG (1987) Pollen evidence for the postglacial origins of Nova Scotia's forest. *Can J Bot* 65:1163–1179

Green DG, Dolman GS (1988) Fine resolution pollen analysis. *J Biogeogr* 15:685–701

Hall RI, Smol JP (1993) The influence of catchment size on lake trophic status during the hemlock decline and recovery (4800 to 3500 BP) in southern Ontario. *Hydrobiologia* 269/270:371–390

Huntley B (1993) Rapid early–Holocene migration and high abundance of hazel (*Corylus avellana* L.): alternative hypotheses. In Chambers FM (ed) *Climate Change and Human Impact on the Landscape* pp 205–215 Chapman and Hall London

Jackson ST (1994) Pollen and spores in Quaternary lake sediments as sensors of vegetation composition: theoretical models and empirical evidence. In Traverse A (ed) *Sedimentation of Organic Particles* pp 253–286 Cambridge University Press Cambridge

Kaland PE (1986) The origin and management of Norwegian coastal heaths as reflected by pollen analysis. In Behre K–E (ed) *Anthropogenic Indicators in Pollen Diagrams* pp 19–36 Balkema Rotterdam

Korhola AA, Tikkanen MJ (1991) Holocene development and early extreme acidification in a small hilltop lake in southern Finland. *Boreas* 20:335–356

Korhola A, Virkanen J, Tikkanen M, Blom T (1996) Fire induced plant rise in a naturally acid hill-top lake, southern Finland: a palaeoecological study. *J Ecol* (in press)

Korsman T, Renberg I, Anderson NJ (1994) A palaeolimnological test of the influence of Norway spruce (*Picea abies*) immigration on lake–water acidity. *The Holocene* 4:132–140

Legendre P (1993) Spatial autocorrelation: trouble or new paradigm? *Ecology* 74:1659–1673

Lotter AF, Birks HJB (1993) The impact of the Laacher See Tephra on terrestrial and aquatic ecosystems in the Black Forest, southern Germany. *J Quat Sci* 8:263–276

MacDonald GM, Larsen CPS, Szeicz JM, Moser KA (1991) The reconstruction of boreal forest fire history from lake sediments: a comparison of charcoal, pollen, sedimentological and geochemical indices. *Quat Sci Rev* 10:53–71

Manly BFJ (1991) *Randomization and Monte Carlo methods in biology.* Chapman and Hall London

Martens H, Næs T (1989) *Multivariate calibration.* J Wiley & Sons Chichester

Miller AJ (1990) *Subset Selection in Regression*. Chapman & Hall London

Montgomery DC, Peck EA (1992) *Introduction to linear regression analysis*. J Wiley & Sons New York

Odgaard BV (1992) The fire history of Danish heathland areas as reflected by pollen and charred particles in lake sediments. *The Holocene* 2:218–226

Odgaard BV (1994) The Holocene vegetation history of northern West Jutland, Denmark. *Opera Bot* 123

Patterson WA, Edwards KJ, Maguire DJ (1987) Microscopic charcoal as a fossil indicator of fire. *Quat Sci Rev* 6:3–23

Peglar SM (1993) The mid–Holocene Ulmus decline at Diss Mere, Norfolk, UK: a year–by–year pollen stratigraphy from annual laminations. *The Holocene* 3:1–13

Peglar SM, Birks, HJB (1993) The mid–Holocene *Ulmus* fall at Diss Mere, south–east England – disease and human impact? *Veg Hist Archaeobot* 2:61–68

Renberg I (1990a) A 12600 year perspective of the acidification of Lilla ØresjØn, southwest Sweden. *Phil Trans Roy Soc London B* 327:357–361

Renberg I (1990b) A procedure for preparing large sets of diatom slides from sediment cores. *J Paleolimnol* 4:87–90

Renberg I, Korsman T, Anderson NJ (1993a) A temporal perspective of lake acidification in Sweden. *Ambio* 22:264–271

Renberg I, Korsman T, Birks HJB (1993b) Prehistoric increases in the pH of acid–sensitive Swedish lakes caused by land–use changes. *Nature* 362:824–826

Rhodes TS, Davis, RB (1995) Effects of late Holocene forest disturbance and vegetation change on acidic Mud Pond, Maine, USA. *Ecology* 76:734–746

Sugita S (1994) Pollen representation of vegetation in Quaternary sediments: theory and method in patchy vegetation. *J Ecol* 82:881–897

Swain AM (1973) A history of fire and vegetation in northeastern Minnesota as recorded in lake sediments. *Quat Res* 3:383–396

Swain AM (1978) Environmental changes during the past 2000 years in north–central Wisconsin: analysis of pollen, charcoal, and seeds from varved lake sediments. *Quat Res* 10:55–68

Tarapchak SJ, Wright HE (1986) Effects of forest fire and other disturbances on wilderness lakes in northeastern Minnesota I. Limnology. *Arch Hydrobiol* 106:177–202

ter Braak CJF (1987a) *Unimodal models to relate species to environment*. Doctoral thesis University of Wageningen

ter Braak CJF (1987b) Ordination. In Jongman RHG, ter Braak CJF, van Tongeren OFR (eds) *Data analysis in community and landscape ecology* pp 91–173 Pudoc Wageningen

ter Braak CJF (1988) Partial canonical correspondence analysis. In Bock HH (ed) *Classification and Related Methods of Data Analysis* pp 551-558 Elsevier Science Publishers Amsterdam ter Braak CJF (1990) *Update notes: CANOCO version 3.10.* Agricultural Mathematics Group Wageningen

ter Braak CJF (1992) Permutation versus bootstrap significance tests in multiple regression and ANOVA. In Jöckel K–H, Rothe G, Sendler W (eds) *Bootstrapping and related techniques* pp 79–86 Springer Verlag Berlin

ter Braak CJF (1994) Canonical community ordination. Part I: Basic theory and linear models. *Ecoscience* 1:127–140

ter Braak CJF (1995) Non–linear methods for multivariate statistical calibration and their use in palaeoecology: a comparison of inverse (K–nearest neighbours, partial least squares and weighted averaging partial least squares) and classical approaches. *Chemometr Intell Lab Syst* 28:165–180

ter Braak CJF, Juggins S (1993) Weighted averaging partial least squares regression (WA–PLS): an improved method for reconstructing environmental variables from species assemblages. *Hydrobiol* 269/270:485–502

ter Braak CJF, Juggins S, Birks HJB, van der Volt H (1993) Weighted averaging partial least squares regression (WA–PLS): definition and comparison with other methods for species–environment calibration. In Patil GP, Rao CR (eds) *Multivariate Environmental Statistics* pp 525–560 Elsevier Science Publishers Amsterdam

ter Braak CJF, Prentice IC (1988) A theory of gradient analysis. *Adv Ecol Res* 18:271–317

ter Braak CJF, van Dam H (1989) Inferring pH from diatoms: a comparison of old and new calibration methods. *Hydrobiol* 178:209–223

ter Braak CJF, Wiertz J (1994) On the statistical analysis of vegetation change: a wetland affected by water extraction and soil acidification. *J Veg Sci* 5:361–372

Verdonschot PFM, ter Braak CJF (1994) An experimental manipulation of oligochaete communities in mesocosms treated with chlorpyrifos or nutrient additions, multivariate analyses with Monte Carlo permutation tests. *Hydrobiol* 278:251–266

Walker D (1982) The development of resilience in burned vegetation. In Newman EI (ed) *The Plant Community as a Working Mechanism* pp 27–43 Blackwell Scientific Publications Oxford

Walker IR, Smol JP, Engstrom DR, Birks HJB (1991) An assessment of Chironomidae as quantitative indicators of past climatic change. *Can J Fisheries Aqu Sci* 48:975–987

Whittaker J (1984) Model interpretations from the additive elements of the likelihood function. *Appl Stat* 33:52–64

Wright HE (1981) The role of fire in land/water interactions. In Mooney HA, Bunnickson JM, Christensen NL, Lotan JE, Reiners WA (eds) *Fire Regimes and Ecosystem Properties US Forest Service* General Technical Report WO–26 pp 421–444

Wright RF (1976) The impact of forest fire on the nutrient influxes to small lakes in northeastern Minnesota. *Ecology* 57:649–663

The Sedimentary Record of Fire in Montane Meadows, Sierra Nevada, California, USA: a Preliminary Assessment

R. Scott Anderson and Susan J. Smith[1]
Center for Environmental Sciences and Education
and Quaternary Studies Program
Box 5694
Northern Arizona University
Flagstaff, Arizona 86011, U.S.A.

Summary

Examination of the charcoal particle stratigraphy in sediment cores from seven meadows from the Sierra Nevada of California allows reconstruction of regional fire periods and their relationship to climate during the Holocene. With the exception of the period of 8700 to 9200 yr BP, charcoal is less frequently encountered in early Holocene sediments than it is in sediments deposited after 4500 yr BP. We hypothesize a climatic control. Empirical and modeled data suggest that the early Holocene was xerothermic in the Sierra Nevada. Pollen evidence suggests forests were more open than today, thus producing less tree biomass for fuel. Intensification of El Niño during the middle Holocene may have been important in changing the character of the Sierran forests as well.

Introduction

The relationship between climate, vegetation type, and fire occurrence has attracted considerable interest from ecologists, land managers, and, more recently, from climate and vegetation modellers. For instance, land managers have developed models for prescribed burning to reduce fire hazards in temperate forests that are based upon weather patterns, atmospheric humidity, and remnant fuel moisture (e.g. Rothermel 1983). Ecologists have long recognized that certain species, such as lodgepole pine (*Pinus contorta* var. *latifolia*), and certain plant communities, such as California chaparral and giant sequoia (*Sequoiadendron giganteum*)-mixed conifer forests, are adapted to periodic fire (Keeley 1986, Swetnam 1993) (nomenclature follows Hickman 1993). Paleoecologists have often noted charcoal particles in sediments as part of their analysis of vegetation and climate during the Holocene (Waddington 1969, Anderson et al. 1986, Clark 1988, MacDonald et al. 1991), although interpretation of charcoal data has often been limited by methodological and taphonomic problems (Patterson et al. 1987).

---

[1] Bilby Research Center, Box 6013, Northern Arizona University, Flagstaff, Arizona 86011, USA.

NATO ASI Series, Vol. I 51
Sediment Records of Biomass Burning
and Global Change
Edited by James S. Clark, Hélène Cachier,
Johann G. Goldammer, and Brian Stocks

Direct evidence for strong linkage between climatic factors and the frequency of forest-fire disturbance has been demonstrated by several recent studies using historical and tree-ring records. Simard et al. (1985) compared historical records of fire activity and El Niño in the United States that showed decreased fire activity subsequent to El Niño years. Swetnam and Betancourt (1990, 1992) extended the record farther back in time by analysis of fire-scarred trees (chronologies back to 1700 A.D.). They demonstrated that small areas burn after wet springs (El Niño years), while large areas burn after dry springs (La Niña years). Less direct but equally compelling evidence was provided recently by Overpeck et al. (1990) using pollen data to model increased future disturbance with increasing global temperatures. Each of these studies has demonstrated the importance of regional and global climatic patterns on ecosystem processes.

Examination of charcoal and plant remains in stratigraphic sediments can extend the record of fire disturbance and ecosystem effects to more ancient times. Construction of these longer term records of fire–vegetation interactions, and comparison with independently derived records of climate change, should provide additional confirmation of the relationships outlined by Simard et al. (1985) and Swetnam and Betancourt (1990), at least at the local level, thereby increasing confidence in future climate scenarios. Ultimately, however, our goal should be to resolve climate-fire patterns and predictions at the landscape level.

The approach we used to examine the long-term history of fire consists of documenting charcoal particle deposition in sediment cores of montane meadows from the Sierra Nevada of California. Our comparisons of sedimentary charcoal, pollen, and plant macrofossil data from these cores strongly suggests that climate is the driving force for change in Holocene fire regimes. Perhaps surprisingly, our data suggest that burning rates, as measured by charcoal deposition, during the warm, dry early Holocene were lower than during the latter Holocene, a period of greater effective precipitation. We hypothesize that, among other factors, climate has influenced fire regimes in the Sierra Nevada through biomass accumulation (in the form of fine fuels) in the understory, with greater biomass accumulation during the late Holocene than that of earlier time periods.

Regional setting

The Sierra Nevada is a westward tilting granitic batholith, bounded by metamorphic crystalline rocks (Bateman and Wahrhaftig 1966) in eastern California (Fig. 1). The northwest–southeast oriented range is 575 km long and 100 to 125 km wide. The steep eastern escarpment is the western terminus of the Great Basin physiographic province, and it separates the floristically diverse cis-montane Californian Floristic Province from the less diverse Great Basin Floristic Province.

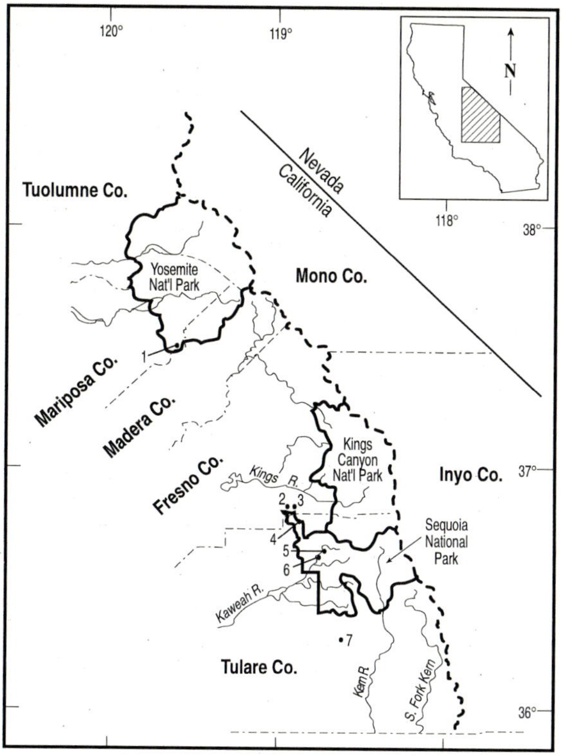

Figure 1. Location of the Sierra Nevada and meadow sites mentioned in the text. 1 = JB Swale; 2 = Hightop; 3 = Huckleberry; 4 = Weston; 5 = Long; 6 = Circle; 7 = Dogwood.

Modern climate of the region is Mediterranean, with hot, arid summers and cool, humid winters (Major 1988). Dominant winter storms enter the region from the west, and lose moisture on the west side of the range. For any given elevation, precipitation is often twice as high on the western side of the crest as on the east. Convective thunderstorms provide some summer moisture as do sporadic storms originating in the south (Gulf of California) or southwest (Pacific Ocean). Precipitation increases with elevation; most precipitation above 2200 m falls as snow, with large spatial variability (Armstrong and Stidd 1967).

Modern vegetation of the range has been described in terms of "zones" or "formations" that generally parallel the axis of the range. All meadow sites described here lie in the sequoia-mixed conifer forest, occurring at 1200 to 2200 m asl. Dominant tree species include white fir (*Abies concolor*), sugar and ponderosa pines (*Pinus lambertiana* and *P. ponderosa*), incense-cedar (*Calocedrus decurrens*), and, locally, giant sequoia (*Sequoiadendron giganteum*). Above this forest type at 2200-2750 m is the upper montane assemblage, which includes lodgepole and western white pines (*Pinus contorta* var. *murrayana* and *P. monticola*), sugar pine, red fir (*Abies magnifica*), and Sierra juniper (*Juniperus occidentalis* ssp. *australis*). The subalpine forest (2750-3300 m asl) consists of lodgepole pine with mountain hemlock (*Tsuga mertensiana*), western white pine, whitebark pine (*Pinus albicaulis*), and Sierra juniper. The alpine zone above 3300 m is dominated by herbs and subshrubs.

Below the mixed conifer forest is chaparral (915-1225 m), with a diverse array of shrubs including buckbrush (*Ceanothus* spp.), manzanita (*Arctostaphylos* spp.), mountain mahogany (*Cercocarpus betuloides*) and chamise (*Adenostoma fasciculatum*). Below 915 m is the blue oak (*Quercus douglassi*)–digger pine (*Pinus sabiniana*) woodland.

Study sites

The seven sites for which charcoal data are available occur between 1786 and 2206 m elevation on the west flank of the Sierra Nevada (Table 1). Six of the sites are elongated meadows, varying in length from 500 to 1500 m and completely surrounded on the sideslopes by closed mixed conifer forest (Fig. 2). Three of these sites (Hightop, Circle, Dogwood) occur within giant sequoia groves, while the other three (Huckleberry, Weston, Long) are outside the groves. JB Swale, the northernmost site, is a small, circular meadow also surrounded by mixed conifer forest.

Meadow genesis and sedimentation pattern varies among sites (Anderson and Smith 1994). Hightop, Long, and Dogwood each contain single small streams that flow seasonally along the long axes of meadows. By late summer the streams are largely dry. Huckleberry and Weston Meadows appear to be fed by small springs at the head of each meadow. Circle Meadow lies in a crescent-shaped bedrock depression, and thus is fed by slopewash from the gentle surrounding slopes. JB Swale is a small depression; sediment collects behind a granitic rampart that partially blocks a small drainage.

Table 1. Location of sites mentioned in the text.

Meadow	County	Elevation (m)	Latitude (° N)	Longitude (° W)
JB Swale	Mariposa	1786	37° 30'	119° 37'
Hightop	Fresno	1908	36° 48'	118° 57'
Huckleberry	Fresno	1987	36° 46'	118° 57'
Weston	Tulare	2036	36° 43'	118° 53'
Long	Tulare	2206	36° 35'	118° 44'
Circle	Tulare	2085	36° 34'	118° 45'
Dogwood	Tulare	1987	36° 12'	118° 40'

Methodology

All meadows in this study were cored between July 1987 and June 1991. We used a Livingstone sediment corer to obtain each sediment record. The Livingstone corer removes 1-m long × 5-cm diameter core sections, with successive sections retrieved from the same hole. In the field the cores were wrapped in plastic wrap and aluminum foil, stored in wooden boxes, and transported to the Laboratory of Paleoecology (Northern Arizona University) for cold storage.

Detailed core description and pollen, macrofossil, and charcoal–particle analyses were performed on each core. We concentrate here on the methodology for the charcoal particle analysis, but we also provide a general methodology for the other analyses. We used the sediment thin–section procedure to produce our charcoal (fire) records, a methodology developed by Clark (1988) and refined for meadow sediments by Smith and Anderson (1995). Long segments of cores were impregnated with epoxy resin before production of semi-transparent thin-sections. This method produces a continuous stratigraphic record and has the advantage of allowing *in situ* study of the entire record, which can include subtle and gradational transitions. Sediments are dehydrated by replacing interstitial water with acetone and then impregnated with epoxy. Application of vacuum enhances the replacement and evacuates gas bubbles from the epoxy and sediments. The resulting epoxied sediments are cured 24 to 36 hours at 40°C. Thin-sections were prepared by standard procedures using a quick-setting epoxy to bond the chip to the slide.

Charcoal particles for each 1 mm depth increment were identified and measured at 100× using a Reichert microscope. All particles having one dimension > 100 μm were measured with the aid of a reticle gridded in 100 μm increments. Cumulative particle area was determined for each transect.

Figure 2. Photograph of Circle Meadow, Giant Forest, Sequoia National Park, California. Note triangular-shaped "cat face" fire-scar at the base of the giant sequoia extreme right of the photograph.

Pollen samples were prepared by standard techniques (Faegri and Iversen 1989). One ml sediment subsamples were treated with dilute warm KOH (to disaggregate organics), 10% HCl (to remove carbonates), acetolysis solution (to digest organics), and HF (to remove silicates). Resulting pollen residues were stored in silicone oil. A minimum of 300 grains (exclusive of spores) were identified; difficult grains were checked against modern pollen reference specimens. Pollen data are not reported here. Rather, a generalized pollen stratigraphy is discussed (see below). Core chronologies were estimated from conventional bulk sediment radiocarbon dates obtained from Beta Analytic, Inc.

Results

Chronology

A total of 29 radiocarbon dates were obtained from the eight cores, including two locations in Circle Meadow (Table 2). Three to five conventional radiocarbon dates were obtained for each core with the exception of a single date from Circle Meadow core #5. Records from each meadow exceeded 9400 years in length with the longest record (Weston Meadow) extending to 10400 yr BP. Charcoal records for several of the cores did not span the entire length of core (see below). Chronology of sedimentary events for each core was determined by interpolation between radiocarbon dates.

Stratigraphy of sediment

Stratigraphies of five sites (Hightop, Weston, Long, Circle, and Dogwood) are summarized by Anderson and Smith (1994); stratigraphies of two additional sites, JB Swale and Huckleberry Meadow, are similar to these other sites. Details vary among records but general patterns are apparent. Basal sediments of each core are coarse sands or gravels. A transition to colluvium or finer sands occurs by 9500 yr BP. Early to middle Holocene sediments consist of bands of silt with variable amounts of peat or coarse sands. Late Holocene sediments are peats, rarely interspersed with sandy colluvium. The transition time from predominantly sandy colluvium to predominantly peat varies slightly among sites, but generally occurs between 4500 and 3000 yr BP.

Charcoal records

Figure 3 details individual charcoal records for the seven meadows, including two records from Circle Meadow. The meadows are arrayed from the northernmost site

(JB Swale in Yosemite National Park) to the southernmost site (Dogwood Meadow in Mt. Home State Forest). Each charcoal record consists of a series of "spikes". Each spike represents charcoal particle area deposited over a 25 ^{14}C-yr period as determined from interpolation between ^{14}C dates (shown as dots on the records). The number of data points for each 25-yr period decreases from the most recent periods to those at 10000 yr BP (see below).

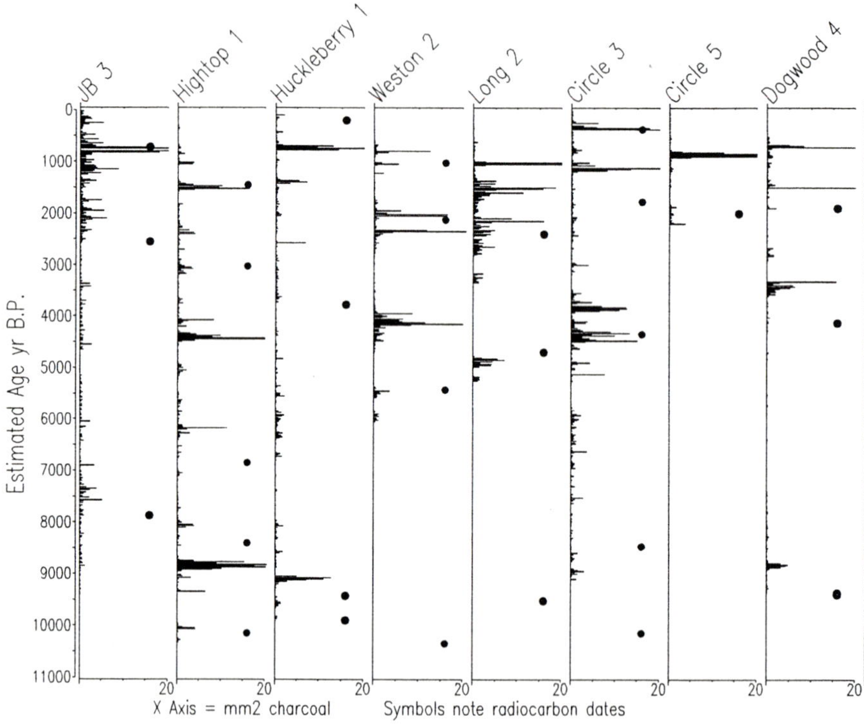

Figure 3. Individual charcoal records from eight cores within the Sierra Nevada. Each bar on the bar graphs is the sum of charcoal deposited over a 25-radiocarbon year period. The dots represent locations of radiocarbon dates.

Individual fires or perhaps fire periods are recorded as major increases in charcoal abundance (large "spikes") well above the background level for each record. Hightop Meadow has one of the longest and most continuous records of charcoal accumulation. The Hightop record shows major charcoal peaks at approximately 8800, 6200, 4400, and 1400 yr BP, with secondary peaks near 10000, 9400, 8000, 4100, 3100, 2300, and 1000, yr BP. A similar record is found in the long core from Circle Meadow (core # 3). Major charcoal peaks there occur at 5100, 4400, 3800, 1000, and

300 yr BP, with secondary peaks at 9100, 6000, 3000, and 700 yr BP. Each of the remaining records show slightly different occurrences of charcoal "spikes".

Table 2. Radiocarbon dates for sites mentioned in the text.

Meadow	Core number	Depth (cm)	Age (^{14}C yr BP)	Laboratory number
JB Swale	3	25-30	730 ± 100	Beta-34999
	3	75-85	2560 ± 70	Beta-35000
	3	106-114	7870 ± 110	Beta-31990
Hightop	1	65-75	1440 ± 60	Beta-34997
	1	155-162	3020 ± 90	Beta-34998
	1	214-225	6840 ± 80	Beta-62502
	1	250-259	8410 ± 100	Beta-62503
	1	280-286	10160 ± 90	Beta-31989
Huckleberry	1	30-40	200 ± 60	Beta-52698
	1	81-95	3770 ± 110	Beta-62504
	1	126-135	9430 ± 90	Beta-52699
	1	220-230	9890 ± 500	Beta-46163
Weston	2	55-60	1030 ± 80	Beta-29030
	2	95-100	2130 ± 60	Beta-62507
	2	134-144	5430 ± 90	Beta-29031
	2	247-256	10360 ± 70	Beta-27602
Long	2	85-92	2410 ± 80	Beta-62505
	2	198-206	4690 ± 140	Beta-62506
	2	393-401	9530 ± 80	Beta-27597
Circle	3	54-60	370 ± 70	Beta-42421
	3	65-81	1780 ± 90	Beta-43961
	3	117-122	4340 ± 70	Beta-42422
	3	170-175	8480 ± 130	Beta-42423
	3	269-279	10150 ± 140	Beta-31987
	5	89-94	2000 ± 70	Beta-46004
Dogwood	4	85-99	1900 ± 70	Beta-34995
	4	155-160	4130 ± 70	Beta-34996
	4	175-180	9370 ± 80	Beta-62501
	4	211-220	9400 ± 70	Beta-31988

Inspection of charcoal records reveals consistencies. Four of the five records that stretch back into the earliest Holocene (Hightop, Huckleberry, Circle, Dogwood) have charcoal peaks between 8700 and 9200 yr BP. Four records (Hightop, Huckleberry, Weston, Circle) also reveal somewhat elevated charcoal amounts centered near 6000 yr BP. Major charcoal peaks occur in several records from 3800 to 4500 (Hightop, Weston, Circle, Dogwood), 2000 to 2300 (JB, Weston, Long), 1400 to 1500 (Hightop, Huckleberry, Long, Dogwood), 1000 to 1100 (JB, Hightop, Weston, Long, Circle), and 700 to 800 yr BP (JB, Huckleberry, Weston, Circle, Dogwood). Repeated occurrences of charcoal peaks in these records is striking, because the sites are distant from one another (Fig. 1), and ages of charcoal peaks are interpolations between radiocarbon dates and not dates directly obtained on the charcoal bands themselves (Fig. 3).

Combining records allows us to examine the pattern of burning in Sierran meadows during the entire Holocene. Figure 4 shows cumulative area for each 25-yr period for the eight records and "corrected charcoal" (area per core). Patterns in the individual records (Fig. 3) are enhanced in the composite record (Fig. 4a). The composite record shows little charcoal in sediments deposited during the first millennium of the Holocene, although a fire event may have occurred at 10000 yr BP, and at least one is recorded in the interval from 8700 to 9100 yr BP. Low background charcoal with some small peaks occur in the interval from 6300 to 8700 yr BP. The period from 4500 to 6300 yr BP shows slightly increased charcoal. After 4500 yr BP, however, charcoal increases substantially and, with the exception of a short period centered around 3000 yr BP, remains high until the most recent century of fire suppression.

Discussion

The charcoal accumulation patterns from each of seven meadow sites differ slightly in detail, as would be expected from locally derived records. However the general synchroneity over a large portion of the Sierra Nevada, as shown by the composite charcoal record (Figs. 4a and 4c), suggests a broader-scale forcing mechanism such as climate. We believe that climate has influenced the long-term natural fire regime in the Sierra Nevada by two or more factors acting individually or in combination. These two factors include (1) a dry early Holocene, which produced forests with too little biomass to carry significant fires, and (2) development of El Niño conditions in this part of North America during the middle Holocene.

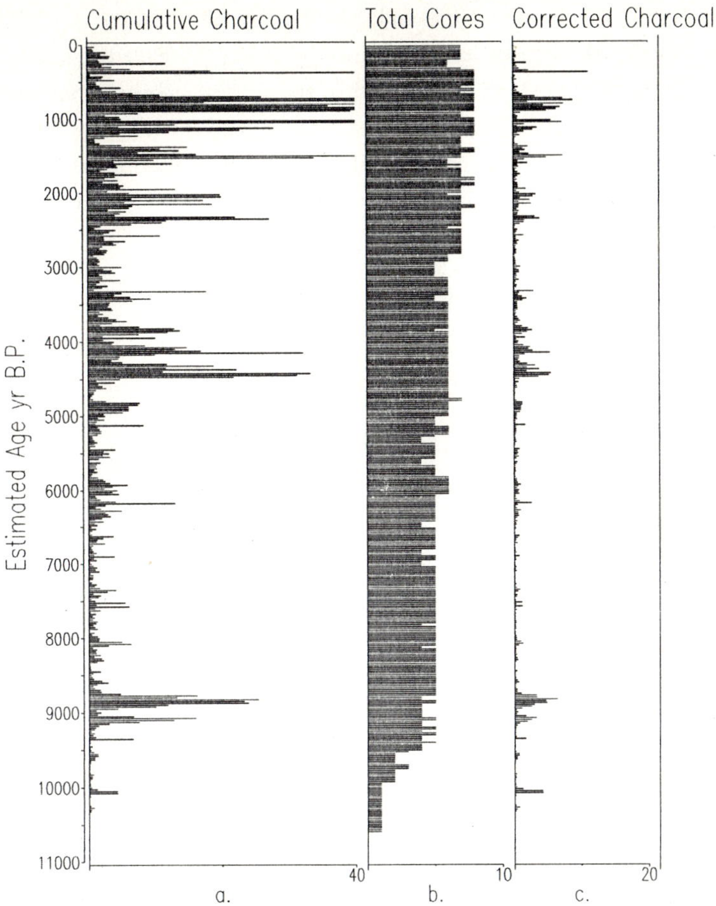

Figure 4. (a) Composite charcoal record constructed by summing the 25-year values for each core in Fig. 3; (b) Number of cores with data for each 25-year period; (c) value in 4(a) divided by the corresponding value in 4(b).

The biomass hypothesis

The vegetation history of the Sierra Nevada has been summarized recently in several papers (Davis et al. 1985, Anderson 1990, Anderson and Smith 1994). Anderson and Smith (1994) used a pollen–indicator–species approach to deduce vegetation and paleoclimatic change within the range. They produced a series of pollen time-space diagrams for climatically sensitive pollen types, including sagebrush (*Artemisia*), bush chinquapin (*Chrysolepis*), oak (*Quercus*), mountain hemlock (*Tsuga mertensiana*), fir (*Abies*), and giant sequoia (*Sequoiadendron giganteum*) (Fig. 5). Of these six species, sagebrush, bush chinquapin, and oak were most abundant during the early Holocene. Sagebrush declined in importance by 6000 to 7000 yr BP, oak by 6300 yr BP, and bush chinquapin by 5500 yr BP. Most species of sagebrush characteristically grow on dry slopes, sandy soils, or exposed open locations where soil moisture may be limited (Hickman 1993). Bush chinquapin, a montane chaparral shrub, prefers dry open forest (Hickman 1993), while many species of oak in the Sierra also prefer open conditions.

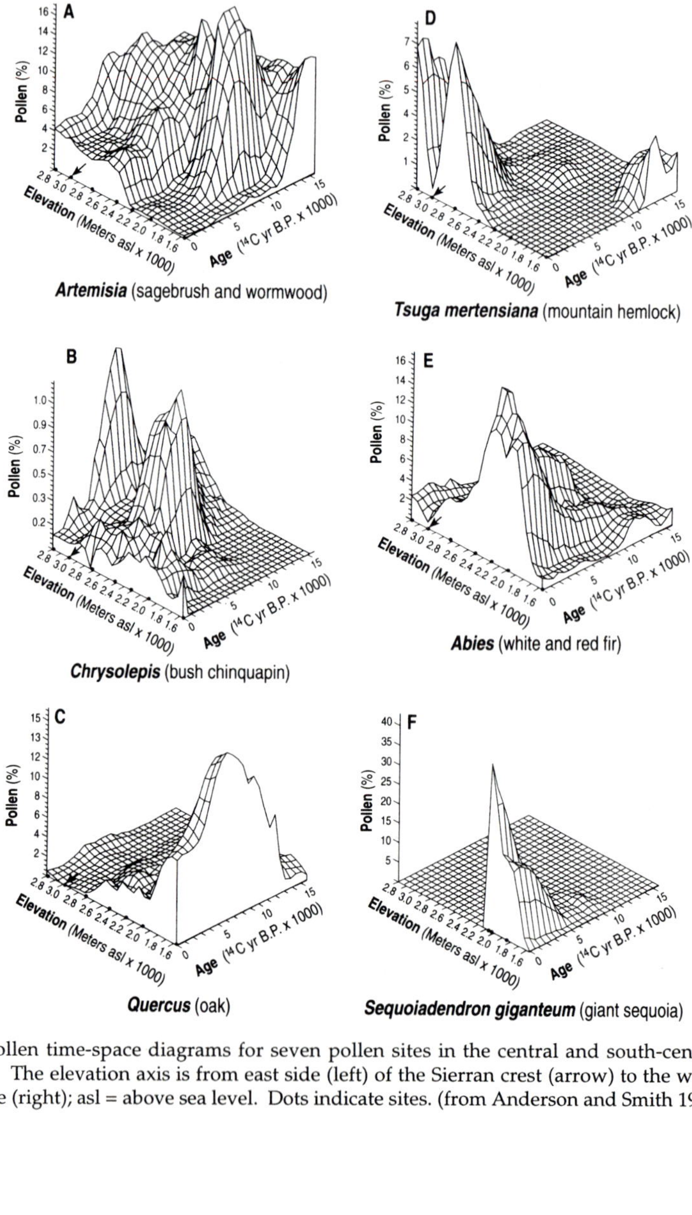

Figure 5. Pollen time-space diagrams for seven pollen sites in the central and south-central Sierra Nevada. The elevation axis is from east side (left) of the Sierran crest (arrow) to the west side of the range (right); asl = above sea level. Dots indicate sites. (from Anderson and Smith 1994).

During the latter half of the Holocene tree species dependent upon abundant soil moisture increased (Fig. 5). Red fir (*Abies magnifica*) depends upon a deep winter snowpack, which provides sufficient soil moisture to last through the summer dry season (Barbour et al. 1991). Although fire–dependent, giant sequoias are limited by soil moisture (Harvey 1980). Mountain hemlock, confined to subalpine forests, also prefers moist soils.

Our data suggest that forests in the Sierra were more open prior to 5000 to 6000 yr BP than they were during the latter Holocene. This condition suggests less biomass, especially coniferous biomass that tends to carry fire. The change toward cooler or wetter conditions after 5000 yr BP allowed for increases in the coniferous species mentioned above, which produce the abundant fine fuels that burn readily, especially during extended periods of summer drought.

The El Niño-Southern Oscillation (ENSO) hypothesis

Swetnam and Betancourt (1990, 1992) documented correspondence between high Southern Oscillation Index (SOI— La Niña phase) and large area burned, suggesting a second possible explanation for the Sierran fire regime. At present California lies at the northernmost extension of ENSO conditions (Meko 1992; Null 1992). Yet our understanding of the antiquity of El Niño in North America is poor. Geological and archaeological evidence suggests near-modern ENSO conditions originated in South America as early as the mid-Holocene. Changes in mulluscan fauna from warm- to cold-water species occurred along coastal Peru by 5000 yr BP, suggesting that the boundary between the warm Panamic Province and the cold Peruvian Province shifted northward by that time (Rollins et al. 1986). Wells (1990) noted an increase in large floods along coastal Peru after 3200 yr BP. Lake–level fluctuations in coastal South America and fossil plant assemblages suggestive of increased moisture and increased drought and fire frequency indicated to McGlone et al. (1992) an intensification of El Niño by 4000 yr BP.

Intensification of El Niño conditions (winter/spring precipitation) in western North America would increase soil moisture during the summer growth season and inhibit fire activity during the dry summer. It would also allow for greater biomass production so that during the alternate La Niña years fire would be promoted. Although direct evidence is lacking, the fire record from the Sierra Nevada may be a proxy for El Niño conditions in North America.

Although Holocene climatic regime may be the overriding factor controlling fire occurrence within the Sierra Nevada human impact on the landscape, primarily by aboriginal burning, cannot be ruled out as a cause for the distinct change in fire regime beginning 4500 years ago. For instance, Anderson and Carpenter (1991) used paleoecological and ethnographic evidence to infer protohistoric Miwok use of fire to manipulate the late-Holocene vegetation of Yosemite Valley in Yosemite National Park California. However the archaeological record of the middle Holocene is sparse and the available data suggest that cultural instability rather than long-term permanent settlement predominated in the Sierra (Moratto 1984).

Conclusions

Charcoal extracted from montane meadows in California's Sierra Nevada clearly document the occurrence of natural fire throughout the Holocene. Each of the seven sites shows a slightly different pattern of major fires, which is to be expected based upon local site characteristics and our reliance on interpolation between radiocarbon dates. The composite record of fire from the meadows demonstrates a distinctive pattern of burning, which, when considering the wide geographic area of the study, is strongly suggestive of a climatic cause. Human impact on these events cannot be discounted.

We believe that the temporal pattern of fire can be explained most simply by the occurrence of an early to middle Holocene xerothermic period in which forests were too open to carry large fires. With the climatic changes that accompanied the late Holocene driven by increases in winter storms from the north Pacific or an intensification of El Niño–like conditions the modern mixed conifer forest developed allowing for greater biomass accumulations thus encouraging greater fire occurrence.

Acknowledgments

We thank Bill Peachey, David Parsons, David Graber, Jan Van Wagtendonk, Peter Koehler, and Nate Stephenson for assistance. Research was funded by NPS contracts CA 8000-7-0001 and CA 8013-8-0002 to R.S. Anderson. Laboratory of Paleoecology Contribution # 55.

Literature references

Anderson RS (1990) Holocene forest development and paleoclimates within the central Sierra Nevada California. *J Ecol* 78:470-489

Anderson RS, Carpenter SL (1991) Vegetation change in Yosemite Valley Yosemite National Park California during the protohistoric period. *Madrono* 38:1-14

Anderson RS, Davis RB, Miller NG, Stuckenrath R (1986) History of late- and post-glacial vegetation and disturbance around Upper South Branch Pond northern Maine. *Can J Bot* 64:1977-1986

Anderson RS, Smith SJ (1994) Paleoclimatic interpretations of meadow sediment and pollen stratigraphies from California. *Geology* 22:723-726

Armstrong CF, Stidd CK (1967) A moisture-balance profile on the Sierra Nevada. *J Hydrol* 5:258-268

Barbour MG, Berg NH, Kittel TGF, Kunz ME (1991) Snowpack and the distribution of a major vegetation ecotone in the Sierra Nevada of California. *J Biogeog* 18:141-149

Bateman PC, Wahrhaftig C (1966) Geology of the Sierra Nevada. In Bailey EH (ed) *Geology of northern California California Div Mines Geol Bull* 190:107-172

Clark JS (1988) Stratigraphic charcoal analysis on petrographic thin sections: application to fire history in northwestern Minnesota. *Quat Res* 30:81-91

Davis OK, Anderson RS, Fall P, O'Rourke MK, Thompson RS (1985) Palynological evidence for early Holocene aridity in the southern Sierra Nevada California. *Quat Res* 24:322-332

Faegri K, Iversen J (1989) *Textbook of pollen analysis*. Wiley Chichester

Harvey HT (1980) Giant sequoia reproduction survival and growth. In Harvey HT, Shellhammer HS, Stecker RE (eds) *Giant sequoia ecology US Dept Int Nat Park Serv Sci Monogr* 12:41-68

Hickman JC (ed) (1993) *The Jepson manual*. University of California Press Berkeley

Keeley JE (1986) Resilience of Mediterranean shrub communities to fires. In Dell B, Hopkins AJM, Lamont BB (eds) *Resilience in Mediterranean-type Ecosystems* pp 95-112 Dr W Junk Dordrecht Netherlands

MacDonald GM, Larsen CPS, Szeicz JM, Moser KA (1991) The reconstruction of boreal forest fire history from lake sediments: a comparison of charcoal pollen sedimentological and geochemical indices. *Quat Sci Rev* 10:53-71

Major J (1988) California climate in relation to vegetation. In Barbour MG, Major J (eds) *Terrestrial Vegetation of California* California Native Plant Society Spec Pub 9:11- 74

McGlone MS, Kershaw AP, Markgraf V (1992) El Niño/Southern Oscillation climatic variability in Australasian and South American paleoenvironmental records. In Diaz HF, Markgraf V (eds) *El Niño* pp 435-462 Cambridge University Press Cambridge

Meko DM (1992) Spectral properties of tree-ring data in the United States Southwest as related to El Niño/Southern Oscillation. In Diaz HF, Markgraf V (eds) *El Niño* pp 227- 242 Cambridge University Press Cambridge

Moratto MJ (1984) *California archaeology*. Academic Press Orlando

Null EJ (1992) *A climatology of San Francisco rainfall 1849-1991*. MA thesis San Jose State University San Jose

Overpeck JT, Rind D, Goldberg R (1990) Climate-induced changes in forest disturbance and vegetation. *Nature* 343:51-53

Patterson III WA, Edwards KJ, Maguire DJ (1987) Microscopic charcoal as a fossil indicator of fire. *Quat Sci Rev* 6:3-23

Rollins HB, Richardson III JB, Sandweiss DH (1986) The birth of El Niño: geoarchaeological evidence and implications. *Geoarch* 1:3-15

Rothermel C (1983) How to predict the spread and intensity of wildfires. *USDA For Ser Intermount Res Sta Gen Tech Rep* INT- 93

Simard AJ, Haines DA, Main WA (1985) Relations between El Niño/Southern Oscillation anomalies and wildland fire activity in the United States. *Agri For Meteor* 36:93-104

Smith SJ, Anderson RS (1995) A method for impregnating soft sediment cores for thin-section microscopy. *J Sed Res* A65:576-577

Swetnam TW (1993) Fire history and climatic change in giant sequoia groves. *Science* 262:885-889

Swetnam TW, Betancourt JL (1990) Fire-Southern Oscillation relations in the southwestern United States. *Science* 249:1017-1020

Swetnam TW, Betancourt (1992) Temporal patterns of El Niño/Southern Oscillation - wildfire teleconnections in the southwestern United States. In Diaz HF, Markgraf V (eds) *El Nino* pp 259-270 Cambridge University Press Cambridge

Waddington JCB (1969) A stratigraphic record of the pollen influx to a lake in the Big Woods of Minnesota. *Geol Soc Am Spec Pap* 123:263-283

Wells LE (1990) Holocene history of the El Nino phenomenon as recorded in flood sediments of northern coastal Peru. *Geology* 18:1134-1137

Late Quaternary Climate, Fire, and Vegetation Dynamics

Marjorie Green Winkler
IES-Center for Climatic Research
University of Wisconsin-Madison
1225 West Dayton Street
Madison, WI 53706, USA

Summary

Analysis of the abundance of microscopic charcoal in radiocarbon-dated lake and bog sediment cores from midwestern and northeastern U. S. sites provide climate-driven regional fire histories which can be compared to the regional vegetation changes since deglaciation determined from pollen. Charcoal abundances change at or immediately preceding pollen zone boundaries which delineate plant community change. Vegetation stability in the northern conifer forest, coastal pine barrens, and the midwestern oak-savanna requires high fire incidence and highest charcoal abundances are found while these plant communities dominate the landscape. Low charcoal abundances are found when closed-oak, mesophytic hardwood, floodplain, and northeastern boreal forests are inferred from regional pollen. Stable carbon isotope analysis of the sediment charcoal suggests the proportions of C3 or C4 plants on the landscape during the late-glacial and the Holocene and provides evidence for initiation of C4 prairie vegetation in the midwestern U. S. These results suggest that, as with windborne pollen, a regional microscopic windborne charcoal rain exists. The charcoal results demonstrate that fire disturbance was as important a factor in plant community dynamics in the past as it is today. Regional charcoal analysis can be used to reconstruct paleo-environments and climatic variables, and also to validate climate and vegetation models.

Background

In this study I use regional composites of analyses of lake sediment core charcoal to describe the effect of fire on regional vegetation dynamics and landscape stability since deglaciation in the coastal northeast and the midwestern United States (Fig. 1). Quantification of the microscopic charcoal found in lake sediments and further analysis of the charcoal by mass spectrometry, provides a history of regional fires to compare with regional vegetation changes from pollen and also evidence independent of pollen for changes in climate, the C3/C4 signature of the vegetation, and the landscape structure (forest or prairie).

Charcoal from both local and regional fires is preserved in lake sediments and has many points of origin. Products of fire are distributed widely in the troposphere and locally by runoff (Clark 1988a). In the troposphere, fire ejecta are transmitted in latitudinal bands by winds (Herring 1985, Garstang pers. comm.) and by high vertical velocities that can also send particles into the stratosphere (Kimball 1917,

NATO ASI Series, Vol. I 51
Sediment Records of Biomass Burning
and Global Change
Edited by James S. Clark, Hélène Cachier,
Johann G. Goldammer, and Brian Stocks

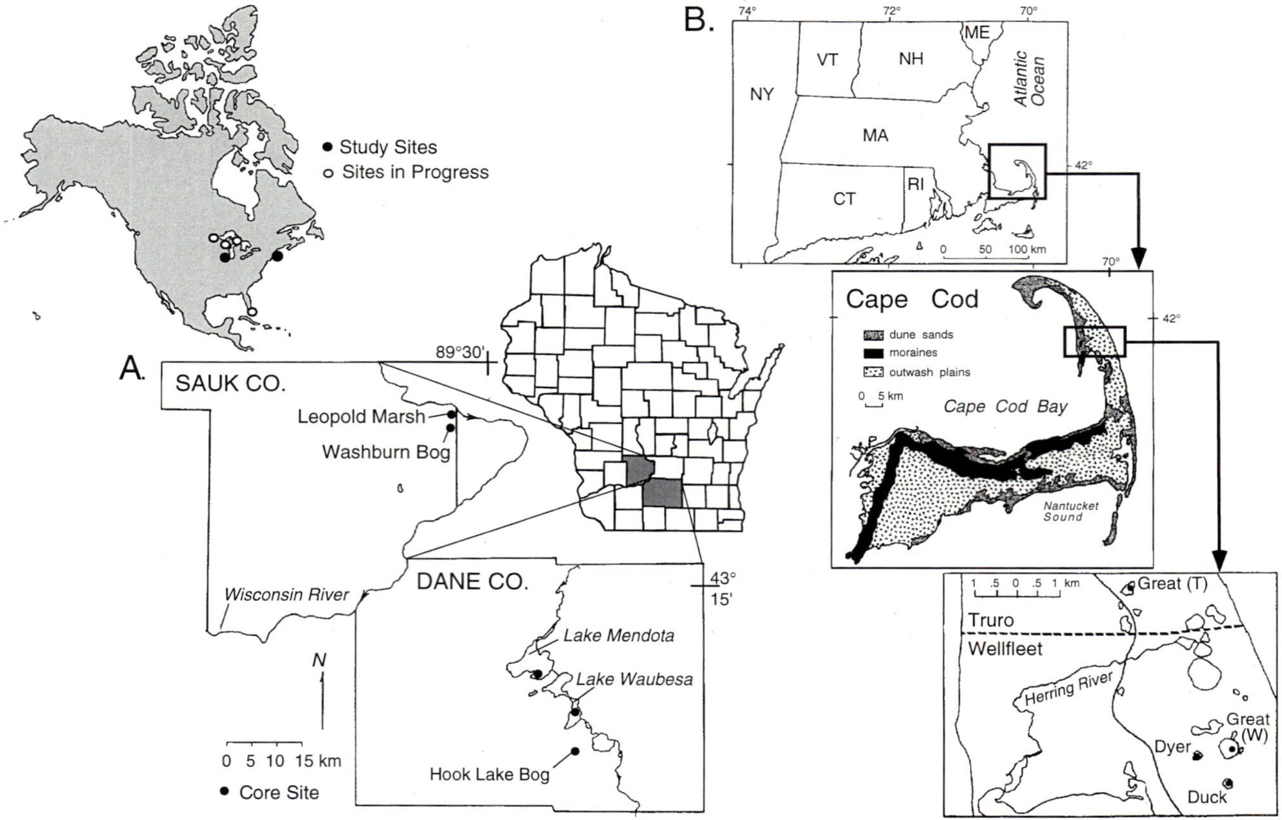

Figure 1. Map of North America (top left) with regional charcoal study sites and in progress studies indentified. Inset A locates the 5 sites in south central Wisconsin. Inset B locates the 4 coastal Massachusetts sites.

Cuniff 1955). Removal processes for charcoal particles from air masses include washout or dryfall which deposit them directly in lakes, bogs, and oceans, and onto snowfields, ice sheets, and ice caps, with similar transport processes and constraints as pollen and other aerosols. Charcoal particles can also be transported from their source by indirect means such as redeposition from the soil via erosion or runoff from the land to streams and lakes.

Evidence that a microscopic charcoal rain exists is found in several diverse studies. Optical depth measurements of stratospheric aerosols taken in Madison, Wisconsin, from July 1910 to December 1918 (Goodman 1984) show three peaks caused by forest fires burning at that time hundreds of miles to the west and northwest in central U. S. and Canada. Charcoal is a component of the particulate fraction in ice cores (Thompson et al. 1986, Holdsworth et al. 1988, Legrand et al. 1992, Cachier 1994) and ocean sediment cores (Smith et al. 1973, Herring 1985, Suman et al. 1994, Verardo et al. 1994). Analysis, therefore, of windborne microscopic charcoal found in lake and bog sediment cores (even at relatively coarse resolution) can provide a fire history for a large region just as analysis of windborne pollen provides regional vegetation history. Here, regional charcoal data for south-central Wisconsin and for coastal Massachusetts is analyzed and compared to regional vegetation changes from pollen at 500-year time intervals since deglaciation.

Method of Charcoal Data Analysis

Lake sediment charcoal was measured by chemical assay (Winkler 1985a) for five sites in south-central Wisconsin and four sites in south-eastern Massachusetts that had records dating from deglaciation. The lakes vary in size and depth and one of the midwestern lakes is now a marsh (Winkler 1992) while two other of the midwestern lakes are now bogs (since about 3000 yr B.P.) (Winkler 1988). Cores were retrieved from the deepest part of each lake. The charcoal in these cores and in the 1 cc sediment aliquot used for analysis, is essentially microscopic. All charcoal in the 1 cc subsample is analyzed chemically, so that size of the particles does not enter in the outcome of the analysis. Since the sediment is not examined separately for macroscopic charcoal, the average size of the charcoal analyzed is hard to calculate. Charcoal particles in these samples were probably less than 100 μm in size with most particles less than 20 μm. Larger aggregates of charcoal would be seen, if present, during stratigraphic examination of the sediment, or subsequent subsampling and processing of samples for other paleoecological analyses (pollen, diatoms, etc.).

Large macroscopic charcoal layers were not seen during laboratory subsampling of any of these cores.

The chemical assay of the sediment subsamples for charcoal utilizes nitric acid digestion and loss-on-ignition (at 500°C) techniques and isolates the charcoal from the whole lake sediment. The technique measures all the charcoal in a sample and provides the percent charcoal in the dry weight of the sample. The percent charcoal value has been compared for several sites to more standard microscope slide counting techniques (Winkler 1985a). In general, the two techniques demonstrated similar intervals of increased charcoal. Charcoal values per se cannot be compared between lakes because the range of values is different for large lakes and small lakes. Charcoal distribution is dependent upon basin morphometry. In large lakes the charcoal is distributed over a larger area and therefore charcoal values are generally not higher than 2-4 percent. In smaller lakes, especially those with funnel-shaped depositional areas, charcoal values can be as high as 15-20 percent. Therefore, the comparison of actual charcoal values as they vary over time in the sediment core can only be made for the individual site. Between sites, however, I use relative charcoal measurements (i.e., high, low) within the same time interval for comparison.

To arrive at a relative scale, charcoal values from each site were ordered from highest to lowest and separated into thirds representing the percent of charcoal values found to be high, medium, and low within 500-year time intervals from 14,000 years ago. For example, there are 38 charcoal values for Lake Mendota, Hole C, that range from 2.43 to 0.42 percent. By ordering the 38 values from highest to the lowest and dividing them into thirds, values 2.43 to 1.27 percent are high, 1.25 to 0.81 percent are medium, and below 0.81 percent are low. For Washburn Bog, there are 35 charcoal values ranging between 12.11 and 0.52 percent. For Washburn, percents from 12.11 to 5.36 are high, 5.08 to 2.39 are medium, and 2.16 to 0.52 are low. The time scale is derived from the radiocarbon chronology at each site. By combining the number of high, medium, and low charcoal values within each time interval from several sites in a region, a composite regional charcoal stratigraphy is constructed for comparison with regional pollen changes within the same time interval. From these analyses a pattern of regional fire history emerges which correlates with other indicators of changing regional hydrology through time in the Midwest (Winkler et al. 1986, Winkler 1988, Chumbley et al. 1990, Baker et al. 1992) and in the northeast (Winkler 1982, Webb 1990, Winkler and Sanford 1995).

Regional Charcoal Results

In this study I interpret high charcoal abundances to indicate more fires in a region during a designated time period than low charcoal abundances. Sampling interval of cores at each site determines the scale of interpretation. Due to relatively coarse sampling interval, emphasis in this study was placed on century to millennium scale changes in charcoal abundance in the lake sediments and not on individual fires. Good inter- and intra-site comparison of the charcoal and pollen data could be made because all cores were relatively well-dated and all sites within a region underwent generally synchronous major pollen changes. Each lake sediment core had between 20 and 60 charcoal analyses and from three to nine radiocarbon dates. Pollen analyses and interpretation of pollen zones and vegetation changes at each site and in each region were done independently from the charcoal analyses. Additional charcoal analyses and mass spectrometry of the charcoal from nearby sites in both areas of the United States reinforce the patterns defined in this study suggesting that fire history sensed by charcoal input to small and medium-sized lakes in these regions is representative of a larger area in each case.

Southeastern Massachusetts (Fig. 2)

Vegetation changes in southeastern Massachusetts since deglaciation were interpreted from pollen analysis of radiocarbon-dated lake sediment cores and have been published elsewhere (Winkler 1985b, Winkler and Sanford 1995). The pollen zones and vegetation formation results (Fig. 2), which here are compared with the charcoal results (Fig. 2), were determined independently of the charcoal analysis. Comparison of the regional charcoal analysis with the vegetation changes in the coastal northeast (Outer Cape Cod) demonstrates several temporal fluctuations.

1. Charcoal abundances were low before about 11,000 yr B.P. when a spruce-tundra parkland was the dominant vegetation on the Cape.

2. Jack pine increased while spruce declined during a period of increased charcoal abundances at about 10,500 yr B.P. Fire-dependent pitch pine then increased in relation to jack and white pine as the climate became warmer and drier in south-eastern New England and fire became more frequent.

3. The period from 9500 to 5000 yr B.P. had continuous high charcoal abundances and the highest fire occurrence within the Holocene in this region (Fig. 2). This is when pine and oak barrens vegetation became established on the Cape Cod coastal plain, the Cape shrubs and herbs such as

Myrica/Comptonia, Ilex, and Rosaceae increased; and Cape heathlands (Ericaceae) which had been abundant before about 10,000 yr B.P. decreased dramatically.

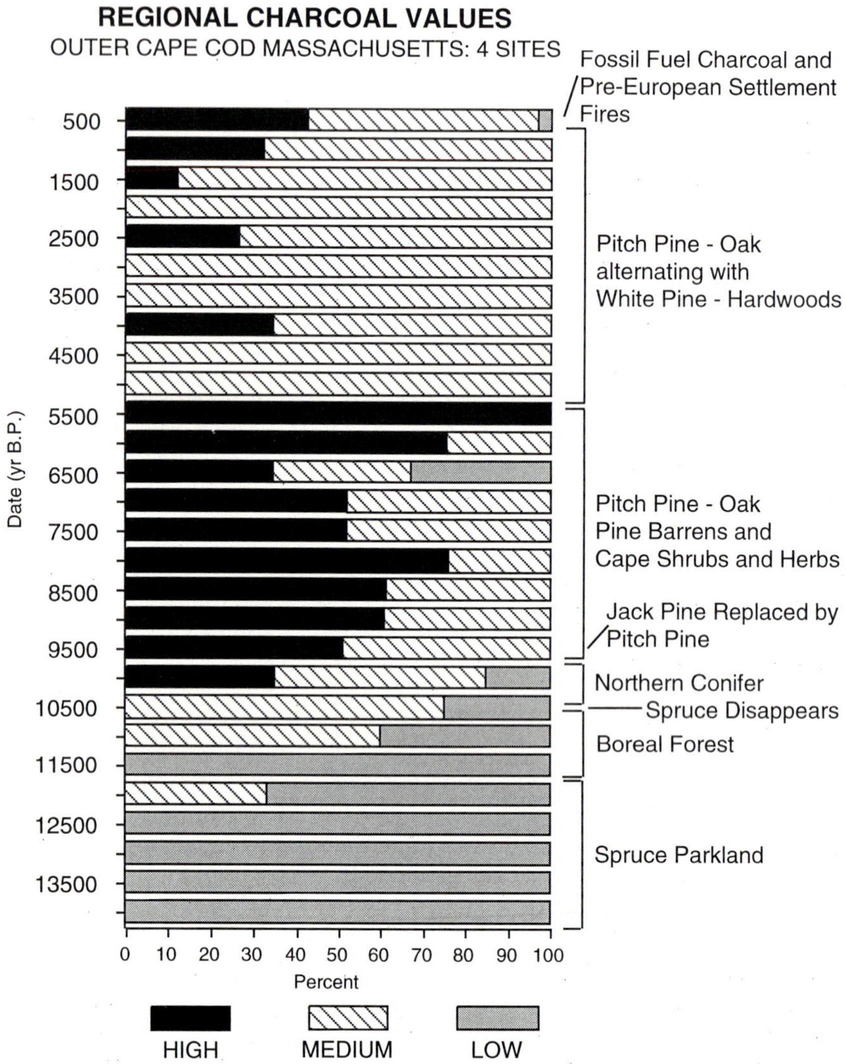

Figure 2. Regional charcoal since deglaciation, Outer Cape Cod, Massachusetts. A composite of high, medium, and low charcoal values from four sites. The timescale is derived from radiocarbon chronologies. Regional plant community changes are interpreted from pollen analyses (Winkler 1982, 1985b, Winkler and Sanford 1995).

4. Many medium charcoal abundances were evident after 5000 yr B.P. but there were no low values, suggesting that regional fires were only somewhat less frequent than before 5000 yr B.P. as the pine barrens alternated with more mesophytic forests as the dominant vegetation.

Southcentral Wisconsin (Fig. 3)

Vegetation changes in this region of the U. S. Midwest have been interpreted from pollen analysis of radiocarbon-dated lake and bog sediment cores (Winkler 1985c, 1988, 1992, Winkler et al. 1986). The results from comparison of the regional charcoal analysis (Fig. 3) with the vegetation changes in south-central Wisconsin since deglaciation (Fig. 3) demonstrate several temporal fluctuations in this region as well.

1. Charcoal values were relatively low before 12,500 yr B.P. when tundra-spruce parkland vegetation gave way to spruce-black ash swamps in south-central Wisconsin.

2. Charcoal was abundant while boreal forest was dominant from about 12,000 to 10,500 yr B.P. High charcoal values also accompanied the regional decline of spruce. Concurrent with the spruce decline, pine increased and then decreased as the climate became warmer and more moist in south-central Wisconsin and charcoal frequencies declined as well.

3. There were many high and medium charcoal values between 10,500 and 8000 yr B.P. when a northern pine-hardwood forest was dominant in the region. A charcoal peak for the region between about 9000 and 8000 yr B.P. suggests more frequent regional fires. This charcoal peak was coincident with an increase in grass pollen and a decrease in pollen from mesophytic hardwoods at this time.

4. There were lower charcoal values when a mesophytic forest dominated by elm and ironwood was prevalent from 8000 to 6500 yr B.P.

5. The period from 6500 to 3500 yr B.P. had continuous high charcoal abundances and highest fire frequency within the Holocene in this region. At that time grass and oak pollen percentages dominate indicating an oak-savanna landscape and eastward expansion of prairie in the U.S. Midwest. Stable carbon isotope analysis of the microscopic sediment charcoal from this time interval shows a shift from values of $\delta^{13}C$ less than -26 per mil, indicative of C3 plants (the signature of most midwestern trees), to values as high as -22 per mil, more representative of an oak-savanna landscape including a higher proportion of C4 prairie grasses (Winkler, 1987, 1994) (Fig. 4).

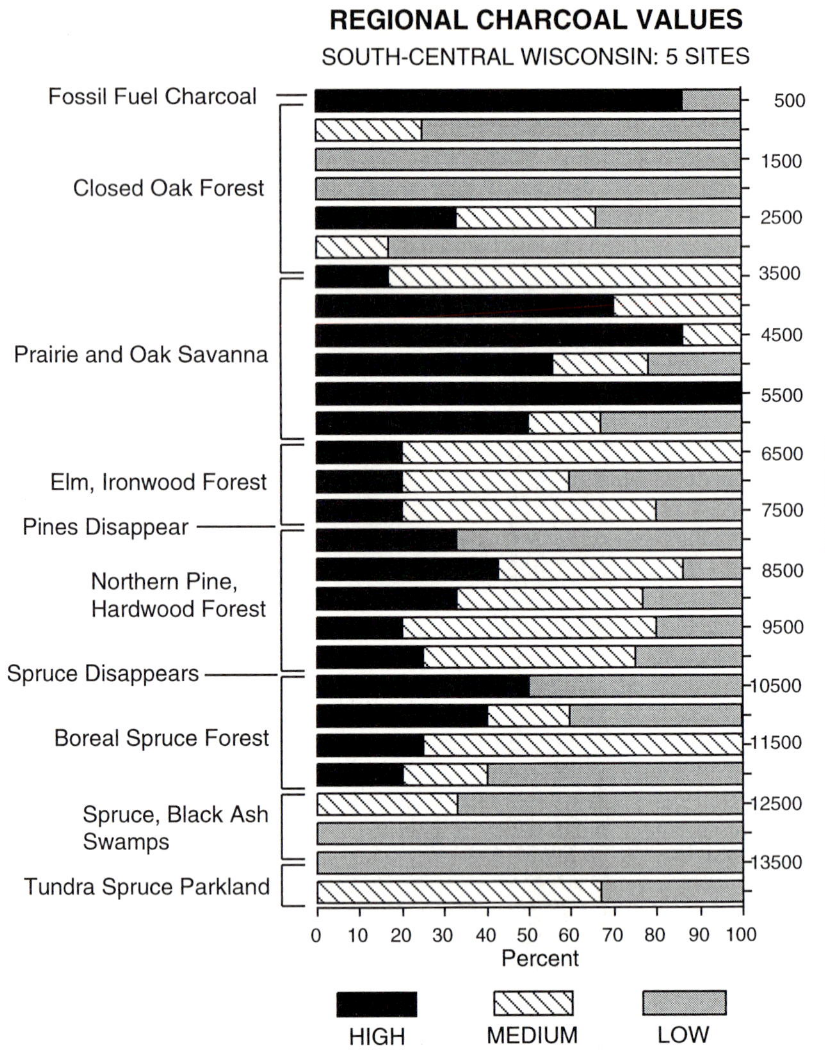

Figure 3. Regional charcoal since deglaciation, southcentral Wisconsin. A composite of high, medium, and low charcoal values from five sites. The timescale is derived from radiocarbon chronologies. Regional plant community changes are interpreted from pollen analyses (Winkler 1985c, 1988, 1992).

6. Charcoal values were low as closed oak forests dominated the region during the last 3500 years. The return to an arboreal landscape dominated by C3

plants is also indicated by the charcoal stable carbon isotope data ($\delta^{13}C$) which again falls below -26 per mil (Fig. 4).

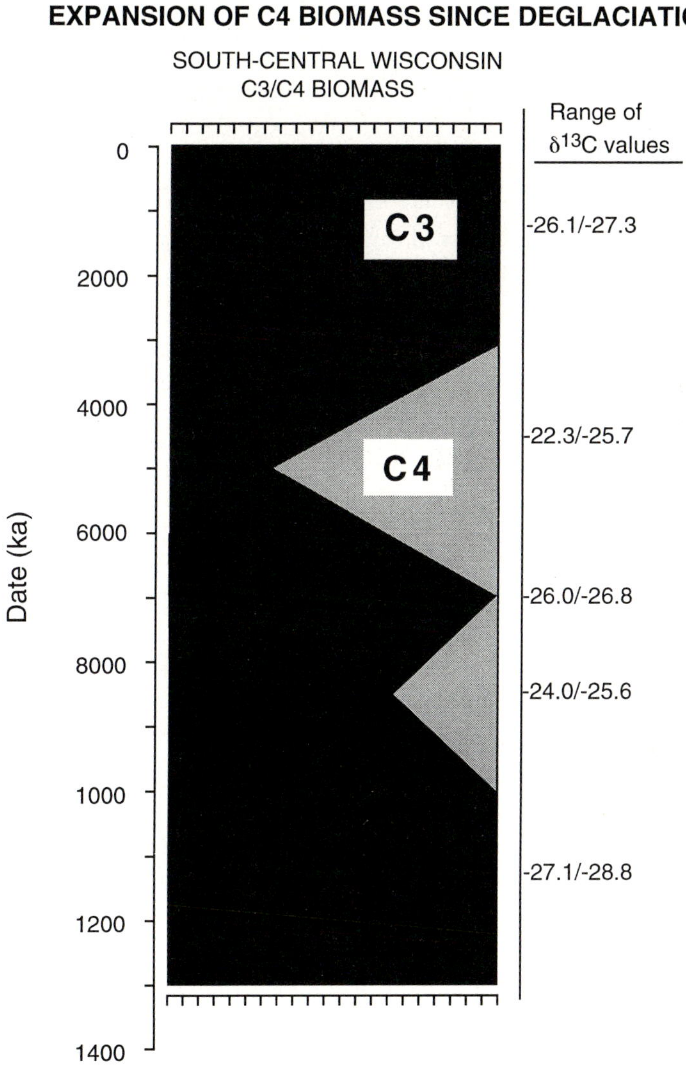

Figure 4. Representation of expansion of C4 plants on the landscape in south-central Wisconsin since deglaciation suggested by stable carbon isotope analysis of microscopic lake sediment charcoal from Lake Mendota. Range of $\delta^{13}C$ values (per mil) measured within pollen zones (see Winkler 1987, 1994).

Post-European Settlement: Both Regions

The plotting interval of 500 years used in this study does not permit interpretation of decadal or century-scale fire frequency. However, closer interval analysis of cores from both regions indicates that high modern (post-European settlement) values in both regions (Figs. 2 and 3) are masked in the most recent 500-year time interval because there were low percent charcoal values from sediments in the 500 to 200 yr. time range. (European settlement occurred 350 years ago in the coastal northeast and 150 years ago in the upper midwest). In fact, charcoal abundances had begun to rise before European settlement, suggesting fires from Native American settlements in both regions. High charcoal values in early post-settlement are the result of land-clearance and post-logging fires, and the burning of wood and coal for heating and industrial uses in the more recent post-European settlement. In the Midwest after European settlement, fire suppression policies produce a dramatic decline of charcoal in some regional lakes that is detected by charcoal analyzed at closer intervals and counted directly by microscope examination (Swain 1978, Gajewski et al. 1985, Clark 1988b).

Discussion

Timing of Regional Vegetation Change

The two regions are generally not synchronous in their fire disturbance pattern as represented by the charcoal abundances measured in this study nor are the regions synchronous in their pattern of vegetation change. However there are some similarities. In both regions rapid vegetation change occurred between 10,500 and 9000 yr B.P. At this time, mesophytic hardwood species (such as beech, birch, chestnut, hickory, maple, and elm) and the conifer, hemlock, appeared, and some of the most abundant early coniferous species disappeared (spruce, jack pine, fir). Accompanying these changes, charcoal in sediments became more abundant in the northeast and more variable in the Midwest. This period of vegetation instability in both regions coincides with a period of enhanced seasonality caused by changes in the earth's orbital parameters (Kutzbach 1981, Kutzbach and Otto-Bliesner 1982, Wright et al. 1993). The coincidence suggests that orbital changes may have influenced biotic responses. A similar pattern of rapid change in vegetation in the late-glacial and the early Holocene and relative stability after the middle Holocene has also been inferred from pollen evidence in northwest Canada (Ritchie 1986), other areas of the northeastern U. S. (Jacobson et al. 1987), in Yunnan Province of China (Walker 1986, Winkler and Wang 1993), and elsewhere.

After these somewhat synchronous rapid changes in vegetation the timing of change in the two regions diverged. The coastal vegetation became stabilized by 9500 yr B.P. while the vegetation in the Midwest continued to change dramatically until the middle Holocene. Possibly a land-sea coastal circulation developed as sea-level rose, stabilizing the climate in the northeast coastal plain by the early Holocene. In the Midwest, the retreating ice sheet continued to affect the climate until the middle Holocene.

Fire/Plant, Fire/Climate Relationships

In both regions charcoal abundances change at or immediately preceding pollen zone boundaries marking vegetation changes in the late-glacial and the early Holocene (Figs. 2 and 3). The fact that the regional pollen zones were derived independently of the charcoal stratigraphies leads to the conclusion that fire disturbance is an important factor in plant community dynamics and change. For many plant communities fire mediates the climate change/vegetation change dynamic. However, it is also possible that each taxon has evolved individualistic adaptations to a range of disturbance regimes and the dominant class of plants in a region at any one time is related to their genetic fire sensitivity or fire dependence.

Plants have a range of reactions in relationship to fire and can be classified in terms of the fire/plant relationship. Fire *sensitive* plants are susceptible to fire and are altered by fire. This class of plants is exemplified by trees such as maple, basswood, and elm which are killed by fire. Fire *tolerant* plants such as bur oaks with thick, corky bark, are able to withstand or endure fires. Fire *adapted* plants such as jack pine have adapted to fire by the development of serotinous cones which ensure reproduction of this fast-growing species after fire. Jack pine also flowers early (at 3 years) and develops viable seed as early as five years (Fowells 1965). Fire *dependent* plants rely on recurrent fire for survival. Pitch pine is fire (or other ground-clearing disturbance) dependent but also has adaptations which ensure survival after fire. These adaptations include serotinous cones, but, more importantly, dormant buds throughout trunk and limbs which enable resprouting from charred stumps soon after fire.

The regional charcoal evidence indicates that in the past, as well as today, fire is required by some plants (or plant communities dominated by fire-sensitive plants [boreal forest, northern hemlock-hardwoods]) to lower vegetation thresholds and initiate change (i.e., overcome "vegetation inertia", (Grimm 1983, Cole 1985, Davis et al. 1991) or to maintain plant community stability as in the tundra, taiga, pine or blueberry barrens, savannas, and grasslands—i.e., those ecosystems which are dominated by fire-dependent plants. High charcoal values are continuous throughout the regional early to middle Holocene dry period on the east coast (9000 to 5000 yr B.P.) when the pine barrens and the Cape herb/shrub vegetation was initiated and when lakes in the coastal region were lower than present (Winkler 1982, 1985b, Webb 1990). Charcoal is also relatively high (only high and medium values) in the sediments on the coast in the years after 5000 yr B.P. while pine barrens vegetation alternates with more mesophytic forests. Highest charcoal abundances also occur between 6500 to 3500 yr B.P. in the Midwest, indicating that this charcoal analysis technique can similarly detect the middle Holocene dry period in that region, identified from other evidence, when oak-savanna covered the landscape, lake-levels were lower than modern levels, mean annual precipitation was between 10 - 20% lower, and mean July temperatures were at least 0.5°C warmer than modern (Bartlein et al. 1984, Winkler et al. 1986, Winkler 1988, Chumbley et al. 1990, Baker et al. 1992).

Fire disturbance frequency, on a regional scale, is ultimately related to climate, with the type and amount of fuel, soils, and topography of a site having secondary regional, but crucial local, importance. High fire disturbance frequency at single sites has been statistically related to drought periods/cycles or dry years within wetter decades in the last centuries (Clark 1988c, 1989, Lorimer 1984, Habeck and Mutch 1973), to times when the relative humidity is less than 40 percent (Lorimer and Gough 1988), to ENSO (El Niño Southern Oscillation) events (Swetnam and Betancourt 1990), and also to subcontinental scale droughts (Cwynar 1977, Frissell 1973, Clark 1989).

The fossil charcoal data from this study indicate that different plant communities have different disturbance tolerances and requirements. This fact has been recognized for modern vegetation assemblages (Kozlowski and Ahlgren 1974, Heinselman 1981, Wright and Bailey 1982, Frelich and Lorimer 1991) and from charcoal and pollen for the northern conifer-hardwood forest in the past 1000 years at individual sites (Swain 1973, 1978, Cwynar 1978, Clark 1990). My study provides documentation from charcoal analysis for the importance of fire disturbance in the past for long-term changes in regional vegetation assemblages over an even longer timespan and for many vegetation assemblages.

I suggest from these charcoal data that as the climate warmed after deglaciation, a high fire frequency accelerated the decline of the boreal vegetation in the Midwest (from about 12,000 to 10,500 yr B.P.). It appears, however, that the late-glacial boreal forest (defined by pollen) in the two regions studied differed in several ways. Although fire was important in the decline of the boreal forest on the northeast coastal plain, fewer fires were evident as spruce, jack pine, and alder flourished. The boreal forest in the coastal region (from about 11,500 to 10,500 yr B.P.) probably grew in a coastal climate that was wetter than the continental climate of the boreal forest in the Midwest (even though sea level was then about 75 meters lower than today [Dyke and Prest 1987]). The boreal forest on the northeast coast may have also differed in structure and composition from that in the Midwest. Possibly jack pine was less abundant and the landscape more open with less organic fuel for large fires. Abundant pollen of the shrub *Hudsonia* provides some evidence for this interpretation (Winkler 1985b). Thus structural differences in the boreal forest plant community may have resulted in a different plant community/fire relationship in each region. In the Midwest fire appears to have mediated a climate/boreal forest relationship while in the northeast, the patchiness of the vegetation modified the fire regime in that regional boreal forest. Possibly, the period of enhanced seasonality at the late-glacial/Holocene transition also affected vegetation response to the changing climate, stimulating or limiting growth of individual boreal-associated taxa in each region and consequently modifying the plant/fire relationship.

Differences in the relationship to fire of the boreal forest plant community in the two regions examined suggest that this plant formation is controlled by more than a fire-mediated climate/vegetation link. For the North American paleo-boreal forest which extends across many regions (and probably for the larger hemispheric circumboreal forest as well) there is climatic, structural, topographic, edaphic, and genetic interactions that alters the vegetation/fire relationship within this forest formation. Circumboreal plant communities most probably exhibit large regional differences that vary with climate, plant composition, and landscape structure. Additional pollen and charcoal study may aid in identification of regional differences in the modern boreal forest as well as in the boreal forests of the past and will help determine the fire/plant relationships caused by regional placement of similar plant communities.

Fire Disturbance and Vegetation Stability

Vegetation associations identified by pollen abundances and containing fire sensitive or fire dependent dominant plants appear to have been regulated by fires in the past, as similar associations are today. For example, since 10,500 yr B.P., a high fire frequency initiated and maintained other plant associations such as northern conifers, pine savannas and oak savannas in the Midwest, and northern conifers and pine and scrub oak barrens in the northeast. Although vegetation structure and composition and topography may control some aspects of fire dynamics (i.e. by smaller, less intense, burns such as interpreted here for the northeast boreal forest compared to the Midwest boreal forest, or through vegetation feedbacks that modify the microclimate under some tree stands as described by Davis et al. 1991 for hemlock), a warm and/or dry climate drives high fire frequency, which forces vegetation associations to break down and contributes to changes in vegetation stability.

From pollen and charcoal studies of the past 1000 years at sites in conifer-hardwood forests of the Northeast and the Midwest, high vegetation stability is noted during periods of high disturbance frequency (short fire recurrence intervals) (Heinselman 1981, A M Swain, pers. comm.). However, when the dynamics of other plant communities (boreal forest, northern conifer forest, floodplain forest, mesophytic forest, oak-savanna, or pine barrens) are studied as they form and collapse in these two regions over the last approximately 14,000 years, several patterns emerge. Stability in closed boreal forests (especially in the Midwest), northern-conifer forests, oak-savannas, and pine barrens is maintained by high fire frequency as interpreted from abundant charcoal. In the oak-savanna and the pine barrens continuous fires seem to be essential for survival of the vegetation formation. A low fire disturbance frequency (as inferred from low charcoal abundances) is required for stability in closed-oak forests, floodplain forests, and mesophytic hardwood forests.

Vegetation assemblages, then, can maintain stability during both high and low fire disturbance regimes. As noted above, however, different plant communities are stable within each of these states. Resilience (persistence of the plant community on the landscape) of high fire disturbance assemblages, such as oak-savanna and pine barrens, is a function of the high fire adaptation of the dominant plants. Low fire disturbance assemblages are resilient because of a variety of different adaptations such as clone and sprout regeneration (floodplain forests), seedling growth in low light (mesophytic hardwoods), drought adaptation (tap root growth) and animal distribution of seed in small openings and at edges (closed-oak forests), and others.

Fire Disturbance and Predicted Global Climate Change

The similarity of ancient and modern plant community/fire relationships in some cases should give us more confidence to interpret internal dynamics of past vegetation. Fire disturbance history of plant communities is an important element in the definition of modern global climate change problems. Recent climate model simulations of vegetation change caused by doubled CO_2 in the atmosphere (Overpeck et al. 1990) suggest that changes in forest taxa would be more rapid with an increase in disturbance. Increased disturbance accelerated rates of vegetation change by about 40 to 220 years, depending on the taxon studied. Therefore, knowledge of the fire history of plant communities may help in the management of natural areas in the future by predicting the future regional displacement and replacement of vegetation types. Fire can also be used as a tool to stimulate or delay vegetation change.

Conclusion

Plant community change in both the coastal northeast and the Midwest during the late-glacial and the Holocene appears to have been accelerated by changes in fire frequency. High charcoal abundances precede some vegetation changes (i. e., from boreal forest to northern-conifer forest, from mesophytic forest to pine barrens or to oak-savanna), while a striking decrease in charcoal abundance precedes or accompanies other plant community changes (i. e., pine barrens to more mesophytic forest, oak-savanna to closed oak forest). Different plant communities, therefore, are stable within both high and low fire disturbance regimes.

There is a regional microscopic charcoal rain that at coarse interval analysis represents the past climate/fire/vegetation history of a region. The method of constructing regional composites from individual lake sediment charcoal stratigraphies can augment other paleo-environmental information. Charcoal analysis can be used to identify intervals of drought or high precipitation, to identify the fire sensitivity of dominant plants in past assemblages, to indicate the proportion of C3/C4 biomass on the landscape, to define the spatial extent of a temporal climatic change, and to provide a chronology of estimates of the atmospheric carbon contribution from paleo-biomass burning as well as estimates of the amount of carbon removed by depositional processes.

Fire is one agent by which climate, the primary forcing mechanism, effects spatial and temporal landscape changes. Regional climates can be reconstructed from regional charcoal through derivation of the climate-C4 relationship, the climate-fire-sensitive plant relationship, and the albedo changes as landscape is opened or closed. Temporal changes in paleoair-mass trajectories can be estimated from landscape changes. Regional climate and landscape history from regional charcoal can be used to validate climate and vegetation models while regional disturbance history can contribute to the construction of realistic dynamic vegetation models.

While this study is suggestive of a long-term (Holocene) climate/fire/vegetation equilibrium, samples were analyzed on a relatively coarse scale. Analyses of regional sites at finer temporal resolution are needed for more definitive interpretation of the relationship between climate (± landscape), fire, and vegetation and the feedbacks between these systems. Although the results of this study suggest that source areas for microscopic, windborne pollen and charcoal may be similar, more experimental data is needed about charcoal source areas. The feedbacks between an individual taxon, fire frequency, and interspecies competition (sensu Davis et al. 1991) also needs further study. Further pollen and charcoal study of the vast modern and late-glacial circumboreal forests would help determine the variables that define fire/plant community relationships for this plant association that is continental rather than regional in scale.

Acknowledgments

I would like to thank S. Barta, P. Sanford, and M. Huh for laboratory assistance and M. Kennedy, M. Woodworth, P. Behling, and J. Kruepke for graphics and other logistical support. P. Sanford, J. Clark, and two unidentified reviewers provided helpful critical review. This research was supported by grants from NARO-NPS, Water Resources Division of NBS, the NBS Global Change Research Program, and the National Science Foundation.

Literature References

Baker RG, Maher LJ, Chumbley CA, Van Zant KL (1992) Patterns of Holocene environmental change in the midwestern United States. *Quat Res* 37:379-389

Bartlein PJ, Webb III T, Fleri E (1984) Holocene climatic change in the northern Midwest: pollen-derived estimates. *Quat Res* 22:361-274

Cachier H (1994) Ice-core records of fires. In *Sediment records of biomass burning and global change* Abstract p 4 NATO Advanced Research Workshop Algarve Portugal

Chumbley CA, Baker RG, Bettis III EA (1990) Midwestern Holocene paleoenvironments revealed by floodplain deposits in northeastern Iowa. *Science* 249:272-274

Clark JS (1988a) Particle motion and the theory of charcoal analysis: source area, transport, deposition, and sampling. *Quat Res* 30:67-80

Clark JS (1988b) Stratigraphic charcoal analysis on petrographic thin sections: application to fire history in northwestern Minnesota. *Quat Res* 30:81-91

Clark JS (1988c) Effect of climate change on fire regimes in northwestern Minnesota. *Nature* 334:233-235

Clark JS (1989) Effects of long-term water balances on fire regime, north-western Minnesota. *J Ecol* 77:989-1004

Clark JS (1990) Fire and climate change during the last 750 yr in northwestern Minnesota. *Ecol Mono* 60:135-159

Cole K (1985) Past rates of change, species richness, and a model of vegetational inertia in the Grand Canyon, Arizona. *Amer Nat* 125:289-302

Cuniff CV (1955) Variations in the intensity of solar radiation at normal incidence on the surface of the earth. *Mon Wea Rev* 83:101-103.

Cwynar LC (1977) The recent fire history of Barron Township, Algonquin Park. *Can J Bot* 55:1524-1538

Cwynar LC (1978) Recent history of fire and vegetation from laminated sediment of Greenleaf Lake, Algonquin Park, Ontario. *Can J Bot* 56:10-21

Davis MB, Sugita S, Calcote RR, Frelich L (1991) Effects of invasion by *Tsuga canadensis* on a North American forest ecosystem. *Proceedings: First European Ecosystem Conference,*Firenze Italy 1991

Dyke AS, Prest VK (1987) Late Wisconsin and Holocene history of the Laurentide Ice Sheet. *Geographie Physique et Quaternaire* 41:237-263 and *Geological Survey of Canada Maps* 1702A and 1703A

Fowells HA (1965) *Silvics of Forest Trees of the United States.* Agriculture Handbook 271 USDA Forest Service Washington DC

Frelich LE, Lorimer CG (1991) Natural disturbance regimes in hemlock-hardwood forests of the Upper Great Lakes Region. *Ecol Mono* 61:145-164

Frissell SS Jr (1973) The importance of fire as a natural ecological factor in Itasca State Park, Minnesota. *Quat Res* 3:397-407

Gajewski K, Winkler MG, Swain AM (1985) Vegetation and fire history in northwestern Wisconsin: Comparing the data from three varved lakes for the past 900 years. *Rev of Palaeobot and Palynol* 44:277-292

Goodman BM (1984) *The climatic impact of volcanic activity.* PhD Thesis University of Wisconsin Madison

Grimm EC (1983) Chronology and dynamics of vegetation change in the prairie-woodland region of southern Minnesota, USA. *New Phytol* 93:311-350

Habeck JR, Mutch RW (1973). Fire-dependent forests in the Northern Rocky Mountains. *Quat Res* 3:408-424

Heinselman ML (1981) Fire and succession in the conifer forests of northern North America. In West DC, Shugart HH, Botkin DB (eds) *Forest Succession* Springer-Verlag New York

Herring JR (1985) Charcoal fluxes into sediments of the North Pacific Ocean: The Cenozoic record of burning. In Sundquist ET, Broecker WS (eds) *The Carbon Cycle and Atmospheric* CO_2*: Natural Variations, Archean to Present Geophysical Monograph* 32:419-442 American Geophysical Union Washington DC

Holdsworth G, Krouse HR, Peake E (1988) Trace-acid ion content of shallow snow and ice cores from mountain sites in western Canada. *Annals of Glaciol* 10:57-62

Jacobson GL Jr, Webb III T, Grimm EC (1987) Patterns and rates of vegetation change during the deglaciation of eastern North America. In Ruddiman WF, Wright Jr HE (eds) *North America and Adjacent Oceans During the Last Deglaciation The Geology of North America* Vol K-3 pp 277-288 Geological Society of America Boulder CO

Kimball HH (1917) The smoke cloud and the high haze of 1916. *Mon Wea Rev* 45:49-52

Kozlowski TT, Ahlgren CE (eds) (1974) *Fire and Ecosystems*. Academic Press New York

Kutzbach JE (1981) Monsoon climate of the early Holocene: Climate experiment using the Earth's orbital parameters for 9000 years ago. *Science* 214:59-61

Kutzbach JE, Otto-Bliesner BL (1982) The sensitivity of the African-Asian monsoonal climate to orbital parameter changes for 9000 years B. P. in a low-resolution general circulation model. *J Atmos Sci* 39:1177-1188

Legrand M, DeAngelis M, Staffelbach T, Neftel A, Stauffer B (1992) Large perturbations of ammonium and organic acids content in the Summit-Greenland ice core. Fingerprint from forest fires? *Geophys Res Lett* 19:473-475

Lorimer CG (1984) Development of the red maple understory in northeastern oak forests. *For Sci* 30:3-22

Lorimer CG, Gough WR (1988). Frequency of drought and severe fire weather in northeastern Wisconsin. *J Envir Mgt* 26:203-219

Overpeck JT, Rind D, Goldberg R (1990) Climate-induced changes in forest disturbance and vegetation. *Nature* 343: 51-53

Ritchie JC (1986) Climate change and vegetation response. *Vegetatio* 67:65-74

Smith DM, Griffin JJ, Goldberg ED (1973) Elemental carbon in marine sediments: A baseline for burning. *Nature* 241:268-270

Suman D, Kuhlbusch T, Verardo D (1994) Sediment records of combustion in the ocean. In *Sediment records of biomass burning and global change NATO Advanced Research Workshop* Abstract p 12 Algarve Portugal

Swain AM (1973) A history of fire and vegetation in northeastern Minnesota as recorded in lake sediments. *Quat Res* 3:383-396

Swain AM (1978) Environmental changes during the past 2000 years in north-central Wisconsin: Analysis of pollen, charcoal and seeds from varved lake sediments. *Quat Res* 10:55-68

Swetnam TW, Betancourt JL (1990) Fire-southern oscillation relations in the southwestern United States. *Science* 249:1017-1020

Thompson LG, Mosley-Thompson E, Dansgaard W, Grootes PM (1986) The Little Ice Age as recorded in the stratigraphy of the tropical Quelccaya Ice Cap. *Science* 234:361-364

Verardo DJ, Ruddiman WF, Freedland C (1994) High resolution record of charcoal deposition in late Pleistocene marine sediments of the tropical Atlantic. *EOS* (AGU Annual spring meeting Abstracts)

Walker D (1986) Late Pleistocene-early Holocene vegetational and climatic changes in Yunnan Province, southwest China. *J Biogeo* 13:477-486

Webb RS (1990) *Late Quaternary water-level fluctuations in the Northeastern United States.* PhD Thesis Brown University Providence RI

Winkler MJ (1982) *Late-glacial and Postglacial vegetation history of Cape Cod and the Paleolimnology of Duck Pond, South Wellfleet, Massachusetts.* MS thesis IES-Land Resources Univ of Wisconsin Madison

Winkler MG (1985a) Charcoal analysis for paleoenvironmental interpretation: A chemical assay. *Quat Res* 23:313-326

Winkler MG (1985b) A 12,000-year history of vegetation and climate for Cape Cod, Massachusetts. *Quat Res* 23:301-312

Winkler MG (1985c) *Late Glacial and Holocene Environmental History of South-Central Wisconsin: A Study of Upland and Wetland Ecosystems.* PhD thesis Univ of Wisconsin Madison

Winkler MG (1988) The effect of climate on the development of two *Sphagnum* bogs in south-central Wisconsin. *Ecology* 69(4):1032-1043

Winkler MG (1992) Since the retreat of the ice sheet: A Wisconsin wetland. In Janelle DG (ed) *Geographical Snapshots of North America International Geographical Congress Publication* pp 197-203 The Guilford Press New York

Winkler MG (1987) Sensing upland plant community changes from $\delta^{13}C$ analyses of charcoal in lake sediment cores. *Proceedings: International Union for Quaternary Research* XII International Congress 290 Abstract

Winkler MG (1994) Sensing plant community and climate change by charcoal-carbon isotope analysis. *Ecoscience* 1(4):340-345

Winkler MG, Sanford PR (1995) Coastal Massachusetts pond development: edaphic, climatic, and sea level impacts since deglaciation. *J Paleolimnol* 14:311-336

Winkler MG, Swain AM, Kutzbach JE (1986) Middle Holocene dry period in the Northern Midwestern United States: Lake levels and pollen stratigraphy. *Quat Res* 25(2):235-250

Winkler MG, Wang PK (1993) The late-Quaternary vegetation and climate of China. In Wright Jr HE, Kutzbach JE, Webb III T, Ruddiman WF, Street-Perrott FA, Bartlein PJ (eds) *Global Climates Since the Last Glacial Maximum* pp 221-261 University of Minnesota Press Minneapolis

Wright HA, Bailey AW (1982) *Fire Ecology*. John Wiley and Sons

Wright Jr HE, Kutzbach JE, Webb III T, Ruddiman WF, Street-Perrott FA, Bartlein PJ (eds) (1993) *Global Climates Since the Last Glacial Maximum.* University of Minnesota Press Minneapolis

Holocene records of fire from the boreal and temperate zones of Europe

R.H.W. Bradshaw, K. Tolonen[1], and M. Tolonen[2]
Southern Swedish Forest Research Centre
Swedish University of Agricultural Sciences
Box 49
S-23053 Alnarp, Sweden

Summary

Fire has long been associated with boreal forest systems, but its natural role in European temperate forest is poorly known. Fire regimes in Europe have been significantly influenced by human activities for varying lengths of time. We examine the late Holocene records of fire as recorded by charcoal fragments in lakes, peatlands, and small forest hollows in a variety of European forest types. A gradient of vegetation response to fire is apparent, ranging from fire-adapted boreal communities to fire-sensitive deciduous forests. Fire frequencies have varied in space and time in association with both climatic and vegetation changes. Human activity has often exploited pre-existing 'natural' fire regimes, having both stimulated and suppressed fire at different times. Our case studies are drawn from Finland, Sweden, Denmark, and Ireland. We explore the linkage between fire and certain key ecological processes such as paludification, species immigrations, and changes in soil microbiology.

Introduction

The study of fire regimes lies at an interface between neoecology and palaeoecology. A long time perspective is valuable for assessing the effects of fire on vegetation and for analysis of temporal and spatial variations in fire interval and response (Clark and Robinson 1993). The boreal and temperate zones of Europe stretch from northern Scandinavia to the Mediterranean and eastwards to the borders of Asia. Fire has long been associated with boreal forest systems, but its natural role in temperate forest is unclear. One feature common to the entire region is that fire regimes have been significantly influenced by human activities for varying lengths of time; one aim of this paper is to examine the time-scales and consequences of this human impact. We also examine the complex interaction of climate, vegetation type, and fire and address the question of whether fire regimes can control vegetation types over large areas for long time periods. We explore these topics through selected case studies based on our own work and drawn from the literature that combine analyses of charcoal fragments and pollen preserved in lake sediments,

[1] Department of Biology, University of Joensuu, P.O. Box 111, FIN-80101 Joensuu, Finland
[2] Department of Botany, University of Helsinki, P.O. Box 7, FIN-00014 Helsinki, Finland

NATO ASI Series, Vol. I 51
Sediment Records of Biomass Burning
and Global Change
Edited by James S. Clark, Hélène Cachier,
Johann G. Goldammer, and Brian Stocks

peats, and small forest hollows. The record from many archaeological sites is usually not temporally coherent, and such data are not included in our survey. We focus on the north temperate and boreal zones by reviewing data from Finland, Sweden, Denmark, and Ireland (Fig. 1), thus excluding studies from Germany (Clark et al. 1989) and the Mediterranean region (Rego 1992).

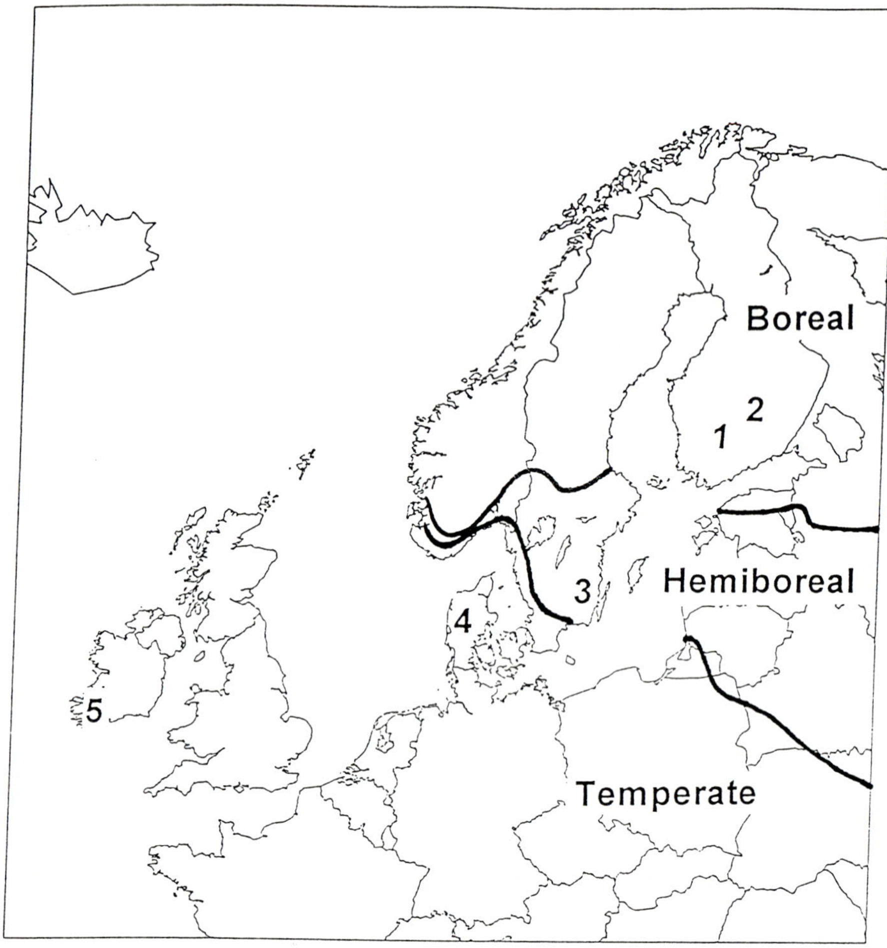

Figure 1. Map showing the sites discussed in the text related to forest zones as defined by Ahti et al. (1968). 1. Lake Ahvenainen, Finland. 2. Kuhmoinen hollow, Finland. 3. Bohult hollow, Sweden. 4. Jutland heathlands, Denmark. 5. Killarney hollow, Ireland.

Site types

Estimation of fire return intervals in the past and comparison of charcoal vs. fire-scar chronologies from trees require good temporal control and high spatial resolution. Varved lake sediments provide the finest temporal control although spatial resolution may be poor. Our first case study is from such material. Peat deposits can also provide a continuous record of charcoal particles (Mehringer et al. 1977, M. Tolonen 1985, Wein et al. 1987). Peat is superior to lake muds in certain respects, because post-depositional transport and mixing is absent; however, dating is much less accurate than it is with varved sediments. Several Finnish studies reviewed here are peat-based analyses. *In situ* charcoal deposits from small forest hollows provide a picture of local fire disturbance of vegetation at and around the study site. These sites collect charcoal at a spatial scale comparable to stand-level studies of fire-scar sequences, and thus provide a link between the broader scales of standard palaeoecological studies and the high spatial resolution of most neoecological studies (Bradshaw 1988). Two examples of such sites are presented in this review.

The range of sites, techniques, and particle sizes examined in these studies shows how the research strategy has evolved independently in many laboratories. It also shows the need for international standards based on the study of charcoal deposition at various types of sedimentary basins after experimental burns.

Case histories

1. Lake Ahvenainen, Finland 61° 02'N, 25° 07'E (K. Tolonen based on M. Tolonen 1978)

Lake Ahvenainen is located in south-east Finland and is the longest sequence of varved lake sediments analysed for pollen and charcoal in the country covering about 7500 years (Fig. 2) (M. Tolonen 1978). Data of this type provide the most reliable estimates of how regional fire frequencies have altered through time, as the sedimentary conditions have remained relatively uniform for thousands of years. Local fire history was also reconstructed for this site by recording large charcoal fragments (> 0.5 mm long). The results presented here show a close correspondence between charcoal deposition, changes in vegetation, and levels of human activity.

Sediments were not sieved, so large particles were included in the analyses. Three size classes of charcoal fragments were recorded (< 50 μm^2, 50 - 400 μm^2, and > 400 μm^2) from pollen slides, and charred fragments > 0.5 mm were counted separately (K. Tolonen 1983). Thus, data differ from the usual "pollen slide charcoal", being more comparable to those obtained from thin sections (Clark 1988, Clark et al. 1989). This only applies, of course, when the time interval between fires (charcoal peaks in sediment) exceeds the sampling intervals.

The proximity of fires was interpreted based on particle size. Occurrence of peak numbers of large charcoal fragments (> 0.5 mm long) suggests local fires. Small charcoal fragments suggest regional fires. It seems likely that the redeposition (and perhaps inwash) of pulses of coarse charred particles takes place over longer time periods than that of fine particles, because peaks of large particles often span more than a single one-cm sample (39 from a total of 43 in "local" fires, but 44 from a total of 58 in "regional" fires).

One-cm increments of sediment were sampled throughout the Lake Ahvenainen core, and, consequently, temporal resolution varied depending on the rate of sedimentation. For the periods from AD 1300 to 1976 and from 0 to AD 1300 each sample represented 8.2 $\pm$ 0.5 yrs (N = 82) and 14.3 $\pm$ 0.6 yrs (N = 137), respectively. Resolution was less good in older sediments, e.g. 25.8 $\pm$ 0.6 (0 - 500 B.C.) and 32.2 $\pm$ 0.6 (500 B.C. - 1500 B.C.). Although the inclusion of charcoal-rich and charcoal-poor layers within the one-cm samples smoothes the charcoal influx curve, the average fire interval was sufficiently large in the main part of the core to permit interpretation. This assumption that fire intervals are long is based on the occurrence of clear and often narrow peaks in the charcoal record. This method suggests fire intervals unusually long compared with contemporary fire regimes.

The lake catchment is today characterized by a dense, mesic *Picea abies* forest (mean age 85-95 years) with scattered *Pinus sylvestris* and some *Betula verrucosa*. No permanent settlements occur within the surroundings or are known from the past.

Briefly, the main stages that emerged in the interpretation of the fire history are as follows (the deepest part of the sediment was not analysed, so about the first 4000 years in the record are missing):

Period 1. Charcoal influx was low (0.02 - 0.27 $\times$ 10^6 $\mu m^2 cm^{-2} yr^{-1}$, mean 0.11 $\pm$ 0.04) during the Atlantic Chronozone (5500 - 4000 BC), when the relative pollen frequencies of Quercetum mixtum (*Quercus, Ulmus, Tilia, Corylus*) comprised up to 24% of the total arboreal pollen. No local fires were detected.

Period 2. Charcoal influx rose by a factor of 33 as *Picea abies* became common in the area (4000 - 3000 BC). It reached values almost as high as those found in the core some 2000 years later during the intensive slash-and-burn period. It seems likely that the abundant fires promoted the spread of spruce in the otherwise dense deciduous forests. Estimated average fire intervals were 358 $\pm$ 130 years (local) and 178 $\pm$ 25 years (regional). These are conservative estimates, because temporal resolution is poor. Slight traces of neolithic impact on the vegetation were found in fossil pollen by the end of the period.

Period 3. Bronze Age and Pre-Roman Iron Age 1500 BC - 0. Average charcoal influx was slightly lower than in Period 2. The inferred average fire interval was 130 - 150 yr. Pollen evidence of small-scale cultivation during the period was recorded. During the period 562 - 164 BC great fluctuations in *Picea* pollen percentages(together with other pollen changes) suggest burning of the woodland at intervals of 75 - 125 years.

Period 4. Iron Age, 0 - 1000 AD. Mean charcoal influx was similar to the previous period, but estimated fire frequencies were higher, 31 - 250 (87 $\pm$ 33) yr for local and 43 - 166 (97 $\pm$ 16) yr for regional fires. This reduction in the fire interval from the previous period is not an artifact, because the deposition time of one-cm samples was about the same in both periods (24 - 26 yr). *Picea* pollen percentages showed cycles with a periodicity of 80 - 100 years, which coincide with a periodicity of the charcoal peaks, representing the slash-and-burn rotation in the vicinity.

Period 5. Late Iron Age and the Medieval Period, 1000 - 1500 AD. Mean charcoal influx increased 1.3 fold compared with that in the previous period ($3.29 \pm 0.42 \times 10^6\ \mu m^2 cm^{-2} yr^{-1}$), and the average fire interval was reduced to 50 - 60 yr. Fluctuations in the charcoal influx values closely correspond to those in the *Picea* pollen percentage curve and also show some similarities to the fluctuations of pollen abundances of the slash-and-burn crops and weeds.

Period 6. Historical time, including the main slash-and-burn period from 1500 - 1750 A.D. Charcoal influx increased to its maximum of 1.2 - 12.0, ($4.07 \pm 0.41 \times 10^6\ \mu m^2 cm^{-2} yr^{-1}$). The close proximity of fires to the lake was suggested by the abrupt 5-fold increase in sediment deposition rate caused by increased erosion from the steeply sloping shores. Fires inferred as local recur at intervals of 22 - 80 (43 $\pm$ 10) yr, and regional fires recur at 12 - 90 (32 $\pm$ 10) yr intervals. Paleolimnological analyses revealed eutrophication as a result of increased nutrient inwash from the catchment.

Period 7. 1750 - 1976. The first 50 yr of the period suggest little fire activity (charcoal influx was only $0.06 \times 10^6\ \mu m^2 cm^{-2} yr^{-1}$). This coincides with the cessation of slash-and-burn practice because of legal restrictions (Soininen 1974). No local fires have been detected from this period. The interpretation of regional fires, and recent fire activity in general, was difficult, because spherical soot particles of industrial origin are included in the charcoal figures and overshadow the forest fire record.

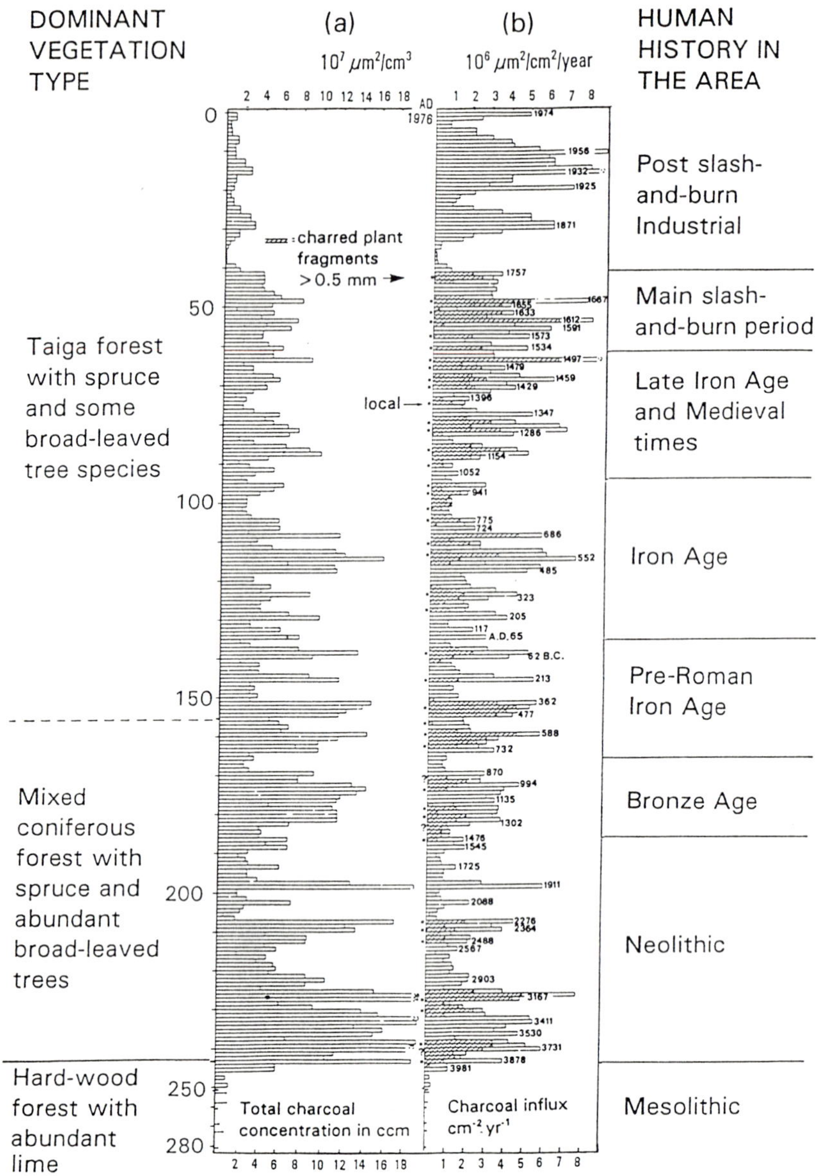

Figure 2. (a) Total charcoal concentrations, and (b) charred plant fragments >0.5 mm (hatched) and <0.5 mm (unshaded) flux to varved sediment from Lake Ahvenainen, South Finland. Inferred 'regional' fire years are indicated with dates in calendar years and 'local' fire years with squares. The dominant vegetation type and period of human history are also indicated. (Redrawn in part from M. Tolonen (1978), Fig. 9).

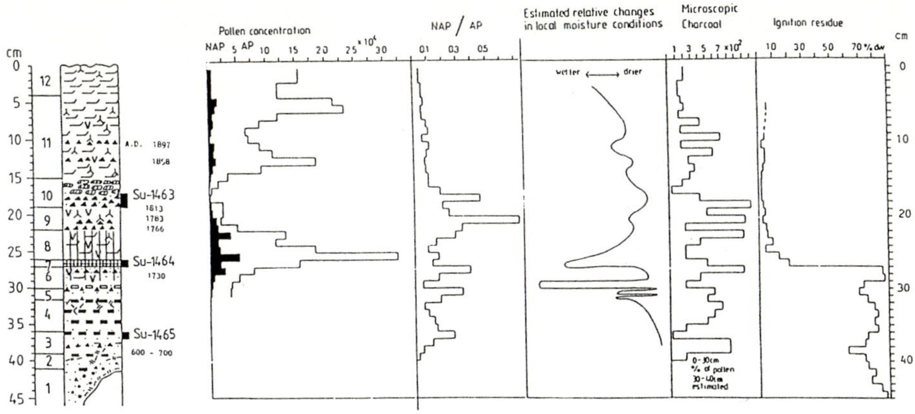

Figure 3. Soil profile from Kuhmoinen, Finland. Stratigraphic units 1-10 described in Table 1. Pollen concentrations are × 10^4 pollen grains cm^{-3}. Nonarboreal vs arboreal pollen ratio (NAP/AP) shows local changes in vegetation. Approximate changes in local moisture conditions are inferred from different bio-indicators in the stratigraphic sequence (remains of algae, mosses and insects); Microscopic charcoal is numbers of fragments in pollen slides, expressed as percentages of total pollen. In the pollen–poor sequence (below 30 cm) the charcoal abundance was estimated relative to samples above.

The lake Ahvenainen record raises the issue of the balance between changes in charcoal source strength and changes in sedimentary conditions in determining peaks and troughs in the charcoal record. The peak values of charcoal influx during period 6 could simply be an example of sedimentary changes. Attention should perhaps be focused more on the relative contrasts within the charcoal record rather than the absolute influx values. This problem is not so acute in peatland sites.

2. Kuhmoinen lake, peatland and hollow, Finland 61° 33'N 25° 12'E (M. Tolonen)

The Kuhmoinen site in Finland demonstrates how different types of sites in the same area record fire regimes with different spatial and temporal resolution. Fire scars on a representative pine stump provide a detailed chronology of the recent fire history. This chronology was correlated with the youngest peat sections within the study area that were not datable by other techniques. Charcoal particles > 10 μm encountered on pollen slides during routine pollen analysis were counted. No sieving was done during sample preparation. Results, expressed as percentages or accumulation rates of pollen, were used to infer the average fire interval and fire intensity.

Table 1. Stratigraphic units recognized from the Kuhmoinen soil profile, Finland (Fig. 3).

Unit	Depth (cm)	Description
12	0 - 4	Recent moss peat with Sphagnum roots and rhizomes of dwarf shrubs
11	4 - 15	Moderately humified light-brown moss peat with stems and roots of ligneous and Ericaceous plants. Charcoal layers at 10 and 13 cm
10	15-18	Wood layer
	18-19	Charcoal layer
9	19-22	Highly humified ligneous peat, dry and friable. Charcoal layers
8	22-26	Brown humified peat
7	26-27	Black-brown, highly decomposed humus, "pollen- mor", greasy and dense
6	27-28	Charcoal layer
	28-31	Red-brown consolidated humus with a relatively high sand content. Macroscopic charcoal visible at 30 cm
5	31	Light grey silt containing slightly leached horizon
4	31-36	Black-dark brown greasy humus containing clay and sand. Very consolidated. Roots and hyphae abundant at 31-32 cm
3	36-39	Grey-brown humic sand. Humic lenses present. Diffuse lower boundary. Charcoal layer at 38-39 cm
2	39-41	Rusty yellow-grey sand/silt with clay
1	41-45	Friable grey sand with stones

The palaeoecological study was undertaken to investigate the Holocene development of vegetation cover and land–use in an area that mostly lacks visible signs of disturbance by fire or man. The hillfort area of Kuhmoinen is located in the mixed coniferous forest region on the midwestern coast of lake Päijänne in the South Finnish Lake District (M. Tolonen 1990). The site is a small paludified hollow atop a rocky hill, with a sparse vegetation cover, so fires would have easily swept over the whole area. Dendrochronological investigations of a representative pine stump, established a local fire chronology for the past three centuries (AD 1630 to 1935). It was possible to trace and date six fires at 1730, 1766, 1783, 1813, 1858, and 1897. The 1813 fire was the most severe, destroying cambium over 30% of the circumference of the tree. The fires occurred at intervals of 36, 17, 30, 45 and 39 years, respectively, from oldest to youngest after the first fire that occurred when the tree was 100 years old. Subsequently, several fires were inferred from analysis of microscopic charcoal fragments in the peaty soil profile (Fig. 3, Table 1). The distance from the soil site to the pine stump was about 70 metres. It seems probable that the charcoal layers identified in soil are contemporaneous with the fire scars.

A 2.5 ha ombrotrophic bog lies 200 m from and 35 m below the hillfort, separated by moraine. A 3.5 m long peat profile (data not presented) spanned the last 9000 yr. When birch, alder, and other deciduous taxa dominated the site (7000 and 4500 yr BP), average fire intervals were long. After spruce immigration, 14 burned layers were recognised from 4800 to 3500 yr BP (calendar year 3000-1700 BC). From 3500 yr BP (calendar yr 2000 - 1700 BC) to 1700 yr BP (AD 200 - 300) 12 peaks in charcoal abundance and large fluctuations in *Picea* pollen indicated severe effects of fire on spruce populations. Pollen types indicating cultivation and forest clearance suggest that the high charcoal concentrations may mean frequent anthropogenic fires from AD 200 - 300 to AD 500 - 750. The fire intervals were shorter than before, but the temporal resolution did not allow identification of individual fires. Historical documents date the impact of land clearance and continuous human activity to 250 yr BP, the time of earliest historical records of the mid-1600s to 1800s. Changes in water-level caused peat growth to increase, perhaps due to climatic conditions of the Little Ice Age.

Microscopic charcoal from non-laminated sediment cores of nearby lakes are comparable to the regional pattern of average fire intervals in South Finnish lake sediments over the last 9000 years (M. Tolonen 1990). The sediment type did not allow fine resolution analysis. In the recent period of intensive cultivation and land use, in spite of fire control, charcoal influx reached maximum values due to domestic burning.

3. Bohult hollow, Sweden 57° 14'N 16° 10'E (Bradshaw)

Analysis of Bohult, a peat-filled small hollow, demonstrates how long-term changes in fire regime are expressed at the scale of single forest stands. Only charcoal fragments larger than 250 μm were analysed. Bohult hollow is located in the mixed boreal-deciduous zone of south-eastern Sweden. The site lies in a reserve that has several documented fires during the past 150 years (Granström 1991). The macroscopic charcoal analyses show distinct temporal variations in charcoal deposition (Axelsson 1995) (Fig. 4). The first evidence of fire after 7000 BP is represented by a peak at 5800 yr BP, and fires are rare and sporadic from 6000 to 4000 yr BP. The charcoal record suggests more frequent and severe fires from 4000 to 2500 yr BP and then a period with almost no evidence of fire (2500 to 1900 BP). This time period also had severely reduced charcoal deposition at a forest hollow in Fiby forest, north of Stockholm (59° 53' N, 17° 21' E)(Bradshaw and Hannon 1992), and we believe this episode reflects climatic conditions unsuitable for wildfires. The charcoal record shows more regularity from 1900 yr BP to the present, suggesting that fire was being managed in a systematic manner by humans. An alternative interpretation is a marked decrease in average fire interval after 2000 yr BP compared with the period prior to 2500 yr BP. Similar patterns has been observed in some sites in both Europe and Canada and are suggestive of a fundamental climatic change (Bergeron pers. comm.).

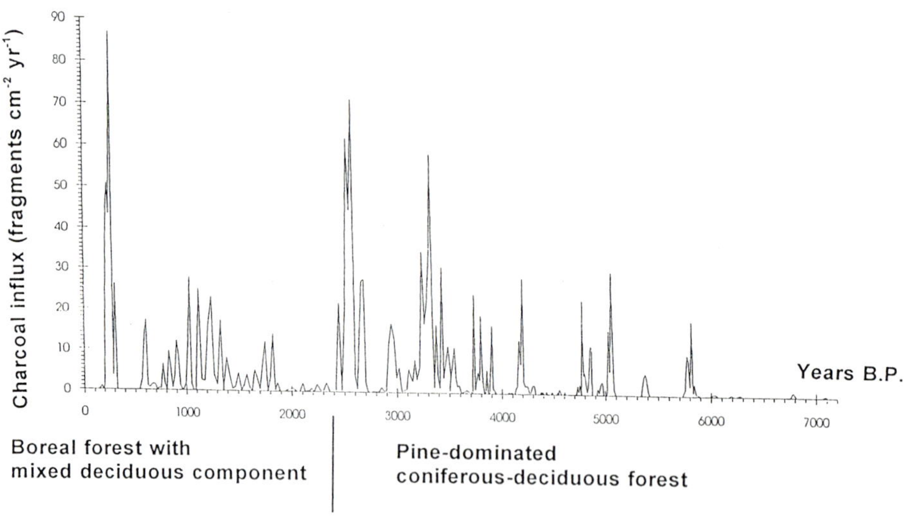

Figure 4. Macroscopic charcoal deposition rate into the forest hollow at Bohult, Sweden during the last 7000 years (after Axelsson and Bradshaw, in prep.).

Pollen analysis shows that *Pinus* was more abundant at Bohult than in areas to the south and west since 4500 yr BP, and these latter areas have far less charcoal in their sediments (Björkman and Bradshaw, in prep.). Fire has been instrumental in maintaining *Pinus* as a substantial forest component up to 1000 yr BP, and remnants of this *Pinus* distribution remain in the present landscape (Björse et al. 1995). The first signs of human influence in the pollen diagram occur about 2000 yr BP, coincident with the change in pattern of charcoal deposition. *Picea* does not become a significant feature of the local vegetation until about 1000 yr BP. In summary, the site demonstrates i) temporal variation in charcoal deposition that may prove to be under climatic control, ii) influence of fire regime on maintenance of *Pinus* populations in the landscape, and iii) human interference with fire regime. The first written records of slash-and-burn agriculture from the region date from the 1300s (Larsson 1980).

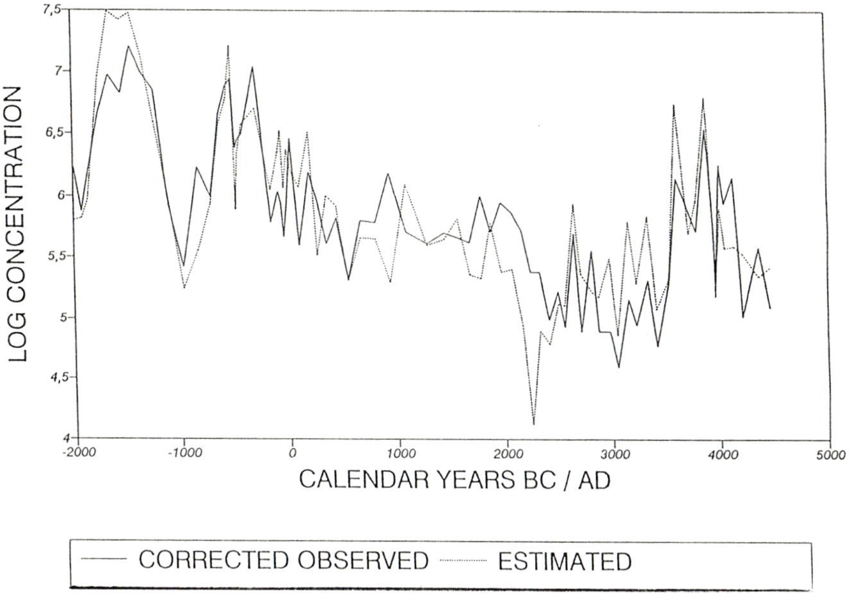

Figure 5. Corrected observed and estimated log concentration of charred particles at Skånsø, Denmark plotted against age. The corrected observed data were calculated from y = -8.62 + 1.3099 x, where y = ln(charred particle concentration) and x = ln(*Calluna* pollen concentration)(after Odgaard 1994).

4. Jutland heathlands, Denmark (Odgaard 1994)

Odgaard's (1994) palaeoecological study of western Denmark heathland shows a close relationship between heathland extent, represented by *Calluna vulgaris* pollen, and charred particles in lake sediments (Fig. 5). Data are from unlaminated lake sediments, but 23 radiocarbon dates provide a reliable chronology, with each sample representing a mean period of 9.1 ^{14}C years Small patches of heathland were recorded from 8000 to 4000 BC, and these could be attributable to natural fires. The first permanent expansion of heathland dates from 3000 BC around lake Skånsø, NW Jutland, and Odgaard assigns this to human exploitation of a pre-existing fire regime, as was done in the Bohult study. Recent relaxation of heathland management has led to forest expansion in some areas, so the long history of human fire management has had a major influence on vegetation.

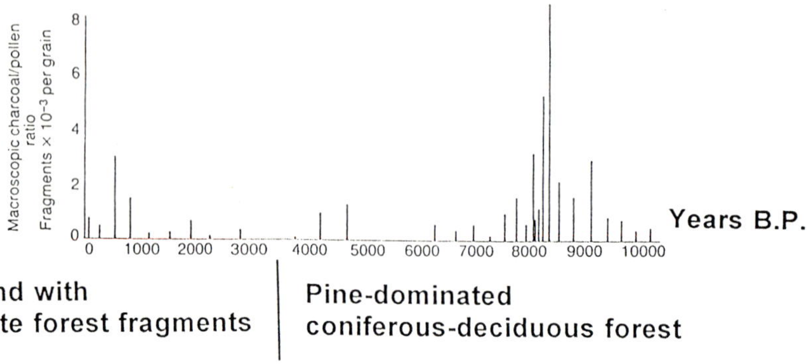

Figure 6. Macroscopic charcoal deposition in a forest hollow at Killarney, Ireland during the last 10000 years (after O'Sullivan 1991).

5. Killarney hollow, Ireland 52° 00'N 9° 35'W (Bradshaw).

The Holocene record of charcoal fragments >150 μm from a tiny woodland hollow in southwestern Ireland shows marked temporal variation (O'Sullivan 1991, Bradshaw 1993)(Fig. 6). There is a peak of charcoal deposition between 9000 and 8000 yr BP, and a second peak during the last 1000 yr. Pollen analysis showed that the forest type during the former period was composed of *Quercus, Alnus, Pinus, Betula* and *Corylus* (O'Sullivan 1991). Climatic reconstructions for this time period indicate greater seasonality than today with warm, dry summers (COHMAP 1988): conditions more suited for wildfire than the present wet, mild conditions. We have interpreted the data as indicating a mixed coniferous-broadleaved forest with an active, natural wild-fire regime. The gradual disappearance of *Pinus* from Ireland correlates with reduction in sedimentary charcoal at many western sites, and the presence of *Pinus* in otherwise broad-leaved forest is probably a result of fire (Bradshaw and Browne 1987).

The second peak in charcoal values dates from the last 1000 years and is related to human pressures and heathland expansion in the Killarney region (Mitchell 1988, Bradshaw 1991). Fires originate more readily in ericaceous heathland than within the *Quercus* woodland of Killarney today, and these fires can gradually erode the edges of the surviving woodland fragments. Charcoal from these fires has traveled to the woodland hollow which is now close to extensive heathland areas.

The dynamics of the populations of the fire-adapted *Arbutus unedo* are closely linked to this recent period of burning that largely results from human activities (Mitchell 1990, Dodson and Bradshaw 1987).

Discussion

Forest types and fire sensitivity

The combined charcoal and pollen data we present here show that a gradient in sensitivity of vegetation to fire can be recognized in Europe stretching from fire-adapted vegetation in the boreal zone to fire-sensitive temperate deciduous forest. We present a conceptual model illustrating this gradient with our case studies placed in their appropriate positions (Fig. 7). The two extremes of this gradient have been thoroughly discussed in the Scandinavian literature (Zackrisson 1977, Granström 1991, Odgaard 1994). In the disturbance-driven, fire-adapted boreal forests, large fires occur frequently, and usually give rise to predictable successions with the vegetation reverting to pre-fire conditions. In the species-poor forests of northern Europe, individual fires generate little vegetational response as sensed by regional pollen diagrams, although local pollen diagrams can capture aspects of the post-fire succession (Bradshaw 1993). By contrast, fire may have a devastating effect on fire-sensitive temperate broadleaved forest, although the scale and intensities of burning are quite different from the boreal zone (Bergeron pers.comm.). Non-anthropogenic fire is a rare event in European temperate forests today, but was probably more common under different climatic conditions in the past. Human exploitation of fire in these regions has resulted in significant local shifts in vegetation type, for example the creation of heathlands. Regular burning maintains heathlands, but their initial creation almost certainly involved a combination of burning and browsing. A close correspondence has been demonstrated between *Calluna* pollen and charcoal fragments in western Denmark, and such a relationship doubtless has also existed in western Ireland and Norway. The heathlands of western Europe are examples of how human activity has intensified the formerly natural fire regime.

The role of fire in the intermediate hemiboreal zone (mixed deciduous-coniferous forest) (Ahti et al. 1968) is of considerable interest (Fig. 7). Fires periodically occur in this forest zone and can have a decisive impact on forest succession at stand level. We have examples of such sites in southern Sweden (Björkman and Bradshaw 1995, Lindbladh and Bradshaw 1995), where the occasional fire can have considerable influence on forest dynamics, for example catalyzing the immigration of a new tree species. This type of fire regime seems to be associated with broad-leaved forests with a limited presence of *Pinus* – a forest type that has been virtually eradicated from southern Sweden during the last 1000 years (Björse et al. 1995), but existed in western Ireland 8000 years ago. Green (1982) studied a comparable but more species–rich mixed conifer-hardwood forest in Nova Scotia, where fire played a similar role. Forest succession in the hemiboreal zone may be particularly sensitive to changes in average fire interval: an increase in interval permitting the establishment of disturbance-sensitive late-successional species such as *Fagus sylvatica* (Bergeron pers. comm.).

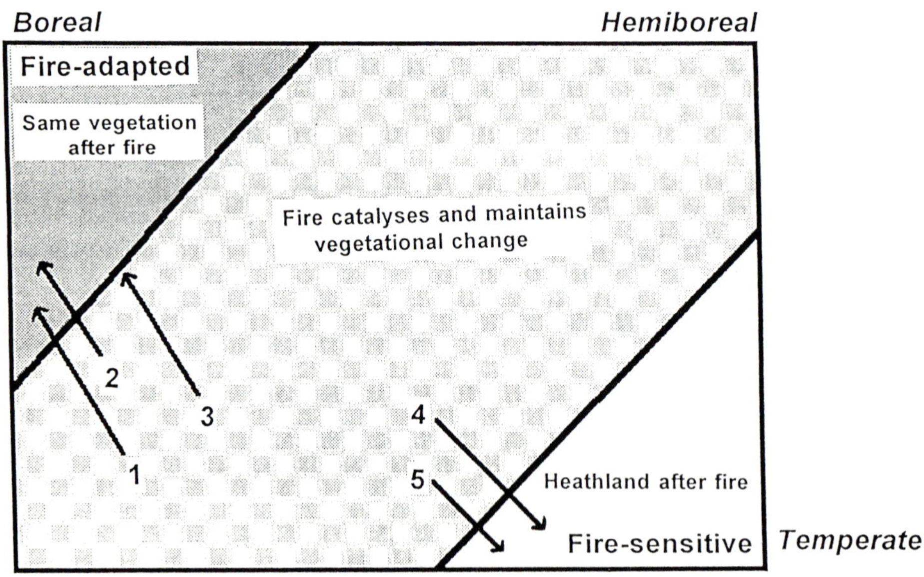

Figure 7. Conceptual model showing the long-term effects of fire in boreal, fire-adapted vegetation (heavy shading), hemiboreal or mixed coniferous-deciduous forest (light shading) and temperate, fire-sensitive vegetation. The approximate trajectories of the five study sites during the investigated time-periods are marked by arrows. Sites are numbered as in Fig. 1.

Our data also document changes in average fire intervals and possibly intensity of burning in northern Europe during the Holocene and suggest possible driving forces controlling this dynamism. Climatic change, vegetational change, and human activities are the three major, inter-connected forces for change. Charcoal influx data from Finland show that forest fires were about two to three times more frequent or intensive between 9000 and 7000 BP than between 7000 and 4500 yr BP.

The same trend is apparent in south-western Ireland (Fig. 6). The increase in average fire interval seen at Bohult 2500 yr ago is an example of another change of continental-scale significance that resulted from climatic change. These broad patterns are consistent with climatic reconstructions from general circulation models, for example, the increased continentality of north-west European climates during the early Holocene (COHMAP 1988). It is clear that climate exercises the ultimate control over fire regimes.

Striking correlations between fluctuations in forest types and charcoal influx records demonstrate the close relationship between fire and forest type, but cannot usually distinguish between cause and effect. The anomalously high abundances of *Pinus* pollen in SE Sweden during the last 2000 years are matched by high charcoal influx values (Fig. 4). *Picea abies* in Sweden is usually disadvantaged by fire to the benefit of *Pinus sylvestris* (Zackrisson, 1977, Bradshaw 1993, Almquist-Jacobson 1994). However the spread of *Picea* into the dense deciduous forests of south-eastern Finland after 4500 BP was apparently assisted by fire. Our working hypothesis is that the fire regime can control forest type at a regional scale, rather than vice versa.

There is little doubt regarding the influence of human activity on fire regimes in northwestern Europe. Both sediment and fire-scar studies from Finland show a halving of average fire interval in recent times. This decreased fire interval occurred during the Iron Age in southern Finland and in late medieval times in the north. Average fire intervals of 122 $\pm$ 26 yr estimated from fire scars in northern Sweden (Zackrisson 1977), were reduced during the period of intensive slash-and-burn agriculture to 12 to 22 yr according to fire-scar and sedimentary charcoal studies (Haapanen and Siitonen 1978, Huttunen pers. comm.). Fire intervals were shorter still in dry forests dominated by *Pinus* (Vakurov 1975). Improved fire prevention was practiced in Scandinavia from the latter part of the 19th century onwards and led to marked changes in fire frequencies. The average fire interval lengthened considerably [in an extreme case from 52 years (in AD 1676 to 1700) to 7025 years (in 1926-1950)], or more generally from about 100 years (1551 to 1875) to about 3500 years (1901 to 1975) in remote northern Sweden (Zackrisson 1977). This important result clearly implies that the fire statistics from the post slash-and-burn period do not correspond to the situation during earlier periods in the Holocene, either as a result of fire control or as a consequence of changing climate after the Little Ice Age (Engelmark et al. 1994).

Average fire interval is only one way of characterizing the fire regime. It is hard to reconstruct intensities from fossil data, but some data exist on the areas burned. Forest fire statistics from Finland and Sweden covering the period AD 1865 to 1922 have shown that individual burns were usually small. Most were less than 5 ha, and the mean size was 25 ha (Saari 1923). Comparable figures for Norway were even smaller with a mean of 7 ha. The mean burn size for Finland during the 1800's was 131 ha, and the proportion of state-owned forest land that burned annually ranged between 0.0013 and 0.38, with more land always burned in the south of the country.

Fire and ecological processes

Fire potentially exerts a major influence over several key ecological processes. Fires physically disrupt dominating vegetation, release nutrients and generate charcoal that has specific physical and chemical properties. Palaeoecological studies can provide important evidence for the long-term effects of fire upon ecosystems. Paludification is often associated with forest fires on many mineral soil sites in northern Europe (Lukkala 1933, Huikari 1956). Others have found that blanket peat initiation was a consequence of the destruction of forest cover by fire, partly through loss of transpiring trees, and partly through reduction in soil porosity by fine charcoal (Moore 1975, Jakobi et al. 1976). Recent studies in northern Sweden have shown the fertilizing effects of fires in species-poor systems with deep layers of acid humus (Bradshaw and Zackrisson 1990) and confirmed the role of charcoal in counteracting the allelopathic effects of *Empetrum hermaphroditum* on tree seed germination (Bradshaw et al. ms.). Laboratory studies demonstrated the inhibitory effect of *Empetrum* exudates on seed germination of *Pinus* and other species from northern Sweden, an effect that was reversed by addition of activated charcoal (Nilsson et al. 1993). Pollen and charcoal analyses suggested that this phenomenon also occurred under field conditions, and, furthermore, charcoal had a stimulatory effect on soil microbial activity (David Wardle pers. comm.) and diversity of ground vegetation. A combination of these physical and chemical effects probably explains the frequently observed correspondence between charcoal and immigration of tree species into new areas (Tallantire 1977, Green 1982, Björkman and Bradshaw 1995).

An important new result has emerged in the study of accumulation rates in long peat sequences from Finland (Tolonen et al. 1992) (Fig. 8). A considerable proportion (up to 50%) of the organic matter was lost in repeated (up to 31) fires several thousand years ago in peatlands, including those at Patvinsuo (Fig. 8). The loss of material can be estimated by comparing individual peat sequences with a large number of other peat sequences with basal dates. At some sites the average annual loss of carbon from fire exceeded 2 g C $m^{-2}yr^{-1}$. This type of data can help quantify a potentially significant and variable part of the global carbon cycle.

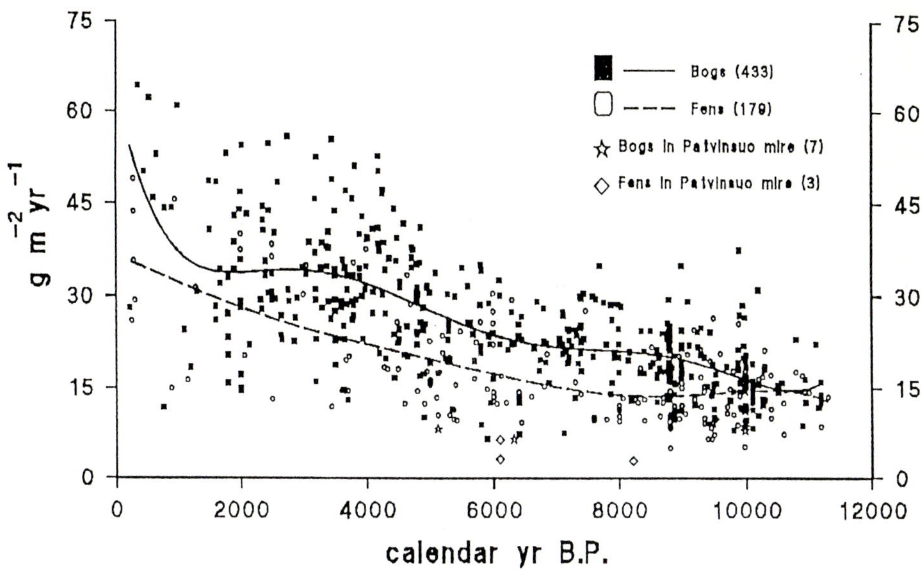

Figure 8. Long-term rates of (apparent) carbon accumulation (g C $m^{-2}yr^{-1}$) against age (calibrated yr BP) of basal peat in dated peat columns in Finland (bulk density measured in about 40 % of cases and estimated in the rest). The two lines are splined regressions for six segments of the age axis. Peat cores with basal dates of 5000 to 10000 yr BP showed anomalously low accumulation rates for bogs and fens from Patvinsuo, eastern central Finland, when compared with similar cores from 622 sites in other parts of Finland. Numerous (up to 35) macroscopic charcoal horizons in the Patvinsuo cores suggest that organic matter losses due to fires were responsible for these low accumulation rates. Original figure from the studies of Tolonen et al. (1994).

The general findings from our review are that humans have exploited and amplified existing fire regimes in Europe. The time-scale is dependent on historical factors but was far earlier in western Denmark than in central Finland. Natural fire regimes have varied in space and time, probably in response to climate-forcing. Average fire interval has a controlling influence on vegetation type and can generate fire-adapted vegetation that succeeds to other forest types when the fire regime is altered. We are only beginning to understand the influence of fire on important ecosystem processes and its role in the global carbon cycle.

Acknowledgments

We would like to thank Anna Axelsson and Aileen O'Sullivan for help with data collection and Yves Bergeron for commenting on the manuscript.

Literature References

Ahti T, Hämet-Ahti L, Jalas J (1968) Vegetation zones and their sections in northwestern Europe. *Ann Bot Fenn* 5:169-211

Almquist-Jacobson H (1994) *Interaction of Holocene climate, water balance, vegetation, fire, and cultural land-use in the Swedish Borderland.* PhD Thesis University of Lund Sweden

Axelsson A (1995) *Brandhistorien i sydsverige under senare holocen* (Fire history of southern Sweden during the late Holocene [English summary]) Undergraduate thesis University of Lund

Björkman L, Bradshaw RHW (1995) The immigration of *Fagus sylvatica* L. and *Picea abies* (L.) Karst. into a natural forest stand in southern Sweden during the last two thousand years. *J Biogeogr* (in press)

Björse G, Bradshaw RHW, Michelson DB (1995) Calibration of regional pollen data to construct maps of former forest types in southern Sweden. *J Palaeolimnol* (in press)

Bradshaw RHW (1988) Spatially-precise studies of forest dynamics. In Huntley B, Webb T (eds) *Vegetation History* pp 725-751 Kluwer Academic Publishers

Bradshaw RHW (1991) Spatial scale in the pollen record. In Harris DR, Thomas KD (eds) *Modelling Ecological Change* pp 41-52 Institute of Archaeology London

Bradshaw RHW (1993) Forest response to Holocene climatic change - equilibrium or non-equilibrium? In Chambers FM (ed) *Climate Change and Human Impact on the Landscape* pp 57-65 Chapman and Hall London

Bradshaw RHW Zackrisson O (1990) A two thousand year history of a northern Swedish boreal forest stand. *J Veg Sci* 1:519-528

Bradshaw RHW and Browne P (1987) Changing patterns in the post-glacial distribution of *Pinus sylvestris* in Ireland. *J Biogeogr* 14:237-248

Bradshaw RHW and Hannon GE (1992) Climatic change, human influence and disturbance regime in the control of vegetation dynamics within Fiby Forest, Sweden. *J Ecol* 80:625-632

Clark JS (1988) Stratigraphic charcoal analysis on petrographic thin sections: application to fire history in northwestern Minnesota. *Quat Res* 30:81-91

Clark JS, Merkt J, Muller H (1989) Post-glacial fire, vegetation, and human history on the northern alpine forelands, south-western Germany. *J Ecol* 77:897-925

Clark JS, Robinson J (1993) Palaeoecology of fire. In Crutzen PJ, Goldammer JG (eds) *Fire in the Environment* pp 193-214 John Wiley and Sons

COHMAP (1988) Climatic changes of the last 18,000 years: observations and model simulations. *Science* 241:1043-1052

Dodson JR, Bradshaw RHW (1987) A history of vegetation and fire, 6600 B.P. to present, County Sligo, western Ireland. *Boreas* 16:113-123

Engelmark O, Kullman L , Bergeron Y (1994) Fire and age structure of Scots pine and Norway spruce in northern Sweden during the past 700 years. *New Phytol* 126:163-168

Granström A (1991) Elden och dess följeväxter i södra Sverige. *Skog Forskning* 4:22-27

Green DG (1982) Fire and stability in the postglacial forests of southwest Nova Scotia *J Biogeogr* 9:29-40

Haapanen A, Siitonen P (1978) Forest fires in Ulvinsalo strict nature reserve. *Silva Fenn* 12:187-199

Huikari O (1956) Studies on the role of primary mire formation in the origin of Finnish peatlands. *Comm Inst Forest Fenn* 46(6):1-79

Jakobi RM, Tallis JH, Mellars PA (1976) The southern Pennine Mesolithic and the ecological record. *J Archaeolog Sci* 1976(3):307-320

Larsson LJ (1980) Svedjebruket i SmÜland. *Kronobergsboken 1979-80* 1-13

Lindbladh M, Bradshaw RHW (1995) The development and demise of a mediaeval forest-meadow system at Linnaeus' birthplace in southern Sweden: implications for conservation and forest history. *J Veg Hist Archaeobot* 4:153-160

Lukkala OJ (1933) Does paludification of forest soils take place at the present time? *Commun Inst Forest Fenn* 19:1-127

Mehringer PJ, Arno SF, Petersen KL (1977) Postglacial history of Lost Trail Pass Bog, Bitteroot Mountains, Montana. *Arctic Alpine Res* 9:345-368

Mitchell FJG (1988) The vegetational history of the Killarney oakwoods, SW Ireland: evidence from fine spatial resolution pollen analysis. *J Ecol* 76:415-436

Mitchell FJG (1990) The impact of grazing and human disturbance on the dynamics of woodland in S.W. Ireland. *J Veg Sci* 1:245-254

Moore PD (1975) Origin of blanket mires. *Nature* 256:267-269

Nilsson M-C, Högberg P, Zackrisson O, Fengyou W (1993) Allelopathic effects by Empetrum hermaphroditum on development and nitrogen uptake by roots and mycorrhizae of *Pinus silvestris*. *Can J Botany* 71:620-628

Odgaard BV (1994) The Holocene vegetation history of northern West Jutland, Denmark. *Opera Botan* 123:1-171

O'Sullivan A (1991) *Historical and contemporary effects of fire on the native woodland vegetation of Killarney, S.W. Ireland.* PhD thesis Trinity College Dublin

Rego FC (1992) Land use changes and wildfires. In Teller A, Mathy P, Jeffers JNR (eds) *Responses of Forest Ecosystems to Environmental Changes* pp 367-373 Elsevier

Saari E (1923) Forest fires in Finland. Acta *Forest Fenn* 26(5):1-155

Soininen AM (1974) Old traditional agriculture in Finland in the 18th and 19th centuries. *J Agric Soc Finland* 46(Suppl):1-459

Tallantire PA (1977) A further contribution to the problem of the spread of spruce (*Picea abies* (L.) Karst.) in Fennoscandia. *J Biogeogr* 4:219-227

Tolonen K (1983) The post-glacial fire record. In Wein RW, MacClean DA (eds) *The Role of Fire in Northern Circumpolar Ecosystems* SCOPE 18:21-44 John Wiley and Sons

Tolonen M (1978) Palaeoecology of annually laminated sediments in Lake Ahvenainen, S. Finland. I. Pollen and charcoal analyses and their relation to human impact. *Ann Botan Fenn* 15:177-208

Tolonen M (1985) Palaeoecological record of local fire history from a peat deposit in SW Finland. *Ann Botan Fenn* 22:15-29

Tolonen M (1990) Pollen analytical evidence of ancient human action in the hillfort area of Kuhmoinen, central Finland. In Taavitsainen JP (ed) *Ancient hillforts of Finland* pp 247-264 SMYA

Tolonen K (1987) Bog profile evidence of fire and vegetation dynamics since 3000 years BP in the Acadian Forest. *Can J Botany* 65:1180-1186

Tolonen K, Vasander H, Damman AWH, Clymo RS (1992) Rate of apparent and true carbon accumulation in Boreal peatlands. *Proc 9th Internat Peat Congr* Uppsala Sweden 22-26 June 1992 Vol 1:319-333

Tolonen K, Turunen J, Vasander H, Junger H (1994) Rate of carbon accumulation in boreal mires. In Kanninen M, Heikinheimo P (eds) *The Finnish Research Programme on Climate Change. Second Progress Report* 1 94:297-302 Publ Acad Finland

Vakurov AD (1975) *Forest Fires in the North.* Ixdatjelstvo Nauka Laboratorija Lesovedenija Moscow (in Russian)

Wein RW, Burzynski M P, Sreenivasa BA, Tolonen K (1987) Bog profile evidence of fire and vegetation dynamics since 3000 years BP in the Acadian Forest. *Can J Botany* 65:1180-1186

Zackrisson O (1977) Influence of forest fires on the North Swedish boreal forest. *Oikos* 29:22-32

Controls on Charcoal Distribution in Lake Sediments: Case Studies from Yellowstone National Park and Northwestern Minnesota

Cathy Whitlock, J. P. Bradbury[1], Sarah H. Millspaugh[2]
Department of Geography
University of Oregon
Eugene, Oregon 97403, USA

Summary

The identification of past fires from sedimentary charcoal requires information on the processes that transport and deposit charcoal into lakes. Two case studies are presented that describe the taphonomy of lake-sediment charcoal. In Yellowstone National Park, charcoal accumulation was measured in eight lakes during the five years following large fires in 1988. Lakes in burned and unburned watersheds received charcoal as airborne fall-out during the fire, and secondary charcoal was transported from the littoral zone in subsequent years. In the fifth year, sites within burned watersheds accrued more charcoal in deep water than those within unburned watersheds. The lag before clear registration of a fire signal suggests that charcoal peaks in the stratigraphic record represent several years of sediment accumulation. At Elk Lake, Minnesota, sediment-trap studies and a transect of surface samples revealed that charcoal deposition in deep water occurred during times of strong spring circulation, when charcoal and other allochthonous material were moved from the littoral to the profundal zone. The transport mechanisms within the lake were not related to the fires that produced the charcoal. A record of sedimentary charcoal from Elk Lake showed that intervals with abundant charcoal generally coincided with decades of regional drought and burning. Both case studies demonstrate the importance of understanding charcoal depositional processes as an important step in fire history investigations.

Introduction

Analysis of charcoal particles in lake sediments has become a widely used tool in studies that examine the role of fire in shaping vegetation and the effect of climate change on disturbance regimes. Two approaches are usually taken in charcoal studies. General variations in the amount of sedimentary charcoal through time are used to infer broad changes in fire frequency and magnitude. For example, the abundance of charcoal particles in lake sediments dating to the early Holocene is often used as evidence that fires were more frequent then than today (Cwynar 1987, Mehringer et al. 1977). A second approach is to assign sedimentary layers with high charcoal concentrations, i.e., charcoal peaks, to fires near the study site. Clark (1990) identified individual fires in small catchments in northwestern Minnesota based on

[1] U.S. Geological Survey, Box 25046, Federal Center, Denver, Colorado 80225, USA
[2] Department of Geography, University of Oregon, Eugene, Oregon 97403, USA

NATO ASI Series, Vol. I 51
Sediment Records of Biomass Burning
and Global Change
Edited by James S. Clark, Hélène Cachier,
Johann G. Goldammer, and Brian Stocks

peaks of macroscopic charcoal in annually laminated (varved) sediment cores. The two approaches provide complementary information: one describes the status of regionwide burning, the other discloses changes in local fire occurrence.

The chronology for a fire history comes from varve counting or radiometric age determinations of the sediments. Although charcoal records lack the specificity of dendrochronologic data in identifying the age or location of particular fires, their strength lies in potentially extending a fire chronology back thousands of years beyond the age of living trees. In addition, charcoal data can be compared with other paleoecologic records, such as pollen, sedimentological, geochemical, and diatom records collected from the same sediment cores, as well as dendrochronologic data from trees within the watershed (Clark 1990, Green et al. 1988, Millspaugh and Whitlock 1995). These records taken together provide information on past vegetation, disturbance regimes, limnology, and perhaps on the sensitivity of ecosystems to past climate changes.

Implicit in fire-history studies are assumptions concerning the transport, deposition, and burial of charcoal in lake sediments following a fire. It is widely assumed that sedimentary charcoal is introduced to lakes primarily as airborne fall-out during the fire event. Secondary or redeposited material is considered a relatively minor contribution to the charcoal record. This assumption is based on theoretical models that describe charcoal transport in terms of the aerodynamic properties of various sized particles (Clark 1988, Patterson et al. 1987) and, to some extent, from empirical studies of the atmospheric transport of particles during modern fires (e.g., Radtke et al. 1990). Support has also come from the close match between known fire events and charcoal peaks in dated sediment cores collected in various regions, including northwestern Minnesota (Clark 1990), Maine (Patterson and Backman 1988), and Yellowstone National Park (Millspaugh and Whitlock 1995).

Other studies have challenged this assumption by noting the oftentimes poor match between known fires and charcoal peaks at particular sites (Cwynar 1978, Green et al. 1988, MacDonald et al. 1991) and the lack of comparison among charcoal records at adjacent sites (Patterson and Backman 1988). A further complication is that charcoal peaks often span several centimeters of lake core that comprise years of sediment accumulation. Broad peaks have been variously interpreted as evidence of multiple fires, accumulation of secondary charcoal following a fire, or bioturbation of charcoal following burial (Patterson et al. 1987). Thus, the presence of low amounts of charcoal even in nonfire years, the sometimes inconsistent correlations between charcoal peaks and times of known fires, and the broad stratigraphic range of many charcoal peaks represent problems in interpreting charcoal data.

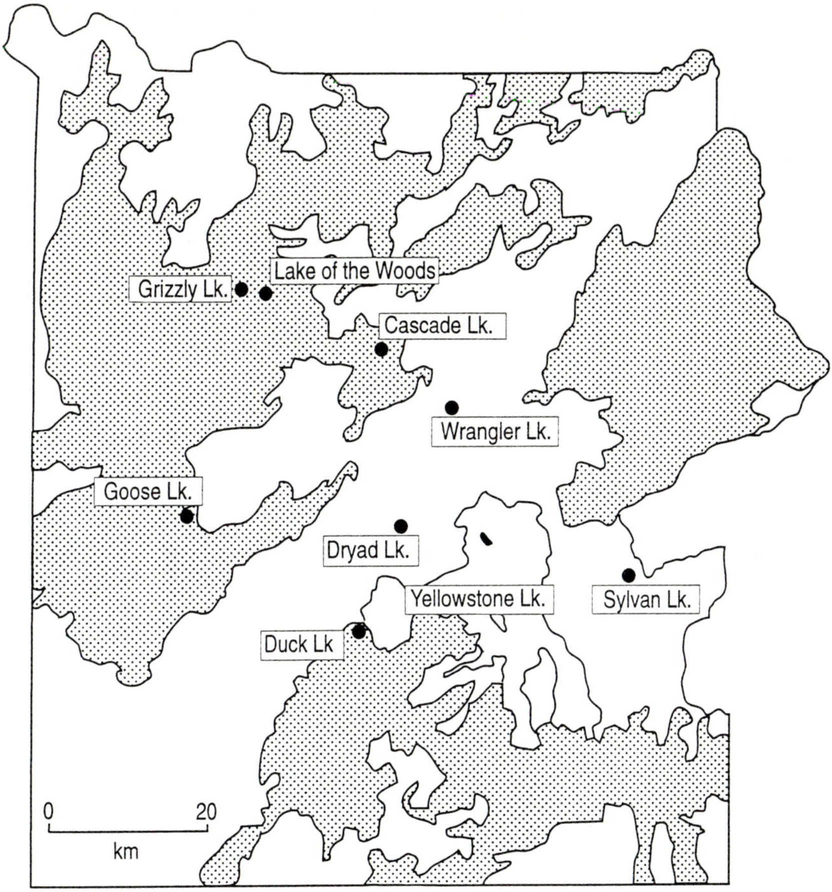

Figure 1. Location of sites for modern charcoal study in relation to 1988 fires (shaded) in Yellowstone National Park.

In this paper, we describe two studies that examine the taphonomy of lake-sediment charcoal. One study examines the record preserved in eight small lakes in Yellowstone National Park (YNP) in northwestern Wyoming following the fires of 1988 (Whitlock and Millspaugh 1996). Surface sediments were collected from these lakes between 1989 and 1993 to provide information on the introduction and accumulation of charcoal in the years after fire. The other study focuses on a 1535-yr charcoal stratigraphy from Elk Lake in Itasca State Park, northwestern Minnesota (Bradbury 1996). The charcoal record at Elk Lake is compared with stratigraphic diatom and phytolith data from the same core to infer mechanisms of charcoal transport. Interpretation of the fossil record at Elk Lake is based on the abundance of charcoal and phytoliths in surface sediments and sediment traps and on recent limnology. Both the Yellowstone and Elk Lake studies provide insights into the relationships between fire, site characteristics, and sedimentary charcoal that form the foundation for most fire history-studies.

Charcoal Deposition In Yellowstone National Park

In the summer of 1988, 44.5 percent (cs. 400,380 ha) of YNP burned in natural and human-caused wildfires. The size and intensity of the fires were attributed to a combination of conditions, including severe winter drought, above-average summer temperatures, strong winds, and the large build-up of fuels in the form of downfallen timber and mature forest stands (Schullery 1989, Balling et al. 1992). The fires created a mosaic in which 56 percent of the area was burned by high-intensity forest canopy fires, 37 percent of the area experienced surface or ground forest fire, and 7 percent was burned meadow and sagebrush grasslands (GYPFRAC-BAST 1988).

Table 1. Yellowstone study sites (from Whitlock and Millspaugh 1996).

Site	Lat., Long., Elev (m)	Surface area (ha)	Maximum depth (m)	Catchment size (ha)	Vegetation type[a]	% catchment burned[b]
Duck Lake	44° 25'N, 110° 32'W, 2357	9.2	19	59	lp, s-f-p	60
Grizzly Lake	44° 26'N, 110° 46'W, 2287	46.5	11	1571	lp, s-f-p	90
Goose Lake	44° 32'N, 110° 50'W 2189	12.1	9	416	lp	95
Cascade Lake	44° 45'N, 110° 31'W, 2432	12.2	8.5	218	lp, s-f-p	84
Lake of the Woods	44° 47'N, 110° 43'W, 2347	10.5	7.5	1286	lp, #,s-f-p	45
Wrangler Lake	44° 41'N, 110° 26'W, 2384	11.4	10	103	lp	0
Dryad Lake	44° 33'N, 110° 31'W, 2531	13.9	8	1243	lp	0
Sylvan Lake	44° 28'N, 110° 10'W, 2564	8.2	7.5	114	lp, s-f-p	0

[a]Vegetation type: s-f-p = spruce-fir-pine forest, lp = lodgepole pine forest.
[b]Data for the 1988 fires from GYPFRAC-BAST (1988).

The study lakes, located in central YNP (Fig. 1), were chosen specifically because they were simple systems with small surface area, conical bathymetry, small- to moderate-sized catchments, and fed by minor or intermittent streams (Table 1). Another criterion of selection was that the catchments had not experienced fires in the last century; this reduced the possibility of 1988 charcoal mixing with material from older fires. The catchments of five lakes (Cascade Lake, Duck Lake, Goose Lake, Lake of the Woods, and Grizzly Lake) were extensively burned in 1988. Of these, the catchments of Grizzly and Goose lakes had the most area burned, and that of Duck Lake was the least affected. Three lakes lay outside the burned area: Dryad and Wrangler lakes were located 7 km downwind of major fire storms, and Sylvan Lake was located 3 km south and upwind of the nearest fire.

Three of the study lakes, Dryad, Goose, and Wrangler, were surrounded by mixed-age stands of *Pinus contorta* in 1988. The catchments of Duck Lake, Grizzly Lake, Cascade Lake, Lake of the Woods, and Sylvan Lake supported *Pinus contorta* forest with *Picea engelmannii, Abies lasiocarpa*, and *Pinus albicaulis*. Cascade Lake was bordered on the northeast by an extensive wetland meadow of grasses, sedges, and herbs. Wrangler Lake and Lake of the Woods had a large mat of *Nuphar* and *Scirpus* in shallow water. The other sites had little or no riparian margin, and forest extended to the lake shore.

The methodology used in this study is described in Whitlock and Millspaugh (1996). Between 1989 and 1993 short sediment cores (6-cm long) were taken from the lakes by use of a gravity sampler that recovered the sediment surface intact. Cores were collected from 1989 to 1992 along shallow- to deep-water transects that paralleled the local wind direction in the catchments during the 1988 fires. In 1993 cores were collected from deep-water sediments only. Cores were extruded from the top of the coring tube at intervals of 2 cm, and the segments were stored in plastic bags and refrigerated. In the lab, subsamples of 5- or 10-ml volume were taken from each segment and soaked in a dilute solution of sodium hexametaphosphate (Calgon) for 3 days. The sediment was wet-sieved through a nested series of metal screens with mesh sizes of 0.500, 0.250, 0.125, and 0.063 mm, and charcoal particles in the different size classes were counted under a stereomicroscope at 375× magnification. Charcoal counts were divided by the subsample volume to obtain the charcoal concentration (number of particles ml^{-1}) of the core segment. Concentration were multiplied by the core-segment length to determine charcoal accumulation (number of particles cm^{-2}), and the accumulations of each core segment was summed to give the total number of particles cm^{-2} (hereafter referred to as charcoal abundance). This measurement is not an accumulation rate or flux (particles cm^{-2} yr^{-1}), which would require precise information on deposition rates or the age of each core segment. Rather, it represents the total number of particles deposited on a given surface area at a particular sampling location over the time interval represented by the core segments. It is assumed that all the charcoal in the uppermost sediments is from the 1988 fires.

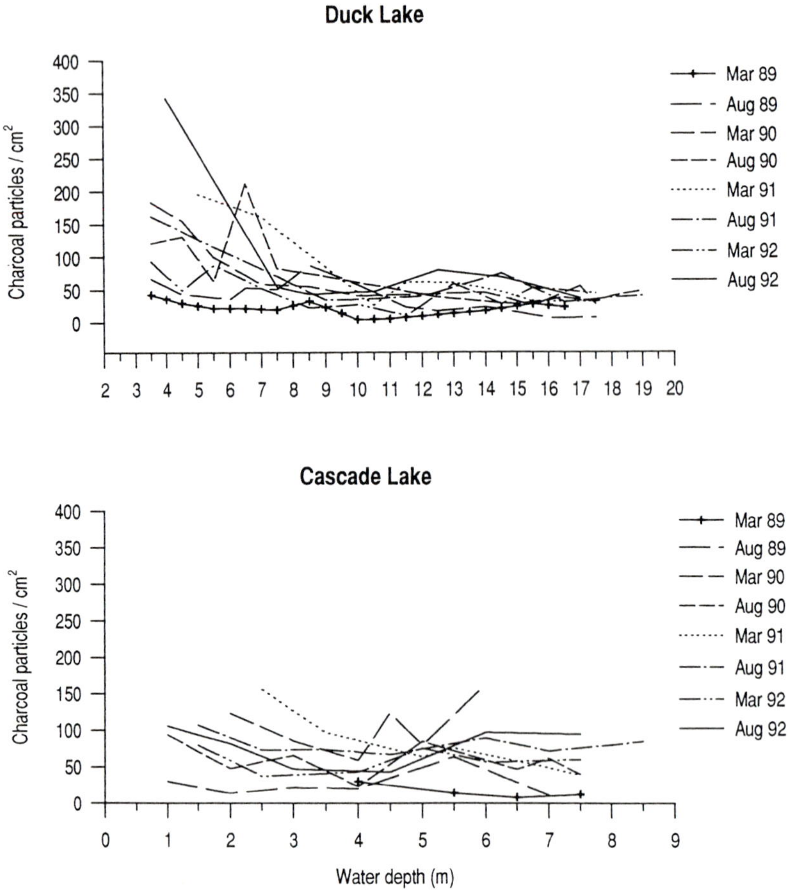

Figure 2. Charcoal abundance in the sediment taken from sampling stations from 1989 to 1992. Duck Lake is a deep lake (>10 m water depth) with a burned catchment; Cascade Lake is a shallow lake (<10 m water depth) with a burned catchment.

The selection of the core length was governed by our desire to analyze enough of the core to include all the charcoal from the 1988 fire that might be buried in the sediments and at the same time to minimize inclusion of pre-1988 charcoal from deeper sediments. Fire-history data from YNP indicated that none of the watersheds burned in the 100-year period prior to 1988 (Romme and Despain unpublished data,

Millspaugh and Whitlock 1995). It was therefore assumed that charcoal in sediments younger than 100 years old came from the 1988 fires. Lead-210 dating of the uppermost sediments in small lakes in YNP indicated sediment deposition times of 5-13 yr cm^{-1} (Engstrom et al. 1991, Millspaugh and Whitlock 1995), suggesting that the upper 8-20 cm spanned the last 100 years. An ash layer between 4-5 cm depth at Grizzly Lake was ascribed to the 1980 eruption of Mt. St. Helens, implying a sediment deposition time of ca. 3 yr cm^{-1}. Test cores collected from the sites in 1989 contained high charcoal values in the upper 2 cm of sediment, low or zero values below that depth. Cores collected from all sites in 1990 and 1993 had high amounts of charcoal in sediments from 0-4 cm sediment depth and low amounts below 4 cm depth. Based on this information, 0-2 cm sediment depth was analyzed for the first year (1989) and 0-4 cm depth was examined for the next four years (1990 to 1993).

The transects revealed considerable movement and addition of charcoal in the sediments between 1989 and 1992. Duck Lake exemplified the spatial and temporal patterns of charcoal accumulation observed in deep lakes (>10 m water depth) with burned catchments (Fig. 2). Along a particular transect, charcoal abundance was highest in shallow-water sediments and decreased with water depth. In successive samplings charcoal generally increased at a given location, and the increase was more gradual in deep-water sediments (Whitlock and Millspaugh 1996).

Cascade Lake, in contrast, was typical of shallow lakes (<10 m water depth) with burned catchments. Charcoal abundance varied irregularly from shallow-water to deep-water sediments in a given sampling and from year to year. The data indicated no trend in space or time that might suggest steady transport of charcoal to deep water.

Because paleoecological studies usually examine cores collected from the center of a lake, the patterns of charcoal accumulation in deep-water sediments from 1989 to 1993 were of particular interest (Fig. 3). Charcoal values generally increased in deep-water sediments at all sites following the first sampling in 1989. The rate of increase was slower in deep lakes than in shallow ones and slowest in unburned sites, and the trends were not unidirectional. Initially, the charcoal values in the deep-water sediments of burned sites were comparable to those of unburned sites, but by 1993 burned sites had charcoal abundances that were higher (Fig. 3). Thus, five years of accrual was required before burned sites could be distinguished from sites that had not burned.

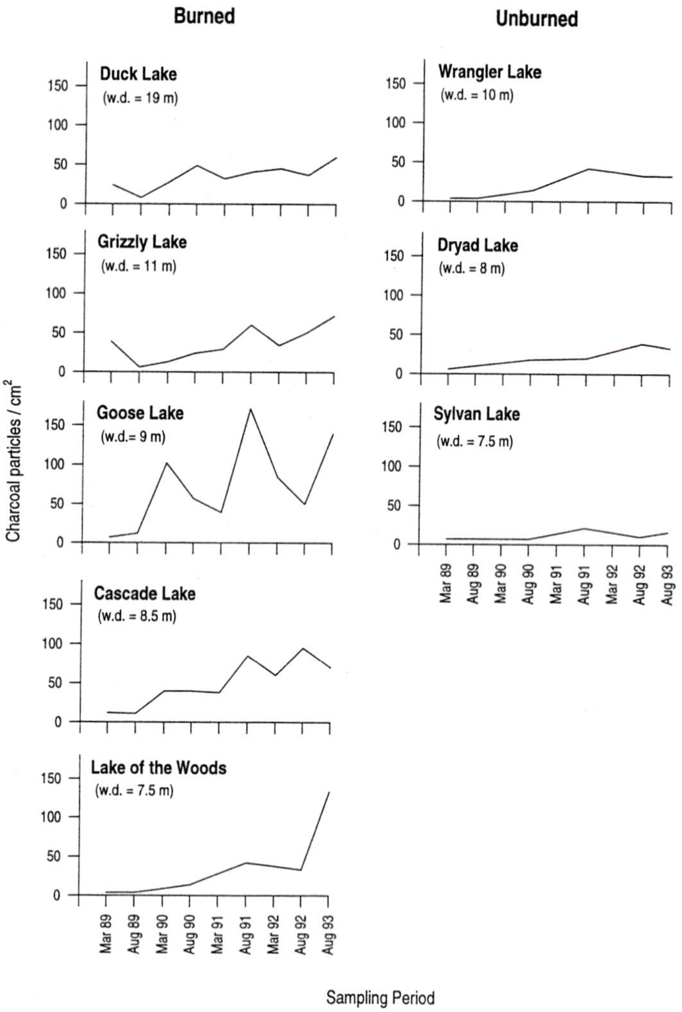

Figure 3. Comparison of charcoal abundance in deep-water sediments from 1989-1993. Lakes are separated between those with burned catchments and those with unburned catchments. Sites are arranged by water depth.

Charcoal, phytoliths, and diatom data from Elk Lake

Elk Lake is part of a complex drainage system that forms the headwaters of the Mississippi River in Itasca State Park (Fig. 4). The lake is approximately 100 ha and has a maximum water depth of 30 m. Before Euro-American forest clearance, in the 20th century, sediment at Elk Lake accumulated as varves of diatom-rich organic matter and calcium carbonate (Anderson 1993). The chronology for the past 1535 years was based on varves deposited before 1903 and an overlying nonlaminated "disturbed" zone that extended to the surface when the core was collected in 1984 (Dean et al. 1994).

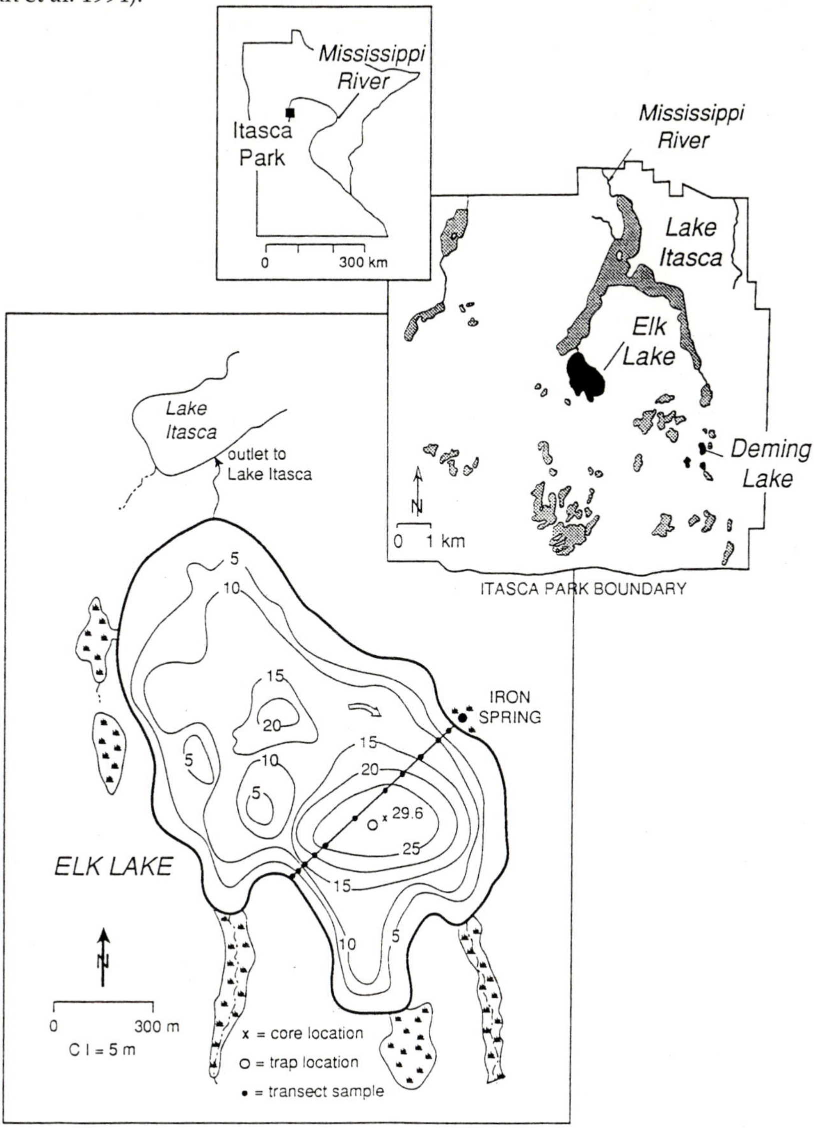

Figure 4. Location of Itasca State Park, Elk Lake, and Deming Lake studied by Clark (1990). Inset shows bathymetric map and location of deep-water core, sediment trap, and the transect of surface-sediment samples discussed in text.

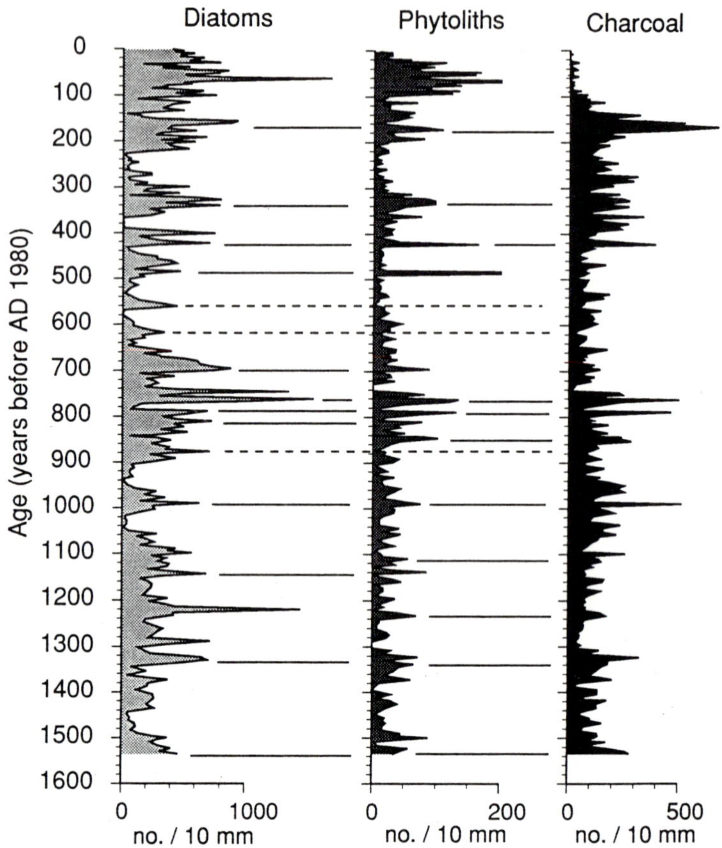

Figure 5. Stratigraphic profiles of diatoms, phytoliths, and charcoal from the 1535-year record of varved sediments at Elk Lake.

Diatom, phytolith, and charcoal analyses were undertaken for contiguous samples spanning approximately five varve years. Phytoliths and charcoal were also analyzed in a transect of surface (upper one cm) sediment samples that probably represent deposition in Elk Lake for the past decade. In addition, analyses were made of sediment collected in an automatic trap located in the deepest part of the lake from July 1983 to July 1984 (Fig. 4; Anderson 1977, Bradbury 1988, Nuhfer et al. 1993). Samples were processed by digestion in concentrated HNO_3 at 100°C, and diatoms, phytoliths, and charcoal particles were tallied from equal volumes of processed material. The resulting, internally comparable concentrations of particles are presented as the number of particles encountered along a 36-mm traverse across a microscope coverslip at 1000X magnification. Typically, charcoal particles enumerated in these samples ranged from 10 to 30 mm in length, and, in many cases, they retained visible cuticular structures.

In the 1535-year record, peaks in the abundance of diatoms, phytoliths, and charcoal were generally correlated (Fig. 5, Bradbury 1996). The diatom peaks indicated periods of lake circulation and nutrient cycling, whereas the phytoliths, which are components of plant tissue, and charcoal were evidence of allochthonous sediment transport. The frequent correspondence between peaks of diatoms, phytoliths, and charcoal suggests that factors that increased lake circulation, nutrient cycling, and thereby diatom production in the past were often associated with sediment transport and phytolith and charcoal deposition.

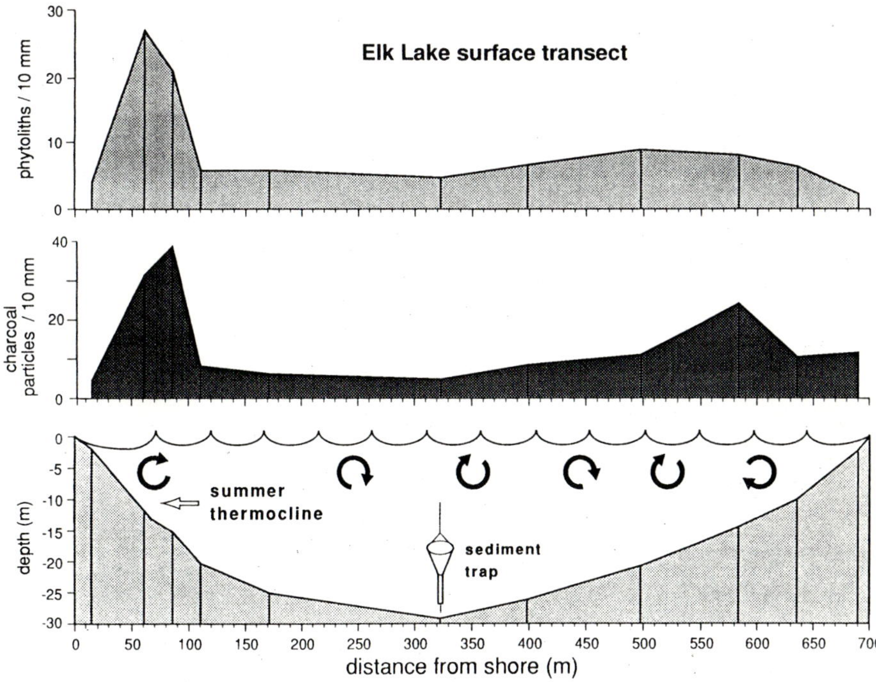

Figure 6. Numbers of phytoliths and microscopic charcoal particles per 10-mm traverse across a microscope coverslip from surface-sediment samples along the Elk Lake transect (Fig. 4). Bottom section shows depth and spatial location of transect samples, the average depth of the summer thermocline, and the location of the sediment trap.

The transect of surface-sediment samples revealed abundant charcoal and phytoliths at water depths between 10 and 15 m, especially on the southwest side of the lake (Fig. 6). This depth range was at or just below the mean depth of the summer thermocline. Apparently, charcoal particles and phytolith-bearing plant fragments were concentrated by a clockwise surface nearshore current gyre generated by the west-northwest winds that prevail in spring. Charcoal particles and plant fragments were subsequently deposited from the quiet waters of the upper hypolimnion as the lake stratified, and over time they accumulated in the sub-littoral zone of the lake.

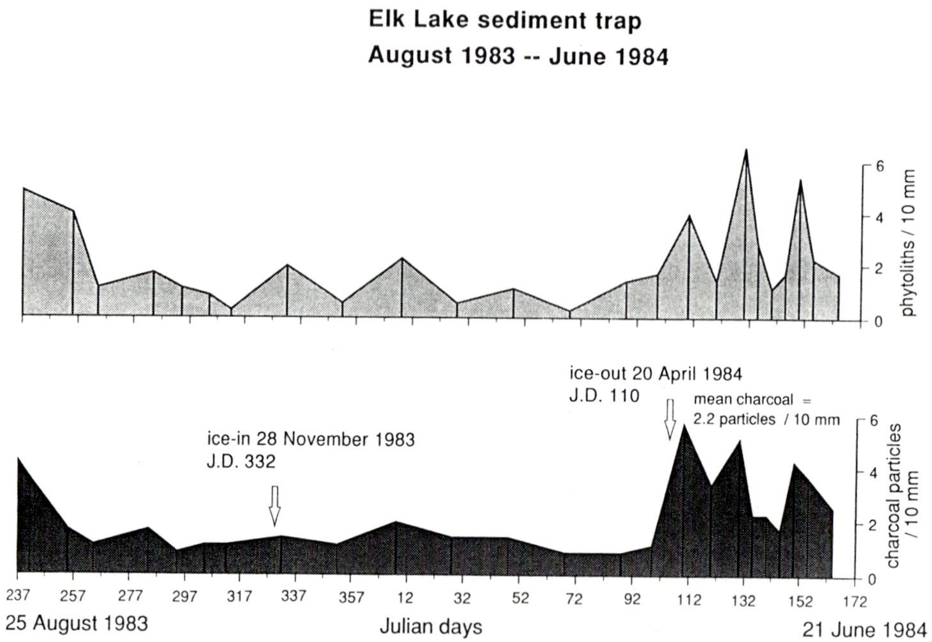

Figure 7. Numbers of phytoliths and charcoal particles per 10-mm traverse across a microscope coverslip from samples deposited in a sediment trap (Fig. 4) installed in Elk Lake between July 1983 and June 1984.

Sediment-trap samples provided a temporal perspective on charcoal and phytolith deposition in the deepest part of the lake (Fig. 7). The highest values of both were found in trap samples collected during spring circulation, shortly after ice-out. These data suggest that when the lake circulated, wind-driven vertical currents moved over the bottom, resuspended particles in the upper hypolimnion, and transported the material to deep water (Nuhfer et al. 1993). Peaks in the abundance of charcoal particles and phytoliths collected between 20 April and 20 June 1984 imply three major circulation events during spring 1984. Daily temperature records from Itasca State Park document three cold episodes during this period between ice-out and lake stratification (Bradbury 1988), and the weather conditions associated with these episodes may have been responsible for stronger-than-normal circulation.

The sediment-trap data also recorded a large bloom of diatoms during spring 1984 (Bradbury 1988). The strong circulation that winnowed and redeposited charcoal and phytoliths from the littoral and sub-littoral zones apparently delivered nutrients to the photic zone and sponsored this bloom. In contrast, autumn 1983 experienced a gradual temperature decline, and the circulation, while strong enough to generate a nutrient flux and diatom bloom, was not sufficient to cause significant charcoal and phytolith redeposition (Bradbury 1988). The more-moderate autumn circulation pattern provides an explanation for those intervals in the core where diatom peaks were not accompanied by increased amounts of charcoal and phytoliths. Apparently, the conditions that allow a large diatom bloom are not always adequate to transport charcoal.

Discussion

The post-fire increase in charcoal in Yellowstone lakes underscores the importance of secondary, redeposited charcoal in the sedimentary record. The accumulation of charcoal in the deep-water sediments increased in all lakes in the five years after the 1988 fires. By 1993, more charcoal had accumulated in lakes with burned catchments than in those with unburned catchments. The secondary charcoal probably had several sources: In burned catchments, surface run-off may have introduced charred material to lakes in the period before vegetation recolonized the slopes. Small debris flows were noted in 1989 in the Duck Lake and Grizzly Lake catchments, although these flows did not reach the shoreline. Other lake-core studies have inferred episodes of post-fire erosion from stratigraphic increases in magnetic minerals and geochemical changes associated with charcoal peaks (Cwynar 1978, Rummery et al. 1979, Millspaugh and Whitlock 1995). In YNP, for example, magnetic susceptibility correlated with charcoal peaks in sediments that were ^{210}Pb-dated to times of catchment fire. Charcoal peaks not associated with high values of magnetic susceptibility were assumed to represent extralocal fires (Millspaugh and Whitlock 1995).

A second source of charcoal in Yellowstone lakes came from burned trees growing near the shoreline. As roots decayed and trees fell into lakes, charcoal was added directly to the littoral zone. Charred wood from standing trees was also introduced to lakes by strong local winds. Lake ice in March 1989 and 1990, for example, contained significant amounts of wind-blown charcoal that was deposited during spring thaw.

A third source of secondary charcoal was the material introduced as atmospheric fall-out during the fires, carried in the water column by wind-generated currents, and deposited in the littoral zone in the first year. Whitlock and Millspaugh (1996) observed in 1989 that charcoal abundances at Goose Lake were greatest in the littoral zone that was in the downwind direction of the fire path. This observation implies that much of the airborne fall-out at Goose Lake was carried in suspension by wind-generated currents and concentrated on the leeward side. In subsequent years, some of this charcoal was apparently moved to deep-water sediments by resuspension during seasonal circulation and perhaps by slumping. The fact that wind-generated currents and wave action are most effective in shallow water (Larsen and MacDonald 1993) may explain why the year-to-year variability in charcoal distribution in shallow lakes was greater than in deep lakes. It should also be noted that this third source of charcoal was the only secondary material available to lakes in unburned catchments and probably accounted for the lower amounts of charcoal accumulation at those sites.

How long will charcoal continue to accumulate following the 1988 fires in YNP? Short-core studies from some of the sites show that charcoal peaks corresponding with known fires span a core interval representing ca. 20 years of sediment accumulation (Millspaugh and Whitlock 1995). The stratigraphic thickness of these peaks probably results both from bioturbation, which tends to blur the sharpness of the peak, and from the accumulation of secondary charcoal after a fire. The 20-year span represented by past fire events in YNP suggests that rates of charcoal accrual from the 1988 fires will likely decrease in the next ten years and approach background values.

The importance of seasonal circulation in subaqueous charcoal redeposition is also evident from the sediment transect data and sediment trap data from Elk Lake (Bradbury 1996). As in YNP, charcoal introduced during and after a fire was probably concentrated on the leeward side by regional winds (Fig. 8). This material was later moved by horizontal currents to marginal areas where it sank in near-shore quiet water (below the thermocline) as stratification developed. During subsequent spring circulation events, some charcoal was picked up from the sub-littoral zone and redistributed to deeper parts of the basin.

Major fires have not occurred in Itasca State Park since 1922, and no fires have reached the Elk Lake shoreline since 1885 (Frissell 1973). Nonetheless, the stratigraphic record indicates some accumulation of charcoal in deep-water sediments during this period. Surface-sediment and sediment-trap data help to explain the charcoal accumulation. The transect data reveal that charcoal in the presumed sub-littoral source region for that year was approximately 40 particles/10 mm traverse (Fig. 6). Charcoal abundance in the sediment trap during spring 1983 was ca. 0.5 particles/10 mm traverse (Fig. 7). Between 1850 and 1890, when several fires occurred in the region, the average charcoal abundance in the deep-water core of Elk Lake was ca. 1 particle/10 mm traverse. After 1900, when fires were rare, charcoal values dropped to ca. 0.5 particles/10 mm traverse, an amount comparable to the maximum value in the sediment trap data in 1983 (Fig. 7). Perhaps this low value (0.5 particles/10 mm traverse) should be considered a background level for charcoal accumulation under the climate conditions of the last century.

The highest accumulation of charcoal in the last 200 years of the Elk Lake record occurred between 1810 and 1830 (ca. 6 particles/10 mm)(Fig. 9). This was a period of drought in northeastern Minnesota (Meko 1992), and dendrochronologic data indicate that 80% of Itasca State Park burned between 1800 and 1849 (Frissell 1973). The abundant charcoal at Elk Lake is consistent with evidence of dry conditions and fires in the early 19th century, despite the indirect way in which charcoal accumulated.

The Elk Lake and climate data were compared with a record of macroscopic charcoal (>0.060 mm particle size) from Deming Lake, a small varved-sediment lake in Itasca State Park (Clark 1990, Fig. 4). When charcoal accumulation rates calculated by Clark (1990) were averaged for the same time intervals as Elk Lake (Fig. 9), the Deming Lake record did not register high charcoal amounts in the early to mid 19th century. The dissimilarities between the two records may reflect inaccuracies in varve counting, different methods of analysis and enumeration, or different processes affecting both primary and secondary deposition of charcoal. Because both lakes record low charcoal amounts in the early 20th century, the different methodologies of charcoal tabulation or varve chronologies seem unlikely to be responsible for the discrepancies between the two profiles. More probably, the difference may be caused by the fact that the two sites portray fire history on different spatial scales. Elk Lake, with a large catchment, probably collects charcoal from a large part of the Itasca region. The macroscopic charcoal at Deming Lake provides a history of local fires in or near its small catchment (Clark 1990). Whether a small catchment burns depends not only on the regional climate conditions but also on the availability of an ignition source, favorable weather for fire spread, and the characteristics of the age-class within the catchment. Thus, the local fire history described by the charcoal record at Deming Lake may have been only indirectly controlled by drought cycles that affected the Itasca region and were registered at Elk Lake.

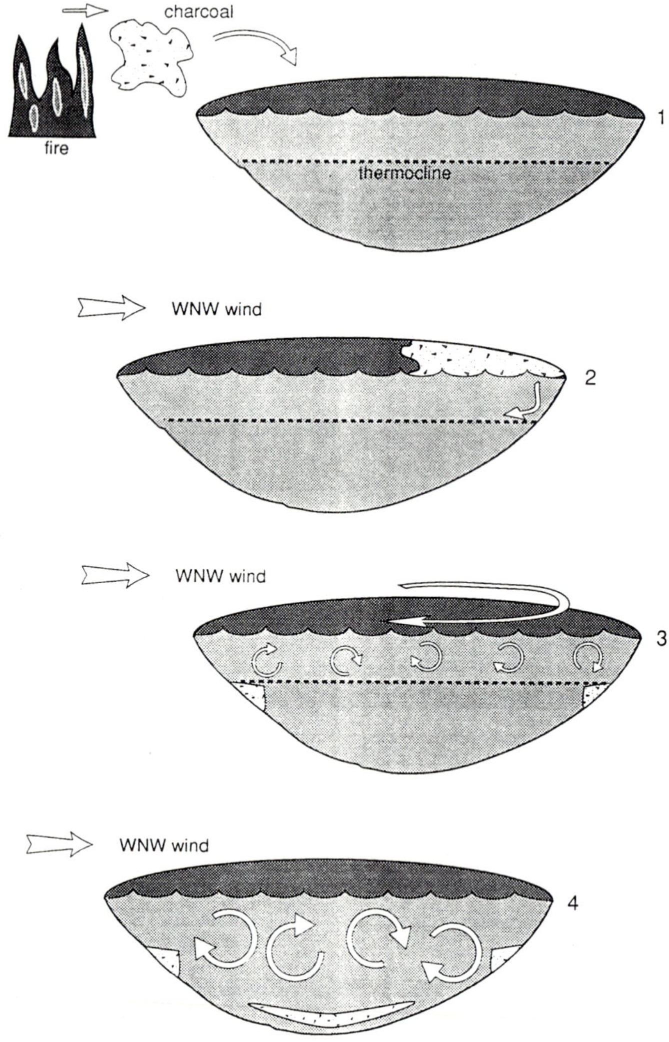

Figure 8. Sequence of events that control deposition of charcoal particles to the deep-water sediments of Elk Lake. (1) Introduction of charcoal onto the water surface. (2) Concentration of floating charcoal particles to the leeward side of the lake by winds. (3) Horizontal current gyres developed by WNW winds that move charcoal around the lake and sedimentation of charcoal fragments in near-shore, quiet water during summer stratification. (4) Entrainment of charcoal particles from sub-littoral sites of deposition during subsequent periods of circulation and redeposition of the particles in the profundal region of the lake.

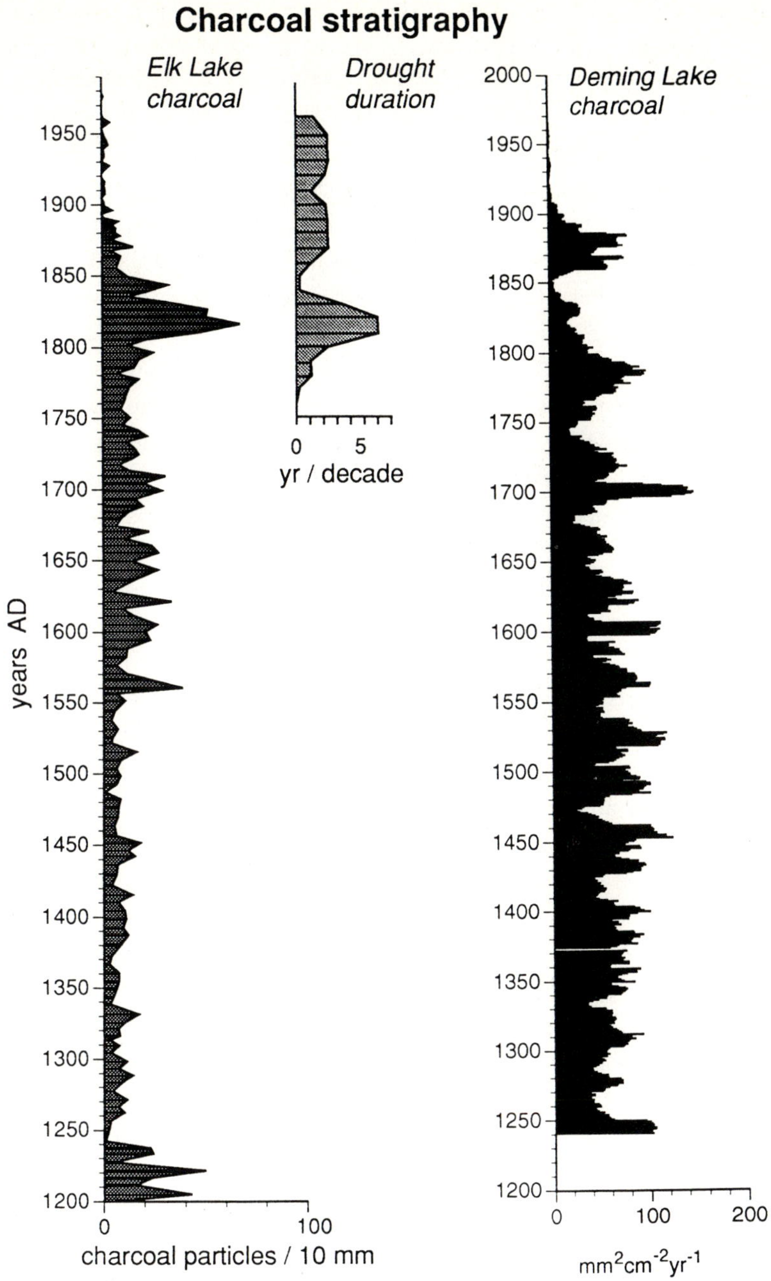

Figure 9. The charcoal stratigraphies of Elk and Deming lakes, back to AD 1200, based on their respective varve chronologies (Dean et al. 1994, Clark 1990), are compared with the number of drought years per decade based on tree-ring studies in northeastern Minnesota (Meko 1992).

Conclusions

Modern charcoal deposition in small lakes in YNP and a 1535-year charcoal record from Elk Lake provide some empirical data that can be used to refine key assumptions in charcoal analysis. Both studies suggest that much of the charcoal found in deep-water sediments is secondary rather than primary material deposited during a fire. The source of secondary charcoal is apparently from burned slopes and trees and also from airborne material deposited in the littoral and sub-littoral zones during or soon after a fire. Shallow-water charcoal is incorporated into deep-water sediments during periods of wind-driven currents, strong wave action, and episodes of slumping. In YNP, the year-to-year variability in the distribution of charcoal in the sediments was greater in shallow lakes than in deep ones. Charcoal abundance in the deep-water sediments increased in the five years following the 1988 fires at all sites, but in 1993 burned sites had notably more charcoal than unburned sites. At Elk Lake, diatom and phytolith data provided an independent means of inferring the time and mechanisms of charcoal transport and deposition. Strong circulation in spring accounted for the increase in charcoal in deep-water sediments and in the sediment trap during non-fire years, even after fires had ceased for 70 years.

The characteristics of a study site are of unquestionable importance to processes that control charcoal deposition, and these characteristics should be well understood before reconstructing a fire history from lake-sediment records. Varved-sediment lakes offer an opportunity for a refined chronology (Swain 1973, 1978, Cwynar 1978, Gajewski et al. 1985, Clark 1990, Tolonen 1986). However, the dissimilarity of fire histories from lakes in Itasca State Park suggests that sedimentological and limnological variations may be significant even between varved-sediment sites. A lake with broad littoral and sub-littoral zones potentially adds a source of secondary charcoal that would not affect a lake with steep bathymetry. Similarly, a large lake with a sizable catchment probably collects charcoal from a larger source area than a small lake; such a site would be better for inferring regional burning and drought cycles. A small lake may record local fires within a small catchment but be less sensitive to regional climate conditions.

In conclusion, both the records from YNP and Elk Lake indicate that the post-fire patterns of charcoal accumulation are complex. Variations in fire size, intensity, and proximity undoubtedly influence the production of charcoal and the amount that is received in a lake as airborne fallout. Other factors come into play after a fire that are equally important in determining how much charcoal accumulates in the sediments. Reconstruction of fire intensity or extent would be difficult without first understanding the charcoal taphonomic processes of a site. Lags in charcoal accumulation may account for the broad peaks that often represent a single fire in the stratigraphic record. They may also explain the sometimes poor match between charcoal peaks and well-dated fire events. The shape of a charcoal peak quite likely tells more about the sedimentary history of charcoal within the lake following a fire than about the amount of biomass that was burned within the catchment.

Acknowledgments

James S. Clark and Peter Kershaw provided helpful comments on the manuscript. The Yellowstone research was supported by grants from the University of Wyoming-National Park Service Research Center, the National Science Foundation (ATM-9096230, SES-9305436), and Sigma Xi. Roger Y. Anderson and Walter E. Dean were instrumental in developing the chronology of the high-resolution varved record at Elk Lake. The Elk Lake research is part of the U.S. Geological Survey Climate Change Program.

Literature references

Anderson RY, Dean WE, Bradbury JP (1993) Elk Lake in perspective. In Bradbury JP, Dean WE (eds) Elk Lake, Minnesota: Evidence for rapid climate change in the north-central United States: Boulder, Colorado *Geological Society of America* Special Paper 276:1-6

Anderson RY (1977) Short term sedimentation response in lakes in western United States as measured by automated sampling. *Limnology and Oceanography* 22:423-433

Balling Jr RC, Meyer GA, Wells SG (1992) Climate change in Yellowstone National Park: is the drought related risk of wildfires increasing? *Climatic Change* 22:34-35

Bradbury JP (1988) A climatic-limnologic model of diatom succession for paleolimnological interpretation of varved sediments at Elk Lake, Minnesota. *J of Paleolimn* 1:115-131

Clark JS (1988) Particle motion and the theory of stratigraphic charcoal analysis: source area, transport, deposition, and sampling. *Quat Res* 30:67-80

Clark JS (1990) Fire and climate change during the last 750 yr in northwestern Minnesota. *Ecol Mono* 60:135-159

Cwynar LC (1978) Recent history of fire and vegetation from annually laminated sediment of Greenleaf Lake, Algonquin Park, Ontario. *Can J of Botany* 56:10-12

Cwynar LC (1987) Fire and forest history of the North Cascade Range. *Ecology* 68:791-802

Dean WE, Bradbury JP, Anderson RY, Bader LR, Dieterich-Rurup K (1994) A high-resolution record of climate change in Elk Lake, Minnesota for the past 1500 years. *US Geological Survey* Open File Report 94-578:127

Frissell SS (1973) The importance of fire as a natural ecological factor in Itasca State Park, Minnesota. *Quat Res* 3:397-407

Gajewski K, Winkler MG, Swain AM (1985) Vegetation and fire history from three lakes with varved sediments in northwestern Wisconsin (USA). *Rev of Palaeobot and Palynol* 44:277-292

Green DC, Singh G, Polach H, Moss D, Banks J, Geissler EA (1988) A fine resolution palaeoclimatology and palaeology from southeastern Australia. *J of Ecol* 76:790-806

GYPFRAC-BAST (Greater Yellowstone Post-Fire Resource Assessment Committee, Burned Area Survey Team) (1988) Preliminary burned area survey of Yellowstone National Park and adjoining national forests. *NPS/USFS/NASA/Montana State University, Yellowstone National Park, Wyoming*

Larsen CPS, MacDonald GM (1993) Lake morphometry, sediment mixing and the selection of sites for fine resolution palaeoecological studies. *Quat Sci Rev* 12:781-792

MacDonald GM, Larsen CPS, Szeicz JM, Moser KA (1991) The reconstruction of boreal forest fire history from lake sediments: a comparison of charcoal, pollen, sedimentological, and geochemical indices. *Quat Sci Rev* 10:53-72

Meko DM (1992) Dendroclimatic evidence from the Great Plains of the United States. In Bradley RS, Jones PD (eds) *Climate since AD 1500* pp 312-330 Routledge London

Mehringer PJ, Arno SF, Petersen KL (1977) Postglacial history of Lost Trail Pass Bog, Bitterroot Mountains, Montana. *Arctic and Alpine Research* 9:345-368

Millspaugh SH, Whitlock C (1995) A 750-yr fire history based on lake sediment records in central Yellowstone National Park. *The Holocene* 5:283-292

Nuhfer EB, Anderson RY, Bradbury JP, Dean WE (1993) Modern sedimentation in Elk Lake, Clearwater County, Minnesota. In Bradbury JP, Dean WE (eds) Elk Lake, Minnesota: Evidence for rapid climate change in the north-central United States Boulder Colorado *Geological Society of America* Special Paper 276: 75-96

Patterson III WA, Backman AE (1988) Fire and disease history of forests. In Huntley B, Webb III T (eds) *Vegetation History Handbook of Vegetation* pp 603-632 Science Kluwer Academic Publishers Dordrecht

Patterson III WA, Edwards KJ, MacGuire DJ (1987) Microscopic charcoal as a fossil indicator of fire. *Quat Sci Rev* 6:3-23

Radke LF, Lyons JH, Hobbs PV, Hegg DA, Sandberg, DV (1990) Airborne monitoring and smoke characterization of prescribed fires on forest lands in western Washington and Oregon. *US Dept of Agri For Serv* Pacific Northwest Research Station Portland Oregon General Technical Report PNW-GTR-251

Rummery TA, Bloemendal J, Dearing J, Oldfield F, Thompson R (1979) The persistence of fire-induced magnetic oxides in soils and lake sediments. *Annales de Geophysiqua* 35:103-107

Schullery P (1989) The fires and fire policy. *Bioscience* 39:686-695

Swain AM (1973) A history of fire and vegetation in northeastern Minnesota as recorded in lake sediments. *Quat Res* 3:383-396

Swain AM (1978) Environmental changes during the past 2000 yr in north-central Wisconsin: analysis of pollen, charcoal and seeds from varved lake sediments. *Quat Res* 10:55-68

Tolonen K (1986) Charred particle analysis. In Berglund BE (ed) *Handbook of Holocene Palaeoecology and Palaeohydrology* John Wiley and Sons Ltd New York

Whitlock C, Millspaugh SH (1995) Testing the assumptions of fire-history studies: an examination of modern charcoal accumulation in Yellowstone National Park, USA. *The Holocene* 6:7-15

Reconstruction of Fire Disturbance and Forest Succession from Fossil Pollen in Lake Sediments: Potential and Limitations

S. Sugita, G. M. MacDonald[1,2] and C. P. S. Larsen[3,4]
Department of Ecology, Evolution and Behavior
University of Minnesota
100 Ecology, 1987 Upper Buford Circle
St. Paul, MN 55108 U.S.A.

Summary

The size of a fire, its proximity to a lake and the size of the lake all affect the way the fire is represented in pollen records. To quantify how these three factors affect pollen representation of disturbance events such as fires, computer simulation models of pollen deposition on a lake were applied. As a model input, we used both a vegetation map of the boreal forest around Rainbow Lake A (3 ha in size), Alberta, Canada, and an hypothetical map of patchy vegetation, which consists of circular patches of aspen and spruce in a matrix of pine, with a circular lake (3 ha) at the center.

We verified our models by comparing modern pollen assemblages observed at Rainbow Lake A with modern pollen assemblages predicted by the models. Then, the impact of a fire that burned around the lake in AD 1943 was simulated to determine how this fire would have changed pollen deposition in the lake, assuming that the pre-fire vegetation was the same as the modern vegetation. The fire in 1943 burned an area ca. 400 ha within a 2.5 km radius of the lake, but did not burn within 400 m of the lake. Our simulation results showed that the 1943 fire should not have caused significant changes in pollen loadings and percentages just after the fire; the actual pollen records from the laminated sediment of the lake shows no apparent record of this fire, supporting the simulation results.

Experiments, which simulated disturbances of different sizes (4 ha, 100 ha and 2500 ha) with various distances to a lake in the hypothetical, patchy landscape, provided additional insights into how the size and proximity of fires would influence pollen signals of disturbance. The results demonstrated that fossil pollen most likely records disturbances which are at least eight times larger than the lake and are located within a few hundred meters of the lake shore. Thus, past fires and their effects on vegetation can be detected under limited, but identifiable, conditions. We suggest that the simulation models similar to the one we use here are a useful tool in choosing sites from which paleo-fire records could be obtained using fossil pollen.

---

[1]Department of Geography, McMaster University, Hamilton, Ontario L8S 4K1, Canada
[2]Current Address: Department of Geography, University of California, Los Angels, Los Angels, CA 90095-1524 U.S.A.
[3]Department of Geography, McMaster University, Hamilton, Ontario L8S 4K1, Canada
[4]Current Address: Department of Geography, University at Buffalo, Buffalo, NY 14261-0023 U.S.A.

NATO ASI Series, Vol. I 51
Sediment Records of Biomass Burning
and Global Change
Edited by James S. Clark, Hélène Cachier,
Johann G. Goldammer, and Brian Stocks

Introduction

Fire is one of the most important agents of disturbance in many forested ecosystems, and post-fire succession has been a focus of attention by forest ecologists for over 80 years (e.g. Cooper 1913). Fire recurrence intervals and post-fire development of temperate and northern forests can vary in length from decades to centuries (Heinselman 1981), making it difficult to produce observational records of succession for individual sites. Although choronosequences are often used to infer temporal patterns of post-fire succession (e.g. Dix and Swan 1971, Johnson 1981, Romme 1982, DeGrandpre et al. 1993), such studies suffer from an inability to resolve the impact of spatial heterogeneity within and between sites (Pickett 1989). In contrast, retrospective studies using fossil pollen and charred particles from lake sediments can document long-term history of fire and post-fire development of vegetation from individual sites (e.g. Tsukada and Deevey 1967, Waddington 1969, Swain 1973, 1978, Cwynar 1978 1987, Green 1981, 1982, Sugita and Tsukada 1982, Gajewski et al. 1985, Patterson and Backman 1988, Clark 1990, MacDonald et al. 1991). These paleo-records have provided important insights into fire regimes and their relationships to changes in vegetation.

There are, however, methodological problems about the interpretation of disturbance records from fossil pollen. How does lake size affect sensitivity of a pollen record to fire disturbance and estimates of fire frequency? Is it possible to distinguish local post-fire vegetation changes from more distant ones? In general, such questions on the spatial resolution of fossil pollen records remain an important question to palynologists (e.g. Jacobson and Bradshaw 1981, Prentice 1988, Jackson 1990, Sugita 1993). Anemophilous (wind dispersed) pollen taxa are generally the most abundant in lake sediments and most of the key taxa used to reconstruct post-fire succession are of this type. However, a significant proportion of the anemophilous pollen in lake sediments comes from sources that are tens of kilometers distant (Jacobson and Bradshaw 1981, Prentice 1988, Jackson 1990, Sugita 1993). Therefore, patchy and episodic disturbances near lakes may only be detectable in the pollen record if the signal, in the form of changes in pollen deposition due to the local disturbance, can be separated from the background pollen arriving at the lake from more distant sources.

There is no consensus among paleoecologists on how significantly pollen loadings and assemblages are influenced by the size or proximity of disturbance. There must be critical limits at which a disturbed area is too small, or too far away, to significantly change the pollen signal recorded in a lake of a certain size.

The relatively small changes in fossil pollen assemblages after the presumed disturbance events studied by Cwynar (1978) and Swain (1973) may reflect such limitations. Empirical studies remain too sparse to provide general conclusions on this topic. Simulation experiments specifically designed to examine the effects of changes in critical spatial factors (e.g. lake size, proximity and size of fire disturbance) on pollen deposition in lakes can provide important guidelines for detecting past disturbances in fossil pollen.

Modeling pollen dispersal and deposition has been a relatively recent phenomenon in palynology. Prentice's model (1985) and Sugita's model (1993, 1994) have provided reasonable explanations of the effects of lake size on pollen representation of vegetation with a homogeneous or patchy landscape. Calcote (1995) has indirectly verified Sugita's model by showing that the estimated source area of pollen for small forest hollows in northern Michigan matched the model's predicted source area of pollen. There have been, however, no attempts to verify the models by comparing directly predicted and observed pollen assemblages. It is critical to have verified models to better understand pollen representation of vegetation and to increase the reliability of simulation results.

In this paper, we use simulation experiments to investigate how the location and size of a disturbance patch relative to the size and location of a lake can affect the pollen representation of disturbance and post-disturbance development of plant communities. First, to verify the models for pollen dispersal and deposition for simulation experiments, the surface pollen assemblages from Rainbow Lake A, Alberta, Canada (MacDonald et al. 1991) are compared with the pollen assemblages predicted from the models (Sugita 1993, 1994, this paper). Second, changes in pollen deposition on Rainbow Lake A after a hypothetical fire, similar to the 1943 fire in the Rainbow Lake region reconstructed using dendrochronology, are simulated to compare with the observed pre- and post-fire changes in pollen assemblages in the lake. Third, to better understand how pollen records disturbance in general, we use a simulated landscape of patchy vegetation and create disturbance patches in the landscape to simulate the pollen records that would result in lakes of different sizes following fires of different sizes and proximities to the lakes. Note that we did not simulate charcoal deposition on lakes, because mechanisms of charcoal emission and dispersal in air differ from those of pollen transportation (Prentice 1985, Clark 1988).

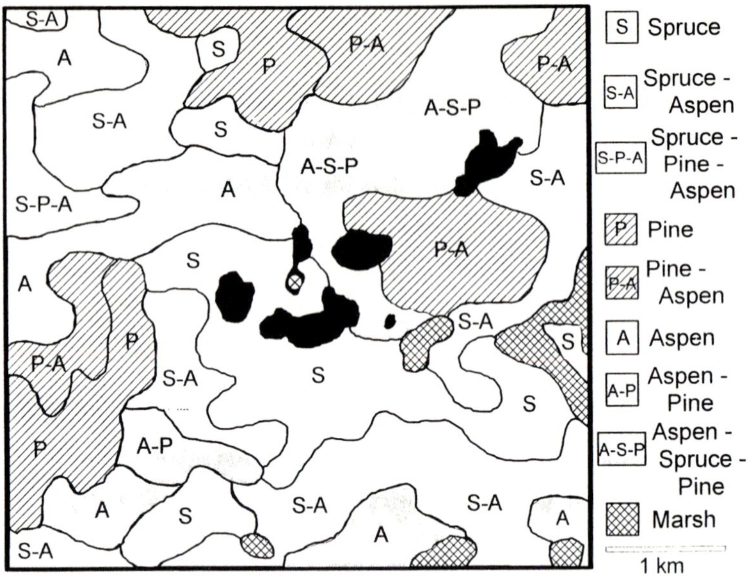

Figure 1. Forest cover map in the Rainbow Lakes region, Alberta, Canada (redrawn from MacDonald et al. 1990, and Larsen and MacDonald, in prep.) Black portions represent lakes, and *f* symbol shows Rainbow Lake A, from which pollen data were obtained.

Study Site

Rainbow Lake A (59°48' N, 112°10' W; 2.7 ha in area) is located in Wood Buffalo National Park, Alberta, Canada. This park is 45,000 km^2 in size. Forest cover map around the lake is shown in Fig. 1. Rainbow Lake A is a solutional sink hole formed in gypsum karst. It is presently surrounded by a mature white spruce (*Picea glauca*)-dominated forest with smaller amount of aspen (*Populus tremuloides*), paper birch (*Betula papyrifera*), tree alder (*Alnus incana*), jack pine (*Pinus banksiana*) and black spruce (*Picea mariana*). Fires in the region are typically large. More than 99% of the area burned in the park over the last 39 years has been the results of fires greater than 100 km^2 in size, with fires of up to 1800 km^2 having been observed in the park.

In the vicinity of Rainbow Lake A, fire scarred jack pine trees found within a 1 km radius of the lake evidence the occurrence of six fires over the past 150 years. Scar evidence was found for only two fires, at 110 and 125 years before 1986, within a 100 m radius of Rainbow Lake A. Notably, the pollen record only appears to reflect the occurrence of a fire that occurred 125 years before 1986 (Fig. 2). The pollen evidence is in the form of a decline in white spruce, followed by subsequent peaks in grass, willow, aspen and white spruce (MacDonald et al. 1991). This pattern of change in pollen abundance is consistent with observational data on post-fire succession in the boreal forest (e.g. Kelsall et al. 1977, Van Cleve et al. 1986). The lack of a clear pollen response to the other fires, including a 408 ha fire that burned to within 400 m of the lake in 1943 (Fig. 3), suggests that the pollen record from Rainbow Lake A is only sensitive to fires hat burn within 100 m of the lake shore.

Method

Models for Pollen Dispersal and Deposition on a Lake Surface

The model for pollen dispersal and deposition proposed by Prentice (1985, 1988), and modified by Sugita (1993), describes a relationship between pollen loading on a sedimentary basin and four factors influencing pollen loading: pollen productivity, pollen dispersal characteristics, spatial distribution of source plants, and size of a sedimentary basin. Pollen loading is defined as the number of pollen grains deposited onto the surface of a bog or lake water. Prentice (1985, 1988) estimated pollen deposition at a point at the center of a basin, which makes his model appropriate for estimating pollen deposition in a bog, where post-deposition movement of pollen is negligible. Sugita (1993, 1994) modified Prentice's model to estimate pollen loading over the entire surface of a basin. Since sediment mixing and focusing redistribute pollen originally deposited on the lake surface (Davis and Brubaker 1973, Davis et al. 1984), Sugita's model is more appropriate for pollen assemblages in lake sediment.

Sugita's model (1993, 1994) is used for our simulation experiments and is described in detail in Appendix 1. Assumptions of the model are: (1) The sampling lake is a circular opening in the forest canopy. (2) The dominant agent of pollen transport is wind above the canopy. (3) Pollen productivity is constant for each taxon. (4) Spatial distribution of each taxon is expressed as a function of distance from a point at the center of the lake. This means that the directions of pollen source do not influence the pollen loading on the lake. (5) Arc-wise pollen deposition is expressed as a function of distance from a point source, derived from a diffusion model of small particles from a ground-level source

(Sutton 1953, Chamberlain 1953). The shape of the function is mainly determined by wind speed and taxon-specific deposition velocity (v_g) of pollen.

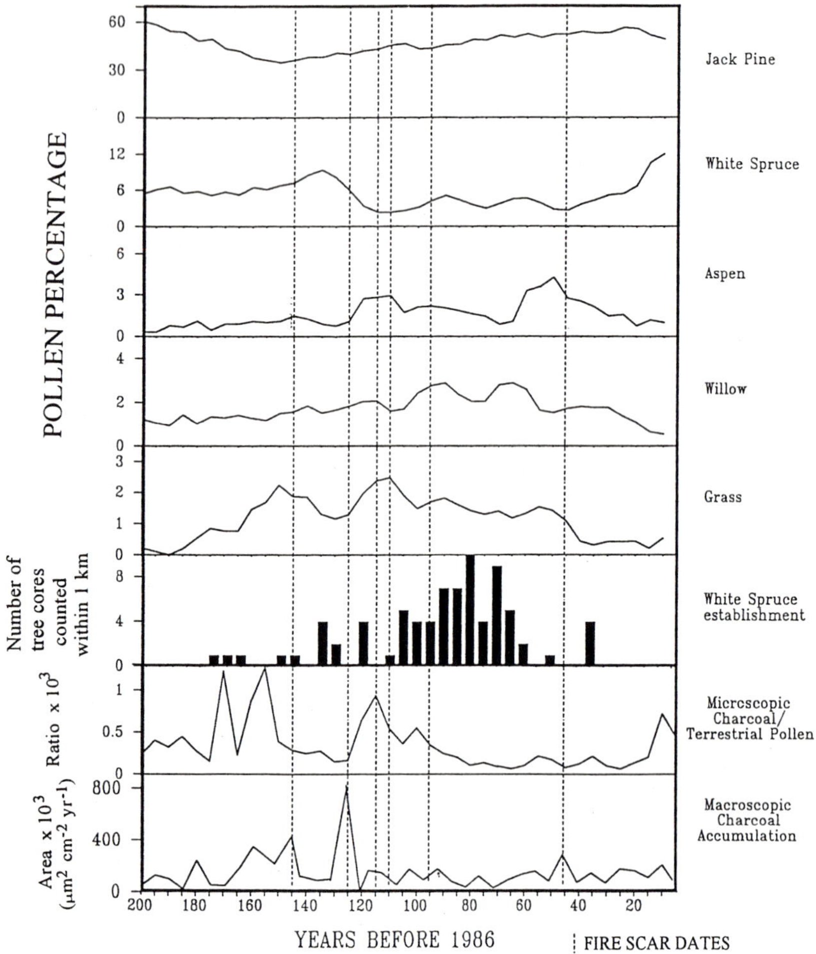

Figure 2. Changes in the pollen percentages of selected taxa following a series of fires near Rainbow Lake A. The timing of fires and the establishment of the present spruce trees around the lake are based on dendrochronological analysis. The pattern of decline in spruce pollen ~140 years ago, followed by sequential peaks in grass, aspen, willow and finally spruce is consistent with post-fire succession in the boreal forest. A small fire that occurred in the 1943 (Fig. 3) did not extend to the shores of the lake and appears to have had no impact on the pollen record (after MacDonald et al. 1991). Pollen records are smoothed with three-term, equal-weight moving average. The broken lines represent dates of fire scars found within 1 km radius of Rainbow Lake A.

In essence, Sugita's model (Sugita 1993, 1994) is a one-dimensional model, as is Prentice's model (Prentice 1985, 1988). In order to apply this one-dimensional model for estimating pollen loading on the entire lake surface, Sugita (1993) assumes that the ratio of pollen deposition on a lake to pollen deposition between concentric rings of radii z-R and z+R (z = distance between a point at the center of a basin and a point source of pollen; R = radius of a basin) is the same as the ratio of the area of a lake to the area between concentric rings of radii z-R and z+R (Appendix 1). Critics may argue that a lake is a two-dimensional object, and thus, a two-dimensional model of pollen deposition would be more appropriate than a model proposed by Sugita (1993, 1994). Appendix 2 shows a finite-line source model of pollen dispersal and deposition on a lake, which is a two-dimensional model. This model is hereafter referred to as FLS model. Pollen assemblages predicted by Sugita's model and FLS model using modern vegetation data around Rainbow Lake A are compared to demonstrate that these two models provide nearly identical results.

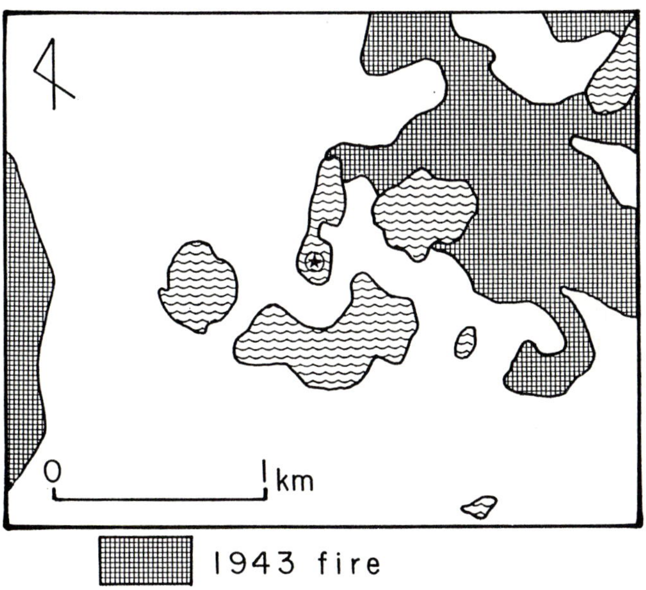

Figure 3. The area burned in 1943 around Rainbow Lake A (after MacDonald and Larsen, unpublished).

Parameter Estimates Required for Model Calculations

Four parameters are included in Sugita's model and FLS model: the mean plant abundance of each taxon at a given distance from the lake, terminal velocity of pollen and pollen productivity of each taxon, and the mean wind speed. The mean summer wind speed of the lake, 4.5 m sec-1, was estimated from the weather station at Ft. Smith, located 25 km northeast of Rainbow Lake A (1951-1980 climate normals, Environment Canada, 1982).

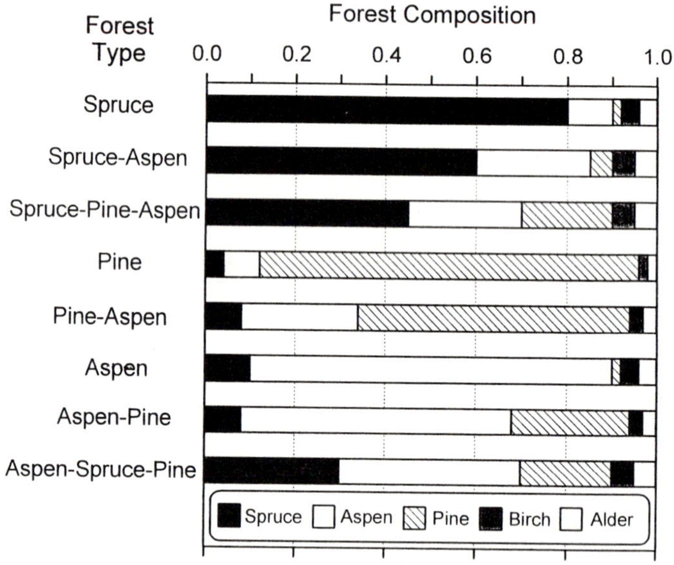

Figure 4. Preliminary estimates of percent cover of forest composition in the eight forest types within the area of forest cover map of Fig. 4 (Larsen and MacDonald, in review). Five major taxa were included.

Mean plant abundance. Plant abundances of five major taxa (spruce, aspen, pine, birch and alder) around Rainbow Lake A were estimated from a detailed vegetation map (Fig. 1). MacDonald et al. (1991) classified eight forest types around the lake; these types are spruce, spruce-aspen, spruce-pine-aspen, pine, pine-aspen, aspen, aspen-pine, and aspen-spruce-pine. Species composition for each forest type was estimated by Larsen from ground surveys and aerial photographs (Fig. 4). Areas occupied by each forest type were calculated around the lake in 100-m-wide concentric rings out to 2500 m, using GIS (ARC/INFO). Mean plant abundances of five taxa were then estimated within each concentric ring (Fig. 5). The mean abundance for each taxon between 2000 and 2500 m radius was used as the estimate of regional plant abundance beyond 2500 m up to 2000 km, which was required for model calculations.

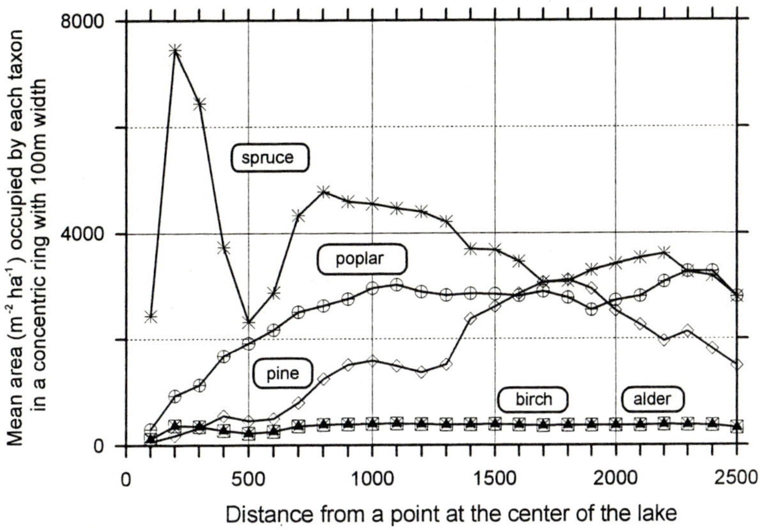

Figure 5. Mean areas (m^2 ha^{-1}) occupied by five taxa in a concentric ring with 100 m width at different distances from the lake. These estimates were calculated from the areas occupied by individual forest types and forest composition in each of forest types.

Terminal velocity of pollen grains. Sugita's model and FLS model require estimates of the deposition velocity (v_g) of pollen (Appendices 1 and 2). The deposition velocity of pollen at the ground can be assumed to be equivalent to the terminal velocity (v_s) of pollen (Gregory 1973, Chamberlain 1975, Prentice 1985, Okubo and Levin 1989). Even so, reliable estimates of the terminal velocity of pollen are not available for most North American taxa. Thus, we compare model predictions based on two sets of estimates of terminal velocity. One set was taken from Eisenhut (1961); 0.056 (m sec^{-1}) for spruce, 0.025 for aspen, 0.031 for jack pine, 0.024 for birch, and 0.021 for alder. These estimates are for congeneric species in Europe, except jack pine. Another set includes estimates for spruce (0.087 m sec^{-1}), pine (0.025), birch (0.025), and alder (0.021) from Gregory (1973) and aspen (0.025) from Eisenhut (1961). The terminal velocity of spruce pollen listed in Gregory (1973) is significantly greater than the estimates from Eisenhut (1961). Spruce is one of the major taxa in the Rainbow Lakes area, so these two sets of estimates provide a good test for sensitivity of the models to different terminal velocity estimates.

Pollen productivity. Pollen productivity of major tree species in North America has been estimated by comparing pollen percentages in surface sediment samples with plant abundance around the study sites (Webb et al. 1981, Bradshaw and Webb 1985, Prentice and Webb 1986, Jackson 1990, Calcote 1995). Several authors have tried to correct for pollen dispersal bias in such estimates by applying weighting procedures (Webb et al. 1981, Prentice and Webb 1986, Schwartz 1989). However, they ignore differences in pollen dispersal characteristics among taxa, so they underestimate pollen productivity for heavy, poorly dispersed types such as spruce and fir, while overestimating pollen productivity of light pollen types such as pine and oak (Sugita and Calcote, in prep.)

We used correction methods to minimize the effects of pollen dispersal bias on pollen productivity estimates for major taxa around Rainbow Lake A. (Appendix 3; Sugita and Calcote, in prep.). The correction was applied to data in the literature that estimate relative pollen productivity using the maximum likelihood method (e.g. Prentice and Webb 1986, Prentice et al. 1987, Calcote 1995, Jackson et al. 1995). Pollen productivity of each taxon was calculated relative to birch. Maximum likelihood estimates of productivity for spruce and aspen were taken from Prentice and Webb (1986) and for pine and birch from Calcote (1995). There are no reliable estimates for alder, so its pollen productivity was assumed to be the same as birch. Corrected productivity for spruce, aspen, pine, birch and alder based on Sugita's model are 0.78, 0.02, 1.57, 1.00 and 1.00, respectively, and, based on FLS model, 0.81, 0.02, 1.57, 1.00 and 1.00, respectively.

Observed Pollen Assemblages from the Surface Sediment of Rainbow Lake A

We used the mean pollen percentages of individual taxa over the last ten years at Rainbow Lake A (MacDonald et al. 1991; Fig. 2) to verify the pollen assemblages predicted by Sugita's model and FLS model. The 99 % confidence intervals of the observed percentage values were calculated using a formula proposed by Mosimann (1965). We did not calculate confidence intervals for the predicted percentages, because we lack error estimates for plant abundance of each taxon surrounding the lake, terminal velocity of pollen, and pollen productivity.

Simulation of Pre- and Post-Fire Pollen Assemblages in Rainbow Lake A

The second experiment was to simulate changes in pollen assemblages in Rainbow Lake A just before and after a hypothetical fire in the modern vegetation. The area burned by a hypothetical fire was assumed as the same area burned in 1943. Figure 3 shows a portion of the area burned in 1943. For the simulation, the area burned in 1943 was estimated up to 2500 m from the lake shore of Rainbow Lake A. In total, 408 ha was burned within 2500 m from the lake. Closest distance between the lake and the fire was 400 m. The post-fire plant abundance of five taxa (spruce, aspen, pine, birch, and alder) in 100-m-wide concentric rings out to 2500 m from the lake was estimated from the modern forest-cover map (Fig. 1), assuming that all trees were removed in the burned area. Mean plant abundances of these five taxa beyond 2500 m from the lake was assumed to be the same as those in the modern vegetation. Wind speed was set to 4.5 m sec^{-1}, and fall speed of pollen and pollen productivity for each taxon were the same as described in "*Parameter estimates required for model calculations*". Both Sugita's model and FLS model were used to estimate pollen loadings on Rainbow Lake A after the fire.

Effects of Disturbance on Pollen Loadings and Percentages in a Simulated Patchy Vegetation

We used a simulated landscape of patchy vegetation similar to the landscapes used by Sugita (1994). Three taxa (spruce, aspen and pine) occupied a 50-km x 50-km simulated landscape (Fig. 6). Spruce formed large, randomly distributed circular patches, which did not overlap each other. The mean size of spruce patches was 50 ha (ca. 800 m in diameter) with a standard deviation of 20 ha. Overall percentage cover of spruce patches in the simulated landscape was 40%. The matrix outside of spruce patches was occupied by aspen. Small circular patches of pine were also distributed randomly and occupied 20% of the area in the plot. The mean and standard deviation of pine patches were 5 ha (ca. 250 m in diameter) and 2.5 ha, respectively. Pine patches did not overlap each other, but overlapped with spruce patches and the aspen matrix. A mixed forest composition of spruce and pine or aspen and pine was assumed with equal abundances of each taxon in the overlapped areas.

Two sizes of lakes were used: medium size (radius R = 100m; 3.1 ha) which is similar to the size of Rainbow Lake A (MacDonald et al. 1991; Larsen and MacDonald, in review), and large size (R = 1000 m; 314.2 ha). Lake locations were selected randomly within a 4-km x 4-km cell at the central portion of the simulated landscape. We used pollen assemblages from nine lakes for analyzing the effects of size and location of a disturbed patch and size of a lake on pollen representation of disturbance.

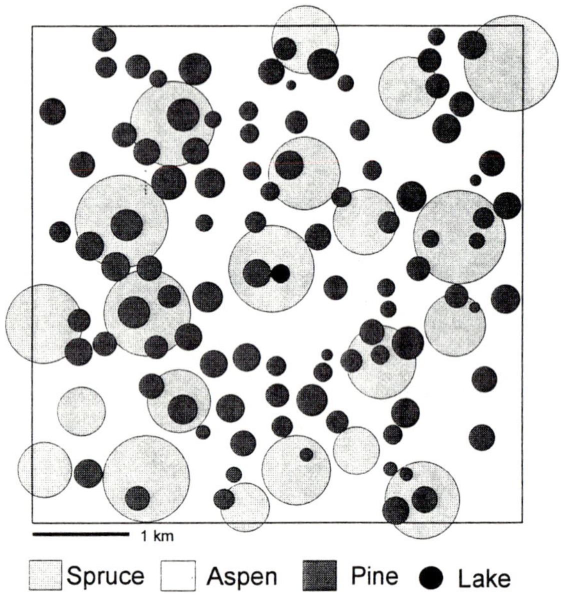

Figure 6. Schematic map of a simulated patchy landscape. A lake is shown at the center in black, large patches of spruce in light shading, aspen matrix in white, and small patches of pine in darker shading. For simulations, the plot size was fixed to 50-km x 50-km.

Two different shapes of patchy disturbance were simulated (Fig. 7): one was the disturbance at the lake shore, which completely surrounded a lake, and the other was a square patch of disturbance which was placed at various distances from a lake (0 m - 10000 m). The canopy was assumed to be completely destroyed in each disturbed patch. Patch sizes of disturbance varied from 4 ha to 2500 ha.

Considering the sampling and counting errors associated with pollen analysis, only large changes in pollen loading or pollen percentage are likely to be detected as significant changes in fossil pollen spectra. Therefore, interpretation of simulation results has to be realistic and practical as well. We tentatively set a value of 10% change in pollen loading or percentage values between time periods as the threshold for "detectable" change.

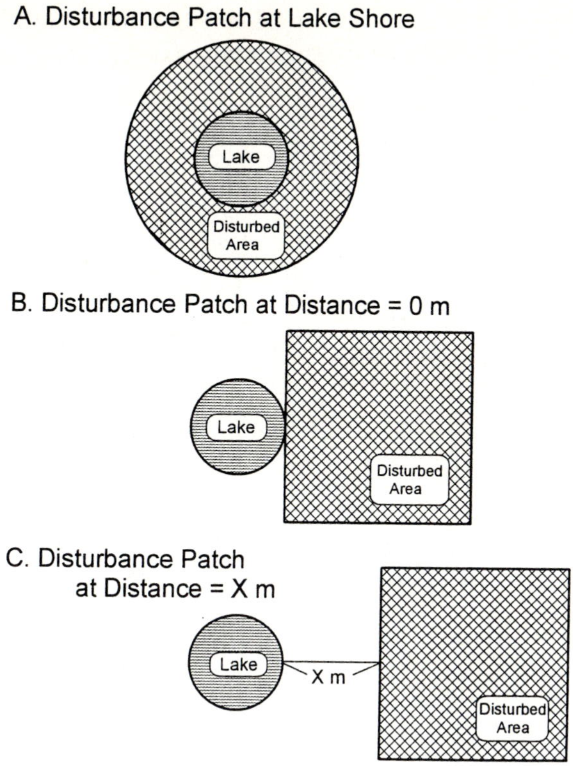

Figure 7. Simulation designs of a lake and a disturbance patch in a simulated landscape. A disturbance patch at lake shore surrounded completely a lake (A). A disturbance patch at a given distance was defined as a square disturbance patch east of a lake with that distance (0 m in B, and X m in C).

Results

Comparison between the Predicted and Observed Pollen Assemblages in the Surface Sediment of Rainbow Lake A

Overall, the predicted pollen assemblages match the observed pollen assemblages well (Fig. 8). Using a set of Eisenhut's estimates of terminal velocity, the pollen assemblages predicted by both Sugita's model (Fig. 8-A) and FLS model (Fig. 8-B) are particularly similar to the observed assemblages. The predicted percentages of spruce, aspen, pine, and alder fall within the 99% confidence intervals of the observed values, and the predicted birch percentages are at the lower confidence limits of the observed. Similarity of the results in Figs. 8-A and 8-B indicates that the prediction of pollen assemblages is robust to the differences between the two models for pollen deposition on the entire surface of a basin.

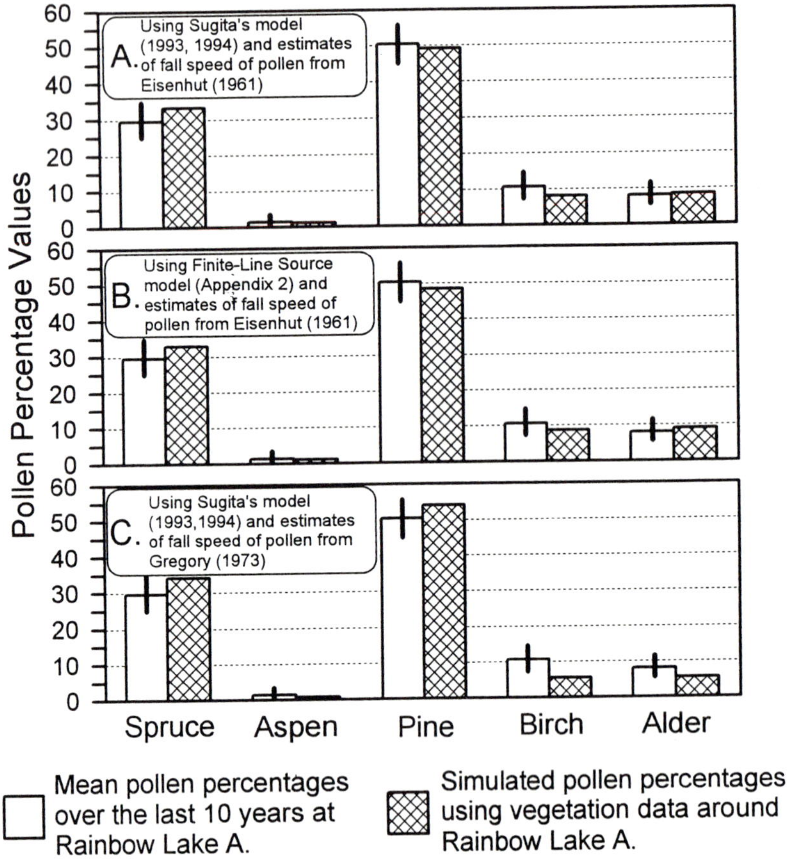

Figure 8. Comparison between the observed and predicted pollen assemblages from Rainbow Lake A. The observed pollen percentages were the means over the last ten years (Larsen and MacDonald, in review). The 99% confidence intervals are shown for the observed values. The predicted pollen assemblages were based on (A) Sugita's model (1993, 1994, Appendix 1) with a set of terminal velocity estimates from Eisenhut (1961), (B) Finite-Line Source model (Appendix 2) with a set of terminal velocity estimates from Eisenhut (1961), and (C) Sugita's model with a set of terminal velocity estimates from Gregory (1973) except for aspen (Eisenhut 1961). Wind speed was set to 4.5 m sec-1, and the pollen productivity for each taxon was estimated according to the methods described in Appendix 3.

The pollen assemblages, predicted by Sugita's model using a set of estimates of terminal velocity compiled by Gregory (1973), are less similar to the observed assemblages (Fig. 8-C) than the other two results. The predicted percentages of birch and alder are outside of the 99% confidence intervals of the observed. These results suggest that differences in the estimates of terminal velocity of pollen have larger effects on predictions of pollen assemblages than the differences in the models. Nonetheless, considering the simplicity of the models, uncertainties of pollen analysis (e.g. sampling and counting errors) and the uncertainties of parameter estimates, the predictions based on vegetation data are remarkably similar to the actual pollen percentages.

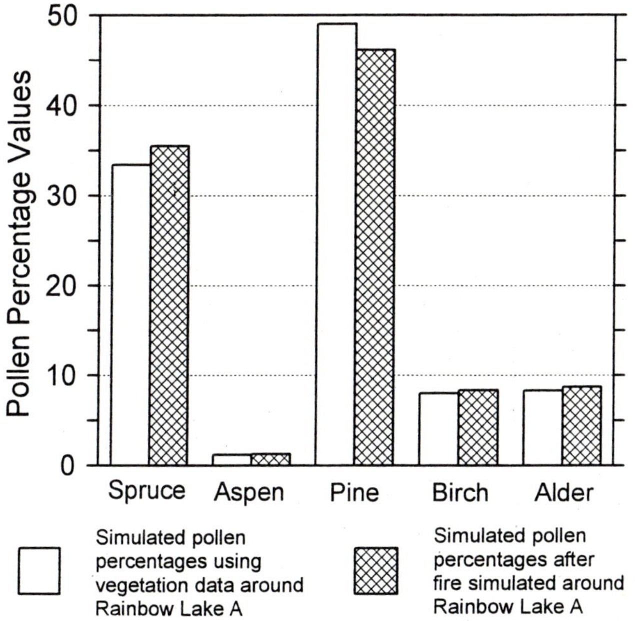

Figure 9. Simulated changes in pollen assemblages in Rainbow Lake A just before and after a hypothetical fire in the modern vegetation. The area burned by a hypothetical fire is assumed as the same area burned in 1943 (Fig. 3). The simulated pollen assemblages in modern conditions and after the fire using Sugita's model (Appendix 1) are shown. FLS model (Appendix 2) provides nearly the same results.

Pre- and Post-Fire Pollen Assemblages in Rainbow Lake A

Assuming a hypothetical fire that occurs in the modern forest with the same area as the 1943 fire, Sugita's model predicts that changes in pollen percentages of all the taxa (spruce, aspen, pine, birch and alder) in Rainbow Lake A are less than 5 % relative to the pre-disturbance conditions (Fig. 9). FLS model provides the nearly same results (Sugita, unpublished). Of course, it is difficult to compare directly the simulation results with the actual changes in pollen assemblages from the lake (Fig. 2), because the forest composition and forest cover before the 1943 fire is not well known. Nevertheless, even in a real forested vegetation, little change in pollen percentages and pollen accumulation rates would be expected after the 1943 fire or fire events with the similar size and location from the lake.

Simulation Experiments on Changes in Pollen Assemblages Immediately after Disturbance

The simulation results on pollen representation of disturbance in the simulated landscape using Sugita's model are presented in Figs. 10 and 11. The results using FLS model are nearly identical to the results shown in the figures. Relative changes in pollen loadings and in pollen percentages immediately after disturbance are shown in Figs. 10 and 11, respectively. For example, the relative change of spruce in pollen loading after disturbance at lake shore (lake radius 100 m; disturbed area 2500 ha) is -0.6, which corresponds to a 60% decline relative to the pre-disturbance level (Fig. 10-1).

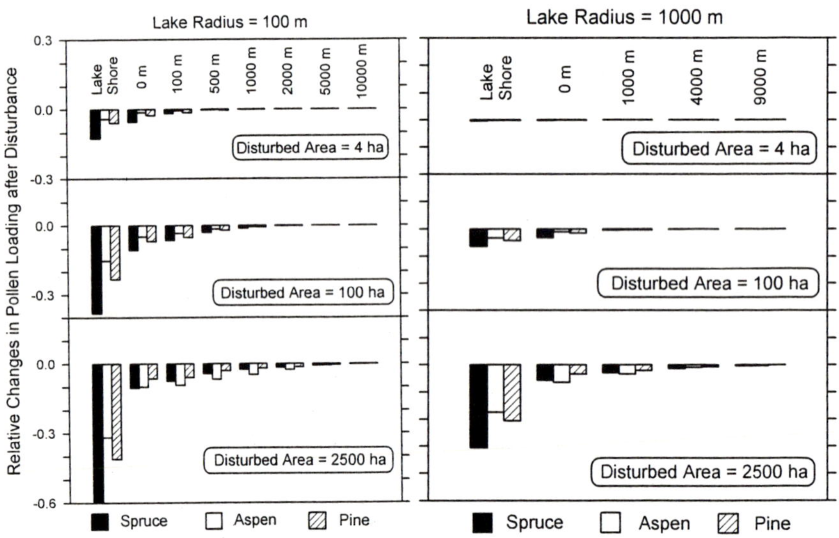

Figure 10. Changes in simulated pollen loading after disturbance, relative to the pre-disturbance levels. Changes are expressed in proportion. Figure 10-1 shows the simulation results for a lake 100 m in radius, and Fig. 10-2 for a lake 1000 m in radius. Simulations for the three different sizes of the disturbed area (4 ha, 100 ha, and 2500 ha) were presented in the top, middle and bottom rows, respectively, in each figure. Each column presents the relative changes in pollen loading for each taxon for a disturbance at various locations from the lake. The mean change in pollen loading at nine sites are calculated for each taxon. "Lake shore" is concentric area around the lake.

Simulations show that **pollen loadings** just after disturbance decline in all cases (Figs. 10-1 and 10-2). However, disturbance events only a few hundred meters away from a lake 100 m in radius do not cause detectable changes in pollen loadings of any taxon (Fig. 10-1). A disturbance patch of 4 ha, which is equivalent to the size of the lake, generates less than a 10% decline of pollen loadings, except for spruce in the case of lake shore disturbance. A disturbance patch of 100 ha, ca. 32 times larger than a lake, causes a significant decline of pollen loadings (>20%) when it occurs at lake shore. When the distance between the lake and disturbance patch becomes 100 m or greater, however, the decline of pollen loadings is less than 10%. When a large disturbance patch (2500 ha) occurs at the lake shore, a decline of 30-60% of pollen loading takes place; however, a 2500 ha disturbance causes less than a 10% decline in pollen loading when it occurs only 100 m away from the lake. Other simulations indicate that pollen loadings on a lake 100 m in radius do not change significantly when a disturbance patch smaller than 25 ha occurs at the lake shore (Sugita, unpublished). When pollen loadings on a lake 1000 m in radius are considered, the relative decline after disturbance is smaller than 10% at all cases, except for a large disturbance (2500 ha) at the lake shore (ca. 20-30% decrease) (Fig. 10-2). The area of this disturbance patch is ca. 8 times larger than the lake. For pollen to detect disturbance events in the simulated landscape, therefore, the area of a disturbance patch at lake shore needs to be at least 8 times larger than the lake.

Temporal changes in **pollen percentage** do not necessarily reflect the direction of changes in forest composition in the disturbed area (Figs. 11-1 and 11-2). Pollen percentages do not show a simultaneous decline of all of the three species after disturbance simply because of a closed universe of percentages. Spruce pollen percentages always decline after disturbance, whereas aspen and pine increase for most of the simulation results. This trend is especially clear when a disturbance at the lake shore is simulated. This phenomena can be explained by the differences in pollen dispersal characteristics among the three taxa. Spruce pollen travels a much shorter distance than aspen and pine pollen, so that, once disturbance removes all the canopy in the patch, spruce pollen coming from outside of the disturbed area is much less abundant than aspen and pine. Thus, spruce pollen percentages tend to decrease after disturbance, whereas pollen percentage of pine and aspen tend to go up.

When a lake 100 m in radius is used (Fig. 11-1), relative changes in pollen percentages do not exceed 5% of the pre-disturbance levels except for disturbances 100 ha and 2500 ha in area at the lake shore. A disturbance 4 ha in area at lake shore does not cause changes greater than 10% in pollen percentages. When the lake radius is 1000 m (Fig. 11-2), relative changes in pollen percentage were well below 2% of the pre-disturbance percentages, except for a large disturbance (2500 ha) at lake shore (5-10% changes).

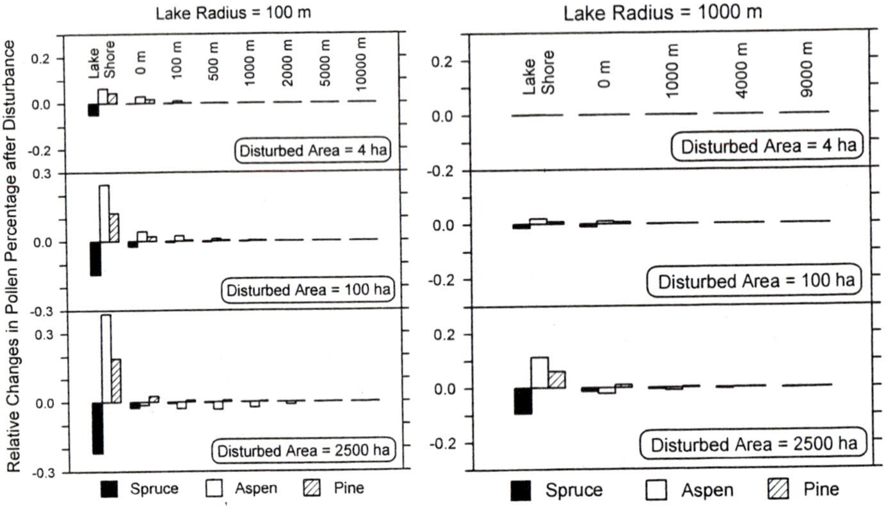

Figure 11. Changes in simulated pollen percentage after disturbance, relative to the pre-disturbance levels. Changes are expressed in proportion. Figure 11-1 shows the results for a lake 100 m in radius, and Fig. 11-2 for a lake 1000 m in radius. Other explanations of the figures are the same as in Fig. 10.

Discussion and Conclusions

Our simulations show clearly that for pollen to record severe disturbance of the local vegetation, the disturbance patch must be significantly larger than a lake (at least eight times larger than the lake in the simulated landscape) and it must be located at, or very near, the lake shore. Even a 2500 ha disturbance that was a few hundred meters away did not cause detectable changes in pollen loading or pollen percentage. This conclusion is valuable for interpreting fossil pollen records and for reconstructing vegetation dynamics. When abrupt changes in pollen assemblages, unrelated to regional pollen stratigraphy, are detected, a reasonable interpretation is that a the sizable disturbance event took place very close to the lake. Simulation results also imply that the estimates of disturbance frequency based on pollen from large lakes could be lower than the estimates from smaller lakes, because large lakes may record only infrequent disturbances that occur over hundreds of square kilometers which tend to occur less frequently than small scale disturbances do. More empirical studies are required to assess the effects of lake size on estimating fire frequency.

Our simulation results support the conclusion of MacDonald et al. (1991) that the pollen record is capable of coarsely resolving successional stages when stand-clearing fires burned most of the drainage-basin of Rainbow Lake A. Large fires at the lake shore, such as are common in the boreal forest around Rainbow Lake A, should be represented fairly well in the pollen record from lakes of similar size. For example, a fire 125 years before 1986, which took place within a 100 m radius of Rainbow Lake A (MacDonald et al. 1991), and its effects on vegetation are clearly recorded in both pollen and charcoal profiles in the lake (Fig. 2). To obtain more detailed information on the areas burned and spatial heterogeneity of post-disturbance successional patterns, however, a network of study sites within a small area will be required. Even a medium-size disturbance, such as 408 ha disturbance in 1943 (Fig. 3), only a few hundred meters away does not create dramatic changes in pollen percentages and pollen loading as shown in real pollen data from Rainbow Lake A (Fig. 2) and in simulated pollen assemblages (Fig. 11). The sizes of lakes also should be carefully chosen according to the spatial scales of disturbance investigators are interested in. For example, large lakes (e.g. 300 ha lakes) would not be able to detect changes in forest composition after disturbances 100 ha in size (Figs. 10-2 and 11-2). Careful site selection is a key for successful research projects, and simulation experiments like ours can provide insights into efficient research designs.

It should be noted that the Interpretation of pollen percentage data, such as used in this study, can be problematic. Simulations demonstrate clearly that the direction of changes in pollen percentages for individual species do not necessarily reflect the direction of shifts in forest composition. Fagerlind (1952) first pointed out this problem, and cautioned palynologists on the danger of misinterpreting pollen percentage data on the direction of plant succession. Percentage data, thus, may not be appropriate for reconstructing the successional pathways after disturbance. Although pollen percentages are not ideal, they can still be useful with careful inspection. For example, MacDonald et al. (1991) found a similar pattern of post-fire changes in pollen percentages and PARs from Rainbow Lake A. Our simulations used only three arboreal taxa. In reality, other early-successional and indicator species may indicate the types of forest changes in post-disturbance vegetation. Also, when a large data set of surface pollen assemblages and vegetation survey is available in the same region, pollen assemblages in percentages could be compared with modern analogs for interpretation at the forest stand level. Nevertheless, other independent evidence of fire and other disturbance, such as charcoal and pollen of early-successional species, is critical for detecting disturbance and its effects on forest development.

In conclusion, simulation experiments suggest that fossil pollen records can portray past disturbances (fire or other agents) and post-disturbance succession under restricted but identifiable circumstances. Research designs aided by simulations and based on networks of fossil pollen sites can provide estimates of location and magnitude of disturbances and recurrence intervals of fire and other disturbance agents.

The sedimentary records from Rainbow Lake A (Fig. 2) show a correspondence between macroscopic charcoal and pollen for the fire 125 years before 1986. If only the charcoal evidence had been used, another local fire might have been noted 145 years before 1986. A combination of pollen and charred particles analyses may best assure that only local fires are recorded in the sediments, forming a basis for reliable sedimentary records of past biomass burning.

Acknowledgments

Sugita is grateful to M.B. Davis and R.R. Calcote for stimulating discussion and encouragement. An earlier version of the manuscript was greatly improved by the comments from R.R. Calcote, J.S. Clark, M.B. Davis, S.T. Jackson, C. Whitlock, and two anonymous reviewers. Financial support was provided for Sugita by grants from NSF (DEB9221371) and the Mellon Foundation to M.B. Davis, for MacDonald by NSERC Operating and Equipment Grants, and for Larsen by McMaster Northern Studies Grants and McMaster Geography Departmental Scholarships.

Literature References

Bradshaw RHW, Webb III T (1985) Relationships between contemporary pollen and vegetation data from Wisconsin and Michigan, USA. *Ecology* 66:721-737

Calcote R (1995) Pollen source area and pollen productivity: evidence from forest hollows. *J of Ecol* 83:591-602

Chamberlain AC (1953) *Aspects of the Travel and Deposition of Aerosol Clouds and Vapours.* Atomic Energy Research Establishment Report HP/R 1261 pp 1-35

Chamberlain AC (1975) The movement of particles in plant communities. In Monteith JL (ed) *Vegetation and the Atmosphere* v 1 pp 155-203 Academic Press New York/London

Clark JS (1988) Particle motion and the theory of charcoal analysis: source area, transport, deposition, and sampling. *Quat Res* 30:67-80

Clark JS (1990) Fire and climate change during the last 750 yrs in northwestern Minnesota. *Ecol Mono* 60:135-159

Cooper WS (1913) The climax forest of Isle Royal, Lake Superior, and its development. *Botanical Gazette* 55:1-44 55:115-140 55:189-235

Cwynar LC (1978) Recent fire and vegetation from annually laminated sediment of Greenleaf Lake, Algonquin Park, Ontario. *Can J of Botany* 56:10-21

Cwynar LC (1987) Fire and the forest history of the North Cascade Range. *Ecology* 68:791-802

Davis MB, Brubaker LB (1973) Differential sedimentation of pollen grains in lakes. *Limnol and Oceanogr* 18:638-646

Davis MB, Moeller RE, Ford J (1984) Sediment focusing and pollen influx. In Haworth Y, Lund JWG (eds) *Lake Sediments and Environmental History* pp 261-293 University of Leicester Press Leicester

DeGrandpre L, Gagnon D, Bergeron Y (1993) Changes in the understory of Canadian southern boreal forest after fire. *J of Veg Sci* 4:803-810

Dix RL, Swan JMA (1971) The role of disturbance and succession in upland forest at Candle Lake, Saskatchewan. *Can J of Botany* 49:657-676

Eisenhut G (1961) *Untersuchungen über dei Morphologie und Ökologie der Pollenkörner heimischer und fremdländischer Waldbäume.* Paul Parey Hamburg (Translated by Jackson ST, Jaumann P, 1989)

Environment Canada (1982) Climatic Normals. *Atmospheric Environment Service* Ottawa

Fagerlind F (1952) The real signification of pollen diagrams. *Botaniska Notiser* 105:185-224

Gajewski K, Winkler MG, Swain AM (1985) Vegetation and fire history from four lakes with varved sediments in northwestern Wisconsin (USA). *Rev of Palaeobot and Palynol* 44:277-292

Green DG (1981) Time series and postglacial forest ecology. *Quat Res* 15:265-277

Green DG (1982) Fire and stability in the postglacial forests of southwest Nova Scotia. *J of Biogeogr* 9:29-40

Gregory PH (1973) *The Microbiology of the Atmosphere* (3rd ed). Leonard Hill Aylesbury

Heinselman ML (1981) Fire and succession in the conifer forests of northern North America. In West DC Shugart HH Botkin DB (eds) *Forest Succession* pp 374-405 Springer-Verlag NewYork

Jacobson GL, Bradshaw RHW (1981) The selection of sites for paleoenvironmental studies. *Quat Res* 16:80-96

Jackson ST (1990) Pollen source area and representation in small lakes of the northeastern United States. *Rev of Palaeobot and Palynol* 63:53-76

Jackson ST, Webb III T, Prentice IC, Hansen JE (1995) Exploration and calibration of pollen/vegetation relationships: a PC program for the extended R-value models. *Rev of Palaeobot and Palynol* 84:365-374

Johnson EA (1981) Vegetation organization and dynamics of lichen woodland communities in the Northwest Territories, Canada. *Ecology* 70:1335-1345

Kelsall JP, Telfer ES, Wright TD (1977) The effects of fire on the ecology of the boreal forest, with particular reference to the Canadian north: a review and selected bibliography. *Can Wildlife Srv* Occasional Paper 32

Larsen CPS, MacDonald GM (1996) A one thousand year record of boreal forest fire and succession. (in review)

MacDonald GM, Larsen CPS, Szeicz JM, Moser KA (1991) The reconstruction of boreal forest fire history from lake sediments: a comparison of charcoal, pollen, sedimentological and geochemical indices. *Quat Sc Revs* 10:53-71

Mosimann JE (1965) Statistical methods for the pollen analyst: multinomial and negative multinomial techniques. In Kummel B, Raup D (eds) *Handbook of Paleontological Techniques* pp 636-673 WH Freeman San Francisco

Okubo A, Levin SA (1989) A theoretical framework for data analysis of wind dispersal of seeds and pollen. *Ecology* 70:329-338

Patterson III WA, Backman AE (1988) Fire and disease history of forests. In Huntley B Webb III T (eds) *Vegetation history* pp 603-632 Kluwer Academic Publisher Dordrecht

Pickett STA (1989) Space-for-time substitutions and an alternative to long-term studies. In Likens GE (ed) *Long-term Studies in Ecology* pp 110-135 Springer-Verlag New York

Prentice IC (1985) Pollen representation, source area, and basin size: toward a unified theory of pollen analysis. *Quat Res* 23:76-86

Prentice IC (1988) Records of vegetation in time and space: the principles of pollen analysis. In Huntley B Webb III T (eds) *Vegetation history* pp 17-42 Kluwer Academic Publishers Dordrecht

Prentice IC, Berglund BE, Olsson T (1987) Quantitative forest-composition sensing characteristics of pollen samples from Swedish lakes. *Boreas* 16:43-54

Prentice IC, Webb III T (1986) Pollen percentages, tree abundances and the Fagerlind effect. *J of Quat Sc* 1:35-43

Romme WH (1982) Fire and landscape diversity in subalpine forests of Yellowstone National Park. *Ecol Mono* 52:199-221

Schwartz MW (1989) Predicting tree frequencies from pollen frequency: an attempt to validate the R value method. *New Phytologist* 112:120-143

Sugita S (1993) A model of pollen source area for an entire lake surface. *Quat Res* 39:239-244

Sugita S (1994) Pollen representation of vegetation in Quaternary sediments: I. Theory and methods in patchy vegetation. *J of Ecol* 82:881-897

Sugita S, Tsukada M (1982) The vegetation history in western North America. I. Mineral and Hall Lakes. *Jap J of Ecol* 32:499-511

Sutton OG (1953) *Micrometeorology*. McGraw-Hill New York

Swain AM (1973) A history of fire and vegetation in northeastern Minnesota as recorded in lake sediments. *Quat Res* 3:383-396

Swain AM (1978) Environmental changes during the past 2000 years in north-central Wisconsin: analysis of pollen, charcoal and seeds from varved lake sediments. *Quat Res* 10:55-68

Tsukada M, Deevey EJ (1967) Pollen analysis from four lakes in the southern Maya area of Guatemala and El Salvador. In Cushing EJ, Wright Jr HE (eds) *Quaternary Paleoecology* pp 303-331 Yale University Press New York

Van Cleve K, Chapin III FS, Flanagan PW, Viereck LA, Dyrness CT (eds) (1986) *Forest Ecosystems of the Alaska Taiga*. Springer-Verlag New York

Waddington JCB (1969) A stratigraphic record of the pollen influx to a lake in the Big Wood of Minnesota. *Geological Society of America*,Special Paper 123:263-283

Webb III T, Howe SE, Bradshaw RHW, Heide KM (1981) Estimating plant abundances form pollen percentages: the use of regression analysis. *Rev of Palaeobot and Palynol* 34:269-300

Appendix 1. Sugita's model for pollen dispersal and deposition (1993, 1994). Definitions and dimensions of mathmatical symbols are listed at the end of this appendix.

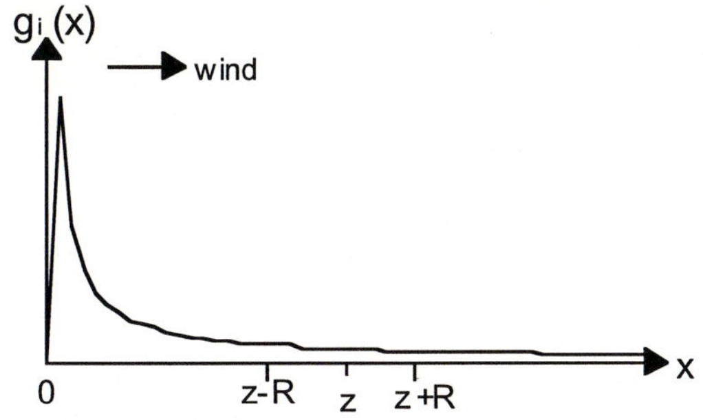

Let's assume a lake (R m in radius), the center of which is located at distance z from a point source of pollen. Using assumption (4) in the text, the pollen source strength of species i at distance z in a landscape is expressed as:

$$Q_{0,i}(z) = P_i \cdot X_i(z) \cdot 2 \cdot \pi \cdot z \cdot dz \qquad 1).$$

Total pollen deposition between distances z-R and z+R can be calculated as $\int_{z-R}^{z+R} Q_{o,i}(z) \cdot g_i(x)dx$, where $g_i(x)$ represents a function describing arc-wise pollen deposition.

Sugita's model assumes that the ratio of pollen deposition on a lake to pollen deposition between concentric rings of radii z-R and z+R is the same as the ratio of the area of a lake to the area between concentric rings of radii z-R and z+R. Then, total pollen deposition on a lake is approximated as:

$$Y_i(z) = \frac{R}{4 \cdot z} \cdot Q_{0,i}(z) \cdot \int_{z-R}^{z+R} g_i(x)dx \qquad (2).$$

Combining eqs. (1) and (2), total pollen loading on a lake, which comes from between lake shore and Zc (Zc > R) from the center of a lake, thus, becomes,

$$Y_i(R,Zc) = \int_R^{Zc} \frac{\pi \cdot R \cdot P_i \cdot X_i(z)}{2} \cdot \left[\int_{z-R}^{z+R} g_i(x)dx \right] dz \qquad (3).$$

The dimension of $Y_i(R,Zc)$ is [grains], which is appropriate for total pollen loading on a lake with radius R. Mean pollen loading on a lake is then defined as $Y_i(R,Zc) / (\pi R^2)$, and its dimension is [grains m^{-2}].

The dimensions of the symbols are defined as follows:

P_i : pollen productivity of species i . [grains · kg^{-1} (or other units of biomass)] or [grains · m^{-2} (or other units of area)].
X_i (z): average abundance of species i at distance z (m) from a point at the center of a lake. [kg (or other units of biomass) · m^{-2}].
b_i: a species-specific constant, defined by wind speed and fall speed of pollen (Prentice 1988). [$m^{-1/8}$].

$$b_i = 4 \cdot v_g \Big/ \left(n \cdot u \cdot c_z \cdot \sqrt{\pi} \right)$$

R: radius of a circular lake. [m].
u: mean wind speed. [m sec^{-1}].
c_z: diffusion coefficient for vertical direction, [$m^{1/8}$].
c_y: diffusion coefficient for lateral direction. [$m^{1/8}$].
v_g: deposition velocity of pollen. [m sec^{-1}].
z: distance from a point at the center of a lake. $z \geq R$. [m].
x: distance from a source. [m].
$g_i(x)$: arc-wise pollen deposition function of species i at distance x (m). [m^{-1}]. Prentice (1985) and Sugita (1993, 1994) define $g_i(x)$ as,

$$g_i(x) = b_i \cdot \gamma \cdot x^{\gamma 1} \cdot exp(-b_i \cdot x^{\gamma}). \quad (\gamma=1/8)$$

Appendix 2. Finite-Line Source (FLS) model for pollen dispersal and deposition (Sugita, in preparation).

Definitions and dimensions of mathmatical symbols are the same as in Appendix 1. Assumptions (1)-(4) in the text apply to this model. Main difference from Appendix 1 is that FLS model is a two-dimensional model, assuming a finite line source with distance z between the line and a point at the center of a lake.

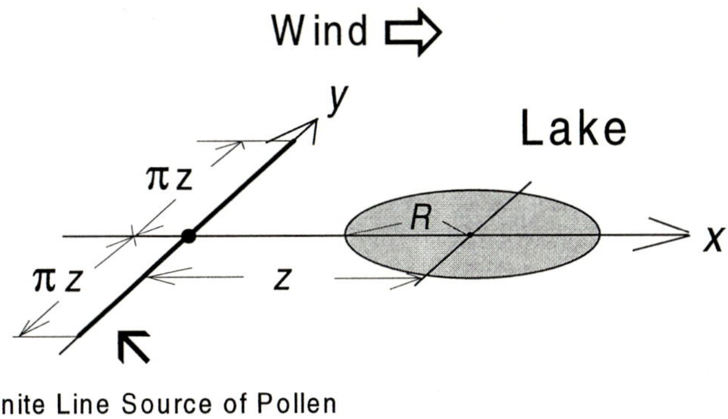

In FLS model, the length of the line source equals the circumference a circle with radii z, which is $2 \cdot \pi \cdot z$. The mean source strength of the line $Q_{0,i}(z)$ has the dimension of [grains m^{-1}] (Sutton 1953). Total pollen production (Ti(z)) of species i at distance z equals $P_i \cdot X_i(z) \cdot 2 \cdot \pi \cdot z \cdot dz$. In this model, $Q_{0,i}(z)$ is defined as:

$$Q_{0,i}(z) = \frac{T_i(z)}{2 \cdot \pi \cdot z} = P_i \cdot X_i(z) \cdot dz \quad \text{.......(1).}$$

Dimension of $Q_{0,i}(z)$ is [grains m^{-1}], appropriate for the model. Total pollen loading on a lake from a finite line source at distance z can be described from Sutton (1953) and Chamberlain (1953) as:

$$Y_i(z) = Q_{0,i}(z) \cdot \int_{z-R}^{z+R} \frac{b_i \cdot n}{4 \cdot x^{1-\gamma}} \cdot e^{-b_i \cdot x^{\gamma}} \cdot \int_{-\sqrt{R^2-(z-x)^2}}^{\sqrt{R^2-(z-x)^2}} \left\{ erf\left(\frac{\pi \cdot z - y}{c_y \cdot x^{1-\gamma}}\right) + erf\left(\frac{\pi \cdot z + y}{c_y \cdot x^{1-\gamma}}\right) \right\} dy \cdot dx \quad \text{....(2).}$$

Combining eqs. (1) and (2), total pollen loading of species i on a lake, which originates between the lake shore and a given distance Z_c, becomes,

$$Y_i(R,Z_c) = P_i \cdot \int_{R}^{Zc} X_i(z) \cdot \int_{z-R}^{z+R} \frac{b_i \cdot n}{4 \cdot x^{1-\gamma}} \cdot e^{-b_i \cdot x^{\gamma}} \cdot \int_{-\sqrt{R^2-(z-x)^2}}^{\sqrt{R^2-(z-x)^2}} \left\{ erf\left(\frac{\pi \cdot z - y}{c_y \cdot x^{1-\gamma}}\right) + erf\left(\frac{\pi \cdot z + y}{c_y \cdot x^{1-\gamma}}\right) \right\} dy \cdot dx \cdot dz \quad \text{...(3).}$$

erf represents the error function, defined as $erf(x) = \frac{2}{\sqrt{\pi}} \cdot \int_0^x e^{-t^2} dt$. Dimension of $Y_i(R,Z_c)$ is [grains], appropriate for total pollen loading on the entire surface of a lake.

Appendix 3. Correction methods for pollen productivity estimate (Sugita and Calcote, in preparation).

Pollen productivity of species i (P_i) relative to pollen productivity of species j (P_j) can be estimated using the maximum likelihood estimates of pollen representation coefficients for species i (α_i) and species j (α_j) (Prentice and Parsons 1983, Prentice and Webb 1986, Sugita 1994, Calcote 1995, Jackson et al. 1995). Homogeneous plant abundance for each species is assumed for all the correction methods. b for species I (b_i) is defined in Appendix 1. R represents radius of a sedimentary basin used for the study, ZL represents the maximum extent of vegetation survey, in meters from a point at the center of a basin, for estimating plant abundance of species i and j.

(1) Correction method, according to Sugita's model (1993, 1994) (Appendix 1):

$$\frac{P_i}{P_j} \equiv \frac{\alpha_i}{\alpha_j} \cdot \frac{\int_R^{ZL}\int_{z-R}^{z+R} g_j(x)\cdot dx\cdot dz}{\int_R^{ZL}\int_{z-R}^{z+R} g_i(x)\cdot dx\cdot dz}$$

$$= \frac{\alpha_i}{\alpha_j} \cdot \left(\frac{b_i}{b_j}\right)^8 \cdot \frac{\Gamma_{b_j\cdot(ZL-R)^{1/8}}(8) - \Gamma_{b_j\cdot(ZL+R)^{1/8}}(8) + \Gamma_{b_j\cdot(2\cdot R)^{1/8}}(8)}{\Gamma_{b_i\cdot(ZL-R)^{1/8}}(8) - \Gamma_{b_i\cdot(ZL+R)^{1/8}}(8) + \Gamma_{b_i\cdot(2\cdot R)^{1/8}}(8)},$$

where $\Gamma_x(a)$ is the incomplete gamma function, defined as $\Gamma_x(a) \equiv \int_0^x e^{-t}\cdot t^{a-1}\cdot dt$.

(2) Correction method, according to Finite-Line Source model (Appendix 2):

$$\frac{P_i}{P_j} \equiv \frac{\alpha_i}{\alpha_j} \cdot \frac{f_j}{f_i},$$

$$\text{where } f_i = \int_R^{ZL}\int_{z-R}^{z+R} \frac{b_i\cdot n}{4\cdot x^{1-\gamma}} \cdot e^{-b_i\cdot x^{\gamma}} \cdot \int_{-\sqrt{R^2-(z-x)^2}}^{\sqrt{R^2-(z-x)^2}} \left\{ erf\left(\frac{\pi\cdot z-y}{c_y\cdot x^{1-\gamma}}\right) + erf\left(\frac{\pi\cdot z+y}{c_y\cdot x^{1-\gamma}}\right)\right\} dy\cdot dx\cdot dz$$

erf is the error function (see Appendix 2) and $\gamma \equiv 1/8$.

The contribution of humans to past biomass burning in the tropics

A.P. Kershaw, M.B. Bush[1], G.S. Hope[2], K.-F. Weiss[3], J.G. Goldammer[3], R. Sanford[4]
Centre for Palynology and Palaeoecology
Department of Geography and Environmental Science
Monash University
Clayton, Vic. 3168 Australia

Summary

Tropical ecosystems, particularly savannas, contribute substantially to global biomass burning which influences the composition of the atmosphere as well as the nature and stability of soils, hydrology and vegetation. The history of burning in the tropics is constructed mainly from the analysis of charcoal preserved in accumulated sediments and within soils derived from a range of tropical environments. There is evidence for burning from the mid Tertiary associated with the development of savannas and drier or more seasonal rainforest types under global drying. The trend towards increased burning appears to have accelerated in more recent geological times due to the activities of people. Some separation of human from natural burning has been possible from a comparison of past burning patterns in areas with different human histories, and particularly from rainforest which is unlikely to experience frequent fires in the absence of people. Although there is evidence that *Homo erectus* used fire as long ago as 1.5 my, it is not until the time of evolution of *H. sapiens* that any relationship between people and biomass burning can be demonstrated. The earliest proposed anthropogenic burning is from at least 40 ky and possibly up to 140 ky BP in Australia where the environment may have been extremely sensitive to impact from an additional ignition source. Dates for increased burning within the rainforests of New Guinea also extend back to about 40 ky. Further increases in tropical burning are recorded by sites in Central and South America from the early Holocene, presumably with the arrival of people, and in most parts of the tropical region from the mid Holocene with the development and spread of agriculture and with the colonization of oceanic islands. Charcoal peaks are frequently associated with times of climate change and it is likely that, regardless of the nature of future fire management strategies, biomass burning will continue to increase in the near future because of the predicted period of climate change.

Introduction

Tropical ecosystems today contribute substantially to global biomass burning (Goldammer 1990, 1992, 1993a, Crutzen and Goldammer 1993). In fact it has been estimated that about 87% of the atmospheric carbon emission from biomass burning

[1] Department of Botany, Duke University, Durham, North Carolina 27708, USA
[2] Division of Archaeology and Natural History, The Australian National University, Canberra, A.C.T. 0200, Australia
[3] Fire Ecology Research Group, Max Planck Institute for Chemistry, c/o Freiburg University, PO Box D-79085, Freiburg, Germany
[4] Department of Biological Sciences, University of Denver, Denver, Colorado 80208, USA

NATO ASI Series, Vol. I 51
Sediment Records of Biomass Burning and Global Change
Edited by James S. Clark, Hélène Cachier, Johann G. Goldammer, and Brian Stocks

comes from the tropics (Andreae 1991). Most sustained burning occurs in savannas, which occupy at least a quarter of the terrestrial tropics. Here a continuous ground cover of grasses combined with a highly seasonal climate allows burning to occur every one to three years in most areas. As the dry season intensifies and rainfall becomes more unreliable, savanna gives way to desert that is either incapable of carrying fire or experiences only occasional burns. On its most humid margins, savanna is replaced by rainforest habitats, which experience lower fire frequency. Here, reduced seasonality and higher rainfall combined with a dense tree canopy that maintains a moist understory and a less-flammable flora, excludes natural fire except, perhaps, during severe droughts. If the restricted montane communities are excluded, the third major tropical vegetation type is seasonal forest, which may be evergreen and sclerophyllous, semideciduous or deciduous. These forests generally occur under similar climatic conditions as savanna but are distinct, because they occupy soils unsuited to the development of a continuous grass cover or they have a different history. These forests experience a variety of fire regimes, depending largely on the nature and degree of human impact, but are intolerant of regular burning.

It is clear that people are an integral part of tropical systems, and there is increasing concern over the extent to which their impact, through burning, is contributing to global environmental changes. Human effects include: i) short term increases in atmospheric CO_2, due to more frequent or geographically more extensive burning, ii) longer term decreases in CO_2, because of carbon sequestration, increased erosion and its impact on hydrology and land use, iii) decreased precipitation by reduction of the tropical rainforest moisture source, iv) higher atmospheric pollution levels, and v) changes in biotic communities, including losses of biodiversity and increased outbreaks of disease. Before policy for remedial action can be formulated, it is necessary to examine the historical role of humans in the production of the present extent and nature of burning patterns.

It is fortuitous that fire leaves a direct indication of its occurrence in the form of black carbon (soot or charcoal). However, it is only recently that measurements of carbon have been undertaken systematically in palaeoecological studies within the tropics. Consequently the presence and impact of fire has generally been inferred from the nature of vegetation changes and, in particular, disturbance indicators within pollen records. Even with black carbon records, it is extremely difficult to determine fire regimes or to separate natural from human produced fires.

This paper focuses on the charcoal in association with other palaeoecological data, where available, to reconstruct the history of burning in tropical regions. The data base is small, and much information is still unpublished and therefore not fully accessible. However, there is some coverage of the major tropical regions that, through their different histories of human colonisation and land use, should allow some separation of human and natural burning and the establishment of general vegetation/land use–fire–charcoal relationships. Sites with known burning records together with the age range of these records and a human cause, where inferred by the original author, are shown on Fig. 1.

Pre-human burning

There has been a great deal of debate over the frequency and extent of biomass burning before the appearance of modern people. Such debate has centred on savanna vegetation—whether savannas are a product of human impact or, if they existed before people, whether fire was a factor in their development. Ecological studies suggest that burning has a long history, because savanna and sclerophyll plants possess adaptations to survive fire. However, many of the adaptations may have evolved in response to other disturbance factors such as climatic variability, grazing, infertile or waterlogged soils.

Ample evidence now indicates that vegetation has been subject to burning since continents were first colonized by land plants. Charcoal in Pacific Ocean cores indicates that there has been a general increase in burning through the Cenozoic, which can be related to global drying (Herring 1985, Suman et al., this volume). The relationship between savanna and fire has been convincingly demonstrated in a marine core from the Niger Delta (Morley and Richards 1993). Here, an increase in grass pollen is followed by the presence of charred grass cuticle fragments around the Early/Mid Miocene boundary with a subsequent sharp rise in both components in the Late Miocene, around 8 my BP. A similar sharp increase in grass pollen occurs about this time (i.e. Late Miocene) in a marine core off northwestern Australia (Martin and McMinn 1993) and, although there is no accompanying charcoal curve, an increase in burning with savanna expansion might be inferred. Synchroneity between these two records suggests development of savanna within the tropics. There are no indications of a relationship between tropical forests and fire during the Tertiary. However, there is evidence for an increase in charcoal within the Latrobe Valley coals of southeastern Australia in the Early Miocene (Kershaw et al. 1991b). That these deposits formed under a subtropical rainforest environment lacking grasses indicates the possibility of fire in tropical forests in the late Tertiary.

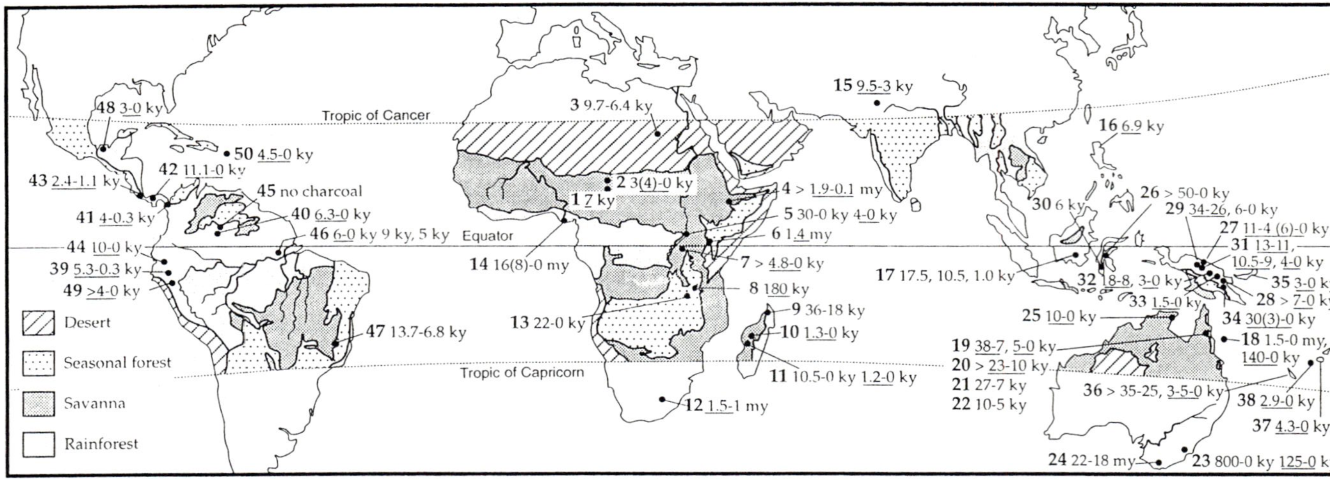

Figure 1. Location of dated burning records in millions of years before present (my) or thousands of years before present (ky), predominantly within the tropics. The records underlined have been considered by the authors to represent mainly anthropogenic burning. 1. Fachi-Dogonbolo (Neumann 1988) 2. Segedim (Schulz 1996) 3. Selima Oasis (Haynes et al. 1989) 4. Omo (Deschamps 1984) 5. Lake Mobutu Sese Seko (Weiss in prep.) 6. Chesowanja (Gowlett et al. 1981) 7. Ahakagyezi Swamp (Hamilton et al. 1986) 8. Kalambo Falls (James 1989) 9. Matsabory Ampozalana Bog (Burney 1987b) 10. Lake Kavitaka (Burney 1987a) 11. Lake Tritrivakely (Burney 1987b) 12. Swartkans Cave (Brain and Sillen 1988) 13. Lake Ishiba Bgandu (Weiss in prep.) 14. Niger Delta (Morley and Richards 1993) 15. Rajasthan Desert (Singh et al. 1974) 16. Mt. Puloz (Goldammer 1992) 17. Kutai National Park (Goldammer and Seibert 1989) 18. ODP site 820 (Kershaw et al. 1993) 19. Lynch's Crater (Kershaw 1986) 20. Strenekoff's Crater (Kershaw et al. 1991a) 21. Humid tropics (Hopkins et al. 1993) 22. Lake Barrine (Walker and Chen 1987) 23. Lake George (Singh and Geissler 1985) 24. Latrobe Valley (Kershaw et al. 1991b) 25. Groote Eylandt (Shulmeister 1992) 26. Wanda (Hope et al. in prep.) 27. Hordorli (Hope and Tulip 1994) 28. Wanum (Garrett-Jones 1979) 29. Kelela and Supulah Hill (Haberle et al. 1991) 30. Ulu Laang (Glover 1977) 31. Telefomin (Hope 1983) 32. Haeapugua (Haberle 1993) 33. Mt Wilhelm (Corlett 1984) 34. Kosipe (Hope 1982) 35. Balof (Hope unpub.) 36. Lac Suprin (Hope unpubl.) and Lac St. Louis (Stephenson unpubl.) 37. Raralevu (Southern 1986) 38. Anietyum (Hope and Spriggs 1982) 39. Lake Ayauchi (Bush and Colinvaux 1988) 40. San Carlos de Rio Negro (Sanford et al. 1985) 41. Lake Wodehouse and Cana Swamp (Bush and Colinvaux 1994) 42. La Yeguada (Piperno et al. 1990, Bush et al. 1992) 43. La Selva (Horn and Sanford 1992) 44. Laguna Llaviuva (Bush unpubl.) 45. Hill of Six Lakes (Bush unpubl.) 46. Lagoa Geral and Lagoa Comprida (Bush unpubl.) 47. Lagoa dos Olhos (De Oliviera 1992) 48. Lago Catemaco (Byrne and Horn 1989, Horn et al. 1992) 49. Junin (Hansen et al. 1994) 50. Laguna Tortuguero (Burney et al. 1994).

Within the Quaternary, high concentrations of grass pollen and charred grass cuticles in the Niger delta core indicate savannas or open woodlands. Oscillations in the concentrations of the charred grass cuticles might relate to glacial/interglacial fluctuations. A marine core from the continental slope adjacent to humid tropics of northeastern Queensland records a predominance of seasonal rainforest pollen, but includes Poaceae, through the last 1.5 my (Kershaw et al. 1993). Charcoal is relatively high throughout the period. Within the humid tropics, a pollen diagram from Lynch's Crater, spanning the last 190,000 yr, indicates near absence of charcoal within more humid (annual rainfall > 2000 mm) phases corresponding with the heights of interglacials, when the site was surrounded by complex rainforest (Kershaw 1994). There is some evidence of burning within the seasonal *Araucaria* forests that were regionally dominant during dry glacials, but the highest charcoal levels occur as peaks at sharp vegetation and climate changes.

The archaeological record and environmental disturbance

The pattern of prehistoric settlement of the world (Fig. 2) and subsequent cultural developments together with evidence for disturbance to the vegetation provide some framework for evaluation of the anthropogenic fire record.

Human settlement

It is generally agreed that people evolved within the present savanna regions of Africa, and substantial evidence for their evolutionary development is preserved in the rift deposits of east Africa (Clark 1980). Fossil records show that the oldest archaeological evidence for *Homo habilis*, the first hominid to use stone tools and therefore perhaps the first to modify his environment to some degree, corresponds with a change to drier and cooler conditions around 2.5 my BP (Bonnefille 1983). It is probable, however, that climatic change was responsible for the development or expansion of savanna vegetation in the region rather than the activities of people. It was not until about a million years later, with the appearance of *Homo erectus*, that there were further substantial cultural developments and a spread of people beyond Africa (Williams et al. 1993). Within the region of interest, the site of Sangiran shows that *H. erectus* had migrated through the Asian tropics to Indonesia by 1.3 my BP.

The oldest evidence for *H. sapiens* is found in southern Africa, dated approximately to the last interglacial. It might be inferred that the world was colonized by modern people from this region, but the human remains suggest that

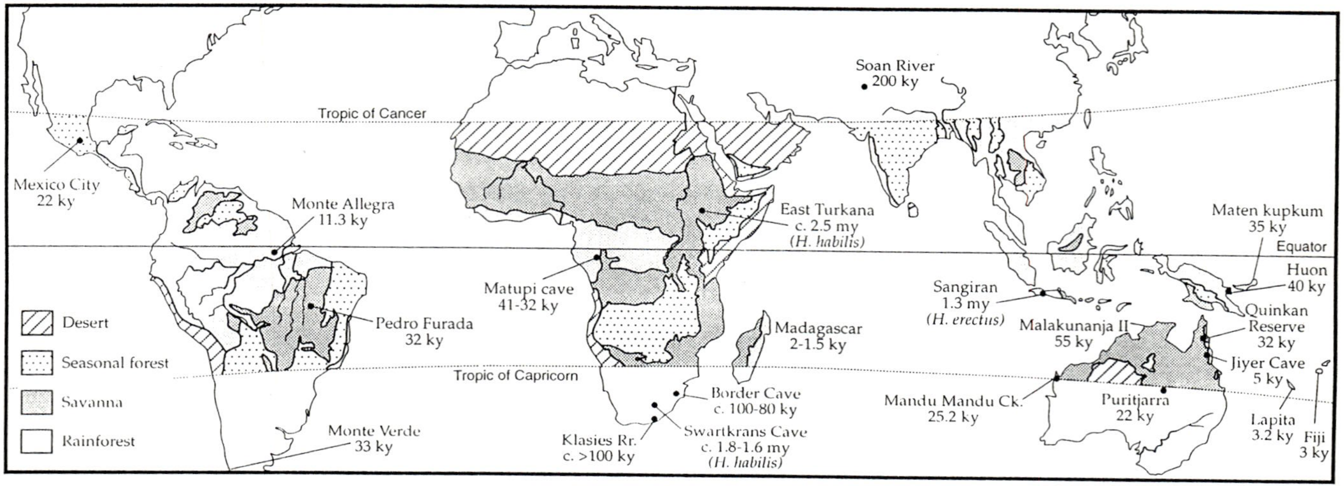

Figure 2. Location of sites showing earliest evidence of people predominantly within tropical regions. Except where indicated, the dates are considered to relate to *Homo sapiens*. Mexico City (Fagan 1989), Monte Allegra (A. Roosevelt pers. comm.), Pedro Furada (Guildon and Delibrias 1986), Monte Verde (Bednarik 1989), Matupi Cave (Van Noten 1977), Klasies River (Deacon 1992), Swartkrans Cave (Brain and Sillen 1988), Border Cave (Deacon 1992), Madagascar (Verin et al. 1970), East Turkana (Fagan 1989), Soan River (Williams et al. 1993), Sangiran (Williams et al. 1993) Mandu Mandu Creek Rockshelter (Morse 1988), Malakunanja II (Roberts et al. 1990), Matenkupkum (Allen 1994), Huon (Groube et al. 1986) Quinkan Reserve (Morwood unpubl.), Jiyer Cave (Horsfall and Hall 1990), Lapita (Enright and Gosden 1992), Fiji (Enright and Gosden 1992).

the population of southern Africa had been isolated for a long period and that mainstream evolution had continued elsewhere. Consequently the actual time and place of evolution of *H. sapiens* is unknown, but is likely to have occurred in the Mid Pleistocene, possibly somewhere in sub-Saharan Africa (Deacon 1992).

Tropical rainforest has been considered unlikely to have contributed to the evolution of modern people. In addition to a paucity of archaeological evidence from such environments, there is controversy over whether rainforest contains the resources to support human populations without the adoption of agriculture (Bird-David 1988, Bailey et al. 1989, Colinvaux and Bush 1991, Hart and Hart 1986). The earliest evidence for people within the present distribution of African rainforest is during the last glacial period when savannas prevailed (Hart and Hart 1986).

There is debate over the time of appearance of *H. sapiens* in both the Americas and New Guinea and Australia, islands that were joined during lower sea level phases of the Pleistocene. It is generally accepted that people entered the Americas during the later part of the Pleistocene. The timing of that entry divides the archaeological and anthropological communities into 'early' and 'late' advocates. 'Early' entry suggested by radiocarbon dated material from cave deposits sets human arrival in South America at about 40-50 ky BP (Guidon and Arnaud 1991). 'Late' entry, the majority view, asserts that humans arrived at about 11-13 ky BP, taking advantage of lowered sea-levels and ameliorating climate to cross Beringia. The 'late' entry school rejects the 'early' dates as being either unrealistic or good but only loosely associated with human artifacts.

It has been argued, from radiocarbon dating of archaeological sites, that people occupied most environments within Australia and New Guinea between 40 and 30 ky BP, with perhaps a somewhat later date for arid regions (Allen 1989) and a much later date (5 ky BP) for tropical rainforest (Horsfall and Hall 1990). However, the application of thermoluminescence (TL) dating to a site in the Northern Territory has indicated an older settlement age of between 55 and 60 ky BP (Roberts et al. 1990). The archaeological debate is centered around the fact that radiocarbon ages are close to the limit of resolution of the method and may reflect an artificial 'radiocarbon boundary', and the questioning of the general reliability of TL dating has yet to be resolved (Allen 1994). However, additional TL evidence from other northern Australian sites (Fullager, personal communication) is leading to a greater acceptance of the older age. The debate is complicated by suggestions, from palynological data, of major changes in the vegetation at earlier times that have been interpreted as a result of Aboriginal burning, and these data will be considered later in relation to the charcoal evidence for human impact. There is much less debate over the timing of subsequent colonization of oceanic islands, essentially the last 4000 to 5000 yr. The degree of cultural sophistication was sufficient to ensure ample archaeological evidence of the presence of people.

Cultural developments and environmental disturbance

The initial colonizers of the major landmasses of the tropics were hunters and gatherers. There is little convincing evidence of widespread impact by these people except perhaps in Australia, where substantial vegetation changes are accompanied by increases in the abundance of charcoal. These will be examined later. Elsewhere, inferred human impact on vegetation by hunters and gatherers has been confined to rainforest, probably because the separation of human from natural impact in open communities that are adapted to disturbance is difficult. An early date of 18 ky for disturbance, in the absence of charcoal evidence, has been tentatively proposed by Maloney (1985) for highland north Sumatra.

It is only with or after the arrival of agriculture that confidence has been given to the allocation of a human cause for disturbance indicators in vegetation. This is apparent in the review of human impact on the world's vegetation from palynological records (Walker and Singh 1991). The clearest, and some of the oldest evidence in the tropics is derived from Central and South America where, in addition to the preservation of food crops in archaeological sites, maize pollen is recorded along with other disturbance indicators in palaeoecological sites. The archaeological record indicates the development and spread of agriculture in central and northwestern South America between about 8 and 4 ky BP. These dates are supported by palaeoecological data with the earliest evidence for maize pollen in highland Colombia dating to about 8 ky BP (Kuhry 1988) and in the Amazon lowlands to about 6 ky BP (Bush et al. 1992).

Similar levels of disturbance are recorded in pollen diagrams from the Southeast Asian to New Guinea region from about the same time, but the lack of identifiable crop pollen has inhibited a direct connection with agriculture (Flenley 1988, Maloney 1985). Some claims for the identification of domesticated rice phytoliths have been made, but it is not yet certain that these can be separated from wild varieties (Chang 1989). A similar problem has confronted archaeologists with claims for rice cultivation beyond 7 ky BP in both southeast Asia and southern China. However, there seems little doubt that rice cultivation was widely practiced in both regions before 5 ky BP. There is some indication of root crop cultivation or intensive collection before this time with the earliest date of about 11 ky BP coming from excavations at Spirit Cave in northeast Thailand (Bellwood 1985). A date almost as old (10.4 ky BP) comes from Kuk Swamp in New Guinea, where an infrastructure indicates the irrigated cultivation of taro or yams (Golson and Gardner 1990). The possibility of much older cultivation comes from the Huon Peninsula of New Guinea where archaeological artifacts, dated to at least 40 ky BP, are likely to have been used for felling trees (Groube et al. 1986). Regardless of the age for the beginning of agriculture, marked disturbance in a number of pollen diagrams from Indonesia and New Guinea strongly suggests widespread forest clearance from after 7.5 ka to 5 ka BP. As previously mentioned agriculture, in association with initial human colonization, then spread through the islands of the Pacific, notably after 4.5 ky BP, and continued until about 1 ky BP.

In contrast to the early evidence for people, and despite the proximity of the tropical regions to the major Middle East centre for an origin of agriculture, obvious signs of widespread disturbance in Africa occur late in the archaeological and palaeoecological record. There is archaeological evidence for cattle raising from about 7 ky BP, including the Sahara, when moisture levels were relatively high. Disturbance indicators in pollen records are evident only one or two thousand years before an intensification of agriculture with the arrival of ironworking about 2 ky BP (Walker and Singh 1991).

Post-occupation burning

A variety of sites for which direct evidence of burning has been recorded, with an indication of whether a human cause has been invoked, are shown on Fig. 1. Except for some older records, almost all sites relate to general environmental burning rather than to those that can be attributed to the domestic use of fire. In addition, full charcoal records are provided for selected studies within the major areas of investigation through the later part of the Quaternary (Figs. 3-6). Charcoal values within these studies have been expressed in a variety of ways. Little attempt has been made to standardise these data, a process which could easily lead to distortion of results. It is felt that individual authors have selected the most appropriate way of presenting their information. As a consequence of this lack of standardization together with the difficulty of relating charcoal values to levels of burning at any site, a point which will be brought up in discussion, no useful comparisons can be made of actual charcoal abundances between sites. The only meaningful comparisons are between values within sites and times of change in values between sites.

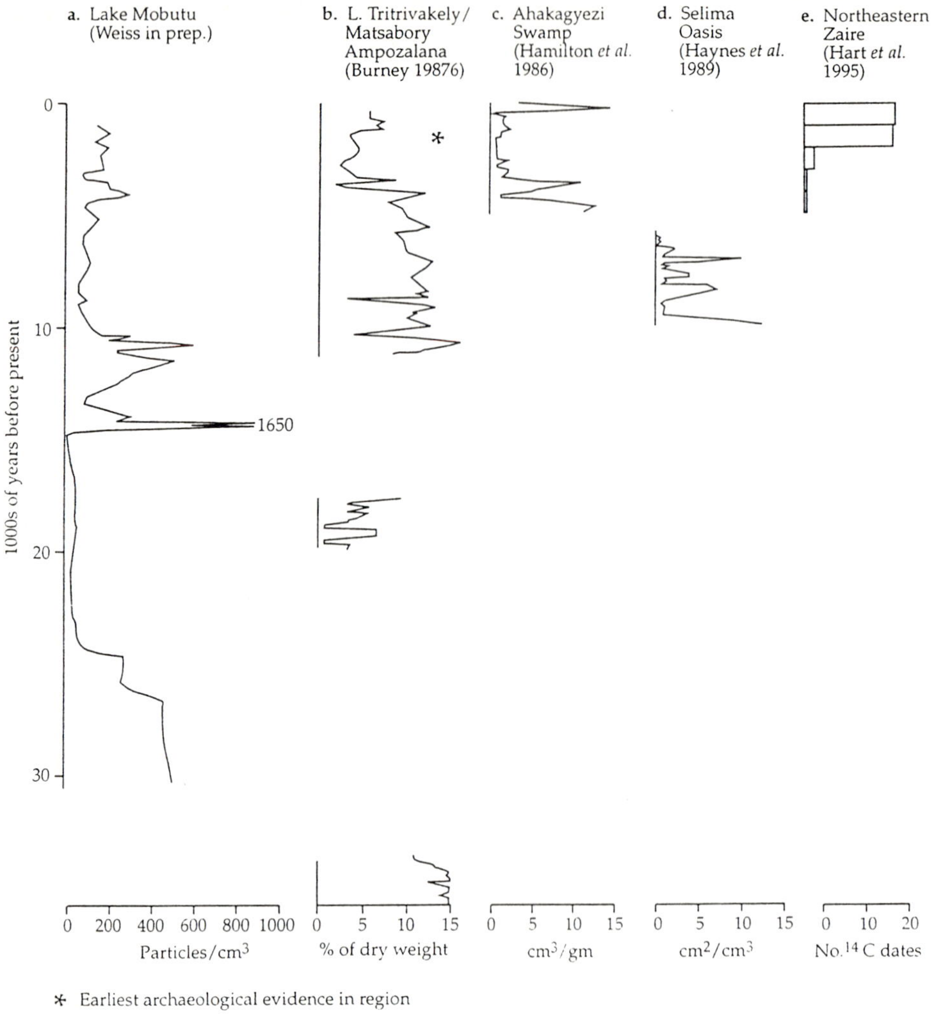

Figure 3. Selected charcoal records from Africa. Site locations are indicated in Figure 1.

Africa

Africa has the oldest records of supposed human burning dating back to about 1.5 my BP (Brain and Sillen 1988). Only archaeological records extend back beyond the last glacial period, and they give little indication of the impact of fire on the landscape. It might be inferred that the ability of people to use fire would have influenced fire regimes and affected the nature and distribution of vegetation types, but the extent of these changes cannot be determined.

Continuous paleoecological sequences, including charcoal curves, are restricted to the last 36 ky, and a number are shown on Fig. 3. The most significant for an assessment of human impact are those from Madagascar, where it had been argued that fire was brought by the first inhabitants about 1.5 ky BP, and that its direct impact and indirect influence on the vegetation were major causes of faunal extinctions. Charcoal records of Burney (1987a, b) clearly indicate that fire and the extensive savanna vegetation were natural. There is some increase in charcoal from low levels around the time of inferred human arrival, but highest charcoal values are recorded between 11 and 4 ky BP and from about 35 to 36 ky BP (Fig. 3b). These data suggest the causes of faunal extinction need reevaluation in light of the time of human arrival.

Most available data from the mainland come from the montane woodland and savanna areas of east Africa, where charcoal has been found throughout existing records. The long continuous record from Mobutu Sese Seko, on the border between Uganda and Zaire (Weiss, in preparation), interpreted in relation to Poaceae pollen, indicates that highest charcoal value occurs when intermediate precipitation levels prevailed from 30 to 25 ky BP and 15 to 10 ky BP (Fig. 3a). At this time there was the most continuous grass cover. Lowest values are recorded in the dry glacial period, between 25 and 15 ky BP, when the vegetation was presumably too sparse to sustain regular burning. Values are also relatively low between 10 and 4 ky BP when high precipitation resulted in an expansion of forest vegetation. Increased charcoal concentrations within the last 4 ky are attributed to an expansion of savanna under dry conditions together with, perhaps, an intensification of human impact, especially agricultural and pastoral activities.

Hamilton et al. (1986) are inclined to interpret presence of charcoal through the last 4.8 ky in Ahakagyezi Swamp as a response to human impact. They consider that charcoal in association with pollen indicators demonstrates an early date for forest clearance and cultivation (Fig. 3c). Although the pollen diagram indicates xeric conditions after about 3.6 ky BP, lack of evidence for crop plants together with the fact that the record does not extend beyond the inferred cultivation phase, must make the human impact interpretation tentative. In this regard, it is interesting to note that the charcoal pattern is similar to that from Madagascar through the recorded period, despite very different human occupation histories.

A recent study of soil charcoal in northeastern Zaire has revealed evidence of rainforest burning through the last 4.2 ky (Hart et al. 1996) (Fig 3e). As in the studies from savanna environments, climate is considered to have exerted the major control over burning patterns with the initiation of charcoal deposition coinciding with the beginning of the drying trend in central Africa and a subsequent charcoal increase around 2 ky BP corresponding with a severe and widespread dry period evident in pollen records within the region (Maley 1992, Schwartz et al. 1990). However, the possibility of human burning through the record is not excluded, and sustained high charcoal levels within the last 2 ky are considered to most likely be the result of human activity.

Other charcoal records come from the dry northern part of Africa. The present Sahara Desert supported areas of savanna during the early to mid Holocene that were associated with fire. There are individual occurrences of charcoal centred on 7-7.5 ky BP (Neumann 1988) and a continuous charcoal sequence from Selima Oasis dated between 9.7 and 6.3 ky B.P (Haynes et al. 1989) (Fig. 3d). Further south, in the present Sahel, charcoal is recorded back to 8 ky BP, but values are highest between 4 ky BP and the present (Schulz 1996). In both areas, fire frequency is related to climate and the ability of the vegetation to carry fire, although human burning is seen as an integral factor. In fact, it is proposed by Schulz (1996) that the Sahel is anthropogenic, created by hunting and stock herding activities within the last 4 ky.

Central and South America

Almost all paleoecological records where charcoal has been systematically counted occur within presently rainforested areas. The longest of these records is from La Yeguada at an elevation of 650 m in Panama (Bush et al. 1992) (Fig. 4a). The site experiences high rainfall (3300 mm annually) with a strong summer maximum. Little charcoal is evident in the xeric, more seasonal phase of the diagram dated between about 14.5 and 11 ky BP. There is then a peak in charcoal, interpreted as a combination of human impact and natural disturbance resulting from an amelioration in climate. Charcoal is subsequently recorded consistently through the diagram, initially as a result of pre-agricultural human disturbance and later from agricultural practises. Despite pollen and phytolith evidence for an intensification of agriculture at about 4 ky BP, there is no accompanying increase in charcoal concentrations. However, agricultural land use and increased charcoal concentrations are evident in humid tropical settings in the Amazon (Sanford et al. 1985, Saldarriaga and West 1986, Bush et al. 1989, Bush and Colinvaux 1994), southern Central America (Bush and Colinvaux 1994) and Mexico (Horn et al. 1992) from the mid-Holocene (see Fig. 4b and c). A similar human origin, presumably the consequence of shifting agriculture, may also be invoked for the presence of soil charcoal in degraded forest soils in southeast Venezuela dating back to 3.5 ky BP (Folster 1992).

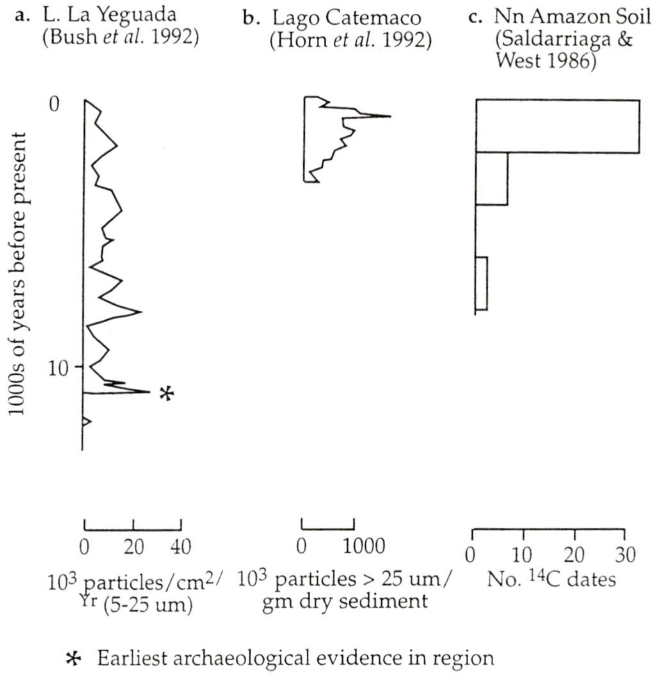

Figure 4. Selected charcoal records from Central and South America. Site locations are indicated in Fig. 1.

Palynological records from the Amazon provide equally convincing evidence of human occupation, with only two out of about 10 analysed lake sequences showing no pre-European contact disturbance histories, and many containing evidence for maize cultivation as well as charcoal (Bush, Colinvaux and Piperno, in prep.). The existence of some natural fire is suggested from a site near Monte Allegra, which has a continuous 6 ky record of inferred human burning. At this site, presently receiving about 1800 mm rainfall per annum, two spikes of charcoal within the last 9 ky suggest a fire intervals of about 4 ky. However, it would appear that natural fire in forests receiving greater than 2500 mm of rainfall is rare. One palaeoecological record, Hill of Six Lakes, covering the last 100 ky, which is near charcoal sites of Sanford et al. (1985), contains no charcoal despite supporting high cerrado rainforest and occurring on an edaphically dry rock outcrop that should attract lightning (Bush, unpublished).

Outside rainforested environments, one site (Laguna Junin), situated above the treeline in the Peruvian Andes, contains charcoal through the whole of the record, which extends beyond 4 ky BP (Hansen et al. 1994). An increase in charcoal around 4 ky BP is attributed to either dry climatic conditions or humans, although a general increase in pollen of disturbance and possibly cultivated plants at this time, corresponding with ethnobotanical data for cultivation in the region, is considered to indicate the beginning of wide-scale plant domestication (Hansen et al. 1994).

An interesting feature of a number of records, including Ayauchi (Bush et al. 1992), Lake Wodehouse, and Cana Swamp (Bush and Colinvaux 1994), where charcoal is clearly linked to maize pollen and phytoliths, is the disappearance of charcoal around 0.3 ky BP. This feature is associated with a disruption of the agricultural system, possibly as a result of the extermination of the indigenous people by disease with the arrival of Europeans.

Most records, unfortunately, do not extend sufficiently far back in time to provide any real assessment of burning during the last glacial maximum when generally drier conditions should have been condusive to frequent fire, and to the expansion of fire tolerant or fire-promoting savanna vegetation. From a limited number of Pleistocene sites revealing savanna vegetation, Absy and van der Hammen (1994) suggest that precipitation within the Amazon Basin may have been as low as 50% of today's levels and that rainforest was restricted to isolated patches. No charcoal data allow an assessment of the possible role of fire in the production and maintenance of these vegetation patterns, but macroscopic charcoal has been recorded in a number of places (van der Hammen, pers. comm.). By contrast, Bush and Colinvaux (1994) and Colinvaux 1987, 1993) suggest that there is no good evidence for a widespread increase in savanna. All their sites show that savanna was either peripheral to the Amazon Basin or currently support savanna/cerrado. The Carajas pollen record actually shows a greater increase in savanna elements during the mid-Holocene than during the Pleistocene (Absy et al. 1991). Bush proposes that a 20% drying is all that is required to satisfy all the palaeoecological data.

The only known palaeoecological site with a significant late Pleistocene occurrence of charcoal, Lagoa dos Olhos, is presently situated on a boundary between cerrados and montane *Araucaria* rainforest (De Oliviera 1992). However, charcoal is concentrated in the early Holocene rather than in the late Pleistocene and may well relate to natural disturbance at the transition between Pleistocene and Holocene conditions.

Southeast Asia, Australia, and the Pacific

This third major tropical region contains the greatest number of charcoal records within the period of human occupation, but almost all are south of the equator. The most notable northern record is from the Rajasthan desert (Singh et al. 1974) though, strictly speaking, this occurs just north of the Tropic of Cancer. This record has similarities to those from the Sahara, in that charcoal occurrence appears to correspond with wettest conditions in the mid-Holocene when the vegetation cover was sufficient to carry fire, yet the sequence has been interpreted in terms of human burning. By contrast, dates on soil charcoal combined with those associated with lignite in presently rainforested parts of eastern Borneo, have been considered to indicate natural burning, probably associated with severe droughts, through at least the last 17.5 ky (Goldammer and Seibert 1989). A recent ENSO (El Niño/Southern Oscillation) event within the region resulted in a 60% reduction in rainfall compared with the annual mean. This drought permitted fires, set by humans, to burn out of control and consume over 5 million hectares of rainforest. Such huge burns may provide useful analogues for past periodic rainforest fires. Within the northern region there are also several archaeological sites, dating back to about 5 ky BP, that show evidence of charred rice indicating some burning in association with food preparation.

South of the tropics, a number of sites in the New Guinea/Indonesian region allow some assessment of burning patterns through a substantial part of the last glacial cycle. Wanda, a small solution hollow formed in ultrabasic rocks in Sulawesi, contains charcoal through the whole of the estimated last 75 ky (Hope et al., in prep.). Burning is likely natural, because the area is regarded as unproductive, even for hunting. Charcoal levels are high between 24 and 13 ky BP, and correspondence with some expansion of grassland at the expense of the presently dominant semi-closed forest suggests an association with drier conditions during the height of the last glacial period. By contrast, the long lowland rainforest record of Hordorli in the northern part of Irian Jaya, estimated to cover the last 60 ky, contains charcoal only in the last 11.4 ky (Hope and Tulip, 1994) (Fig. 5b). Here, the initiation of burning and subsequent increases around 9 and 4 ky BP are attributed to the regional development and expansions of agriculture in the adjacent lowlands together with more local, casual, human disturbance. The disappearance of charcoal in the last 0.5

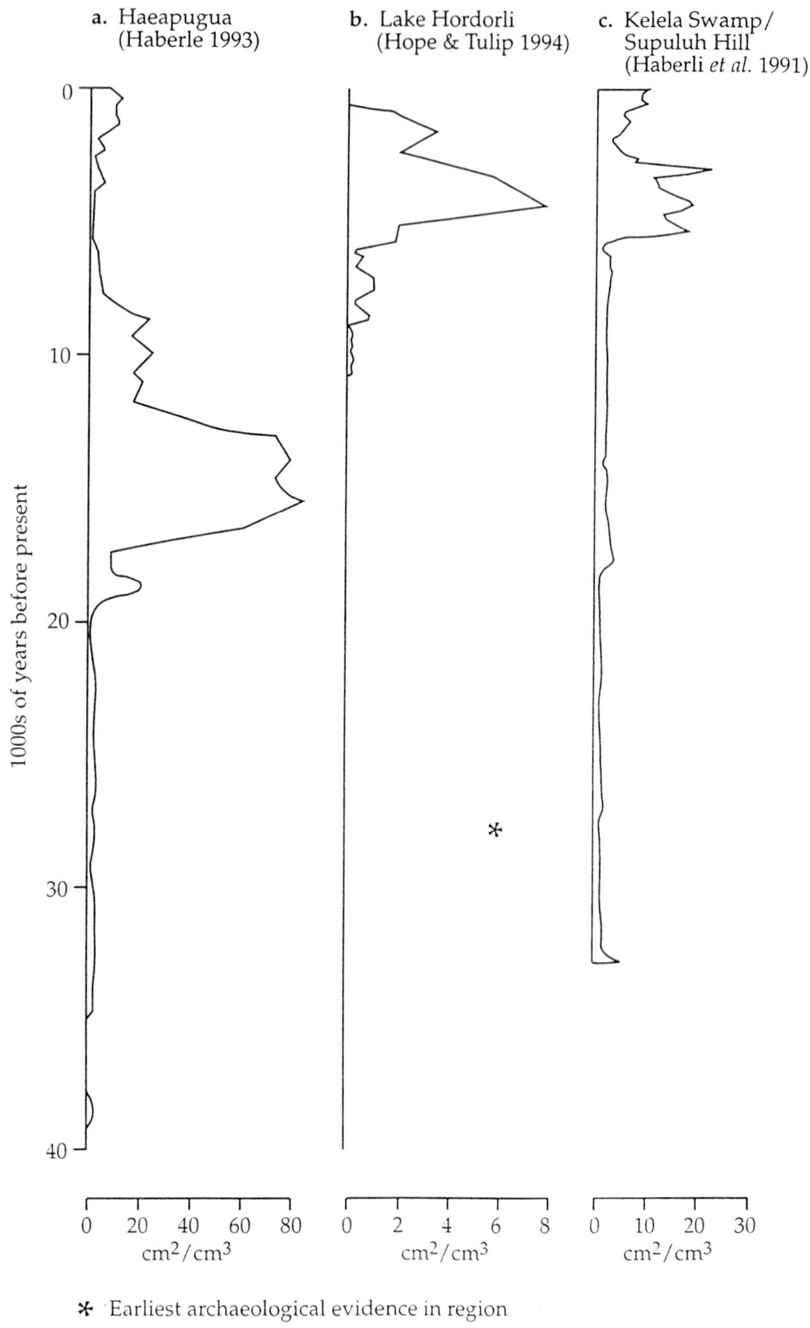

Figure 5. Selected charcoal records from New Guinea. Site locations are indicated in Fig. 1.

ky may be considered to reflect increased silviculture to replace lost forest and decreased dependence on hunted or gathered food. The burning record from Lake Wanum through the last 7-8 ky supports the early date for agricultural disturbance in lowland areas of New Guinea.

Charcoal is present through most of the last 40 ky in sites from highland New Guinea. Haberle (1993) proposes that the beginnings of sustained representation of charcoal, together with a rise in pollen disturbance indicators, demonstrates human influence around Haeapugua from 35 ky BP. High charcoal values between 20 and 9 ky BP, with a peak centred on 15 ky BP, correspond with grassland expansion during the cool, dry glacial (Fig. 5a). Subsequent charcoal increases around 4 and 2 ky BP are attributed to agricultural activities. At Kosipe, charcoal from 30 ky BP is assumed to be anthropogenic, an interpretation supported by the presence of a nearby midden dating back to 26 ky BP. However, some charcoal is present earlier, about 50 ky BP, that is more difficult to attribute to the presence of people. An increase in burning within the last 3 ky indicates intensification of human activity. A similar explanation is provided for the combined sites of Kelela and Supulah Hill, where charcoal from 34 ky BP may be associated with hunting and gathering and around 6 ky BP to agriculture (Fig. 5c).

Other sites in the highland region indicate an initiation of burning with the development of agriculture between about 9 and 3 ky BP. At Telefomin, earlier phases of clearance for agriculture, between 13 and 11 ky BP and 10.5 and 9 ky BP are evident, as well as a more sustained clearance from 4 ky BP. The fine resolution studies of Corlett (1984) show increases in charcoal on Mt. Wilhelm within the last few hundred years with forest destruction and population expansion associated with the cultivation of sweet potato.

The focus of study in Australia has been on the humid tropics of northeastern Queensland, an area containing the largest area of complex rainforest on the continent. Suitable sites for palaeoecological investigation are volcanic crater lakes and swamps on the Atherton Tableland. The longest continuous record is from Lynch's Crater. It covers an estimated 190 ky from 60 m of lake and swamp sediments (Kershaw 1986). As previously mentioned, charcoal occurrences are patchy and low through the earlier part of the sequence, but charcoal levels increase dramatically around 38 ky BP (Fig. 6a). This increase has been attributed to burning by Aboriginal people rather than to climate due, in part, to the fact that no such increase is evident within the previous, climatically analogous, glacial period. Increased charcoal concentrations also correspond well with the earliest archaeological date of 32 ky BP for the presence of people in northeastern

Queensland. At this time moist *Araucaria* forest was replaced by sclerophyll vegetation dominated by the fire-promoting *Eucalyptus* with a grassy understory. It is likely that many forms of seasonal, rainforest-related, forest succumbed to higher intensity or more frequent burns as all have very restricted distributions in Australia today. The burning is also considered to have led to the extinction of at least two genera (*Dacrydium* and *Xanthomyrtus*) (Kershaw 1986, Bohte 1994), which have had a long history on the continent. A similar pattern of vegetation and environmental change has more recently been recorded from nearby Strenekoff's Crater (Kershaw et al. 1991a).

Although charcoal values peak around 38 ky BP within the Lynch's Crater record, the vegetation transition around the site took place over some 12 ky when charcoal, and possibly burning, levels were generally much lower. At the end of this transition soil charcoal from a variety of sites within presently rainforested areas of the humid tropics begins to be recorded (Hopkins et al. 1993) (Fig. 6c). All identifications on charcoal are of *Eucalyptus*, indicating clearly a regional expansion of sclerophyll vegetation. This expansion is considered by Hopkins et al. (1993) to be primarily related to drier conditions, rather than to the impact of Aboriginal people, as its commencement corresponds with the beginning of regionally drier conditions of the last glacial maximum (Hiscock and Kershaw 1992). It is interesting to note that the 38 ky BP charcoal peak has not been recorded in the soil charcoal record, at least to date: perhaps early Aboriginal activity was focused around aquatic sites.

In both the Lynch's Crater and soil records charcoal levels are highest between 15 and 8 ky BP, a period that incorporates the driest conditions in the latter part of the Quaternary within the Australian region (Kershaw and Nanson 1993). Sclerophyll vegetation penetrated some of the wettest areas, which today have rainfall in excess of 3000 mm. These areas include postulated glacial 'refugia' of Webb and Tracey (1981). Although precipitation increased at or slightly before 10 ky BP, sclerophyll vegetation remained the dominant vegetation for at least another 2-3 ky. One explanation for the delay in rainforest recolonisation, at least for the Atherton Tableland, is that it took rainforest some time to migrate from inferred retreats occupied during the glacial period. Thus the actual time of arrival around any one of the six palynological sites within that area depended on its distance from a retreat area (Ash 1988). However, the soil charcoal evidence that brings into question the existence of identifiable retreat areas makes this hypothesis less tenable. More feasible is the proposal that continued fire pressure, that is likely to have had an anthropogenic component, retarded rainforest re-expansion. Time series analysis of a detailed charcoal record from one Tableland site, Lake Barrine (Walker and Chen

1987) (Fig. 6b), suggests that fires during the very early Holocene occurred at an average interval of 10 yr, consistent with regular burning of open sclerophyll woodland. Between 9.3 and 6.8 ky BP the frequency decreased to one fire every 220-240 years, a pattern typical of wet sclerophyll forest dominated by a dense canopy of eucalypts with a rainforest understorey. During this period some rainforest pollen is recorded, but the fire frequency effectively prevented the full establishment of rainforest vegetation.

The period from 6.8 to 5 ky BP shows low levels for charcoal and coincides with the period of maximum rainforest expansion under a wet climate with little seasonal variability (Kershaw and Nix 1988). Charcoal values then rise again, peaking between 4 and 3 ky BP. At Lynch's Crater this increase has been attributed to localised Aboriginal burning, which resulted in the destruction of swamp forest (Kershaw 1983). It is considered that the surrounding complex forest, existing under an estimated mean annual precipitation in excess of 3000 mm, would, most likely, have resisted large-scale burning. The date of about 5ky BP for the initiation of this burning coincides well with the first evidence for human occupation of tropical rainforest within the region (Horsfall and Hall 1990). However, soil charcoal dated between 4 and 2 ky BP probably relates to some expansion of sclerophyll vegetation under lower or more seasonal rainfall. Climatic estimates, combined with the increased occurrence of pollen indicators of more open forest, from Lake Euramoo (Kershaw and Nix 1988) suggest that rainfall became more seasonal after 5 ky BP and drier after 3.6 ky BP, features that may also have resulted from the onset or intensification of El Niño/Southern Oscillation (ENSO) climatic variability (McGlone et al. 1992). A decline in charcoal within the last 2 ky corresponds with evidence for a small increase in precipitation.

The pattern of burning within the Groote Eylandt record (Shulmeister 1992), a sequence from within sclerophyll vegetation, is consistent with much of the data from the Atherton Tableland (Fig. 6d). The site records a gradual increase in precipitation from the beginning of the Holocene to about 7 ky BP. Initially the vegetation was too sparse to support more than occasional fire. Subsequent development of sclerophyll vegetation resulted in frequent fires. Between about 7 and 4 ka BP, precipitation levels higher than today resulted in a reduction in fire frequency. From about 4 ka BP a substantial increase in burning has been attributed to the onset of ENSO (Shulmeister and Lees 1996) with a slight reduction centred on 1.6 ka BP due to some increase in precipitation. Since this time, charcoal levels have been high despite the maintenance of relatively high rainfall. This apparent contradiction is reconciled by the postulation of higher Aboriginal burning levels due to either population increase or a shift in land usage (Shulmeister 1992). Aboriginal burning is identified by a high number of cool early-dry season burns which produce more carbonized particles than ash unlike the hot, late-season burns that are characteristic of the natural fire regimes (Shulmeister 1992).

The interpretation in northern Australia has been complicated recently by pollen and charcoal results from the Ocean Drilling Program core 820 taken from the continental slope adjacent to the Atherton Tableland. As previously mentioned this record shows that charcoal has been produced within this humid tropical region continuously through the last 1.5 million years (Kershaw et al. 1993). However, an increase in burning within the last 140 ky BP has been proposed from higher charcoal levels combined with a sustained decline in dry rainforest relative to sclerophyll vegetation. These changes, which are remarkably similar to those on the Atherton Tableland dated to around 38 ky BP, could indicate that people were present in Australia substantially earlier than suggested from the archaeological record. This date supports a similar suggestion from pollen and charcoal evidence for Aboriginal burning around Lake George in southeastern Australia about 130 ky BP (Singh et al. 1981, Singh and Geissler 1985). Although it is difficult to find a natural cause for these burning increases, archaeologists have argued against the acceptance of such an early Aboriginal occupation (White 1994, Anderson 1994).

On many Pacific Islands, the presence of people is clearly demonstrated by deforestation and abundant artifacts (Flenley 1994). But sediment-based charcoal records often precede archaeological dates, as in Fiji (5-4 ky BP, Southern 1986) and the Cook Islands (2.5 ky BP, Kirch and Ellison 1994), and it has been argued that these records provide the best guide to determination of the age of human settlement (Kirch and Ellison 1994). A similar argument has been put forward for islands elsewhere (Burney et al. 1994). However, as with the much early suggestion for the presence of people in Australia, the use of non-archaeological data has been questioned (Anderson 1994). The majority of palaeoecological records do not go back sufficiently far to determine the existence or extent of natural burning on these islands. One exception is a 13 ky sequence from Lake Tagamaucia in the highlands of Fiji (Southern 1986). No charcoal is recorded at any stage within this rainforest-dominated record despite lower precipitation prior to about 11.4 ky BP. The present day estimated mean annual rainfall of 10,000 mm makes it unlikely that fire was important. By contrast, a record from Plaine des Lacs in southern New Caledonia, which has similar pollen signatures to Lake Tagamaucia (Hope, unpublished), indicates occurrence of intense fires every millennium or so from 40 to 16 ky BP. After this time no sediments have accumulated within the basin. It is considered that natural fires here have been facilitated by a defined dry season, regular cyclones, poor ultrabasic soils, and much lower rainfall (3500 mm) than in the Fijian area. Humans may have been responsible for increased burning from 4 ky that led to a shift from forest to maquis heath vegetation in a shorter record from the region (Hope, unpublished).

Discussion

Sedimentary-based records from the tropics suffer from all the limitations to interpretation of past fire patterns that have been documented in temperate regions (eg. Clark 1983, Clark and Robinson 1993, Patterson et al. 1987). These include assessment of levels of production of charcoal from various fire regimes and vegetation types, differential methods and rates of charcoal transport into basins of deposition, the influences of basin size and nature on charcoal deposition, and the origin and significance of charcoal particle size. Documentation of fire patterns in the tropics is also hindered by a lack of optimal depositional sites. Basins are usually prone to drying-out and oxidation of accumulated organic sediments and to rapid changes in successional status that influence charcoal representation. However, the study of Walker and Chen (1987) demonstrates that, where aquatic conditions and catchments remain relatively constant, statistical methods can be applied to charcoal sequences and reveal meaningful results on fire frequency and intensity.

Some of the limitations of sediment-based studies are overcome by those on soil charcoal. These provide a more precise spatial picture of burning within dryland vegetation. However, preservation of charcoal and its stratigraphic integrity depends on a consistent erosional response to fire events and is limited to particular topographic situations.

In both sedimentary and soil sequences, there are few situations where quantitative charcoal estimates can be related to particular fire regimes or their causes. On Groote Eylandt, Shulmeister (1992) considers that higher charcoal levels in northern Australia are produced from anthropogenic early season fires rather than natural dry season fires. Conversely, Clark in Singh et al. (1981) notes an increase in charcoal representation from fires on Kangaroo Island in southeastern Australia after Aboriginal people had disappeared from the island. In New Guinea, increased charcoal is frequently attributed to agricultural intensification. However, a regional intensification in agriculture, which results in a reduction in hunting and gathering, is proposed as an explanation for a decrease in charcoal within the Lake Hordorli diagram of Hope and Tulip (1994).

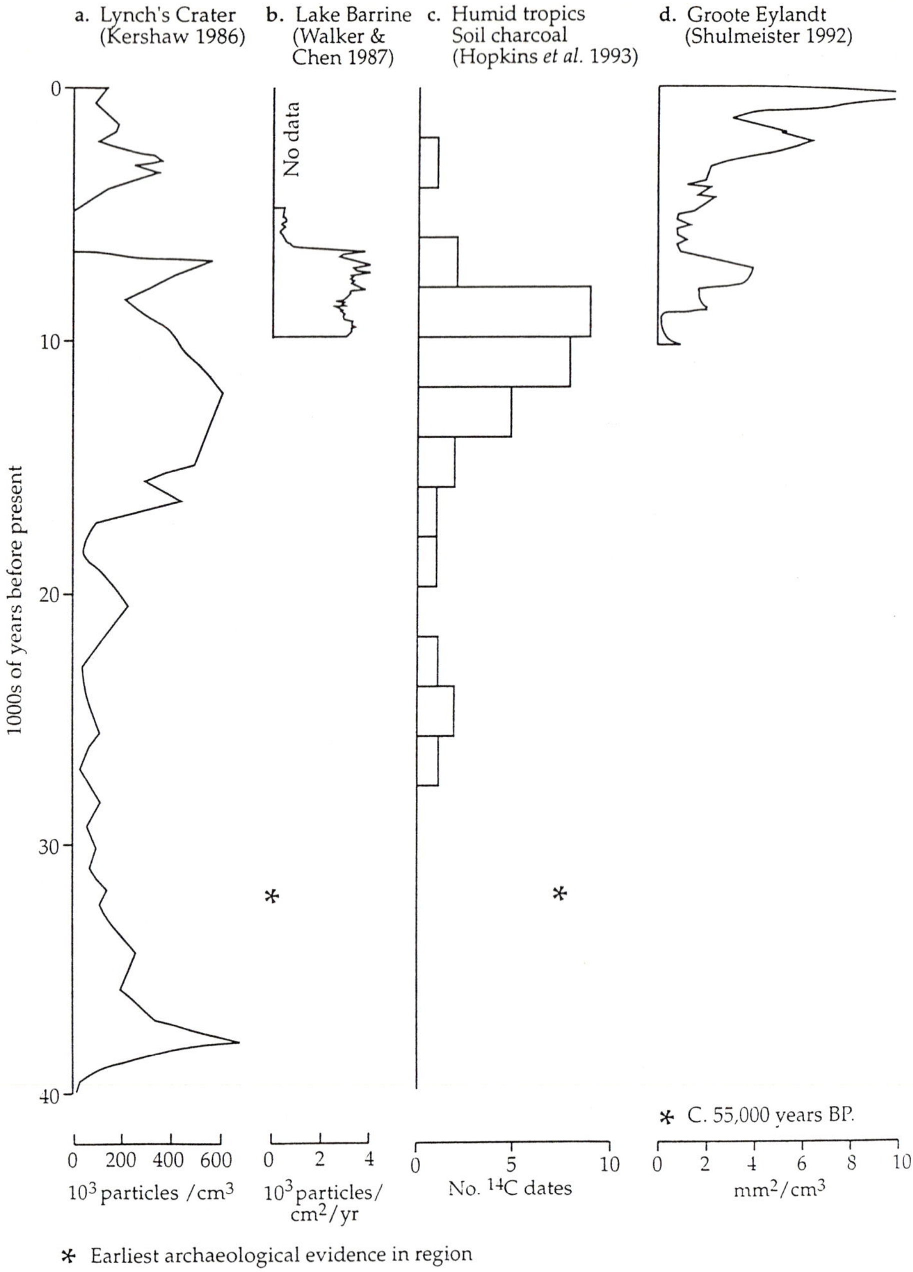

Figure 6. Selected charcoal records from Australia. Site locations are indicated in Fig. 1.

In assessing the human contribution to past biomass burning, the most reliable estimates are probably derived from complex rainforest existing under high, seasonally evenly distributed, rainfall where natural fires are unlikely to occur, and on small oceanic islands where natural ignition sources are limited. The most difficult areas are savannas which, apart from deserts, are least condusive to the production and preservation of continuous sedimentary sequences and which have a demonstrated geological history of natural burning. Separation of human from natural burning is made increasingly difficult towards the present by increasing levels of human disturbance, which produce a more fire-prone environment. Some environments appear to be dry forest or savanna, but, when palaeoecological data are obtained, prove to be 'anthropogenic savannas' (Piperno 1988). One possible example of this is the Sahel of northern Africa as mentioned previously. Another is the Towlan, the once forested district of Central Panama. This area, which has been deforested by human activities for at least the last 3 ky, now resembles scrub savanna and supports a fire regime typical of such an area, though, in reality, it is a human artifact. In Australia, continual Aboriginal burning has probably resulted in the replacement of almost all dry and seasonal rainforests by sclerophyll vegetation that, once established, maintains a higher level of burning after the breakdown of Aboriginal land management. Within wetter forests, the destruction of a moist microclimate by canopy opening would facilitate burning during periods of drought.

A comparison of past charcoal records in regions with different histories of occupation and land use probably provides the most reliable means of identifying the contribution of people to past burning levels. A number of records, e.g. Lynch's Crater and the soil charcoal studies in northeast Queensland, Lake Mobutu in East Africa, and Haeapugua in New Guinea, show high levels of charcoal during the late Pleistocene. These all relate to dry conditions when high fire occurrence might be expected. However, absence of charcoal peaks in other records at this time, particularly from New Guinea, and the absence of a peak at the height of the previous glacial period in Atherton Tableland calls into question a purely natural explanation. It is perhaps most likely that burning was caused mainly by people and was most effective during this drier period. A similar explanation may be offered for changing patterns of burning during the Holocene. There is a broad increase in burning during this period that is generally attributed to intensification of resource use. However, it is probable that increasingly variable climatic conditions during the last 6-4 ky have contributed to a global increase in fire.

The lack of long sedimentary burning records from the Americas means that there are few baseline data against which the predominantly Holocene records can be assessed. However, comparison of soil charcoal records from the Amazon with those from northeastern Australia are particularly informative (Fig. 7). The differences are dramatic. Almost all charcoal from Australia dates to the Pleistocene while all charcoal in Amazonian soils is Holocene and largely restricted to the last 2-4 ky. These patterns suggest strongly that a human presence was probably required to allow substantial burning in moist environments during drier conditions of the last glacial, but that burning during the wetter Holocene was largely restricted to those areas where agriculture was practiced. The limited data from Borneo may be consistent with this pattern: early evidence of burning is restricted to the dry late Pleistocene while the greatest number of charcoal dates are late Holocene, when people are known to have practiced agriculture. The evidence for only recent burning from Zaire is perhaps surprising, considering that people have been present on the continent for such a long period. However, the data are consistent with the generally held view that agriculture developed late within the region.

The available data are too patchy to allow any quantification of changing burning or charcoal sequestration levels during the period of human occupation. The influence of climate change is substantial, but there is no way of determining whether conditions more condusive to burning during drier glacial periods were more than offset by reduced availability of biomass for burning during these times. These data will probably have to await results of long-term elemental soot records from ocean cores (Bird 1996). In general terms, there is little doubt that burning has increased with human evolution and dispersal, and settlement intensification. The trend is most likely to have been an acceleration of that initiated by global drying in the mid-Tertiary.

The earliest documented impact that may be attributed to people occurred within the period 140 to 40 ky in Australia, a continent that, through its general dryness, climatic variability, and high proportion of fire promoting flora, responded dramatically to an additional ignition source. The direct impact of people is likely to have been augmented by the related demise of the megafauna, providing increased fuel (Flannery 1990, Schule 1990). Continued burning resulted in an increasingly fire-promoting vegetation that, in some areas, appears to have supported fire in drier areas or at drier times when the pre-existing vegetation was incapable of supporting fire (Singh et al. 1981). The next major sustained increase in burning was probably initiated by the arrival of people in the Americas and associated decline of the megafauna there around 11 ka. This was followed by intensification of plant utilization and population growth throughout the tropics from the early mid Holocene, and an increase in climatic variability from the mid-Holocene. There is evidence for peaks in burning throughout the whole period that can be related to times of climatic change, when vegetation would have been unstable.

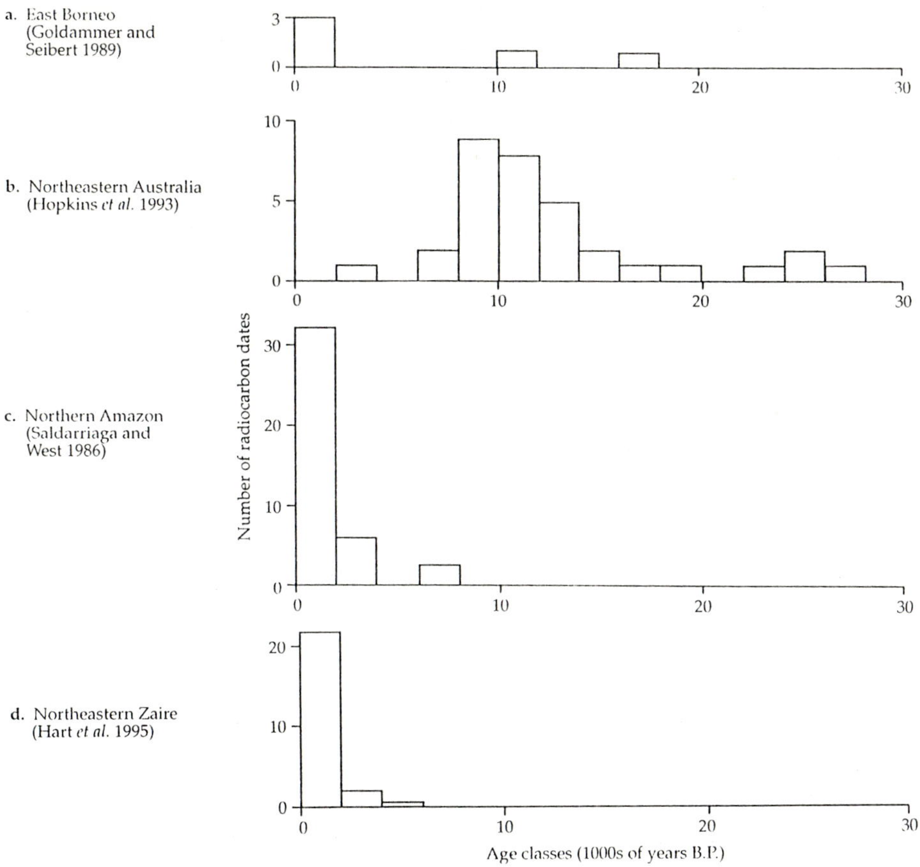

Figure 7. Comparison of soil charcoal date distributions from rainforest regions.

From this history it is possible to make some prediction of future burning levels. There is general concensus that, particularly through rising concentrations of atmospheric CO_2, climate will change rapidly over at least the next 50 years. There is disagreement, though, on the effects of the CO_2 increase on precipitation and hence vegetation cover (Goldammer 1993b). Regardless of the direction of change and of the patterns of climate and vegetation that may emerge, the change itself will undoubtedly result in potentially higher burning levels. Future changes will depend on anthropogenic land use practices and on fire management policies. It has been suggested that, in rainforested areas of South America, a reduction in population levels within the last few hundred years resulted in a reduction of burning. In southeastern Australia, charcoal records indicate a similar reduction in burning with the implementation of fire exclusion policies early this century (Kershaw 1984).

However, in this latter case, fire exclusion led to increased accumulation of fuel which promoted disastrous bushfires and resulted in the implementation of controlled burning practices. This example demonstrates the difficulty of reducing burning for any length of time within a vegetation that has evolved in association with fire. In many parts of the tropics it has been demonstrated that the long association between human-induced fires and vegetation has resulted in a domination of pyrophytic life forms. These will ensure the maintenance of high burning levels within much of the region and a continuation of the trend towards savannization (Goldammer 1993b).

Acknowledgments

We thank Dr Erhard Schulz, Professor Thomas van der Hammen and Dr James Shulmeister for their help with information sources, Dr Jim Clark, Dr Sally Horn and Dr Dan Livingstone for their valuable comments on the manuscript, and Mr Phil Scamp and Mr Gary Swinton for drafting text figures.

Literature References

Absy ML, Cleef A, Fournier M, Martin L, Servant M, Sifeddine A, Da Silva MF, Soubies F, Suguio K, Turcq B, van der Hammen T (1991) Mise en evidence de quatre phases d'ouverture de la foret dense dans le sud-est de l'Amazonie au cours des 60,000 derniers annees. Preiere comparaison avec d'autres regions tropicales. *C R Acad Sci Paris* II 312:673-678

Allen J (1989) When did humans first colonise Australia? *Search* 20:149-154

Allen J (1994) Radiocarbon determinations, luminescence dating and Australian archaeology. *Antiquity* 68:339-343

Anderson A (1994) Comment on J. Peter White's paper 'Site 820 and the evidence for early occupation in Australia'. *Quat Australasia* 12/2:30-31

Andreae MO (1991) Biomass burning: its history, use, and distribution and its impact on environmental quality and global climate. In Levine JS (ed) *Global Biomass Burning:Atmospheric, Climatic and Biospheric Implications* pp 4-21 The MIT Press Cambridge MA

Ash J (1988) The location and stability of rainforest boundaries in northeastern Queensland, Australia. *J Biogeog* 15:619-630

Bailey RC, Head G, Jenike M, Owen B, Zechenter E (1989) Hunting and gathering in tropical rainforest: is it possible? *Amer Anthropologist* 91:59-82

Bednarik RG (1989) On the Pleistocene settlement of South America. *Antiquity* 63:101-111

Bellwood P (1985) *Prehistory of the Indo-Malaysian Archipelago*. Academic Press Sydney

Bird MI (1996) Fire, prehistoric humanity and the environment. *Interdisc Sci Rev* (in press)

Bird-David NH (1988) Hunters and gatherers and other people - a re-examination. In Ingold T, Riches D, Woodburn J (eds) *Hunters and Gatherers 1. History, Evolution and Social Change* pp 17-30 Berg Oxford

Bohte A (1994) *Local vegetation and environments of the last glacial cycle at Lynch's Crater, northeast Queensland.* BSc Honors Thesis Monash Univ Melbourne

Bonnefille R (1983) Evidence for a cooler and drier climate in the Ethiopian uplands, 2.5 Myr ago. *Nature* 303:487-491

Brain CK, Sillen A (1988) Evidence from the Swartkrans cave for the earliest use of fire. *Nature* 336:464-466

Burney DA (1987a) Late Holocene vegetational change in central Madagascar. *Quat Res* 28:130-143

Burney DA (1987b) Late Quaternary stratigraphic charcoal records from Madagascar. *Quat Res* 28:274-280

Burney DA, Burney LP, Macphee RDE (1994) Holocene charcoal stratigraphy from Laguna Toruguero, Puerto Rico, and the timing of human arrival on the island. *J Archaeol Sci* 21:273-281

Bush MB, Colinvaux PA (1988) A 7000-year pollen record from the Amazon lowlands, Ecuador. *Vegetatio* 76:141-154

Bush MB, Colinvaux PA (1994) A paleoecological perspective of tropical rainforest disturbance: records from Darien, Panama. *Ecology* 75:1761-1768

Bush MB, Piperno DR, Colinvaux PA (1989) A 6000 year history of Amazonian maize cultivation. *Nature* 340:303-305

Bush MB, Piperno DR, Colinvaux PA, Krissek L, De Oliviera PE, Miller MC, Rowe W (1992) A 14,300 year palaeoecological profile of a lowland tropical lake in Panama. *Ecol Monog* 62:251-276

Byrne R, Horn SP (1989) Prehistoric agriculture and forest clearance in the Sierra de los Tuxtlax, Veracnz, Mexico. *Palynology* 13:181-193

Chang TT (1989) Domestication and spread of cultivated rices. In Harris DR, Hillman GC (eds) *Foraging and Farming: The Evolution of Plant Exploitation* pp 408-417 Unwin Hyman London

Clark JD (1980) Early human occupation of African savanna environments. In Harris DR (ed) *Human Ecology in Savanna Environments* pp 41-71 Academic Press London

Clark JS, Robinson, J (1993) Palaeoecology of fire. In Crutzen PJ and Goldammer JG (eds) *Fire in the Environment: The Ecological, Atmospheric and Climatic Importance of Vegetation Fires* pp 194-214 John Willey and Sons Chichester

Clark RL (1983) Pollen and charcoal evidence for the effects of Aboriginal burning on the vegetation of Australia. *Archaeol Oceania* 18:32-37

Colinvaux PA (1987) Amazon diversity in the light of the paleoecological record. *Quat Sci Rev* 6:93-114

Colinvaux PA (1993) Pleistocene biogeography and diversity in tropical forests of South America. In Goldblatt P (ed) *Biological Relationships between Africa and South America* pp 473-499 Yale University Press New Haven

Colinvaux PA, Bush MB (1991) The rainforest ecosystem as a resource for hunting and gathering. *Amer Anthropologist* 93:153-160

Corlett RT (1984) Palynological evidence for changing subsistence patterns around Mt. Wilhelm, Papua New Guinea. Singapore . *J Trop Geogr* 5:102-111

Crutzen PJ, Goldammer JG (eds) (1993) *Fire in the Environment: The Ecological, Atmospheric, and Climatic Importance of Vegetation Fires.* John Wiley and Sons Chichester New York

Deacon HJ (1992) Southern Africa and modern human origins. *Phil Trans R Soc Lond B* 337:177-183

Dechamps R (1984) Evidence of bush fires during the Plio-Pleistocene in Africa (Omo and Sahabi) with the aid of fossil woods. *Palaeoecol Africa* 16:291-294

De Oliviera PE (1992) *A palynological record of late Quaternary vegetational and climatic change in southeastern Brazil.* PhD Thesis The Ohio State Univ Columbus

Enright NJ, Gosden C (1992) Unstable Archipelagos–south-west Pacific environment and prehistory since 30,000 B.P. In Dodson J (ed) *The Naive Lands.* Longman Cheshire Melbourne

Fagan BM (1989) *People of the Earth.* 6th ed Scott Foresman Illinois

Flannery TF (1990) Pleistocene faunal loss: implications of the aftershock for Australia's past and future. *Archaeol Oceania* 15:45-67

Flenley JR (1988) Palynological evidence for land use changes in Southeast Asia. *J Biogeog* 15:185-197.

Flenley JR (1994) Pollen in Polynesia: the use of palynology to detect human activity in the Pacific Islands. In Hather JG (ed) *Tropical Archaeology: Applications and New Developments* pp 202-214 Routledge London

Folster H (1992) Holocene autochthonous forest degradation in southeast Venezuela. In Goldammer JG (ed) *Tropical Forests in Transition* pp 25-30 Birkhauser Verlag Basel Switzerland

Garrett-Jones S (1979) *Evidence for changes in Holocene vegetation and lake sedimentation in the Markham Valley, Papua New Guinea*. PhD Thesis Australian National Univ Canberra

Glover (1977) The Late Stone age in eastern Indonesia. *World Archaeol* 9:42-61

Goldammer JG (ed) (1990) *Fire in the Tropical Biota. Ecosystem Processes and Global Challenges*. Springer-Verlag Berlin

Goldammer JG (ed) (1992) *Tropical Forests in Transition*. Birkhauser Verlag Basel Switzerland

Goldammer JG (1993a) *Feuer in Waldokosystemen der Tropen und Subtropen*. Birkhauser Verlag Basel Boston

Goldammer JG (1993b) Historical biogeography of fire: tropical and subtropical. In Crutzen PJ, Goldammer JG (eds) *Fire in the Environment. The Ecological, Atmospheric, and Climatic Importance of Vegetation Fires* pp 297-314 John Wiley and Sons Chichester New York

Goldammer JG, Seibert B (1989) Natural rainforest fires in Eastern Borneo during the Pleistocene and Holocene. *Naturwissenchaften* 76:518-520

Golson J, Gardner DS (1990) Agriculture and sociopolitical organisation in New Guinea highlands prehistory. *Ann Rev Anthropol* 395-417

Gowlett JAJ, Harris, JWK, Walton D, Wood BA (1981) Early archaeological sites, hominid remains and traces of fire from Chesowanja, Kenya. *Nature* 294:125-129

Groube L, Chappell J, Muke J, Price D (1986) A 40,000 year old human occupation site at Huon Peninsula, Papua New Guinea. *Nature* 324:453-455

Guildon N, Delibrias G (1986) Carbon-14 dates point to man in the Americas 32,000 years ago. *Nature* 321:769-771

Guildon N, Arnaud B (1991) The chronology of the New World: two faces of one reality. *World Archaeol* 23:167-178

Haberle SG (1993) *Late Quaternary environmental history of the Tari Basin, Papua New Guinea*. PhD Thesis Australian National Univ Canberra

Haberle SG, Hope, GS, DeFretes Y (1991) Environmental change in the Balien Valley, montane Irian Jaya, Republic of Indonesia. *J Biogeog* 18:25-40

Hamilton A, Taylor D, Vogel JC (1986) Early forest clearance amd environmental degradation in south-west Uganda. *Nature* 320:164-167

Hansen BCS, Seltzer GO, Wright Jr HE (1994) Late Quaternary vegetational change in the central Peruvian Andes. *Palaeo Palaeo Palaeo* 109:263-285

Hart TB, Hart JA (1986) The ecological basis of hunter-gatherer subsistence in African rainforests: the Mbuti of eastern Zaire. *Human Ecol* 14:29-55

Hart TB, Hart JA, Deschamps R, Fournier M, Ataholo M (1996) Changes in forest composition over the last 4000 years in the Ituri Basin, Zaire. *Proc XIV AETFAT Cong*. Wageningen Agric Univ Wageningen Netherlands (in press)

Haynes Jr CV, Eyles CH, Pavlish LA, Richie JC, Rybak M (1989) Holocene palaeoecology of the eastern Sahara; Selima Oasis. *Quat Sci Rev* 8:109-136

Herring JR (1985) Charcoal fluxes into sediments in the North Pacific Ocean: The Cenozoic record of burning. In *The Carbon Cycle and Atmospheric CO2: Natural Variations Archaen to Present Geophys Monogr* Am Geophys Union Washington 32:419-442

Hiscock P, Kershaw AP (1992) Palaeoenvironments and prehistory of Australia's top end. In Dodson J (ed) *The Naive Lands* pp 43-75 Longman Cheshire Melbourne

Hope GS (1982) Pollen from archaeological sites. In Duerden P, Ambrose W (eds) *First Australian Conference on Archaeometry* pp 211-219 Prehistory Dept Australian National Univ Canberra

Hope GS (1983) The vegetational changes of the last 20,000 years at Telefomin, Papua New Guinea. *Singapore J Trop Geogr* 4:25-33

Hope GS, Spriggs MJT (1982) A preliminary pollen sequence from Aneityum Island, southern Vanuatu. *Indo-Pacific Assoc Bull* 3:88-94

Hope GS, Tulip J (1994) A long vegetation history from lowland Irian Jaya, Indonesia. *Palaeo Palaeo Palaeo* 109:385-398

Hopkins MS, Ash, J, Graham AW, Head J, Hewett RK (1993) Charcoal evidence of the spatial extent of the *Eucalyptus* woodland expansions and rainforest contractions in north Queensland during the late Pleistocene. *J Biogeog* 20:357-372

Horn SP, Sanford RL Jr (1992) Holocene fires in Costa Rica. *Biotropica* 24:354-361

Horn SP, Horn RD, Byrne R (1992) An automated charcoal scanner for paleoecological studies. *Palynology* 16:7-12

Horsfall N, Hall J (1990) People and the rainforest: an archaeological perspective. In Webb LJ, Kikkawa J (eds) *Australian Tropical Rainforests: Science - Vaues - Meaning* pp 33-39 CSIRO Melbourne

James SR (1989) Hominid use of fire in the Lower and Middle Pleistocene: a review of the evidence. *Curr Anthropol* 30:1-26

Kershaw AP (1983) A Holocene pollen diagram from Lynch's Crater, northeast Queensland, Australia. *New Phytol* 94:669-682

Kershaw AP (1984) Some applications of studies on vegetation and fire history to forest management. In Ealey EHM (ed) *Fighting Fire with Fire* pp 55-69 Graduate School Environm Sci Monash Univ Melbourne

Kershaw AP (1986) Climatic change and Aboriginal burning in north-east Queensland during the last two glacial/interglacial cycles. *Nature* 322:47-49

Kershaw AP (1994) Pleistocene vegetation of the humid tropics of northeastern Queensland, Australia. *Palaeo Palaeo Palaeo* 109:399-412

Kershaw AP, Baird JG, D Costa DM, Edney PA, Peterson JA, Strickland KM (1991a) A comparison of long Quaternary records from the Atherton Tableland and Western Plains volcanic provinces. In Williams MAJ, DeDeckker P, Kershaw AP (eds) *The Cainozoic of the Australian Region: a Re-evaluation of the Evidence* pp 288-301 Geol Soc Aust Sydney

Kershaw AP, Bolger PF, Sluiter IRK, Baird JG, Whitlaw M (1991b) The nature and evolution of lithotypes in the Tertiary brown coals of the Latrobe Valley, southeastern Australia. *Int J Coal Geol* 18:233-249

Kershaw AP, McKenzie GM, McMinn A (1993) A Quaternary vegetation history of northeastern Queensland from pollen analysis of ODP Site 820. *Proc Ocean Drilling Program Sci Res* 133:107-114.

Kershaw AP, Nanson GC (1993) The last full glacial cycle in the Australian region. *Global Planet Sci* 7:1-9

Kershaw AP, Nix HA (1988) Quantitative palaeoclimatic estimates from pollen taxa using bioclimatic profiles of extant taxa. *J Biogeog* 15:589-602

Kirch PV, Ellison J (1994) Palaeoenvironmental evidence for human colonization of remote oceanic islands. *Antiquity* 68:310-421

Kuhry P (1988) *Palaeobotanical-palaeoecological studies of tropical high Andean peatbog sections, Cordillera Oriental, Colombia.* Thesis Univ Amsterdam (Dissertationes Botanicae 116) Cramer Berlin

Maley J (1992) Mise en evidence d'une pejoration climatique entre ca. 2500 et 2000 ans B.P. en afrique tropical humide. *Bull Soc geol France* 163:363-365

Maloney BK (1985) Man's impact on the rainforests of West Malesia: the palynological record. *J Biogeog* 12:537-558

Martin HA, McMinn A (1994) Late Cenozoic vegetation history of northwestern australia from the palynology of a deep sea core. (ODP Site 765). *Aust J Bot* 42:95-102

McGlone MS, Kershaw AP, Markgraf V (1992) El Niño/Southern Oscillation climatic variability in Australasian and South American paleoenvironmental records. In Diaz HF, Markgraf V (eds) *El Niño. Historical and Paleoclimatic Aspects of the Southern Oscillation* pp 435-462 Cambridge University Press Cambridge

Morley RJ, Richards K (1993) Gramineae cuticle: a key indicator of late Cenozoic climatic change in the Niger Delta. *Rev Palaeobot Palynol* 77:119-128

Morse K (1988) Mandu Mandu Creek Rockshelter: Pleistocene human occupation of north west Cape, Western Australia. *Archaeol Oceania* 23:82-88

Neumann N (1988) Die Bedeutung von Holzkohleuntersuchungen fur die Vegetationsgeschichte der Sahara - das Beispiel Fachi/Niger. *Wurzburger Geogr Arb* 69:71-86

Patterson WA III, Edwards KJ, Maguire DJ (1987) Microscopic charcoal as a fossil indicator of fire. *Quat Sci Rev* 6:3-23

Piperno DR (1988) *Phytolith Analysis: An Archaeological and Geological Perspective.* Academic Press San Diego

Piperno DR, Bush MB, Colinvaux PA (1990) Palaeoenvironments and human occupation in late-glacial Panama. *Quat Res* 33:108-116

Roberts RG, Jones R, Smith MA (1990) Thermoluminescence dating of a 50,000-year-old human occupation site in northern Australia. *Nature* 245:153-156

Saldarriaga JG, West DC (1986) Holocene fires in the northern Amazon Basin. *Quat Res* 26:358-366

Sanford Jr RL, Salarriaga JG, Clark KE, Uhl C, Herrera R (1985) Amazon rain-forest fires. *Science* 227:53-55

Schule W (1990) Landscapes and climate in prehistory: interactions of wildlife, man and fire. In Goldammer JG (ed) *Fire in the Tropical Biota: Ecosystem Processes and Global Challenges* pp 273-318 Springer-Verlag Berlin

Schulz E (1996) The southern limit of the mediterranean vegetation in the Sahara during the Holocene. *Historical Biol* (in press)

Schwartz D, de Foresta H, Deschamps R, Lanfranchi, R (1990) Decouverts d'un premier site de l'age du fer ancien (2110 B.P.) dans le Mayombe congolais. Implications palaeobotanique et pedologiques. *C R Acad Sci Paris* 310:1293-1298

Shulmeister J (1992) A Holocene pollen record from lowland tropical Australia. *The Holocene* 2:107-116

Shulmeister J, Lees BG (1996) Pollen evidence from tropical Australia for the onset of an ENSO dominated climate at circa 4000 BP. *The Holocene* (in press)

Singh G, Geissler, EA (1985) Late Cainozoic history of fire, lake levels and climate at Lake George, New South Wales, Australia. *Phil Trans R Soc Lond B* 311:379-447

Singh G, Joshi RD, Chopra SK, Singh AB (1974) Late Quaternary history of vegetation and climate of the Rajasthan Desert, India. *Phil Trans R Soc Lond B* 267:467-501

Singh G, Kershaw AP, Clark R (1981) Quaternary vegetation and firehistory in Australia. In Gill AM, Groves, RA, Noble IR (eds) *Fire and Australian Biota* pp 23-54 Aust Acad Sci Canberra

Southern W (1986) *The late Quaternary environmental history of Fiji*. PhD thesis Australian National Univ Canberra

Van Noten F (1977) Excavation at Matupi cave. *Antiquity* 51:35-40

van der Hammen T, Absy ML (1994) Amazonia during the last glacial. *Palaeo Palaeo Palaeo* 109:247-261

Verin P, Kottak C, Gorlin P (1970) The glottochronology of Malagasy speech communities. *Oceanic Linguistics* 8:26-83

Walker D, Chen Y (1987) Palynological light on tropical rainforest dynamics. *Quat Sci Rev* 6:77-92

Walker D, Singh G (1991) Eariest palynological records of human impact on the world's vegetation. In Chambers FM (ed) *Climate Change and Human Impact on the Landscape* pp 101-108 Chapman and Hall London

Webb LJ, Tracey JG (1981) Australian rainforests, patterns and change. In Keast A (ed) *Ecological Biogeography in Australia* pp 607-694 Junk The Hague

White JP (1994) Site 820 and the evidence for early occupation in Australia. *Quat Australasia* 12/2:21-23

Williams MAJ, Dunkerley DL, De Deckker P, Kershaw AP, Stokes T (1993) *Quaternary Environments*. Edward Arnold London

Lake Sediment Records of Fossil Fuel-derived Carbonaceous Aerosols from Combustion

B. Lim[1] and I. Renberg[2]
OECD Environment Directorate
2 Rue André Pascal
75016 Paris, France

Summary

The ultimate value of black carbon archives in sediments as paleo-records of past combustion activities depends entirely upon our ability to accurately interpret this information and hinges upon our understanding of black carbon emission, air and water to sediment transport and post-deposition processes. This poses a number of constraints. Nonetheless, present day inventories of black carbon from fossil fuel and biomass burning are available and these could be related to black carbon concentrations in lake sediments. We therefore suggest the use of these relations to extrapolate information on black carbon emissions originating from past biomass burning from black carbon records preserved in ancient lake sediments.

Introduction

On a global scale, present day emissions of black carbon (BC) from fossil fuel combustion and biomass burning are thought to be of similar magnitude (Cooke and Wilson, accepted; Liousse et al. accepted). However, BC concentrations are spatially and seasonally highly variable, particularly over continental source regions (eg. Cooke and Wilson, accepted; Liousse et al. accepted). BC in aerosols is generally assumed to be chemically refractory and is therefore considered to be a conservative tracer of combustion. The deposition and preservation of BC aerosols in lacustrine and marine sediments can thus provide an archive of past atmospheric BC emissions, dependent upon atmospheric dispersion and deposition patterns (e.g. Herring 1985, Wik and Renberg 1995, Smith et al. 1973), type of fuel burnt and the site of combustion. In addition, emission factors for BC and other combustion products such as CO_2 and CH_4 gases are beginning to be quite well documented for both fossil-fuel (Hansen et al. 1989) and biomass fires (Liousse and Cachier 1992). Therefore BC concentrations in lake sediments have

[1] Previous address: Centre de Faibles Radioactivités, CNRS-CEA, Avenue de la Terrasse, 91198 Gif-sur-Yvette Cedex, France

[2] Fire Ecology Group, c/o Freiburg University, Freiburg, Germany

NATO ASI Series, Vol. I 51
Sediment Records of Biomass Burning
and Global Change
Edited by James S. Clark, Hélène Cachier,
Johann G. Goldammer, and Brian Stocks

the potential to provide estimates of gas (CO_2 and CH_4) emissions from past combustion using Gas:BC ratios.

The two principal sources of atmospheric BC are fossil fuel and biomass burning (Fig. 1); both are predominantly anthropogenic activities and fossil fuel emissions are believed to have increased exponentially during recent decades (Herring et al. 1985 and others). Natural lightning-ignited wildfires are thought to be a minor atmospheric source of the BC compared to anthropogenic emissions. Biomass is defined here as all aboveground living or dead, but not yet decomposed vegetation susceptible to fire.

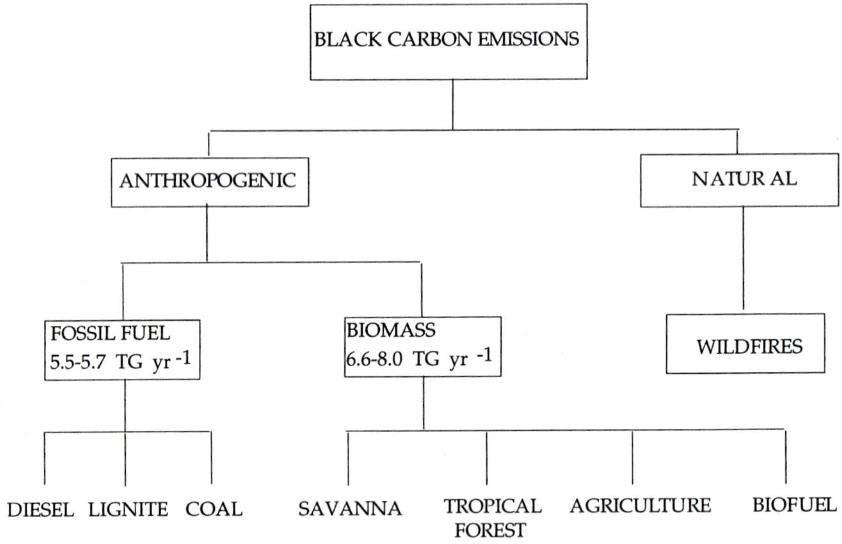

Figure 1. Schematic diagram of black carbon emissions from natural and anthropogenic sources (black carbon emissions are given in Tg yr^{-1}) (range of values are from Cooke and Wilson, accepted; Liousse et al. accepted).

However, before the importance of natural wildfires as emission sources of BC can be assessed, the impact of anthropogenic inputs from fossil fuel and biomass burning upon background BC concentrations must first be evaluated. Recent lake sediments are ideal for such studies, especially if laminated, as they can provide the detailed temporal resolution necessary over the last hundreds of years. On the other hand, the study of deep-sea sediments is more appropriate for obtaining long-term information, as far back as the Cenozoic (Herring 1985). This paper gives a brief introduction to the potential use of BC aerosol records in recent lake sediments as tracers of fossil-fuel and biomass combustion. It focuses upon the concepts rather than analytical methods, but includes an overview of the types of methods available for BC determination.

Definitions of aerosol particles

Combustion processes produce a wide range of particulates whose composition and morphology depend on a number of factors (e.g. fuel type, oxygen flow, temperature). Some of these particles are defined here. Particle sizes are expressed in mm per diameter.

Carbonaceous aerosols are particles which resist decarbonation and contain organic and black carbon, where decarbonation is the chemical treatment of aerosols or particulates with dilute HCl fumes (Cachier et al. 1989).

Black Carbon (BC) is the refractory, experimentally-defined, fraction of carbonaceous aerosols. The cut-off between organic and BC depends critically upon the analytical method used and is discussed in detail elsewhere (Suman et al. accepted).

Charcoal is produced from plant-derived material in low temperature (< 600°C) flames and in limited oxygen supply, in which stratified cell walls are homogenized but recognizable under scanning electron microscopy (SEM) (Griffin and Goldberg 1975) even when fossilized (Jones 1994).

Fly-ash particles are produced from the combustion of coal, oil and other fossil fuels, originating from incompletely burnt fuel and condensation of gases (McCrone and Delly 1973). Fly-ash particles from fossil fuel combustion are often dark and spherical, with particles sizes ranging from <1 mm up to 50 mm. Due to their aerodynamic properties and under certain meteorological conditions, submicron (< 1 mm) and supermicron (> 1 mm) fly-ash particles can typically be transported over thousands of kilometres. Such particles have been found in the remote environment, such as in Arctic aerosols (Rahn et al. 1982), North Atlantic precipitation (Lim et al. 1991, Lim et al. 1994) (Fig. 2a) and snow (Clarke and Noone 1986, Davies et al. 1984), North Pacific deep-sea (Dueser et al. 1983) and lake sediments (Renberg and Wik 1985) (Fig. 2b).

We classify carbonaceous fly-ash particles broadly into two types:

1. Large spherical carbonaceous particles (SCP) are produced mainly by oil and coal burning at high temperatures, having sizes up to about 50 mm (Fig. 2). These particles consist of a carbon matrix with heavy metals and other trace and major elements released from the fuel burnt.

2. Soot spheres, microsoots, chains and chain aggregates are derived from the combustion of a range of fossil fuels. Sizes of individual spheres are generally < 1 mm but aggregate chains are often tens of mms long (Medalia and Heckman 1969). These particles are generated from both domestic and industrial combustion sources, including automobile engines (Ball 1987, Feugier and Gaten 1989).

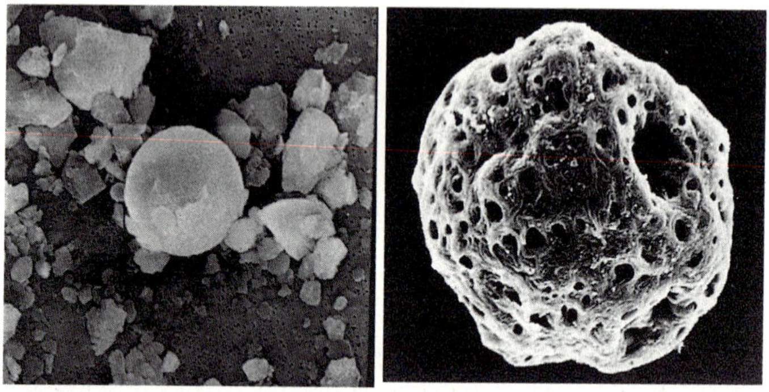

Figure 2. Spherical carbonaceous particles (SCP) found in (a) North Atlantic precipitation and in (b) recent Swedish lake sediments.

It is important to note that soots may also be produced by biomass burning (Cachier 1992, Wolbach et al. 1988, Gaudichet et al. in press) and can only be distinguished from fossil fuel-derived aerosols according to their chemical composition. Soot particles from biomass (Savanna and tropical forest) burning are usually enriched with K (and Cl, P, Cu, Zn) (Artaxo et al. 1993, Gaudichet et al. in press) whereas soots from diesel engine exhausts are often associated with S and K (Gaudichet et al. in press). On the other hand, coal and oil fly-ash particles are typically enriched with trace metals such as Cd, Mn, Pb, V, and Zn (Rahn 1981, Natusch et al. 1973, Wadge et al. 1986). Particle morphology alone is therefore an insufficient criteria for the identification of fly-ash particles but can nonetheless be a powerful tool to broadly classify combustion sources as discussed later.

Analyses of carbonaceous particulate aerosols in sediments

Analyses of BC in sediments fall into two categories, optical (counting) and chemical methods. Optical techniques are time consuming but can provide quantitative and highly specific information on the origin of particles based on their morphology and size (e.g. SCP). As mentioned above, although specific source attribution of combustion-derived particles may be more demanding, microscopy has been used to broadly distinguish between spheres (from coal and oil) and charcoal fragments (from wood burning) (Smith et al. 1973). However, optical techniques may underestimate small particles, and in practice submicron particles are virtually impossible to count. Chemical methods are, by comparison, rapid and all particles, regardless of their size and morphology, should be quantified. The size discrimination of particulates analyzed by optical and chemical methods therefore provide different yet, complementary information. However, bulk methods cannot distinguish between fossil-fuel and wood-derived particles.

Optical methods

Optical analyses of particles from fossil fuel combustion in sediments based on particle counting under microscope were developed by Renberg and Wik (1985) and Rose (1990) for SCP. These methods are based on the removal of organic material and other sediment fractions with chemical digestion and counting of particles on glass Petri dishes using a stereomicroscope (Wik and Renberg 1985) and on microscope slides using a light microscope (Rose 1990). The latter method uses higher magnification and the size limit for identification of SCP is thus about 1-2 mm.

Chemical methods

Chemical methods for the determination of BC in sediments generally consist of two fundamental steps. First, the dried pulverized sediments are treated with a strong acid (e.g. HCl, HF) to remove carbonates and to attack iron hydroxides and silicates. In the second stage, the organic carbon is separated from BC and the residual BC determined in the sediment matrix. Methods of removing the organic carbon include solvent extraction (Wolbach et al. 1985), thermal treatment (Lim and Cachier, accepted), acid (Winkler 1985, Wolbach and Anders 1989) or alkaline oxidation (Smith et al. 1973, Griffin and Goldberg 1975). By far the most widely used method is the latter, using peroxide oxidation. However, as noted earlier, the cut-off between organic and BC is a critical step in the analytical procedure and is experimentally defined. Thus, the chemical method used to determine BC is of paramount importance.

Fossil fuel and biomass burning emissions

Some of the major objectives of this workshop were to evaluate paleo-records of past BC combustions preserved in sediments and to interpret these emissions within the context of past global climate changes. These tasks are complex and involve several aspects of fire history and behaviour, particle formation, deposition, and preservation, which are poorly understood. Nonetheless, in this chapter we focus upon two objectives, namely:

1. How can BC in recent sediments be used to distinguish between fossil fuel and biomass burning ?

2. Can background concentrations of BC from wildfires be estimated from ancient (pre-historic) sediments ?

In the remainder of this paper, we summarize the results of several studies of fossil-fuel derived aerosols in recent lake sediments and suggest how fossil-fuel and biomass BC may be differentiated. Finally, in the conclusions, we speculate upon how BC deposited in recent sediments may be used to extrapolate, and possibly quantify, background concentrations of BC from wildfires that are preserved in ancient pre-historic sediments.

Fossil fuel combustion records in lake sediments

The bulk of combustion-derived particles deposited in sediments are not SCP, but the second category of particles mentioned earlier, the soots. In industrial areas, recent lake sediments are often darker than older sediments and sediments from clean areas. It is possible that this is due to black airborne pollutants that consist of a mixture of combustion products, such as high molecular weight oils and tar from unburned petroleum, charcoal, fly-ash, etc. This blackening of sediments could also be due to the presence of iron sulfides or algal detritus.

The first detailed sediment studies of carbonaceous particles from fossil fuel combustion were made by Griffin and Goldberg (1979, 1981, 1983). They found a temporal correspondence between the BC record in Lake Michigan (analyzed with a chemical method) and fossil fuel combustion history. This relationship has since been found in several other lakes in the USA and in Europe (see review by Wik and Renberg 1995).

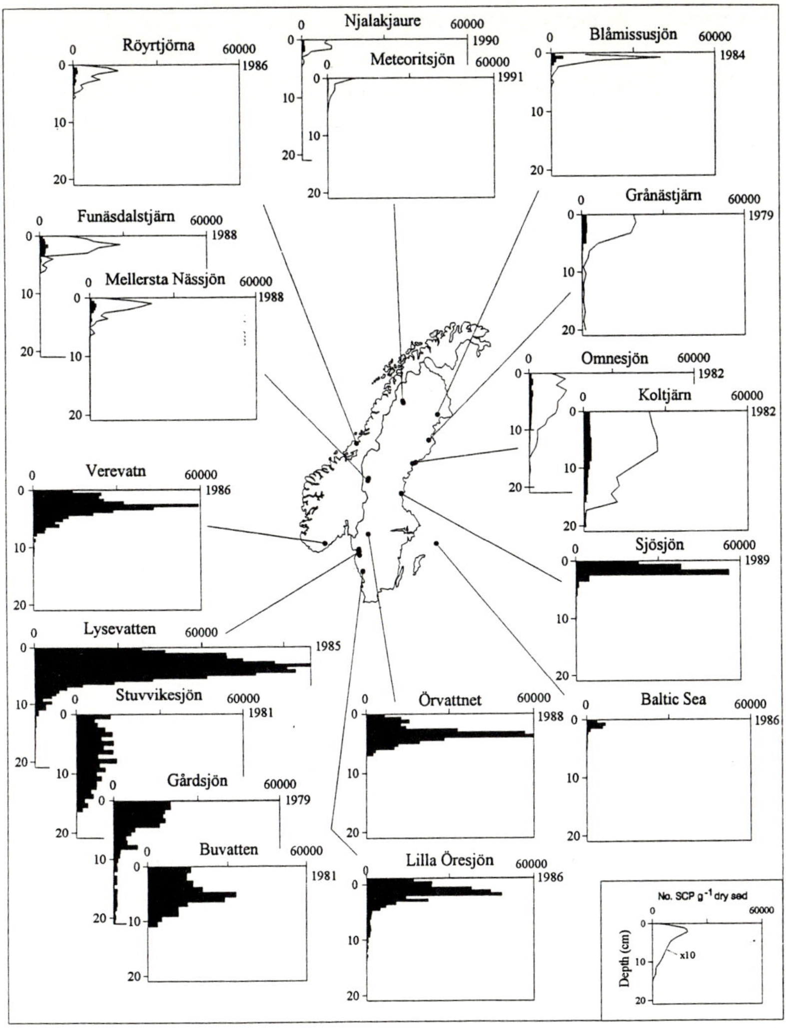

Figure 3. Spherical carbonaceous particles (SCP) concentrations in sediment cores from Swedish and Norwegian lakes and from the Baltic Sea (from Wik and Renberg, accepted)

A series of investigations of lake sediment cores from Sweden revealed a general trend in SCP concentrations (see Wik and Renberg 1995). SCP deposition had already begun during the 18th century, and during the middle of the 19th century the deposition rate slowly began to increase following the industrialization. A marked increase of SCP concentrations in lake sediments occurred after the Second World War coinciding with the expansion of industry in Europe and a rapid increase in fuel consumption. The SCP concentration increased during the 1950´s and 60´s and peaked around 1970 before the so-called "oil crises". The decrease in SCP concentrations after the early 1970s is presumably due to improved air pollution emission controls and decreased fuel consumption. This general pattern can be seen in most lake sediments in severely polluted and clean rural areas as shown in Fig. 3.

There are typically > 10,000 spherical carbonaceous particles per gram dry sediment in recent sediments in areas with high atmospheric deposition of pollution from fossil fuel combustion, such as south-western Sweden. Analyses of surface sediment samples (representing the 1980´s) from about 110 lakes distributed all over Sweden, show a clear south-north gradient, with the most polluted southern areas having more than a hundred times higher SCP concentrations than clean areas in northern Sweden (Wik and Renberg 1991) (Fig. 4). The SCP distribution was similar to the atmospheric deposition pattern of lead and sulfate. Both lead and sulfate are long-range transported pollutants and most of the fall-out over Sweden is considered to be from foreign emissions. How much of the SCP deposition that is due to atmospheric transport from foreign air emission sources is, however, unclear.

These lake studies of fossil-fuel derived particles, together with inventories and emission factors, demonstrate that human activities have significantly perturbed background levels of BC since the 18th century. Following, we discuss the spatial variability of atmospheric BC concentrations.

Fossil fuel versus biomass burning

Earlier estimates of global BC emissions were rather high (Crutzen and Andreae 1990, Penner et al. 1993), but with improving inventory data and emission factors, the best estimates of present BC emissions are now much lower and of the order of 13 Tg yr-1 (Cooke and Wilson, accepted, Liousse et al. accepted).

On average, about 60 % of the total global BC emissions are generated by fossil fuel combustion with the remainder from biomass burning. Sources of BC from the latter sources are predominantly from the Savanna and Cerrado grasslands in Africa and Southern America and tropical forests in Southern America and Central Asia.

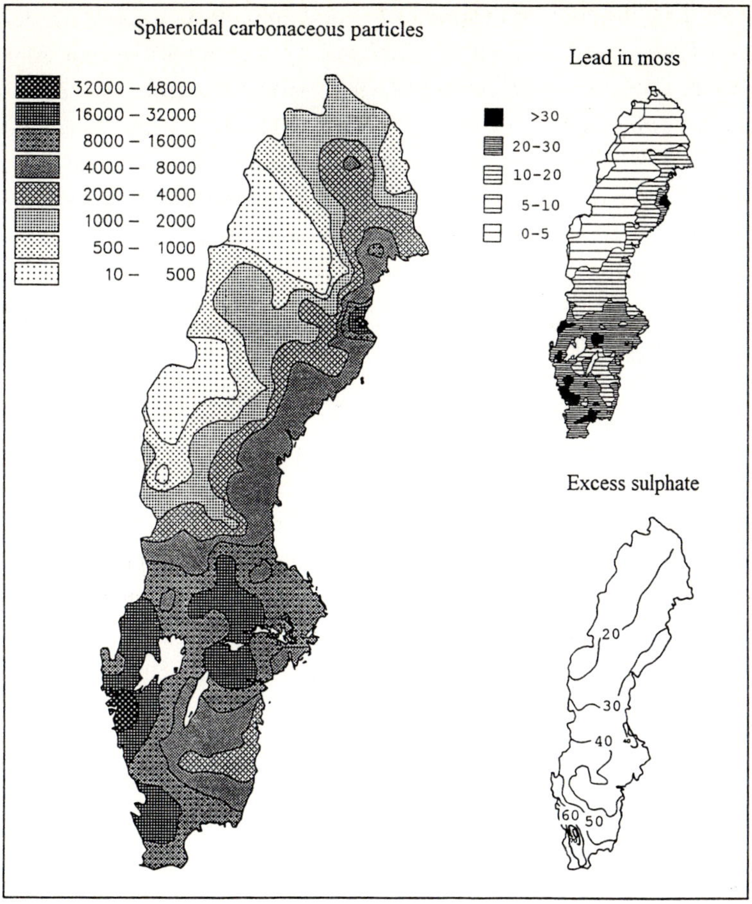

Figure 4. Spherical carbonaceous particles (SCP) concentrations in surface sediment samples (0-1 cm) of Swedish lakes (number g^{-1} dry sediment). The map is based upon analyses of 114 lakes. Inserted maps show results of independent analyses of deposition of airborne pollutants; lead in moss in 1985 (mm g^{-1} dry mass); based upon analyses of >800 samples (from Statens Naturvårdsverk 1987), and annual excess sulfate wet deposition 1983-1987 (meq m^{-2} yr^{-1})(from Granat 1989)(meq are milli equivalents)(bars on figures represent the number of particles per gram of dry sediment and a continuous line is 10 times the number of particles per gram of dry sediment) (from Wik and Renberg 1991).

The three main fossil-fuel sources of BC are primarily hard coal, lignite and diesel fuel. Gasoline, fuel oils, and natural gases are fairly clean fuels with lower emission factors. In general, emission factors of BC from domestic fires are 10 times higher than industrial combustion. These sources and emission factors are reflected in the high atmospheric BC emissions over Europe, Eastern USA and Japan (Fig. 5) due to fossil fuel burning. For biomass burning, highest BC emissions are found over South America, Africa, Central Asia (Fig. 6) and Australia (Cooke and Wilson, accepted). These spatial differences are summarized in Table 1 (For a full discussion of these data, see Cooke and Wilson, accepted). In western Europe, for example, more than 99 % of total BC emissions is from fossil fuel, whereas in Africa it is only about 10 %.

Table 1. Regional comparison of fossil fuel and biomass black carbon emissions (Tg C yr^{-1}) (kind permission of Cooke and Wilson)

REGION	BIOMASS	FOSSIL FUEL
North America	0.07	1.27
Central America	0.03	0.01
Brazil	0.78	0.06
Rest of S. America	0.42	0.04
China	0	1.10
Rest of Asia	0.27	0.85
Western Europe	0.01	1.09
Eastern Europe	0	1.56
former USSR	0.11	1.55
Oceania	0	0.10
Africa	3.81	0.34
Total	5.49	7.97

The main atmospheric deposition pathway of BC is by wet deposition (Suman et al. accepted) and BC deposition maps (Cooke and Wilson, accepted) computed from both wet and dry deposition exhibit a similar pattern to the combined fossil fuel and biomass BC emissions shown in Figs. 5 and 6. Highest atmospheric deposition fluxes are found over industrial America, Europe and China and over the grasslands of the South African and South American continents. Global atmospheric deposition flux estimates (Suman et al. accepted) suggest that about 30-50% of the BC emitted from continental sources is exported and delivered to the open oceans. Therefore, at least half the emitted BC is expected to fall in continental regions. Consequently, we expect a mixture of aerosols from fossil fuel and biomass burning to be deposited to lake sediments. Furthermore, the ratios of Fossil Fuel:Biomass particulates in lake sediments are

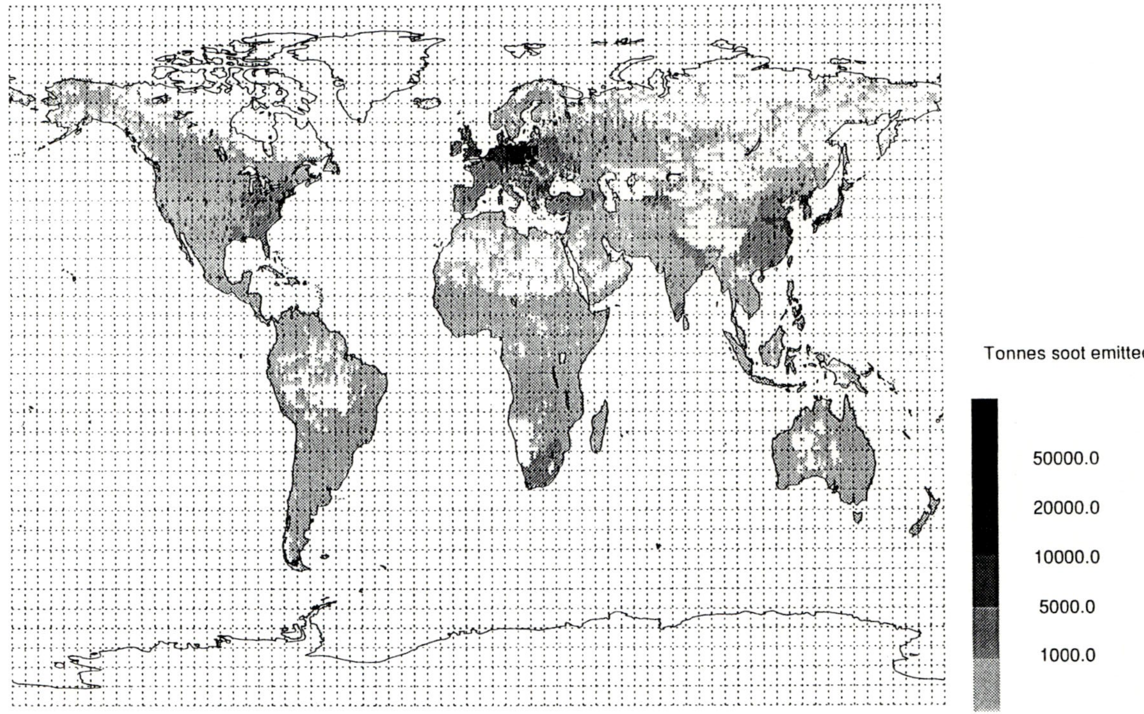

Figure 5. Black carbon emissions from fossil fuel per annum (kind permission of Cooke and Wilson) (Units in tonnes of black carbon emitted per annum)

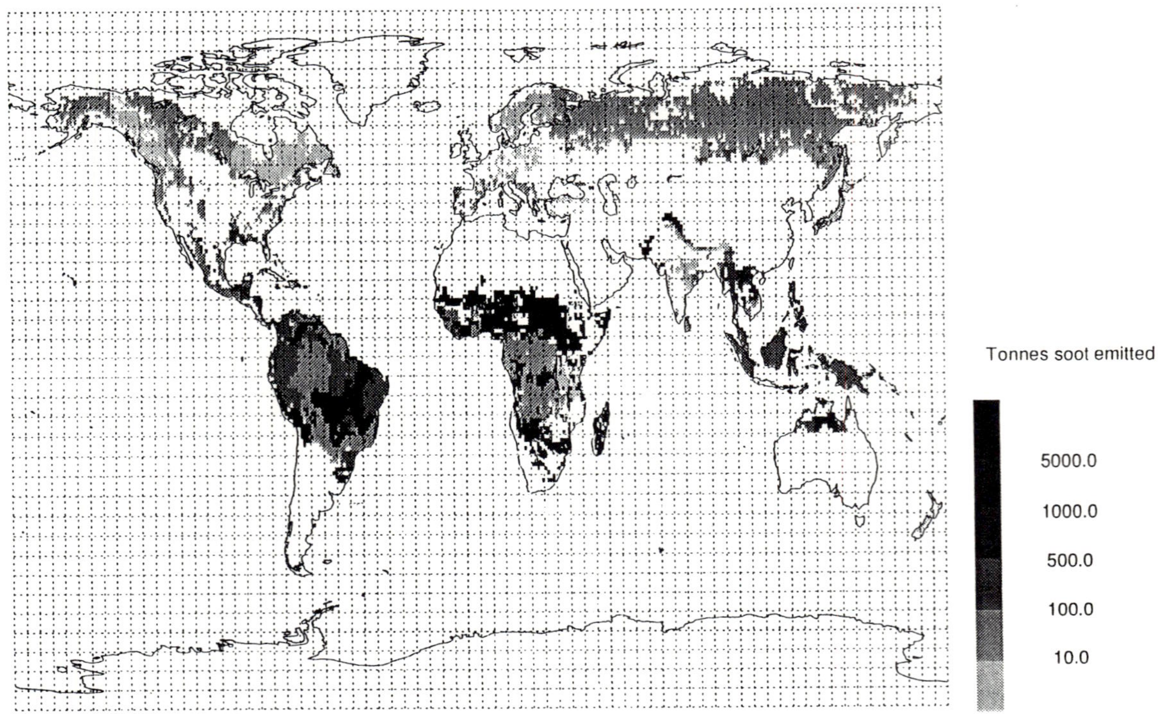

Figure 6. Black carbon emissions from biomass burning per annum (kind permission of Cooke and Wilson) (Units in tonnes of black carbon emitted per annum)

expected to reflect their respective atmospheric ratios, assuming similar atmospheric scavenging efficiencies for the different types of aerosols.

As mentioned earlier, one crucial question is 'How can BC in sediments be used to distinguish between fossil fuel and biomass burning ?' To the best of our knowledge, previous studies of BC in lake sediments are based either on chemical or optical methods. Given that these two methods provide very different, yet complementary, information on the sources of BC, a comparison of the results obtained from these two analytical methods might differentiate between fossil-fuel and biomass BC; thus enabling apportionation of the two particle source strengths.

In theory, a bulk chemical method should determine total (fossil fuel and biomass) BC concentrations. By contrast, fossil fuel BC may be estimated by optical analyses provided that methods for enumeration of microspheres could be developed. The difference between total and fossil fuel BC measured by the two different chemical and optical methods should therefore be equal to BC emitted from biomass burning. In practice, such a distinction may be quantitatively difficult to assess since biomass burning also produces microspheres that may be confused with fossil fuel particles as mentioned earlier. However, in regions where atmospheric inputs are predominantly fossil fuel or biomass, as is the case in Europe or Africa (Figs. 5 and 6), we expect clear patterns to emerge from an intercomparison between chemical and optical measurements. A pilot study is currently underway in our laboratories to evaluate such a hypothesis. The results of such an intercomparison may be used to re-evaluate existing BC data sets of lake sediments and to help distinguish between BC from different fuel emission sources.

Conclusions

Lake sediment records of BC as archives of past fire activities are something of a Pandora's box. Whilst promising to provide insight into past fire activities and global climate changes, the interpretation of BC also raises a number uncertainties. Compared to other archives of combustion like ice-cores and snow, sediment cores are much more complicated to analyze and to interpret due to the chemical complexity of the sample matrix and additional (air:water, surface run-off and water-column) transport and post-deposition processes that must be considered.

Despite these hindrances, recent lake sediment records of fossil fuel-derived aerosols provide an opportunity to further our understanding of past fire activity. Using a combination of optical and chemical analytical methods to determine BC in sediments, we may be able to estimate the BC component generated by biomass burning. Given that present day biomass burning inventories are relatively well known, it should be possible to derive empirical relationships between BC concentrations in recent sediments and combustion source strengths. These present day anthropogenic relationships should enable us to better interpret BC data in ancient sediments and to evaluate the extent of natural fire activity prior to anthropogenic inputs. Using this methodology, we hypothesize that background concentrations of BC due to wildfires may be estimated from BC in pre-historic sediments as functions of several parameters including, distance from fire source, atmospheric and water transport, dispersion and deposition modes.

Further research

As indicated in this paper, the interpretation of BC data in lake sediments should not be oversimplified. Fires produce carbonaceous matter with a range of sizes that undergo different fates in the environment. Field sampling studies suggest that BC aerosols can be dispersed over distances of several thousand of kilometres in the atmosphere, often remote from continental regions, before deposition to the surface waters. The larger residual charcoal fragments are mainly transported by surface runoff and these are probably deposited close to source regions except where long -range fluvial transport is important. These types of dispersion and deposition processes have been mainly observed in the marine sediment record and in coastal sediments where high BC fluxes due to riverine inputs have been found (Suman et al. submitted and others). Intuitively, similar transport and deposition mechanisms of BC to lake sediments are expected. Large lakes remote from combustion sources are likely to contain a strong signal of background aerosol BC concentrations whilst lakes close to fires, e.g. biomass burning, will be dominated by larger charcoal particulates transported by surface runoff (see for example, Griffin and Goldberg 1981).

Further research is needed to provide unravel the complex BC record in lake sediments with confidence and to determine if the above suggestions are justifiable. Progress therefore hinges upon the effective separation of aerosol BC from charcoal in sediments for the development of empirical relationships between SCP and BC. Further development of analytical methods (e.g. isotopic measurements, calibration of chemical with optical methods) are required to separate the atmospheric aerosol and charcoal fingerprints, and indeed the fossil fuel and biomass signals, in lake records.

In addition to the use of different analytical methods to achieve particle size and source separation of BC in sediments, field sampling strategies are also needed to provide further information on the coupling between atmospheric, water and sediment concentrations in lakes. These methods might include simultaneous sediment trapping within the lake water column, as well as at the water/sediment interface.

Acknowledgments

The authors would like to warmly thank Jim Clark and all the folks who so efficiently organized an enjoyable and scientifically productive meeting. We are also indebted to W. Cooke and J.N. Wilson for their invaluable model results. This worked was partly funded by thc Human Capital and Mobility Programme from the European Community. This manuscript was greatly improved by the comments from Richard Zepp, Thomas Kuhlbusch, Tim Jickells, Helene Cachier and Dwight Smith. This is CFR contribution No. 1767.

Literature References

Artaxo P, Yamasoe MA, Martins JV, Kocinas S, Carvalho S, Maenhaut (1993) Case study of atmospheric measurements in Brazil: Aerosol emissions from Amazon Basin fires. In Crutzen PJ, Goldhammer JG (eds) *Fire in the environment. The ecological, atmospheric, and climatic importance of vegetation fire* pp 139-158 Wiley Chichester

Ball DJ (1987) Particulate carbon emissions and diesel engines. In *Vehicle emissions and their impact on European air quality Proceedings of IMechE* 3-5 Nov pp 83-88 London

Cachier H (1992) Biomass burning sources. *Encyclopaedia Earth System Sci* 1:377-385

Cachier H, Bremond MP, Buat-Menard P (1989) Determination of atmospheric soot carbon with a simple thermal method. *Tellus* 41B:379-390

Clarke AD, Noone KJ (1986) Soot in the Artic snowpack: A cause for perturbations in radiative transfer. *Atmos Environ* 19:2045-2053

Cooke WF, Wilson JJN (1996) A global black carbon aerosol model. *J Geophys Res* (accepted)

Crutzen PJ, Andrea MO (1990) Biomass burning in the tropics: Impact on atmospheric chemistry and biogeochemical cycles. *Science* 250:1669-1678

Davies TD, Abrahams PW, Tranter M, Blackwood I, Brimblecombe P, Vincent CE (1984) Black acidic snow in the remote Scottish Highlands. *Nature* 312:58-61

Dueser WG, Emeis K, Ittekkot V, Degens ET (1983) Fly-ash particles intercepted in the deep Sargasso Sea. *Nature* 305:216-218

Feugier A, Gaten P (989) Formation of particulate pollutants in hydrocarbon combustion. *Pollution Atmosphérique* 122:184-189

Gaudichet A, Echalar F, Chatenet B, Quisefit JP, Malingre G, Cachier H, Buat Menard P, Artaxo P, Maenhaut W (1995) Trace elements in tropical African Savanna biomass burning aerosols. (accepted)

Granat L (1989) Luft och nederbördskemiska stationsnätet inom PMK, rapport från verksamheten (1988) Statens Naturvårdsverk SNV PM 3649, 64

Griffin JJ, Goldberg ED (1975) The fluxes of elemental carbon in coastal marine sediments. *Limnol Oceanogr* 20:456-463

Griffin JJ, Goldberg ED (1979) Morphologies and origin of elemental carbon. *Science* 206:563-565

Griffin JJ, Goldberg ED (1981) Sphericity as characteristic of solids from fossil fuel burning in a Lake Michigan sediment. *Geochim Cosmochim* Acta 45:763-769

Griffin JJ, Goldberg ED (1983) Impact of fossil fuel combustion on sediments of Lake Michigan: A reprise. *Environ Sci Tech* 17:244-245

Herring JR (1985) Charcoal fluxes into sediments of the North Pacific ocean: the Cenozoic record of burning. In Sundquist ET, Broeker WS (eds) *The carbon cycle and atmospheric* CO_2*: Natural variations Archean to present* pp 419-442 American Geophysical Union Washington DC

Jones TP (1994) Ultrastructural and chemical studies on Oligocene fossil wood from Bovey Tracey, Devon, UK. *Rev Paleobot and Palynol* 81:279-288

Lim B, Cachier H (1995) Determination of black carbon by chemical oxidation and thermal treatment in recent marine and lake sediments and Cretaceous-Tertiary clays. *Tellus* (accepted)

Lim B, Jickells TD, Davies TD (1991) Sequential sampling of particles, major ions and total trace metals in wet deposition. *Atmos Environ* 25:745-762

Lim B, Jickells TD, Colin JL, Losno R (1994) Solubilities of Al, Pb, Cu, Zn in rain sampled in the marine environment over the North Atlantic Ocean and Mediterranean Sea. *Global Biogeochem Cycles* 8:349-362

Liousse C, Penner JE, Chaung C, Walton JJ, Eddleman H, Cachier H (1995) A Global three-dimensional model-study of carbonaceous aerosols. *J Geophys Res* (accepted)

McCrone WG, Delly JG (1973) *The Particle Atlas, II.* Ann Arbor Science Ann Arbor

Medalia AI, Heckman FA (1969) Morphology of aggregates-II. Size and shape factors of carbon black aggregates from electron microscopy. *Carbon* 7:567-582

Natusch DFS, Wallace JR, Evans CA (1973) Toxic trace elements: Preferential concentrations in respirable particles. *Science* 183:202-204

Penner JE, Eddleman H, Novakov T (1993) Towards the development of a global inventory for black carbon emissions. *Atmos Environ* 27:1277-1295

Rahn KA, Brosset C, Ottar B, Patterson EM (1982) Black and white air episodes: Chemical evolution of Eurasian air masses, and long-range transport of carbon to the Arctic. In Wolff GT, Klimsch RL (eds) *Particulate carbon: Atmospheric life cycle* pp 327-339 Plenum

Renberg I, Wik M (1985) Soot particle counting in recent lake sediments: An indirect dating method. *Ecol Bull* 37:53-57

Rose NL (1990) A method for the extraction of carbonaceous particles from lake sediments. *J of Paleolimn* 3:45-53

Smith D, Griffin JJ, Goldberg ED (1973) Elemental carbon in marine sediments: a baseline for burning. *Nature* 241:268-270

Statens Naturvårdsverk (1987) *Monitor 1987* Liber Förlag Stockholm

Suman D, Kuhlbusch TAJ, Lim B (1995) Marine sediments: A reservoir for black carbon and their use as records of biomass burning. In Clarke JS, Cachier H, Goldhammer JG, Stocks BJ (eds) *Sediment records of biomass burning and global change* Springer Verlag (accepted)

Wadge A, Hutton M, Peterson PJ (1986) The concentrations and particle size relationships of selected trace elements in fly ashes from the UK coal-fired power plants and a refuse incinerator. *Sci Tot Environ* 54:13-27

Wik M, Renberg I (1985) Carbonaceous particles in lake sediments-pollutants from fossil fuel combustion. *Ambio* 14:161-163

Wik J, Renberg I (1991) Recent atmospheric deposition in Sweden of carbonaceous particles from fossil-fuel combustion surveyed using lake sediments. *Ambio* 20:289-292

Wik J, Renberg I (1995) Environmental records of carbonaceous fly-ash particles from fossil-fuel combustion: A summary. *J of Paleolimn* (accepted)

Winkler MG (1985) Charcoal analysis for paleoenvironemental interpretation: A chemical assay. *Quart Res* 23:313-326

Wolbach WS, Lewis RS, Anders E (1985) Cretaceous extinctions: Evidence for wildfires and search for meteoritic material. *Science* 230:167-230

Wolbach WS, Gilmour I, Anders E, Orth CJ, Brooks RR (1988) Global fire at the Cretaceous-Tertiary boundary. *Nature* 334:665-669

Wolbach WS, Anders E (1989) Elemental carbon in sediments: Determination and spectroscopic analysis in the presence of kerogen. *Geochim Cosmochim* Acta 53:1637-1647

PART V:

FUTURE DIRECTIONS

Reports of Working/Discussion Groups at NATO Workshop: Sediment Records of Biomass Burning and Global Change

J. P. Bradbury
U.S. Geological Survey
MS 919 Box 25046 Federal Center
Denver, CO 80225 USA

Preface

On the penultimate day of the workshop the participants were divided into groups to discuss the four key aspects of sediment records of biomass burning. These discussions endeavored to describe the systems and processes pertinent to the production, identification, transportation, and ultimate deposition of biomass burning products. Each discussion group was asked to consider what is known, what needs to be known, what is potentially knowable, and, lastly, to recommend directions and priorities for future research. The group reports below follow this format as closely as possible. The group moderators, rapporteurs, and other contributors to the group reports are listed at the beginning of each section.

I. Sources of Charcoal and Other Products of Combustion From Wildland Fires

J.G. Goldammer, rapporteur and J.B. Kauffman, moderator

Introduction

"Sources" in this paper refer to residual components of vegetation fires and other plant biomass burning (most notably charcoal) that may serve as long-term indicators of fire occurrence in sediments, soils, or geological records. The products of biomass burning provide information concerning fire behavior, biogeochemical cycling, fire ecology, and fire history. A thorough understanding of these disciplines is a prerequisite for determining past climatic, vegetation, and fire regimes through analysis of the paleo/sediment record. The preliminary assumption of this group is that the "sources" used to determine fire history arise from the fire itself. While charcoal and other combustion products may be deposited following a period of transitory storage (soils, water columns or trees,

NATO ASI Series, Vol. I 51
Sediment Records of Biomass Burning
and Global Change
Edited by James S. Clark, Hélène Cachier,
Johann G. Goldammer, and Brian Stocks

etc.), the group focused on variables of interest strictly associated with the actual fire event.

What Do We Currently Know?

In order to interpret the paleo/sediment record, one must have an understanding of current fire regimes. Fire regimes are defined as the pattern or occurrence of fire in a given ecosystem. Fire regimes are described in terms of: 1) fire frequency or mean fire-return interval—the average number of years between fires; 2) fire magnitude—the typical range and aerial extent of fires; 3) severity—the ecological impact on the extant vegetation (i.e., a stand-replacement fire or understory burn); and 4) the seasonality of fires. Fire regimes are described at least in a qualitative sense for most wildland and agricultural systems of the world; less is known concerning historical landscapes.

Important determinants of fire regimes include regional climate, microclimate, topography, ignition sources, and those anthropogenic activities that may alter fuel loads or patterns of ignition. These variables are qualitatively known for most contemporary landscapes, but are uncertain for the historic and prehistoric record.

Total above-ground biomass and/or fuel loads (that portion of the ecosystem exposed to combustion by fire) have been quantified at varying levels of sampling intensity for most global vegetation types. Scientific understanding of fire effects includes ranges in the levels of biomass combustion and fire behavior for many ecosystems, particularly those of temperate North America, boreal forests, tropical savannas, and a limited number of tropical forests. Data exist for only a few ecosystems on the quantities (mass) of ash (defined as vegetative materials that have been thermally transformed through partial or complete combustion). Less information exists on the physical and chemical composition of the post-fire ash component, including the size-class distribution of particles, organic matter content, total carbon concentration, black carbon concentration, and other potential indicators of fire history. Improved quantification of the range in variability of these products of combustion on an ecosystem level basis would facilitate the interpretability of the sediment record.

Finally, terrestrial pools of carbon have been described and partitioned among the various ecosystems of the world. However, partitioning of this global carbon sink into active carbon pools with rapid turnover rates (measured at scales of months to decades) and passive pools with turnover rates measured at scales of centuries or longer are limited. Fires influence the carbon balance of an ecosystem through combustion and transformation of passive organic pools into refractory forms of carbon (elemental, black, inert carbon). Conversely, refractory forms of carbon (charcoal) are susceptible to transformation via combustion processes. To date, studies quantifying these phenomena are limited to only a few ecosystems.

Where Are The Gaps In Our Knowledge, and What Can and Cannot be Addressed?

An in-depth discussion of the gaps in our knowledge of sources is beyond the scope of this contribution. While there is qualitative knowledge of fires regimes in most global vegetation types, there are a great number of ecosystems in which quantitative studies of climate, fire behavior, fire ecology, and fire history have not been conducted. This quantitative knowledge is needed before a better understanding of historic fire regimes can be inferred from the paleo-record. Data on historic fire regimes could be of great utility for models that would predict future fire regimes and carbon dynamics under various scenarios of global change.

A number of basic questions that focus on the variable nature of charcoal production in wildland fires must be addressed. While it is generally assumed that approximately 13% of the biomass consumed by fire is converted to charcoal, few studies have actually quantified this phenomenon. In addition, how the composition and structure of charcoal varies among ecosystems is unknown. Influences of fuels, season of burn, microclimate, and fire behavior on charcoal production and chemistries are in need of quantification. Clearly, improved definitions of charcoal are needed to facilitate field studies of charcoal dynamics. Studies such as these should be carried out in those ecosystems that currently contribute much to global biomass burning (tropical forests, tropical savannas, boreal forests, etc.) and those that might closely resemble past ecosystems under investigation by paleoecologists.

In addition to uncertainties associated with the production of charcoal, there are equally uncertain parameters associated with rates of decomposition or disappearance of charcoal. Studies are needed to quantify turnover rates of surface, sediment, and soil charcoal. These data are needed to interpret the paleo-record in light of global carbon dynamics. Hypotheses related to the importance of refractory carbon (charcoal) as a significant carbon sink should be tested.

Future Research Questions and Directions

Research contributing to understanding the role of fire in global environmental change is critical. It is of great importance to define how fires and land use activities influence the size, composition, and dynamics of global C pools. In order to interpret the sediment record, an understanding of the dynamics of climate, land use, vegetation (fuels) and those combustion processes that determine paleo-records are necessary. Progress towards quantifying the linkages between ecosystem dynamics, global environmental change, and the sediment record can be accomplished by addressing the following research questions:

1. While scientists can measure the products of combustion from fires (e.g. ash, charcoal residual uncombusted debris, etc.) what specifically are those components of biomass burning that are characterized as "charcoal" in paleoecology?

2. What is the quantity, size distribution, and chemistry of charcoal and other products of combustion? With respect to these charcoal components, what is the degree of variability among ecosystems, fire regimes, and land use activities? How will variables, such as season of burn, climate, microclimate, ignition source, and the chemistry and arrangement of fuels influence these components? How can scientists measure and/or model this production for contemporary as well as past and future fire regimes? At what level of specificity can we have a standardized definition for "charcoal"?

3. Are there other products of combustion that may decrease the uncertainty of fire histories? If so, what are the quantities and variability across fire regimes or land use activities?

4. What are the regions or ecosystems in the greatest need of fire ecology studies and fire history (i.e., where do gaps in the data exist with reference to the basic information of fire and influences)?

5. How have humans modified the environment to influence fire regimes (i.e., to what extent has man altered ignition patterns, seasons of fire, and fuels through removal, creation, or active fire suppression)?

6. What is the importance of charcoal, black carbon, or refractory forms of carbon as a sink in the global carbon balance? What is the total pool size of refractory carbon and where is it sequestered? What proportion of the total terrestrial ecosystem carbon pool is comprised of these recalcitrant forms of carbon?

7. What is the global mass balance of charcoal or other refractory forms of carbon? What are the current and historical rates of production? What are the rates of degradation or decomposition of refractory carbon as well as other indicators of fire history?

8. How has/will global environmental and climate change influence the global carbon mass balance (i.e. the changes in fire regimes, or products of combustion)? Through better integration with contemporary fire ecology studies, how much can we learn from the paleo-record that would be used to predict changes that will occur in the future?

9. What are the synergistic relationships between global environmental/climate change and carbon dynamics? If anthropogenic influences and/or changes in climate influence fire regimes, what will be the response of the ecosystem in terms of rates and potentials of carbon sequestration?

II. Characterization of Fire-Altered Materials

G. Dubay, rapporteur and T. Novakov, moderator

Introduction

Biomass burning has affected the earth's environment since colonization by plants. The oldest records of biomass burning are considered to be the sedimentary occurrence of fire-altered material[1] in the lower Carboniferous and transitional Devonian-Carboniferous, from approximately 350 million years ago. At least from that time biomass burning has influenced the development of ecosystems, the atmosphere, and biogeochemical cycles and, thus, the earth's climate. To tackle the problem of correctly evaluating the impact of biomass burning on, e.g. today's atmosphere/climate, there is an urgent need for quantitative information about past and present fire activities. In addition, an understanding of global change requires an appreciation of the nature and yield of pyrogenic substances derived from biomass burning as a function of fuel type, climatic conditions, atmospheric transport, and deposition, including particle sizes, transformations of pyrogenic substances during transport through the

[1] The term "fire-altered material" is used here to summarize all possible tracers in sediments related to biomass burning

atmosphere and water column to sediments, persistence of pyrogenic residues in sediments, sedimentation rates, bioturbation, and other factors that affect the vertical profiles of pyrogenic residues in sediments. Analyses of sediments that contain fire-altered material is one way to address this problem, but others arise, as summarized in three key questions: 1) What is the best tracer for biomass burning in sediments and how can it be differentiated from other material buried with it? 2) How is the tracer changed during transport and residence in the sediments? 3) How can the tracer be related to the source region and its fire activity? Other significant questions go along with these three key questions and have to be solved for an accurate assessment of past fire activity. This can only be solved in a joint, interdisciplinary effort by scientists from various disciplines, e.g. atmospheric chemists, biogeochemists, meteorologists, ecologists, and paleontologists.

What Do We Currently Know?

Black, fire-altered carbonaceous material is the central tracer of biomass burning used for samples from the atmosphere and sediments. Two different analytical techniques are most commonly used for characterization, one of which is based on optical properties of this blackened material. The second uses the chemical/thermal inertness of the carbon compounds in this fire-altered material for differentiation from other organic carbon compounds. Inter comparisons between these two basically different techniques have been conducted for the analyses of atmospheric fire-altered carbonaceous material, but not for sedimentary records which best represent the geochemical record. Inter-comparisons between the two standard methods and any other method in use should be conducted (a) to derive information about the comparability of data sets and to build up one main data set and (b) to identify one or more major tracers for global sedimentary analyses (Suman et al., this volume). Since nearly all data on past fire activity are based on the bulk analyses of carbonaceous fire-altered material, it would be useful to have an independent tracer for the verification of the results, e.g. specific PAH (polycyclic aromatic hydrocarbons) or isotopic signatures.

Current knowledge includes a great deal of information about the residues of biomass burning that can be found in the environment, i.e. sediments, air sampling, etc. This information is scattered among different disciplines and is incomplete. It was the recommendation of this working group that information in the literature be synthesized and maps of fire-altered material be produced. This could then serve as a resource in experimental design and project planning.

The paleo-record indicates that the concentration of gases in the atmosphere has changed as a result of the many factors, including the colonization of the seas and the land by plants and animals. More recently anthropogenic activity has resulted in changes in the gas concentration in the environment. These changes may have a profound impact on global climatic change and on the frequency of biomass burning. There is some information concerning past concentrations of atmospheric gases, including oxygen, nitrogen, carbon dioxide, methane, and carbon monoxide, but it is incomplete. What have been the concentrations of these gases during the past millennia? What impact does the concentration of these gases have on biomass burning? These questions must be answered if we are to understand the interaction of forces that change climate.

Where Are The Gaps In Our Knowledge, and What Can and Cannot be Addressed?

There is a broad information base from a variety of disciplines on characterization of fire-altered materials in sediments, residues, and the atmosphere. Each discipline makes use of different analytical methods and terminologies much of which is inconsistent. We recommend development of a standard terminology for fire-altered materials. This should lead to better understanding of existing data and make this information available to a broader range of scientists.

Terms needing better definition:

1. *Descriptive terms From Atmospheric Research*

·Black Carbon
·PAHs
·Aerosol
soot and micro-soot
small fractions
large fractions
·Trace elements
These latter two headings are linked in the sense that aerosols derived from burning activity often contain high levels of trace elements found in a particular area.

2. *Descriptive Terms From Organic Geochemistry*

·Charcoal
·Humus
·Peat
·PAHs

3. *Descriptive Terms From Paleoecology and Geology*

·Fusinite-Charcoal (fossil & modern)
·Fusain
·Charred organic matter
·Interferences
  minerals
  manganese dioxide
  iron sulfide (pyrite)

4. *Terms Relating to Methods of Characterization*

·Optical Methods:
  "Black"
  "Opaque"
·Reflectance Methods:
  % light reflected
·Chemical Methods:
  Oxidation
    1. HNO_3
    2. HClO
    3. H_2O_2
    4. heat + O_2
·Physical Methods:
  Microscopic Investigation:
    i. examination using a microscope for structural definition.
    ii. determination of cell structure
  Thermogravimetric Analysis (TGA) which involves the loss of mass upon heating in an atmosphere which can be controlled (inert or oxidative)
  $\delta^{13}C$

5. *Terms and Standards to be Used in the Experiments*

Controlled temperature wood "charcoal"
Pine needles/grass
Fossil charcoal
Non-fossil charcoal
Graphite
Aerosol

Future Research Questions and Directions

We began by asking what are the best tracers for biomass burning in sediments, terrestrial residues, and in the atmosphere? This question is not new, but it demands continued attention. W. Chaloner and T. Jones point out that there are many sediment constituents that can be misinterpreted as fire-altered materials. Thus, an inter-laboratory comparison of methods is critical to refining the techniques used to characterize samples. In addition, scientists working in the field should identify and characterize materials, such as fossilized lignin and MnO_2, which can optically masquerade as fire-altered materials. The working group strongly recommends that the major analytical techniques be tested and the limits of these techniques to distinguish fire-altered materials from known interferences be determined.

Atmospheric scientists and those working with sediments have traditionally analyzed different particle sizes. Particle sizes of pyrogenic materials in sediments vary with transport distance. Sample sizes chosen for analysis are critical because large particles (those larger than 10 microns) have short residence times in the atmosphere. Such particles are predominantly deposited in lakes near fires; and they are not transported significant distances. Thus, a technique that measures only large pyrogenic particles in the sediments may only be relevant to local fires. By contrast, more finely divided, sub micron particles come from much more distant sources. These smaller particles move to sediments within larger biological particles such as zooplankton fecal pellets. Biomass burning results in aerosols that can be transported thousands of miles through the atmosphere before deposition. (H. Cashier, this volume) Thus, fine particles of pyrogenic material can provide information about combustion activities over broad geographic areas. Some knowledge of long-range transport characteristics of the atmosphere are required to properly assign the sources of such residues. As this illustrates, particle choice must be compared by the scientists within a discipline and across disciplines so that the data concerning fire-altered materials can be compared.

To understand the impact of biomass fires we must understand what materials remain at the site of the fire as well as what materials are transported various distances from the fire. Materials remaining at the site of the fire may be washed into the local sediment. Recommendations can be summarized as follows:

Area 1. Characterization of Biomass Fires:

a. Determine mass fractions of charcoal left per unit area following wildfire.

b. Establish percentages of charred material that is charcoal and soot from airborne and terrestrial samples.

c. Investigate inert forms of carbon other than charcoal.

d. Determine how mass fluxes of charcoal vary among different ecosystems.

Area 2. Characterization of the Sediments:

a. Quantitative: What is the size of the carbon sink represented by sediments?

b. Qualitative: Which charred materials are stable in sediments at various time scales?

There is currently a need to develop techniques for determining the origin of charred materials found in sediments and other matrices at remote locations. Analytical chemists are using a variety of methods to this end for the combustion products of fossil fuels. Using trace elements, the ratios of stable isotopes for carbon, and polycyclic aromatic hydrocarbons (PAH), they are able to identify the source of the combustion in many cases with reasonable certainty. It is our recommendation that these techniques be applied to the analysis of other samples of fire-altered materials. This is the third area in which advancements in characterization are important, as summarized below:

Area 3. New Methods of Sample Characterization to be Explored

a. Stable isotope analysis for $\delta^{13}C$ should be tested to address problems concerning sources and histories of carbon compounds. Such analysis could be extended to other compounds containing elements with stable isotopes.

b. Trace element profiles should be determined for areas where biomass burning is common to see if any elements or combinations of elements are characteristic of specific sources. If there are elements that appear to be indicative of a particular area, an investigation of the presence of these elements in sediments and aerosols could then be investigated.

c. PAH analysis should be investigated to determine if there are any produced in characteristic quantities from biomass fires. These compounds would then be useful markers of the fire-altered materials from a particular biomass fire.

Extensive PAH analyses have been done on sediments in Europe and the U.S.A., but certain important gaps are present. In particular, there is a lack of data on PAHs in aerosols, lake, and tropical marine sediments, especially in and around boreal forests. Few estimates of emission factors for PAHs from biomass combustion are available. Little or no data exist for isotopic signatures of PAHs. In addition, few careful studies of vertical profiles of sedimentary PAHs with relevant ancillary data, e.g. Pb, Cs, charcoal profiles are available in the literature. Such analyses may yield valuable existence of biomass burning.

III. Particle Transport

Bill Patterson, rapporteur and Helene Cachier, moderator

Introduction

The amount of black carbon (BC) found in sediments depends on source intensity and transport, which includes processes by which material originating from a fire is carried to sites of deposition and ultimately preserved by burial in sediments. Transport will be very different for material transported through the atmosphere and material moved over the ground. A prerequisite for understanding the transport of biomass burn products is an initial classification that considers the differences between aerial and terrestrial transport pathways.

Transport should be studied both in a qualitative way, by identifying and describing the processes that move material away from fire sources, and by quantitative assessments such as particle fluxes (number, mass or volume per unit time) to depositional sites. Transport includes: 1) emission mechanisms at the fire site, 2) factors that affect emission, such as temperature, humidity, and bio-fuel characteristics that determine size, character, and number of particles, 3) transport pathways, both airborne and overland, 4) deposition mechanisms and patterns, and 5) factors that affect the preservation of burned biomass material in the sediment matrix.

Clearly, this definition of transport includes all or most of the elements that were the focus of separate discussion groups and it indicates the willingness of workshop members to view their specialties and problems holistically.

Table 1. Matrix classification illustrating characteristics of biomass-altered material and processes that affect its transport and deposition.

Material	carbon-aerosol	micro-charcoal	macro-charcoal	mega charcoal
Size (μm)	0.01 - 1	1 - 100	100 - 1000	>1000
Emission factor Black carbon[1]	0.4 % flame [2] 3.0% smolder	ash 0.5 - 1.5% [2]		1 - 10 %
Emission factor Organic carbon[1]	2% flame 24% smolder	0.5% - 1.5%		
Transport path	airborne	airborne	air + water	land + water
Transport mechanism	wind	wind + fluvial	saltation, runoff, fluvial	fluvial, gravity
Deposition mechanism	wet (- dry) deposition (9/1)	settling + impaction	air + water settling	water logging, burial
Geographic scale	all scales	all scales	mesoscale or less	local scale
Sampling sites	ocean sediments + polar ice cores	lake + ocean sediments	lake + estuary sediments	paleosols, delta + estuary sediments
Chemical identification	acidic / alkaline oxidative or thermal method	acidic / alkaline oxidative or thermal method	no	no
Morphological identification	microscopy: TEM	microscopy: SEM, TEM, LT	optical microscopy	optical inspection
Chemical characterization	PAH, lipids, waxes, K/BC ratio	no	no	no
Morphological characterization	micro-soot agglomerates	xylem structure	cuticular structure	cell anatomy
Isotopic composition	$\delta^{13}C$, C3-C4 $\delta^{15}N$ (wet/dry)	$\delta^{13}C$, C3/C4 $\delta^{15}N$ wet/dry	$\delta^{13}C$, C3/C4 $\delta^{15}N$ wet/dry	$\delta^{13}C$, C3/C4 $\delta^{15}N$ wet/dry

[1] Emission factors are expressed in grams of carbon per kilogram of dry plant.

[2] Emission factors of carbon aerosols are from Cachier et al. (1995). Emission factors for ash were derived from savanna fires (Kuhlbusch et al. 1995).

What Do We Currently Know?

A table or matrix classification was prepared in order to organize and summarize the working group discussion of the variables that affect emission transport and how transport affects the ultimate deposition of biomass burn products. Included in this table are some chemical and morphological characteristics of biomass burn products that were useful for guiding discussion on the topic.

Where Are The Gaps In Our Knowledge, and What Can and Cannot be Addressed?

Transport is a transient process that cannot be usefully studied after the fact. Monitoring of atmospheric particles and their fallout during natural and experimental fires provides unique information that may be difficult to apply generally or to other specific burn situations. The main problems that frustrate a better understanding of transport processes center on imperfect methods for quantification (flux calculations) and inadequate sophistication of models to provide meaningful insights

Atmospheric pathways

Transport of airborne particles has a strong dependence on meteorological patterns. Convective plumes originating from burns usually do not rise to high altitudes, but atmospheric circulation patterns such as westerly winds and polar jets may induce significant zonal or meridional transport. In addition, the fallout of airborne particles is highly dependent on precipitation patterns. The resulting spatial patchiness is complicated by a large temporal variability of source location and intensity and (or) local meteorological processes that ultimately determine transport direction, distance, and fallout site.

Overland pathways

Similar problems apply to overland transportation of biomass burn products. Sources and emission processes are locally variable even within one burn. The ultimate transport and deposition of charcoal particles has a high dependence on watershed area and vegetative cover, local precipitation and runoff seasonality, runoff volume, and distribution of lakes, marshes, and wetlands as depositional sites.

Future Research Questions and Directions

1. There is a principal need to create a community of scientists working with biomass burning issues that has a well-integrated system of communication and input to mutual problems.

Additional meetings, symposia, workshops and fuller use of electronic communication will promote resolution of this need.

2. Transportation of biomass burn products must be determined by at least decadal monitoring over large spatial scales. Background levels of biomass burn emissions for typical situations (fire-prone boreal forests, tundra, tropical forests, etc.) must be determined by such monitoring programs.

3. Specific studies are needed to relate fluxes of biomass burn products for El Niño/La Niña years, drought and wet cycles, and other short-term climatic phenomena linked to fire frequency and intensity.

4. The physical and chemical transformation of black carbon and particularly charcoal particles at different (long/short) ranges requires additional investigation.

5. A strategic coring program for investigation of transport problems and aerosol/charcoal or airborne/runoff source apportionment should be defined in conjunction with both experimental and natural burns.

6. Particle transport models require review and refinement. There must be an overall effort to adapt diagnostic models (which include clouds) to the specific problem of particle transportation. It is not yet possible to run sensitivity tests for BC deposition.

IV. Deposition

J.P. Bradbury, rapporteur and Glen MacDonald, moderator

Introduction

The group discussing the variables affecting deposition of source materials from fire events was primarily concerned with three major topics:

1. Definition of types of depositional sites and scales of biomass burning (charcoal deposition) they could reasonably be expected to reflect.

2. What are the variety of things produced by fires that are deposited?

3. What are the processes that affect the amount, size, and kind of

biomass burning products once they enter the depositional system?

What Do We Currently Know?

Sites suitable for recording products of biomass burning range from small sites of very local significance to large depositional systems that record only at a global scale. These sites are summarized below in terms of the general approaches identified above.

Fire scar chronologies, soils, peats, small forest hollows and wet meadows were designated as small sites, primarily useful for recording local fires. Whereas global fallout of burned biomass certainly will accumulate at small depositional sites, it is likely to be insignificant in relation to local sources of fire products. However, other small sites, such as hollows and ponds in Arctic tundra or barren polar landscapes or spring mounds in desert areas, may have greater potential for recording global atmospheric deposition of charcoal. Charcoal and charred plant remains, often in very large sizes, dominate the fire products deposited in small sites. Usually there is little difficulty in distinguishing truly burned material and in some cases actual plant taxa may be identifiable.

Soils and hollows do not typically provide an understandable stratigraphic record of fire history, because sediment accumulation is slow, and products from several fires may be mingled together, especially if burrowing animals inhabit the site or if large mammals walk across the wet area. Wet meadows are somewhat better for stratigraphic fire records.

At the next larger scale, stream deposits, and lake sediments are thought to have the potential of recording the products of fires from various parts of the relevant watershed and to have a greater likelihood of recording regional charcoal deposition. Stream deposits are likely to contain large charcoal pieces and even branches and trunks of burned trees. Nevertheless, the stratigraphic record of charcoal in stream deposits may be complicated by scour and redeposition.

Lake sediments are the most used deposits for establishing fire history. Charcoal particles are typically small, especially in large lakes, and usually not identifiable to plant type. Sediment accumulation is generally rapid, and stratigraphic records are excellent. Processes that retain and redistribute charcoal fragments long after fire events can complicate the direct reading of fire history from lake sediments.

River deposits, if they should prove to be suitable for recording charcoal deposition, have the potential for sampling yet larger areas. However, much of the charcoal in fluvial deposits could come from erosion and redeposition of earlier charcoal-rich deposits. Burned logs could be difficult or impossible to evaluate in terms of timing and area of past fires. The same concerns apply to near-shore marine deposits, which receive much of their sediment from fluvial sources.

Open/deep ocean deposits are likely to record charcoal deposition at nearly global scales. Biomass burning products are extremely tiny bits of charcoal and "black carbon" believed to come from fires in vegetation. There may be some potential for establishing sources for these fire products through organic geochemistry, but the terrestrial carbon content of ocean deposits is small, and extraction techniques are time consuming and complicated. The depositional inhomogeneities related to the direction and magnitude of ocean current systems and the biological factors (fecal pellets) responsible for scavenging charcoal and inducing sedimentation have not been systematically evaluated in ocean sediments. Because of erosion, redeposition and scour, fluvial and near-shore marine deposits have stratigraphies that are difficult to unravel and consequently the history of biomass burning is unlikely to be particularly clear in such deposits.

Ice caps have the potential of recording charcoal fallout directly from the atmosphere and can therefore be regarded as the most unambiguous record of global biomass burning. Black carbon and perhaps tiny charcoal fragments form the bulk of terrestrial carbon in ice. Organic chemical techniques, including the analysis of methane from ice, may provide perspectives on the record of biomass burning from ice accumulations. The dynamics of charcoal and aerosol transport from areas of burnable biomass to polar or high mountain ice fields is unknown and it is not possible to fully understand the pattern of global biomass burning from ice records at this time.

Where Are The Gaps In Our Knowledge, and What Can and Cannot be Addressed?

Many of the problems facing insightful interpretation of depositional records of biomass burning relate to problems of the depositional sites themselves. Of the sites listed, only wet meadows, bogs, lakes, deep ocean sediments, and ice caps appear to have deposition rates high enough to produce acceptable stratigraphic records. Data from other systems, such as soils and fluvial deposits, can help amplify the stratigraphic record.

In all sites there is insufficient knowledge about the processes that govern charcoal and black carbon deposition. In all sites, except where burned macroscopic plant material is preserved, there is inadequate knowledge about the identification and fire significance of the presumed fire products. Overall, there is little consistency in the techniques for fire product enumeration and identification between workers.

Future Research Questions and Directions

The deposition of fire products from control burns must be monitored after such sites have been characterized in the non-fire environment. There needs to be much more coordination between fire ecologists and paleolimnologists and paleoecologists who study fire history to refine the dynamic interactions between charcoal production, transportation, deposition, and final burial.

Summary

Workshop working groups discussed the status of knowledge, problems, and recommendations for future research in four broad research categories that focus on the stratigraphic record of biomass burning and its relation to global change. The categories are linked into two fundamental concepts: What does biomass burning produce and how can it be identified on the one hand, and how does it move and where does it wind up on the other hand.

Burning biomass yields smoke, ash, and charcoal. The smoke component has H_2O, CO_2, CO, PAHs and other gases. Also has some fine charcoal and ash that gets deposited somewhere. Of all combustion products, charcoal is the least labile and most likely to remain behind in deposits. Thus charcoal is the starting point from which everything else is inferred. Perhaps investigations of PAHs in sediments can change this.

The first question that we need to know to relate charcoal to global change therefore, is what can we infer from charcoal about the quantity and composition of smoke and, thereby, size and intensity of the fire. This is the main problem of the SOURCES group (Part I). DEPOSITION (Part IV) deals with charcoal, but cannot say anything about global change unless it becomes known what the charcoal means in terms of atmospheric pollutants, greenhouse gases, and the like. The major topic of SOURCES research is to find a way to back-calculate smoke from charcoal. Control fires and lab experiments in different vegetation (biomass) types under different conditions of humidity and temperature, vegetation provide the best way to begin evaluating the connection between SOURCES and DEPOSITION.

However, charcoal itself is a mixture of more-or-less identifiable small black things that have had variable temperature histories, have survived variably oxidative environments, and have come from different plants that give off different gasses in smoke. This is where CHARACTERIZATION (Part II) must play a critical role. Ideally, temperature, oxygen, moisture, and plant source of each charcoal-like particle from a fire could be characterized and thereby provide information about the nature of the smoke that evolved with the fire and consequently the probable impact of those gasses on global change. Minimally, it should be possible to characterize one identifiable charcoal-like particle and know something about the nature of the smoke that produced it. The goal of CHARACTERIZATION is not only to allow particle identification, but, in combination with SOURCES, to develop its significance in terms of the gaseous and fine particulate components of the smoke from which it came. This may be difficult, and further research is needed in this area. PAHs may offer the best hope for achieving these connections.

The research topics alluded to above involve analysis of fires, smoke, and other emissions in as many different environments as possible. Some of this kind of research goes on now, but it needs to become better coordinated with the needs and talents of those working with DEPOSITION of fire-altered material. Depositional sites should be cored in areas where fires have been studied. Dialog must be established between fire and smoke analysts and those who look into the past.

If this approach of dissecting fires into component products and dissecting depositional environments into processes and rates is not realistic, then there has to be some meaningful way to average all "charcoal" in stratigraphic records in terms of "all average fires" that produced it. "Average fires" would give "average smoke" and that might provide some information about global change from fire. We can certainly do better than world averages, at least in some cases. For example, the latitude, longitude, and elevation of a record coupled with the pollen evidence for vegetation would give some idea about what vegetation burned in what probable climatic setting. So one could probably reconstruct paleo "smoke" from spruce/fir forest fires, everglade swamp fires, Amazon jungle fires, grass fires and the like. As with any kind of "average" approach, the more one learns about the system, the less meaning the average will have. The advantage is that "averages" might be tractable enough to get some numbers and maybe some models out fairly soon. Any product of the global average approach will certainly focus questions and generate discussion.

A critical element in the global or regional average approach is to know something about fire size, intensity, and frequency from stratigraphic charcoal. We assume that there is little or no diagenetic change in the evidence of past fire (charcoal). Charcoal is refractory so, what gets deposited is likely to remain unaltered. This probably applies to many depositional cases, including ice, lakes, and probably swamps and oceans. It may not apply to wet hollows and meadows, if they dry out and burn (burning old charcoal in the process), or to forest floors (soil charcoal) for the same reason. Given, however, that most charcoal deposited is also preserved, the question becomes how to quantify past charcoal in terms of fire magnitude, frequency, and intensity. The TRANSPORT (Part III) and DEPOSITION group have related tasks: What does more charcoal mean? More fires? Closer fires? Cooler fires? Wetter fires? Upwind fires? Bioturbation and secondary charcoal deposition will make fire charcoal sources appear longer-lived and larger than they may have actually been. In areas of slow sediment accumulation, such as on forest floors, charcoal of many different ages accumulates together, and an historical sense of change can only be obtained by radiocarbon dating each charcoal fragment.

The problem that makes quantifying stratigraphic charcoal records difficult is that each depositional site requires individual calibration, and that in all probability the calibration will vary through time as the morphometric, climatic, and other environmental boundary conditions change at a given depositional site. For example, lakes in climates of high winds will sample fires from greater distances than will lakes in calm climates. And if climate changes through time, the fire sampling area will also change. Calibrations can be figured out, but to do so would take a fair investment of time and money.

Cooperation of the Paleofire Science Community with Interdisciplinary Fire Research Programs

J.G. Goldammer
Fire Ecology Group
Freiburg University
Freiburg, Germany

In the late 1980's a steadily increasing number of research activities began to investigate the role of fire in the contemporary and historic biosphere and atmosphere. In the context of the role of fire in processes related to "global change", the need was recognized to integrate the approaches and findings of traditionally separated science disciplines into a broader view on global fire. Subsequently, a series of monographs and conferences attempted to formulate the base for a comprehensive global fire science (e.g. Goldammer 1990, Levine 1991, Crutzen and Goldammer 1993, Goldammer 1993, Pyne 1995, Goldammer and Furiaev 1996).

These activities went along with, or were the precursors of, the construction of an organizational framework under which a series of interdisciplinary fire experiments and research programs was initiated. At present the main instrument for international coordination of fire research is under the umbrella of the *International Geosphere-Biosphere Program* (IGBP), a program under the sponsorship of the *International Council of Scientific Unions* (ICSU). One of the first IGBP core projects was the *International Global Atmospheric Chemistry* (IGAC) project, which defined several priority areas of research in atmospheric chemistry and its interactions with the biosphere. Tropical atmospheric chemistry and the role of wildland fires and other biomass burning was identified as one of these priority areas, and a steering committee was set up in 1990 to develop an international activity originally designated *"Impact of Tropical Biomass Burning on the World Atmosphere Project"* (acronym: BIBEX) (Biomass Burning Experiment) (IGAC 1992). In the following years it was recognized that activities of BIBEX should neither be restricted to the tropics, nor to atmospheric chemistry. Subsequently the group was designated as *"Biomass Burning Experiment: Impact on the Atmosphere and Biosphere"* (IGBP 1994b). As with all IGAC Activities, BIBEX utilizes and builds on existing international programs with common goals.

NATO ASI Series, Vol. I 51
Sediment Records of Biomass Burning
and Global Change
Edited by James S. Clark, Hélène Cachier,
Johann G. Goldammer, and Brian Stocks

IGBP also provides a vehicle for research cooperation of paleo sciences. The core project PAGES (*Past Global Changes*) is charged with providing a quantitative understanding of the Earth's past environment and defining the envelope of natural environmental variability within which we can assess anthropogenic impact on the Earth's biosphere, geosphere and atmosphere (IGBP 1994a).[1]

It has been recognized clearly that the scope of research within these two core projects have one common target, the thorough understanding of contemporary and ancient biosphere-atmosphere- climate interactions. As a logic consequence the research programs must be linked with each other. As the mechanisms for conducting joint international and interdisciplinary fire research campaigns have proved successful, they provide a base for future close cooperation with the paleofire research community (e.g. Andreae et al. 1994, Goldammer and Furyaev 1995).

The first example of integrating paleofire research with a large fire experiment has been implemented in FIRESCAN-93 (FIRESCAN Science Team 1994, 1996). In this Siberian campaign a sediment core was recovered from a lake nearby the experimental site, for 14C dating, pollen analysis, and thermogravimetric analysis of sediment charcoal. In addition, ash traps were arrayed around the experimental site in order to obtain information on transport distances of particles. These analyses allowed to reconstruct 4,500 years of fire history and to compare the data with the terrestrial and airborne measurements of the experimental burn (Clark and Hussey 1996, Clark and Richard 1996).

For better estimating the future potential of cooperative research, the ongoing and planned fire research programs as initiated or co-sponsored by BIBEX are described and the joint interests presented (Annex I).

The requirements for the integration of paleofire research into an experimental fire have been elaborated at the workshop; they are summarize in Annex II.

Annex I: Description of ongoing international fire research programs as initiated or co-sponsored by BIBEX and the potential joint interests between the paleofire research community

[1] The link of IGAC with other Core Projects is being built at present, e.g. with *Global Change and Terrestrial Ecosystems* (GCTE) and *Biospheric Aspects of the Hydrological Cycle* (BAHC). Within IGAC the cooperation between BIBEX and other activities is mainly with *Biosphere-Atmosphere Trace Gas Exchanges* (BATGE) and the *Global Emissions Inventory Activity* (GEIA). The intra- and inter-Core Project evaluation of data will follow the mechanisms to be defined by the IGBP Task Force *Global Analysis, Interpretation, and Modelling* (GAIM).

a. Influence of vegetation fires and other biomass burning on tropospheric ozone budgets in the tropics

Campaigns: STARE (the *Southern Tropical Atlantic Regional Experiment*) with the two major components, TRACE-A (*Transport and Atmospheric Chemistry near the Equator - Atlantic*, a sub-project of the NASA *Global Troposheric Experiment*) and SAFARI (*Southern African Fire-Atmosphere Research Initiative*). This experiment (field campaign phase: 1992-95) investigates the impact of biomass burning on the chemical and physical characteristics of the contemporary regional atmosphere (cf. Andreae et al. 1994, Van Wilgen et al. 1996).

In the context of pan-African fire research, the results of STARE will be linked with another international experiment, EXPRESSO (*Experiment for Regional Sources and Sinks of Oxidants*). In this experiment (field campaign phase: 1996-97) the exchange fluxes of reactive trace gases and aerosols between the biosphere and the atmosphere in the tropics will be quantified and the influence of land-use change and biomass burning on biogenic fluxes of reactive or radiatively active trace gases in central Africa will be investigated (EXPRESSO 1994).

Joint Interests: Determine whether elevated levels of O_3 and CO over vast areas of Central Africa and South America, over the tropical Atlantic, and the Indian Ocean are a phenomenon linked to modern fire practices (e.g. driven by large-scale land-use changes). The reconstruction of prehistoric and historic fire regimes will allow to reconstruct the historic and prehistoric atmosphere.

Significance: If elevated ozone concentrations in the lower tropical troposphere have been present in historic and prehistoric times, terrestrial life in mountainous areas of Africa must have adapted to seasonally high levels of ozone. The explanation of mechanisms are of interest for the Northern Hemisphere where plant life suffers from high ozone concentrations caused by industrial emissions.

Methods: Reconstruction of prehistoric and historic African fire regimes through lake sediment analysis (e.g. Weiss et al. 1996).

b. Influence of vegetation fires and other biomass burning on stratospheric ozone

Campaigns: -STARE/TRACE-A/SAFARI (see above), and particularly: -*Fire Research Campaign Asia--North* (FIRESCAN). This research is an effort of institutions and individual scientists jointly working under the umbrella of IGAC-BIBEX and the *International Boreal Forest Research Association* (IBFRA). In the first field phase, a large fire experiment (The Bor Forest Island Fire Experiment) was conducted in Siberia (FIRESCAN Science Team 1994, 1996). In 1997 a set of stand-replacement fires is planned in Canada (Stocks et al. 1995). The aim of the experiments is to investigate behaviour, ecological and atmospheric chemical effects of high-intensity stand-replacement fires in boreal coniferous forests. Trace gas components of interest are those which contribute to the depletion of the stratospheric ozone layer, e.g. methyl bromide (CH_3Br) (Mano and Andreae 1994).

Joint Interests: Determine the role of natural prehistoric and historic wildland fires on the global stratospheric ozone budget, and compare with magnitude and impacts of anthropogenically released compounds that destroy stratospheric ozone (e.g. CFC's).

Significance: Possibility to reconstruct the influence of natural mechanisms determining the pre-industrial stratospheric ozone budget.

Methods: Reconstruction of prehistoric and historic boreal fire regimes through analysis of lake sediments in North America and Eurasia. Determination of depositions of vegetation fire emission products in arctic and Greenland ice cores. Coordinate this research with the IGBP Northern Eurasia Study (Steffen and Shvidenko 1996).

c. Influence of vegetation fires and other biomass burning on the global atmosphere, particularly regional and global transport processes of combustion products from tropical fires

Campaigns: -STARE/TRACE-A/SAFARI, FIRESCAN (see above), and particularly:

-*South East Asia Fire Experiment* (SEAFIRE). In the second half of the 1990's BIBEX research activities will gradually move East from Africa and South from boreal Eurasia into tropical continental and insular South East Asia. The South East Asia Fire Experiment (SEAFIRE) will be designed to investigate this highly complex and diverse fire theatre of SE-Asia (slash-and-burn agriculture, other forest conversion fires, regularly occurring fires in seasonally dry deciduous and semideciduous forests on mainland SE-Asia, regularly occurring fires in the submontane and montane coniferous forests of insular and mainland South Asia, agricultural residue burning, mainly rice straw, throughout the whole region) (SEAFIRE Science Team 1996). Regional pyrogenic smog formation (including imports and export of smoke) will be one of the main objectives of the research. Another objective of SEAFIRE will be the investigation of the role of convective air mass movements ("hot towers"), the Walker Circulation and the El Niño-Southern Oscillation (ENSO) in transport of pyrogenic products from the global "Warm Pool" (SE Asia) to the global system (teleconnections between biomass burning, the Ocean Continent and the global system). The experiment will include non-pyrogenic trace gas and aerosol flux research (terrestrial biogenic, marine, household biomass and fossil fuel burning).

Joint Interests: Determine the role of natural and anthropogenic prehistoric and historic fires in the Warm Pool region on the global atmosphere.

Significance: Possibility to explain the global distribution of pyrogenic and biogenic trace gases in contemporary, historic and prehistoric dimensions.

Methods: Reconstruction of prehistoric and historic tropical SE-Asian fire regimes through analysis of tropical swamps, lakes and ocean sediments of the Sunda shelf.

Annex II: List of requirements for the participation of paleofire research in large fire experiments

1. Minimum size of fire experiment: 10 ha
2. Availability of a lake or peatland nearby
3. Availability of a control area nearby

4. Sedimentation traps along slopes operational before and after the fire

5. Sediment and water samples taken before the fire

6. Instrumentation of the lake

7. Availability of pre-burn baseline information: wet/dry deposition on different target surfaces, general environmental information

8. Possibility to set up wet/dry deposition collectors

Literature References

Andreae MO, Fishman J, Garstang M, Goldammer JG, Justice CO, Levine JS, Scholes RJ, Stocks BJ, Thompson AM, van Wilgen B, STARE/TRACE-A/SAFARI Science Team (1993) Biomass burning in the global environment: First results from IGAC/BIBEX field campaign STARE/TRACE-A/SAFARI-92. In Prinn RG (ed) *Global Atmospheric-Biospheric Chemistry* pp 83-101 Plenum Press New York

Clark JS, Hussey T (1995) A 4500-yr record of fire and vegetation change in Central Siberia based on calibrated sediment charcoal accumulation. *International Union of Forestry Research Organizations* (IUFRO) World Congress 1995 Subject Group S.1-09 Forest Fire Research Tampere Finland August 1995 Abstract Vol p 89

Clark JS, Richard PJH (1996) The role of paleofire in boreal and other cool-coniferous forests. In Goldammer JG, Furyaev VV (eds) *Fire in ecosystems of boreal Eurasia* Kluwer Academic Publisher (in press)

Crutzen PJ, Goldammer JG (eds) (1993) Fire in the environment: The ecological, atmospheric, and climatic importance of vegetation fires. *Dahlem Workshop Reports Environmental Sciences Research Report* 13 John Wiley & Sons Chichester

EXPRESSO Science Team (1994) *Experiment for Regional Sources and Sinks of Oxidants (EXPRESSO).* Proposal for an atmospheric chemistry experiment in the African tropics (June 1994 mimeo)

FIRESCAN Science Team (1994) Fire in Boreal Ecosystems of Eurasia: First results of the Bor Forest Island Fire Experiment. Fire Research Campaign Asia-North (FIRESCAN) *World Resource Review* 6:499-523

FIRESCAN Science Team (1996) Fire in ecosystems of boreal Eurasia: The Bor Forest Island Fire Experiment, Fire Research Campaign Asia-North (FIRESCAN). In Levine JS (ed) *Biomass burning and global change* The MIT Press Cambridge MA (in press)

Goldammer JG (ed) (1990) Fire in the tropical biota. Ecosystem processes and global challenges. *Ecological Studies* 84 Springer-Verlag Berlin Heidelberg New York

Goldammer JG (1993) *Feuer in Waldokosystemen der Tropen und Subtropen.* Birkhauser-Verlag Basel Boston

Goldammer JG (1996) A synthesis of the results of the SAFARI campaign. In van Wilgen B, Andreae MO, Goldammer JG, Lindesay J (eds) *Fire in Southern African savannas. Ecological and atmospheric perspectives* The University of Witwatersrand Press (in press)

Goldammer JG, Furiaev VV (1995) Global change, the boreal forest, and fire: Search for new strategies in science policies and research mechanisms. In Koptyug VA, Klerkx J (eds) *Science Policy: New Mechanisms for Scientific Collaboration between East and West* NATO ASI Series 4, Science and Technology Policy 1:45-61 Kluwer Academic Publishers Dordrecht Boston London

Goldammer JG, Furyaev VV (eds) (1996) *Fire in ecosystems of boreal Eurasia.* Kluwer Academic Publisher (in press)

IGAC (1992) *Biomass Burning Experiment: Impact on the atmosphere and biosphere (BIBEX).* IGAC Document on Activity 2.3 of Focus 2. IGAC Core Project Office Massachusetts Institute of Technology Cambridge

IGBP (1994a) *IGBP in action: Work plan 1994-1998.* IGBP Report 28 Stockholm

IGBP (1994b) *International Global Atmospheric Chemistry (IGAC) Project. The operational plan.* IGBP Report 32 Stockholm.

Levine JS (1991) *Global biomass burning.* The MIT Press Cambridge

Mano S, Andreae MO (1994) Emission of methyl bromide from biomass burning. *Science* 263:1255-1257

Pyne SJ (1995) *World fire. The culture of fire on Earth.* Henry Holt New York

SEAFIRE Science Team (1996) *The South East Asian Fire Experiment (SEAFIRE).* Research prospectus as a sub-component of the Integrated IGBP/IHDP/WCRP Study on Land-use Change in SE Asia (in prep)

Steffen WL, Shvidenko AZ (eds) (1996) *IGBP Northern Eurasia Study. Prospectus for integrated global change research* IGBP Stockholm (in press)

Stocks BJ, Goldammer JG, Cofer WR (1995) 1995 international crown fire experiment in Northern Canada. *International Union of Forestry Research Organizations (IUFRO) World Congress 1995* Subject Group S.1-09 Forest Fire Research Tampere Finland August 1995

van Wilgen B, Andreae MO, Goldammer JG, Lindesay J (eds) (1996) *Fire in Southern African savannas. Ecological and atmospheric perspectives.* The University of Witwatersrand Press (in press)

Weiss KF, Goldammer JG, Clark JS, Livingstone DA, Andreae MO (1996) Reconstruction of prehistoric fire regimes in East Africa by lake sediment analysis. In Levine JS (ed) *Biomass burning and global change* The MIT Press Cambridge (in press)

The ASI Series Books Published as a Result of Activities of the Special Programme on Global Environmental Change

This book contains the proceedings of a NATO Advanced Research Workshop held within the activities of the NATO Special Programme on Global Environmental Change, which started in 1991 under the auspices of the NATO Science Committee.

The volumes published as a result of the activities of the Special Programme are:

Vol. 1: Global Environmental Change.
Edited by R. W. Corell and P. A. Anderson. 1991.
Vol. 2: The Last Deglaciation: Absolute and Radiocarbon Chronologies.
Edited by E. Bard and W. S. Broecker. 1992.
Vol. 3: Start of a Glacial. Edited by G. J. Kukla and E. Went. 1992.
Vol. 4: Interactions of C, N, P and S Biogeochemical Cycles and Global Change.
Edited by R. Wollast, F. T. Mackenzie and L. Chou. 1993.
Vol. 5: Energy and Water Cycles in the Climate System.
Edited by E. Raschke and D. Jacob. 1993.
Vol. 6: Prediction of Interannual Climate Variations.
Edited by J. Shukla. 1993.
Vol. 7: The Tropospheric Chemistry of Ozone in the Polar Regions.
Edited by H. Niki and K. H. Becker. 1993.
Vol. 8: The Role of the Stratosphere in Global Change.
Edited by M.-L. Chanin. 1993.
Vol. 9: High Spectral Resolution Infrared Remote Sensing for Earth's Weather and Climate Studies.
Edited by A. Chedin, M.T. Chahine and N.A. Scott. 1993.
Vol. 10: Towards a Model of Ocean Biogeochemical Processes.
Edited by G. T. Evans and M. J. R. Fasham. 1993.
Vol. 11: Modelling Oceanic Climate Interactions.
Edited by J. Willebrand and D. L. T. Anderson. 1993.
Vol. 12: Ice in the Climate System. Edited by W. Richard Peltier. 1993.
Vol. 13: Atmospheric Methane: Sources, Sinks, and Role in Global Change.
Edited by M. A. K. Khalil. 1993.
Vol. 14: The Role of Regional Organizations in the Context of Climate Change.
Edited by M. H. Glantz. 1993.
Vol. 15: The Global Carbon Cycle.
Edited by M. Heimann. 1993.
Vol. 16: Interacting Stresses on Plants in a Changing Climate.
Edited by M. B. Jackson and C. R. Black. 1993.
Vol. 17: Carbon Cycling in the Glacial Ocean: Constraints on the Ocean's Role in Global Change.
Edited by R. Zahn, T. F. Pedersen, M. A. Kaminski and L. Labeyrie. 1994.
Vol. 18: Stratospheric Ozone Depletion/UV-B Radiation in the Biosphere.
Edited by R. H. Biggs and M. E. B. Joyner. 1994.
Vol. 19: Data Assimilation: Tools for Modelling the Ocean in a Global Change Perspective.
Edited by P. O. Brasseur and J. Nihoul. 1994.

Vol. 20: Biodiversity, Temperate Ecosystems, and Global Change.
Edited by T. J. B. Boyle and C. E. B. Boyle. 1994.

Vol. 21: Low-Temperature Chemistry of the Atmosphere.
Edited by G. K. Moortgat, A. J. Barnes, G. Le Bras and J. R. Sodeau. 1994.

Vol. 22: Long-Term Climatic Variations – Data and Modelling.
Edited by J.-C. Duplessy and M.-T. Spyridakis. 1994.

Vol. 23: Soil Responses to Climate Change.
Edited by M. D. A. Rounsevell and P. J. Loveland. 1994.

Vol. 24: Remote Sensing and Global Climate Change.
Edited by R. A. Vaughan and A. P. Cracknell. 1994.

Vol. 25: The Solar Engine and Its Influence on Terrestrial Atmosphere and Climate.
Edited by E. Nesme-Ribes. 1994.

Vol. 26: Global Precipitations and Climate Change.
Edited by M. Desbois and F. Désalmand. 1994.

Vol. 27: Cenozoic Plants and Climates of the Arctic.
Edited by M. C. Boulter and H. C. Fisher. 1994.

Vol. 28: Evaluating and Monitoring the Health of Large-Scale Ecosystems.
Edited by D. J. Rapport, C. L. Gaudet and P. Calow. 1995.

Vol. 29: Global Environmental Change Science: Education and Training.
Edited by D. J. Waddington. 1995.

Vol. 30: Ice Core Studies of Global Biogeochemical Cycles.
Edited by R. J. Delmas. 1995.

Vol. 31: The Role of Water and the Hydrological Cycle in Global Change.
Edited by H. R. Oliver and S. A. Oliver. 1995.

Vol. 32: Atmospheric Ozone as a Climate Gas.
Edited by W.-C. Wang and I. S. A. Isaksen. 1995.

Vol. 33: Carbon Sequestration in the Biosphere.
Edited by M. A. Beran. 1995.

Vol. 34: Climate Sensitivity to Radiative Perturbations: Physical Mechanisms and Their Validation.
Edited by H. Le Treut. 1996

Vol. 35: Clouds, Chemistry and Climate.
Edited by P. J. Crutzen and V. Ramanathan. 1996.

Vol. 36: Diachronic Climatic Impacts on Water Resources.
Edited by A. N. Angelakis and A. S. Issar. 1996.

Vol. 37: Climate Change and World Food Security.
Edited by T. E. Downing. 1996.

Vol. 38: Evaluation of Soil Organic Matter Models.
Edited by D. S. Powlson, P. Smith and J. U. Smith. 1996.

Vol. 39: Microbiology of Atmospheric Trace Gases.
Edited by J. C. Murrell and D. P. Kelly. 1996.

Vol. 40: Forest Ecosystems, Forest Management and the Global Carbon Cycle.
Edited by M. J. Apps and D. T. Price. 1996.

Vol. 41: Climate Variations and Forcing Mechanisms of the Last 2000 Years.
Edited by P. D. Jones, R. S. Bradley and J. Jouzel. 1996.

Vol. 42: The Mount Pinatubo Eruption: Effects on the Atmosphere and Climate.
Edited by G. Fiocco, D. Fuà and G. Visconti. 1996.

Vol. 43: Chemical Exchange Between the Atmosphere and Polar Snow.
Edited by E. W. Wolff and R. C. Bales. 1996.

Vol. 44: Decadal Climate Variability: Dynamics and Predictability.
Edited by D. L. T. Anderson and J. Willebrand. 1996.

Vol. 45: Radiation and Water in the Climate System: Remote Measurements.
Edited by E. Raschke. 1996.

Vol. 46: Land Surface Processes in Hydrology: Trials and Tribulations of Modeling and Measuring. Edited by S. Sorooshian, H. V. Gupta, and J. C. Rodda. 1997.

Vol. 47: Past and Future Rapid Environmental Changes: The Spatial and Evolutionary Responses of Terrestrial Biota.
Edited by B. Huntley, W. Cramer, A. V. Morgan, H. C. Prentice, and J. R. M. Allen. 1997.

Vol. 48: The Mathematics of Models for Climatology and Environment.
Edited by J. I. Díaz. 1997.

Vol. 49: Third Millennium BC Climate Change and Old World Collapse.
Edited by H. N. Dalfes, G. Kukla, and R. Weiss. 1997.

Vol. 50: Gravity Wave Processes:
Their Parameterization in Global Climate Models.
Edited by K. Hamilton. 1997.

Vol. 51: Sediment Records of Biomass Burning and Global Change.
Edited by J. S. Clark, H. Cachier, J. G. Goldammer, and B. Stocks. 1997.

Printing: Druckhaus Beltz, Hemsbach
Binding: Buchbinderei Schäffer, Grünstadt